D1452884

# TRANSPORT PHENOMENA AND LIVING SYSTEMS

# Transport Phenomena and Living Systems

**BIOMEDICAL ASPECTS OF MOMENTUM AND MASS TRANSPORT**

**E. N. LIGHTFOOT**
*Department of Chemical Engineering*
*University of Wisconsin*
*Madison, Wisconsin*

A WILEY-INTERSCIENCE PUBLICATION

**JOHN WILEY & SONS, New York · London · Sydney · Toronto**

***Library of Congress Cataloging in Publication Data:***

Lightfoot, Edwin N. 1925-
Transport phenomena and living systems.

"A Wiley-Interscience publication."
Includes bibliographies.
1. Biological transport. 2. Membranes (Biology)
3. Rheology (Biology) I. Title. [DNLM:
1. Biological transport. 2. Cell membrane permeability. QH601 L724t 1973]

QH615.L69 574.1'912 73-9541

ISBN 0-471-53515-X

Printed in the United States of America

10 9 8 7 6 5 4 3 2 1

# PREFACE

This book was written primarily to satisfy my curiosity about the utility of biologically oriented transport studies. Hence the first step was to organize the large and frequently contradictory literature in my own mind, and to do this required a considerable narrowing of scope. I decided to concentrate on medically related aspects of momentum and mass transport in this initial effort because these are particularly important and relatively homogeneous—and because of the high quality of work done in these areas. Strong research traditions, built up over many centuries, provided a very strong foundation for the present work.

I also decided at the outset to build on as firm a transport base as possible, and to minimize the use of empiricisms and ill-defined approximations so prevalent in the biological literature. Some sections of the text, particularly III.1, may as a result get off to a slow start and prove difficult to read, but I believe the small extra effort will be repaid. This approach should, in the long run, reduce confusion, clear up apparent contradictions, and improve the chance of gaining new insights. However, because I have a weak biological background, I have tried to stay with relatively simple and well-defined physiological situations. Extension of the text to more complex systems is a worthwhile challenge to the reader.

From the beginning of the writing effort the importance of transport phenomena in applied biology became steadily more apparent. In all living organisms, but most especially the higher animals, diffusional and flow limitations are of critical importance; moreover we live in a very delicate state of balance with respect to these two processes.

Repeatedly it became clear that the greatest need is for better judgement, particularly in adapting problem-solving strategies to the specific physiological and mathematical peculiarities of the problem at hand—and to the inevitable experimental limitations. Strong contrast between the complexity of biological systems and the crudity of biological data remains a fundamental problem.

Thus it becomes particularly important to establish orders of magnitude and to make realistic limiting calculations. Especially attractive for these purposes are dimensional analysis and pharmacokinetic modeling; it seems in fact that these may permit unifying the whole of biological mass transport. Distributed-parameter modeling is also important, but there is real need for more effective interaction between theoreticians, experimentalists, and clinicians. It must be recognized that truly faithful models of any interesting biological system are impossible. To obtain *useful* models one must set rather specific goals and work toward them by systematic alternation of theory and experiment. It is also important to put results in a form readily accessible to potential users, and to focus attention on salient system features.

Of particular specific importance are the estimation of transport properties and the writing of specialized conservation equations, as for ultra-thin membranes. Not enough is known in either area. Both of these situations are in rather strong contrast to those in more classic areas of transport phenomena.

Energy transport is too complex for me to comment on now, and it remains to be organized. Specialization and geometric complexity are formidable problems in generalizing the treatment of biological momentum transfer; this topic seems best left to relatively narrow monographs. Unification of biological mass transport, however, seems both possible and useful; this is an attractive goal for any subsequent text.

Summarizing my writing experience, I feel that this text treats an important area but that it must be considered a transitional effort. Plainly it is built on the "parent text" *Transport Phenomena* (Wiley, 1960), and I hope that it in turn lays a foundation for deeper and more truly unified treatments of biological transport processes. In any event this book owes much to its parent, and the reader of both will find many similarities. Contrasts are also strong, however, and may prove instructive; this work is, for example, far more goal oriented.

Included in this book, as in its parent, are "secret messages." One is simple and specific; the rest are deeply buried. Unacknowledged influences of our close personal and professional associates are of primary importance in all texts, and this is no exception. "Secret messages" from mine are liberally distributed through these pages.

I am most indebted both personally and professionally to my late father, E. N. Lightfoot, and to Professor Olaf A. Hougen. Both put the interests of the next generation ahead of their own and looked far ahead. Both foresaw the development of applied biology much earlier than I and helped me to develop my interests in this area. My colleagues Bob Bird and Warren Stewart provided much of the background in transport theory needed for this effort. In addition they and my late father-in-law James A. Smith set examples of high standards and work performance that I hope will have benefitted this book. My wife, Lila, continues her father's work most effectively, and, like many who write books, I have a devoted, energetic, and creative mother. There are therefore a lot of deeply buried secret messages.

There are also many closer to the surface supplied by my biomedical colleagues. Ken Bischoff of Cornell University and Bob Dedrick of the National Institutes of Health in particular have done much over the years to develop my appreciation for orders of magnitude and to give me a sense of values.

Finally I am grateful to four institutions and their staffs. The University of Wisconsin and its Department of Chemical Engineering have always strongly supported scholarly efforts and encouraged interdisciplinary contacts. They have also helped with the mechanics of manuscript preparation, and I am particularly grateful to Jeanne Lippert, Marcy Sullivan, Diane Peterson, and Celeste Sherven for their efforts. The last stages of the writing took place at Stanford University, the Technical University of Denmark, and University of Canterbury, Christchurch, N.Z. The chance to try out my ideas on both students and staff and the opportunity to work relatively undisturbed are very much appreciated.

**E. N. Lightfoot**

*Madison, Wisconsin*
*May 1973*

# CONTENTS

# TRANSPORT PHENOMENA AND LIVING SYSTEMS

# I INTRODUCTION

## 1. PHILOSOPHY AND PERSPECTIVE

These notes are meant to accompany a senior–graduate-level course for biomedically oriented engineers and for medical scientists with an interest in engineering. Familiarity with the elementary aspects of transport phenomena is assumed, and it is expected that most students will also take courses in physiology and biologically oriented control theory.

This course is based on the fundamental assumption that many biologically important problems can be usefully discussed within the same framework as nonbiological problems. For this reason the organization of the notes is much influenced by the text *Transport Phenomena* (Bird, Stewart, and Lightfoot; 1960; abbreviated *Tr. Ph.* hereafter), which serves as a primary reference. However, we are also much influenced by the fact that we are dealing primarily with the human body and a few rather closely related mammalian systems. We are therefore justified in concentrating our attention on a relatively few specialized aspects of transport phenomena and in extending our discussions to such less well-defined problems as are normally handled in unit operations courses. We further limit ourselves in this initial effort to momentum and mass transfer, in spite of the biological importance of energy transport.

An immediate goal of this course is to assist engineers, medical scientists, and clinicians in solving clinical problems. Many applications of this type, for example, "artificial kidneys" and heart-lung machines, have received a good deal of publicity. However, the material covered should also be useful for those interested in the effects of stress on living organisms. These effects are of obvious importance when unusual environments are involved, as in oceanographic and space research, but they are also important in the so-called normal environments in which we work and live. In addition, it appears likely that the study of biological systems may be of benefit to more conventional industries. Evolution has produced other solutions to many common problems than have engineers (see Bugliarello, 1966). Some of these solutions, such as the echo-location systems of bats and porpoises, are already of practical interest, and others may soon become so.

The biggest biological challenges, outside that of the human brain, are, however, chemical. Among these are the highly selective electrokinetic processes occurring in lipid membranes, and the direct interconversion of chemical, mechanical, and electromagnetic energy. The organization of the intracellular "chemical factories" is also very impressive, as are many of the individual reactions carried out. The chemical and electrochemical achievements of living organisms seem likely to stimulate a large number of important engineering developments.

In recent years observation and explanation have received primary attention in engineering curricula, and this is true in much of the present notes. It should, however, always be kept in mind that these have no engineering relevance except as a basis for invention.

We clearly have a long way to go, and it is of course always difficult to plan for the future. Fortunately, however, bioengineering has a long and distinguished past which can serve as both an inspiration and a guide. The motivations, the successes, and even the failures of earlier workers are of real interest today.

If we define biomedical engineering as the systematic application of existing science and technology to medical problems, our field has its real beginning in the renaissance period (Mendelsohn, in Salkowitz et al., 1968). Its early development was inspired by and dependent upon the simultaneous development of physics; it was based on the belief that living organisms obeyed the same laws as inanimate objects and could be described in terms of these laws. We are still operating on this belief, though on a much more sophisticated level. Consider, for example, the following quotation from Francis Crick (1966):

> The ultimate aim of the modern movement in biology is in fact to explain all biology in terms of physics and chemistry. There is a very good reason for this. Since the revolution in physics in the mid 1920's we have had a sound theoretical basis for chemistry and the relevant parts of physics. This is not to be so presumptuous as to say that our knowledge is absolutely complete. Nevertheless, quantum mechanics, together with our empirical knowledge of chemistry, appears to provide us with a foundation of certainty on which to build biology.

Dr. Crick's optimism may be justified at his level of operation, but engineers would be wise to be more cautious. The experience of the past indicates that bioengineering success is severely limited by the technology on which it is based. Early bioengineers had their greatest successes in applications of Newtonian physics, but were badly misled by their inability to understand chemical transformations and the interconversion of chemical and mechanical energy.

Thus one of the earliest successful bioengineers, Santorio Santorio (or Sanctorius, 1561–1636), a professor of the University of Padua and a contemporary of Galileo, owes his medical fame to his interest in conservation of mass. He found by careful weighing of subjects and their sensible mass inputs and outputs that "insensible perspiration" exceeds all other body excretions combined, but he was of course unable to explain the nature of this weight loss. These experiments, in spite of their limited success, had an enormous influence on medicine and physiology. Sanctorius also developed the first fever thermometer, many years before Fahrenheit (1686–1736). The science of anatomy was founded by Leonardo da Vinci (1452–1519), who also studied the motion of bones and muscles. The laws of mechanics were not well understood in Leonardo's time, however, and the first to treat muscular movements in terms of mechanical principles was the Italian mathematician Giovanni Alphonso Borelli (1608–1679).

The French mathematician René Descartes (1596–1650) was also a "bioengineer" and wrote the first modern textbook on physiology (*De homine*, published posthumously in 1662). Descartes was, however, badly misled in underestimating the limitations of available technology and in insisting on a "rational" explanation of all observed phenomena. Thus he refused to accept William Harvey's experimental observation that muscular contraction of the heart pumped the blood through the circulatory system. He suggested that blood was heated in the heart and that pumping was caused by the resulting thermal expansion—an energy source he understood. His text was largely theoretical and largely discredited by subsequent findings. He did, however, emphasize the importance of the nervous system and its function of coordinating bodily activities.

Similar limitations are found in the applied aspects of bioengineering. The first recognized heart-lung machine was built by Jacobi in 1895. This machine appears to have been a technical success in oxygenating the blood, but a medical failure in terms of patient survival. Even today the most widely used extracorporeal oxygenators cause severe blood damage and cannot be used for more than a few hours. Modern "artificial kidneys" perform only the simplest (though fortunately most vital) of the true kidney's functions: nonselective removal of all low-molecular-weight solutes. The picture is clear: we engineers understand the simple physics of fluid flow and mass transfer to a useful degree, but our knowledge of biochemistry and physiology is very weak. Unfortunately the present notes are concerned almost entirely with "simple physics." The reader should be aware of this limitation at all times. There is no place for the overconfidence of Descartes, no substitute for careful experiment.

On balance, however, we should be more inspired by the immense abilities of our predecessors and by the great progress they made than overwhelmed by the complexities of our field. Accumulated experience indicates that with optimism, careful judgement, and hard work one can solve even very complicated medical problems.

Success is, however, most likely if the engineer works in close continuing cooperation with the appropriate clinicians and medical scientists. The clinician is in the best position to recognize practical problems and needs and to understand the environment in which new solutions must be applied. The basic medical scientist is best able to restate medical problems in terms of questions which can be answered in the laboratory. The proper role of the engineer is in problem simplification, evaluation, and design. Members of any one discipline working alone will normally make only fragmentary contributions to the solution of major problems, and even these are unlikely to be fully utilized.

It should also be kept in mind that the primary concern of the living organism is homeostasis, the maintenance of a viable near steady state. Homeostasis is first and foremost an exceedingly complex exercise in process control, a topic not touched on at all in these notes. Also the anatomical, physiological, and biochemical complexities of living organisms are such that fundamentally mathematical courses in biological transport phenomena and process control are insufficient preparation for biological research. Such essentially descriptive subjects as anatomy, physiology, biochemistry, and practical medicine are still indispensable.

## 2. ORGANIZATION AND MODELING

Since we are primarily interested in the human body it is useful to preface our study with a brief review of its salient characteristics. We begin in Section 2.1 with a brief discussion of levels of organization in the body and then review in Section 2.2 some of the simplified models that have been found useful in engineering analyses.

### 2.1. Levels of Organization within the Mammalian Body

A living mammal is much too complex an organism to consider in all its detail, and it is therefore always necessary to take a simplified view suitable for solving the problem at hand. This in turn may be done at many different levels of organization:

(a) The body as a unit.
(b) The body as a network of interconnected substructures: the organs.
(c) The organs themselves.
(d) The individual cells of which the organs are composed.
(e) The structural components of individual cells, in particular the cell membranes.

We could proceed even further, to the molecular level, and in fact do so very briefly at intervals. For the most part, however, we take a continuum viewpoint. This is primarily because of the limited background of the author and for lack of time. Biochemistry and molecular biology are, however, very important and fast-moving fields, and the prudent bioengineer will make every possible effort to acquaint himself with developments in them.

It is also desirable for the student to provide himself with his own references in physiology and biochemistry, and it may be useful to purchase a medical dictionary. Much of the medical and biochemical background provided in these notes was taken from Guyton (1967), Mahler and Cordes (1966), and White, Handler, and Smith (1968).

### (a) *THE BODY AS A UNIT*

As a focus for attention and to establish orders of magnitude we begin by defining a "standard man" (male) in Table 2.1.1. These figures are by their very nature suspect. Women, for example, are smaller and also differently proportioned: they have a much higher proportion of fat. The orders of magnitude are, however, useful. For example, it is important to keep in mind that the mean blood circulation time is of the order of one minute.

At the simplest level we can consider the body as a thermodynamic system absorbing energy from its environment and in turn releasing heat and doing mechanical work. The body may therefore be considered as a fuel cell. It is in fact much more efficient than any of the so-called biological fuel cells produced to date and is also far more efficient than any heat engine operating within the same temperature limits. An approximate energy balance is shown in Table 2.1.2.

Note that chemical energy is converted to mechanical work through the formation and dissolution of high-energy phosphate bonds. The primary source of this chemical energy is adenosine triphosphate (ATP), which yields its energy on hydrolysis to adenosine diphosphate (ADP) and inorganic phosphate.

Most of the chemical processes of interest take place in the fluid portions of the body, and in fact water is the principal constituent of the

**Table 2.1.1.** "Standard American Male"

*Age:* 30 yr
*Height:* 5 ft 8 in. or 1.86 m
*Weight:* 150 lb or 68 kg
*External surface:* 19.5 $ft^2$ or 1.8 $m^2$
*Normal body temperature:* 37.0°C
*Normal mean skin temperature:* 34°C
*Heat capacity:* 0.86 cal/(g)(°C)
*Capacities:*
Body fat: 10.2 kg or 15%
Subcutaneous fat layer: 5 mm
Body fluids: ca. 51 l or 75%

| | Volume (l) | Fract. of total (%) |
|---|---|---|
| Intracellular: | 27.2 | 40 |
| Interstitial and lymph: | 20.4 | 30 |
| Plasma: | 3.4 | 5 |
| | 51.0 | 75 |

Blood volume: 5.0 l (includes formed elements, primarily red cells, as well as plasma)
Hematocrit = 0.43
Lungs:
Total lung capacity: 6.0 l
Vital capacity: 4.2 l
Tidal volume: 500 ml
Dead space: 150 ml
Mass transfer area: 90 $m^2$
*Mass and energy balances at rest:*
Energy conversion rate: 72 kcal/hr or 1730 kcal/day [40 kcal/($m^2$)(hr)]
$O_2$ consumption: 250 ml/min
(respiratory quotient = 0.8)
$CO_2$ production: 200 ml/min
Heart rate = 65/min
Cardiac output = 5.0 l/min (rest)
= $3.0 + 8M$ in general
($M = O_2$ consumption in liters per minute)
*Systemic blood pressure:* (120 mm Hg)/(80 mm Hg)

**Table 2.1.2.** Distribution of Ingested Food Energy

| | |
|---|---|
| Input: | 100% |
| Losses: | |
| $T\Delta S$ | 5% |
| Biochem inefficiency | 50% |
| Metabolic activity | 45–20% |
| Mechanical work (skeletal muscle contraction) | 0–25% |

One gram-mole glucose

| Aerobic combustion | Controlled biological oxidation |
|---|---|
| 686 kcal heat | 38 moles ATP at ca. 9 kcal free energy per mole: 340 kcal free energy |

body. Water content varies between individuals, from about 45 to 75%, and depends mainly on fat content. It is usually remarkably constant for a given individual. This water is commonly thought of as being distributed between three (or from a geometrical standpoint, three sets of) "compartments" or "spaces": plasma, interstitial fluid, and intracellular fluid. The plasma compartment is the smallest of these and the intracellular the largest. The volume distribution is indicated in Table 2.1.1. Note that the red-blood-cell interiors are part of the intracellular compartment and distinct from the plasma in which they are suspended. The compartments are defined in terms of their availability to various solutes. As an example, the serum proteins (the soluble proteins in blood plasma) occur primarily in the plasma. For most other solutes the distinction between compartments is more ambiguous. Many small solutes—for example, urea and most other small test solutes removed by artificial kidneys during hemodialysis—are distributed throughout all the compartments. Even for these, however, the concentration is not, in general, uniform throughout the body. In addition there may be a distribution lag when these solutes are being removed. Urea, for example, tends to be retained longer in the brain than in most other organs during hemodialysis. As a result, rapid dialysis can cause osmotic pressure to rise in the brain and produce headaches or more serious disorders. The rate of equilibration between compartments has as yet been studied only superficially, from an engineering standpoint, and this is an important area for research.

The fluid composition in the various compartments varies strikingly as shown in Table 2.1.3. Note that the proteins are behaving as anions. That is, they are on the alkaline side of their isoelectric points and thus contain free carboxyl ions:

$$\begin{array}{l} \mathrm{R} - \mathrm{COO}^- \\ | \\ \mathrm{NH_2} \end{array}$$

It is of particular interest to note the very different compositions of plasma and interstitial fluid on the one hand and intracellular fluid on the other. It is also important to note that many solutes are bound in reversible complexes. Many drugs, for example, are bound to serum albumin.

### (b) *THE BODY AS A NETWORK OF INTERCONNECTED ORGANS*

From a geometric and flow standpoint the body may be considered as a complex of highly specialized and interconnected organs, such as the heart, liver, and kidneys. These are connected by the blood stream, which serves to transport oxygen and other nutrients and thus helps to overcome the large diffusional resistance that would otherwise be offered by such a large

**Table 2.1.3.** Ionic Composition of Body Fluids (meq/l)

| Ion | Plasma | | Interstitial | Intracellular |
|---|---|---|---|---|
| $Na^+$ | 138 | | 141 | 10 |
| $K^+$ | 4 | | 4.1 | 150 |
| | | 149 | | |
| $Ca^{2+}$ | 4 | | 4.1 | ... |
| $Mg^{2+}$ | 3 | | 3 | 40 |
| $Cl^-$ | 102 | | 115 | 15 |
| $HCO_3^-$ | 26 | | 29 | 10 |
| $PO_4^{3-}$ | 2 | | 2 | 100 |
| | | 149 | | |
| $SO_4^{2-}$ | 1 | | 1.1 | 20 |
| Organic acids | 3 | | 3.4 | ... |
| Protein | 15 | | 1 | 60 |

body. This key role of the circulating blood is shown schematically in Fig. 2.1.1.

The number of distinct organs is large, and the blood flow path is complex. Hence it is often desirable to focus attention on a smaller group of organs such as the circulatory system shown schematically in Fig. II.3.1.1.

The circulatory and respiratory systems are of particular interest to us in this course. The total circulating blood volume is about 76 ml/kg or about 5.3l for a 70-kg man, and at rest the blood circulates about once a minute.

## (c) *THE INDIVIDUAL ORGANS OF THE BODY*

Many of the individual organs perform well-defined tasks of importance from our point of view. A good example is provided by the lungs, which

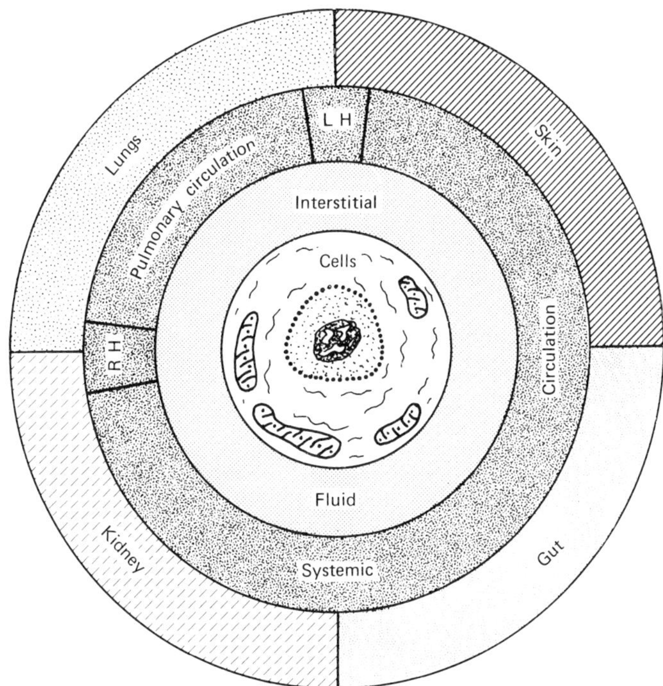


**Figure 2.1.1.** Diffusional topology of the mammalian body. Convective transport takes place in the blood and interstitial fluid, tangentially in the diagram, and between the four "external" organs and the environment. Diffusional transport is primarily at the cellular level, radially in the two inner rings in this diagram. Inspired by Dr. Peter Abbrecht, University of Michigan.

**Table 2.1.4.** Partition of Left Ventricular Output and Pulmonary Oxygen Intake in Man at Rest Under Basal Conditions[a]

| Region | Mass (kg) | Blood Flow (ml/min) | Blood Flow (ml/(100 g)(min)) | A-V $O_2$ Difference (ml/l) | Oxygen Use (ml/min) | Oxygen Use (ml/(100 g)(min)) | Resistance[b] Mean A.P. [in mm Hg] / B. F.[in ml/(100g)(min)] | % of Total Cardiac Output | % of Total Oxygen Use |
|---|---|---|---|---|---|---|---|---|---|
| 1. Hepatic-portal | 2.6 | 1500 | 57.7 | 34 | 51 | 1.96 | 1.56 | 27.78 | 20.40 |
| 2. Kidneys | 0.3 | 1260 | 420.0 | 14 | 18 | 6.00 | 0.214 | 23.34 | 7. |
| 3. Brain | 1.4 | 750 | 53.6 | 62 | 46 | 3.30 | 1.67 | 13.83 | 18. |
| 4. Skin | 3.6 | 462 | 12.8 | 25 | 12 | 0.33 | 7.0 | 8.55 | 4.8 |
| 5. Skeletal muscle | 31.0 | 840 | 2.7 | 60 | 50 | 0.16 | 33.3 | 15.55 | 20.0 |
| 6. Heart muscle | 0.3 | 252 | 84.0 | 114 | 29 | 9.66 | 1.0 | 4.67 | 11.6 |
| 7. Residual tissue | 23.8 | 336 | 1.4 | 129 | 44 | 0.18 | 64.3 | 6.23 | 17.60 |
| Entire body | 63.0 | 5400 | 8.6 | 46 | 250 | 0.40 | 10.6 | 100.0 | 100.00 |

[a] Weighs 63 kg, has a surface area of 1.8 $m^2$, a mean arterial pressure of 90 mm Hg, a cardiac output of 5400 ml, and an oxygen consumption of 250 ml/min.

[b] A.P. = arterial pressure; B.F. = blood flow.

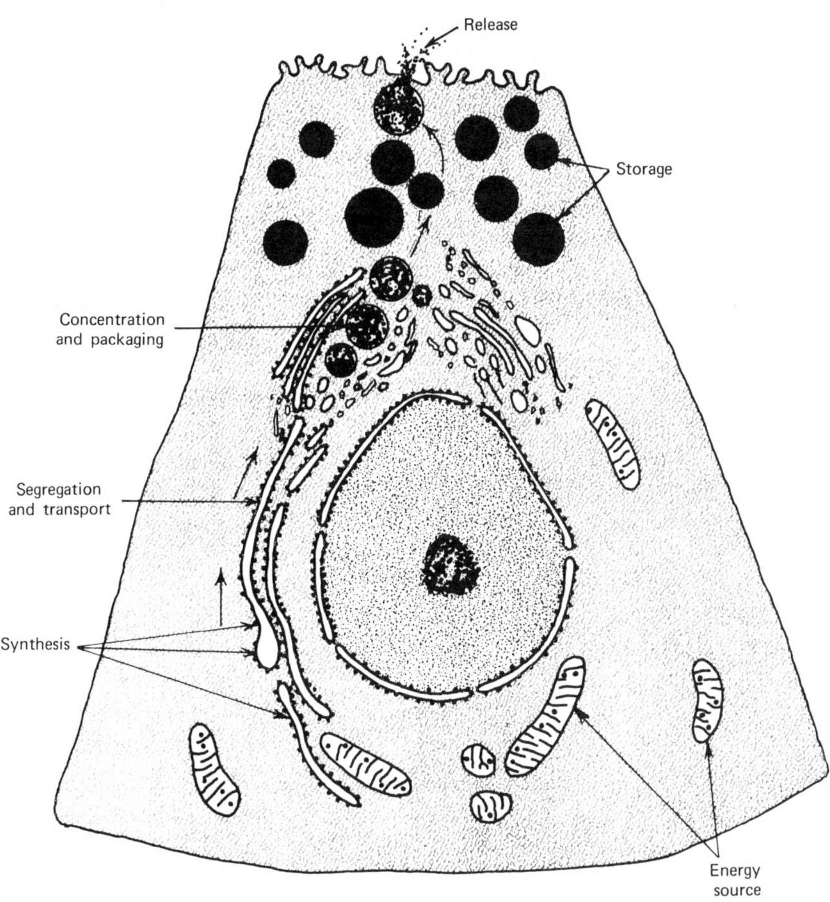


**Figure 2.1.2.** Structure and organization of a characteristic cell. Taken from Mahler and Cordes (1966). Redrawn from D. W. Fawcett, "Structural and Functional Variations in the Membranes of Cytoplasm," in S. Seno and E. V. Cowry, Eds., *Intracellular Membrane Structures*, Jap. Soc. for Cell Biol., Okayama, Japan (1963). Shown here is a cross section of a pancreatic exocrine cell magnified about $10^4\times$.

exchange oxygen, carbon dioxide, and water vapor with the inhaled air and serve to control the pressures of oxygen and carbon dioxide as well as acidity in the blood. The mass-transfer surface of the alveolar sacs in the lungs is of the order of 90 $m^2$, and these surfaces are supplied with a bed of capillaries (vascular bed) of comparable area. Other important organs of mass transfer include the kidneys and gut.

Sometimes, as when studying *pulmonary ventilation*, it is necessary to consider the entire organ, as shown schematically in Figs. 6.3.3 and 6.3.4. The diffusional transfer in the lungs takes place primarily in the alveolar sacs, however, and when studying it we sometimes need consider only a single representative sac. One situation where this simplification is thought to be reasonable is the modeling of respiratory gas transfer in the lung (see Ex. III.6.3.2). At other times the dispersion in behavior of individual units is important; such an example is the "splay" in the transport maximum for glucose in the kidney (see Ex. III.6.3.3). Similarly the highly selective mass transfer in the kidney can be understood in terms of an individual nephron, or even subdivisions of a nephron.

Some of the gross characteristics of individual organs are shown in Table 2.1.4. Much more information of this type is available in such standard references as the *Handbook of Physiology*.

### (d) *FINER SUBDIVISIONS*

We shall have frequent occasion later in these notes to consider the substructures of individual organs and do not discuss them in detail here. However, it is worth noting that a high level of organization exists even at the cellular level, as shown by the sketches of a pancreatic cell in Fig. 2.1.2. It can be seen that this single cell shows a striking parallel to a self-contained chemical plant. We shall see later that even individual cell membranes are surprisingly complicated both in structure and function.

## 2.2 Modeling of Organs and Organ Systems

One of the major current activities of biologically oriented engineers is the development of useful approximate models for such complex physiological systems as the systemic or pulmonary circulation—or even the entire body. Goals of such research range from attempts at the deepest possible conceptual understanding to searches for the simplest reliable guides to diagnosis and treatment of ailments.

Since these notes are devoted in large part to providing a sound basis for modeling, it is clearly premature to discuss this subject in detail in the introductory chapter. It is, however, useful to look ahead to see what sorts of results have been obtained and to assess their utility. It is hoped that such a preview can put the rest of the text in perspective and make some of the detailed discussions more meaningful.

### (a) *FLOW MODELS OF THE CARDIOVASCULAR SYSTEM*

Interest in blood flow received a major impetus with Harvey's discovery of continuing recirculation in 1628, and current research is aimed primarily at improved understanding of the complex pressure-flow relationships observed since that time. Complex conceptual models developed by fundamental researchers are now, however, beginning to be simplified and

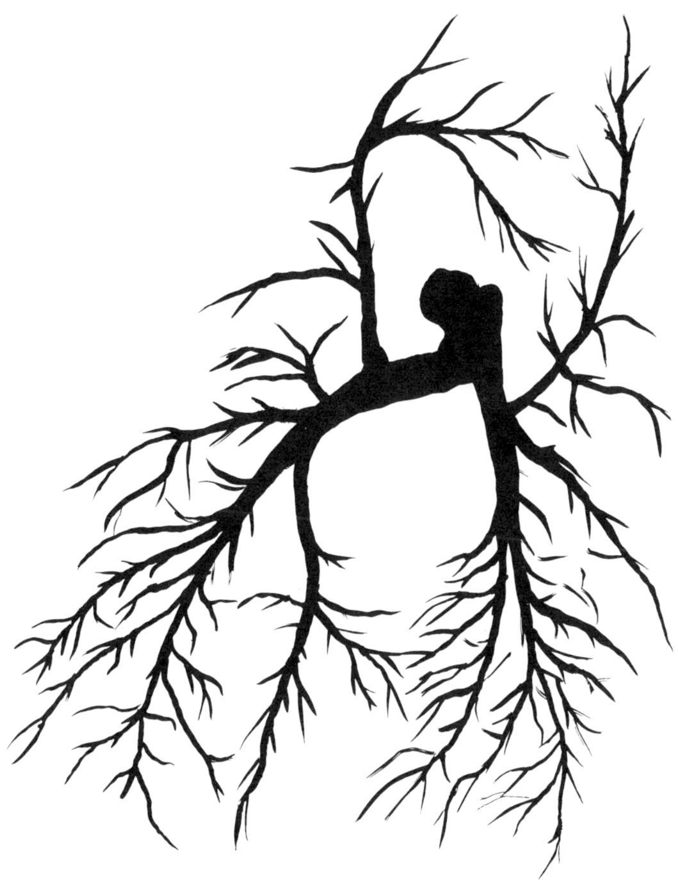

**Figure 2.2.1.** Cast of the pulmonary arterial tree. Drawn from a photograph in E. O. Attinger, *Circ. Res.*, **12**, 623 (1963).

adapted for clinical purposes. The history of circulatory research is described in the monograph of Fishman and Richards (1964), and modeling of the complete circulation is reviewed by Skalak (1971).

The chief characteristics of the circulatory system are its geometric complexity and the peculiar flow characteristics of individual vessel segments. A good model should therefore account for

1. The taper and frequent branching of the system, shown for example by the cast of the pulmonary arterial "tree" of Fig. 2.2.1.
2. The viscous flow resistance and wall elasticity of individual blood vessels, and the inertia of the flowing blood.

In addition the model must recognize the complex pulsatile nature of the blood flow indicated in Figs. 2.2.2 and 2.2.3, and most of those seriously considered do this. They differ very considerably in their geometric complexity and their treatment of flow behavior, however.

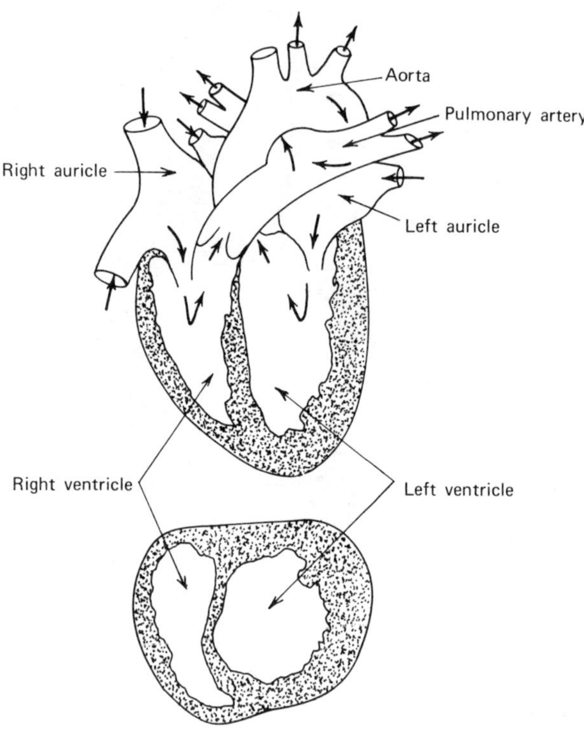


**Figure 2.2.2.** Pressure-flow relations in the heart. (*a*) Schematic diagram of a longitudinal and an horizontal cross-section of the heart.

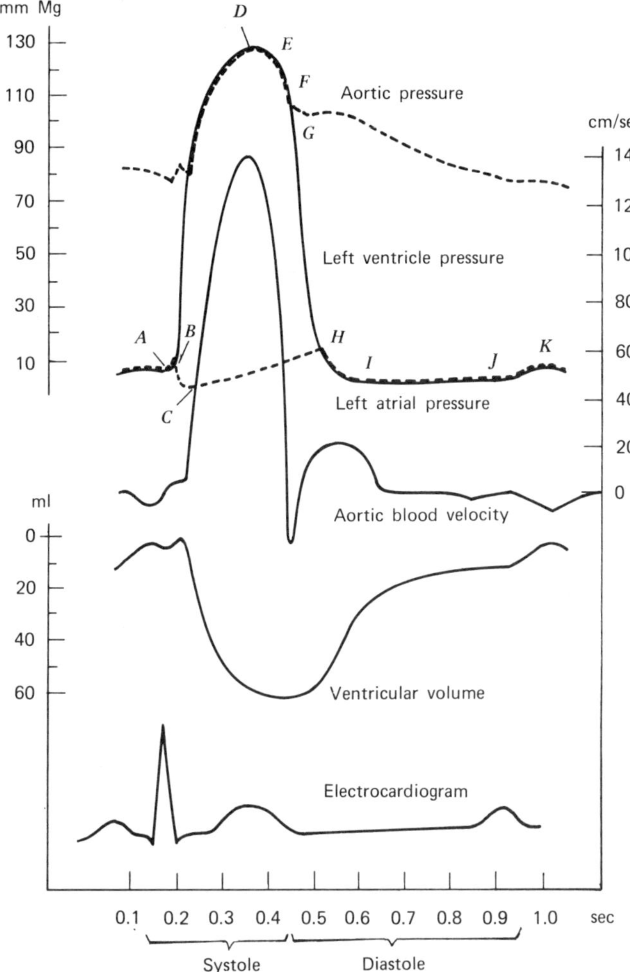


**Figure 2.2.2** (continued). (*b*) Correlation of left ventricular, aortic, and left atrial pressure pulses with ventricualr volume curve, aortic flow, and electrocardiogram.

Skalak (1971) classifies cardiovascular models into four levels depending on their description of flow behavior:

1. Pure resistance models.
2. Lumped-parameter models consisting of networks of discrete resistant, compliant and inertial elements.
3. Distributed linear models.
4. Distributed nonlinear models.

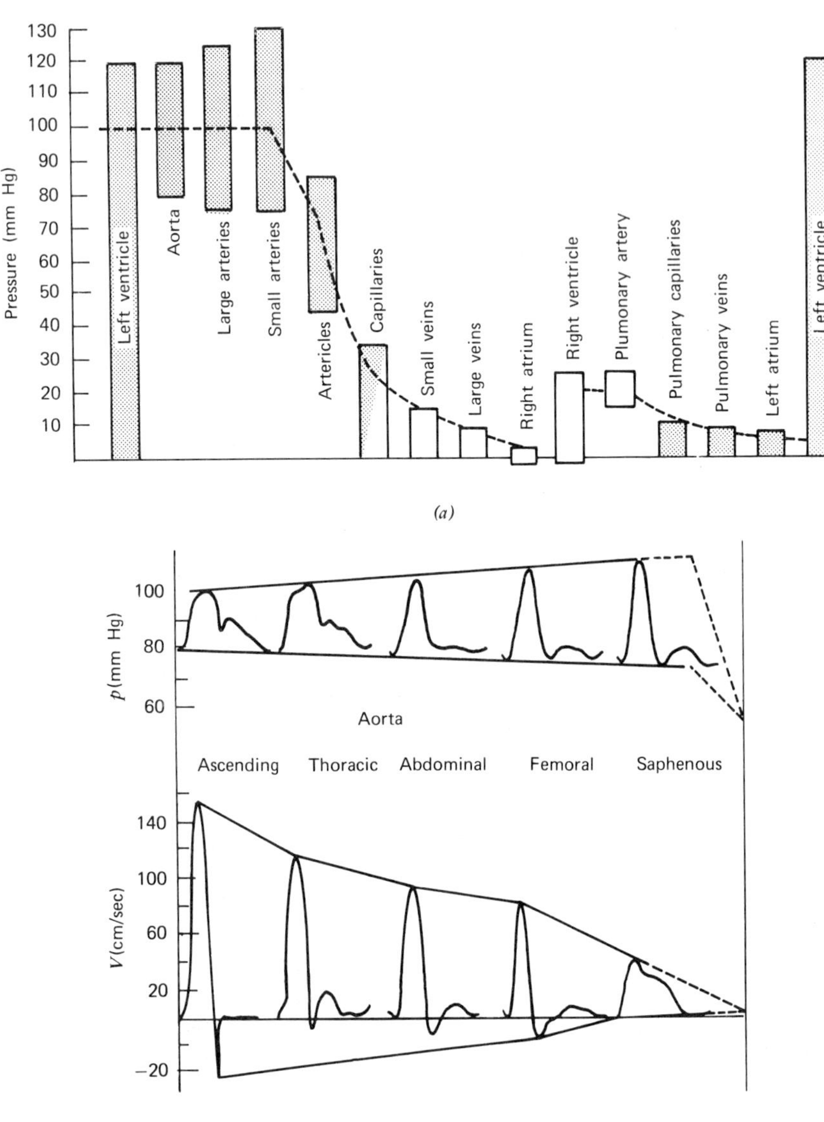


**Figure 2.2.3.** Pressure-flow relations in the systemic and pulmonary circulations. (*a*) Pressure relations in the various segments of the vascular system. The dashed line represents the mean pressure. The light bars represent those parts of the system containing deoxygenated blood. (*b*) Diagrammatic comparison of the behavior of the pressure and flow pulses in the systemic arteries during their peripheral travel. Note that the pressure pulse increases and flow pulse decreases. From D. A. McDonald, *Blood Flow in Arteries*, Edward Arnold (Publishers), Ltd., (1960).

A choice between these successively more complex models depends on the kind of information desired and the relative importance of completeness and simplicity.

The pure-resistance models are limited in principle to steady flows, because they completely neglect inertial effects, and they all take the same simple form

$$R=\frac{p_A-p_V}{Q} \tag{2.2.1}$$

where $R$ = hemodynamic resistance
$p_A, p_V$ = arterial and venous pressures for the system under consideration, and
$Q$ = volumetric blood flow ("perfusion") rate.

When the entire systemic circulation is being measured, $R$ is known as the "total peripheral resistance." These models are most meaningful for exploring the effects of blood rheology and wall distensibility by steady perfusion of isolated limbs or organs. However, the total peripheral resistance obtained by using time-averaged pressure drops and flow rates is often used as a clinical indication of physiological condition. This resistance generally differs somewhat from that obtained by steady perfusion because of a variety of nonlinearities in the flow equations—which is discussed later.

The measured resistance usually decreases with increase of flow rate, as shown in Fig. 2.2.4, or pressure level. This is principally because of the distensibility of blood vessels and the increase in the number of functioning capillaries. In addition the resistance offered by red cells decreases with increase in velocity. Resistance is also affected by a wide variety of physiological control mechanisms, which should not be ignored even though we cannot discuss them here.

The first lumped-parameter model, proposed by Frank in 1899, treated the arteries as a single elastic surge chamber (*Windkessel*) and the rest of the vascular system as a pure resistance in series with the arteries. The resultant pressure-flow relation is

$$Q_i=\frac{1}{k}\frac{dp}{dt}+\frac{p}{R} \tag{2.2.2}$$

where $Q_i$ = rate of blood flow in from the heart,
$p$ = pressure in the *Windkessel* (arteries), and
$k$ = elastic modulus of the *Windkessel*

The venous pressure is considered constant at zero, and the flow through

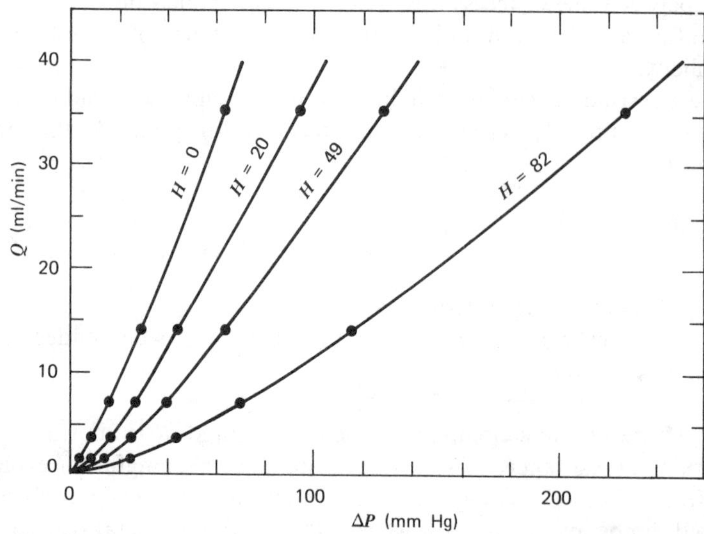


**Fig. 2.2.4.** The effect of pressure on total peripheral resistance. (Taken from Skalak.) The blood flow rate $Q$ is plotted against the pressure head $\Delta P$ for various values of the hematocrit H.

the resistive portion of the system is given by Eq. 2.2.1. This simple model is an improvement over the assumption of a pure resistance in providing for periodic blood storage in the arteries and a time-dependent pressure, and it provides a useful description when the arterial system is stiff and short. Such a situation occurs in the arterial networks of birds and is approached in humans under the influence of vasoconstrictor drugs. This model is equivalent to a combined capacitance and resistance, as indicated in Fig. 2.2.5. The principal fault of the *Windkessel* model is its assumption that the entire arterial system is pressurized at once and its consequent inability to describe wave propagation. This failing is most serious early in systolic ejection where $Q_i \gg p/R$, and Eq. 2.2.2 predicts that the rate of pressure rise, $dp/dt$, is proportional to the inflow rate $Q_i$. In fact, however, the pressure itself is proportional to flow, and this is predicted by wave-propagation theory. Frank himself was well aware of wave-propagation effects and proposed a simple modification of his theory to take them into account. This and later modifications have proven useful and are reviewed by Skalak (1971).

The primary recent emphasis in lumped-parameter modeling has been on adding additional elements, and including inductances as well as resistances and capacitances. The inductive elements are used to represent

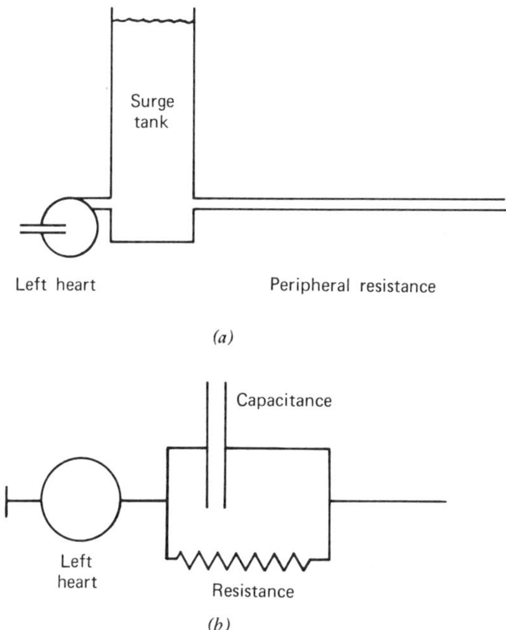


**Fig. 2.2.5.** Hydraulic and electrical equivalents of the Windkessel model of the systemic circulation. (*a*) Hydraulic analog, (*b*) Electrical analog.

the inertia of the flowing blood. The increasing availability of large analog and hybrid, as well as digital, computers has given considerable impetus to this development. A number of rather elaborate programs are now available.

A typical approximation for an arterial section is shown in Fig. 2.2.6. Such a model approaches the actual flow behavior of a Newtonian fluid in a long elastic-walled tube for a long-continued sinusoidal flow as the number of *RLC* elements is increased. The effective resistance and inductance per unit length are, however, frequency dependent even for this idealized case. In practice it is not always necessary to use a large number of elements to describe gross system behavior (see for example Rideout and Schaeffer).

Distributed-parameter models are, however, to be preferred for detailed studies of such phenomena as wave propagation, and these may be either linear or nonlinear. The bulk of the wave-propagation literature deals with linear analyses because these combine simplicity with the ability to de-

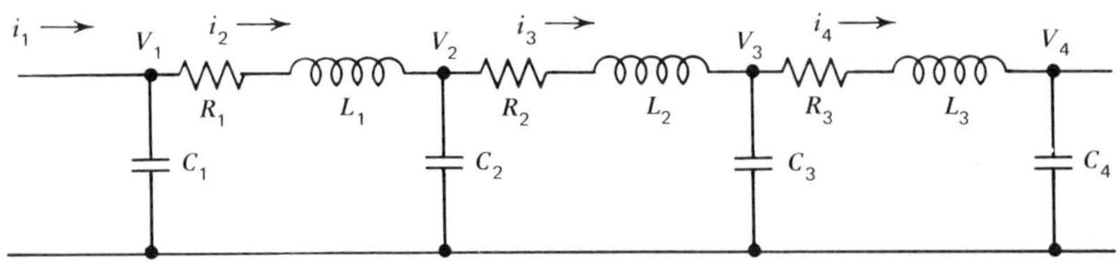


**Fig. 2.2.6.** A linear lumped-parameter model of an elastic artery allowing for viscous flow resistance, elastic expansion of the duct lumen, and blood inertia. Here (1) Electrical current $i$ is analogous to volumetric blood flow rate $Q$; electrostatic potential $V$ is analogous to hydrostatic pressure $p$; (2) viscous flow resistance and blood inertia are approximated for any segment by resistance $R$ and an inductance $L$ in series; wall elasticity is allowed for by a parallel capacitance $C$. Individual $RLC$ elements for short sections, three of which are shown here, are connected in series to approximate the distributed-parameter flow situation being modeled.

scribe the salient features of pressure and flow waves. They have been found very useful for obtaining insight into the nature of arterial flows in particular, and we study them in some detail in subsequent chapters. It is, however, becoming increasingly clear that accurate description of the observed progressive changes in wave forms with distance from the heart requires considering at least some nonlinear effects. The most important of these effects appear to be nonlinear wall rheology, convective acceleration, and the change in vessel lumen (flow cross section) with pressure. Among the more striking manifestations of nonlinearity are the shock waves producing "pistol-shot" sounds characteristic of regurgitation, as in aortic insufficiency.

Current nonlinear models are cruder than their linear counterparts in that they depend upon one-dimensional flow approximations and essentially empirical relations for the position and pressure dependence of flow cross section. Nevertheless they provide information not otherwise available—for example, Rockwell's description of sharpening pressure pulses shown in Fig. 2.2.7. Shown here is the computed dependence of pulse pressure with distance from the heart. This simulation produces many features of the natural pulse including the incisura, growth, and subsequent decay of the pressure wave with distance and the development of the dichrotic wave. The model also predicts observed steepening of the wave front, which appears to depend on the nonlinear terms.

We have now proceeded in our discussion from the simple pure-resistance models through increasingly elaborate lumped- and then distributed-parameter approximations to the blood circulation. Skalak suggests that each of the four major steps in this sequence offers about an equal

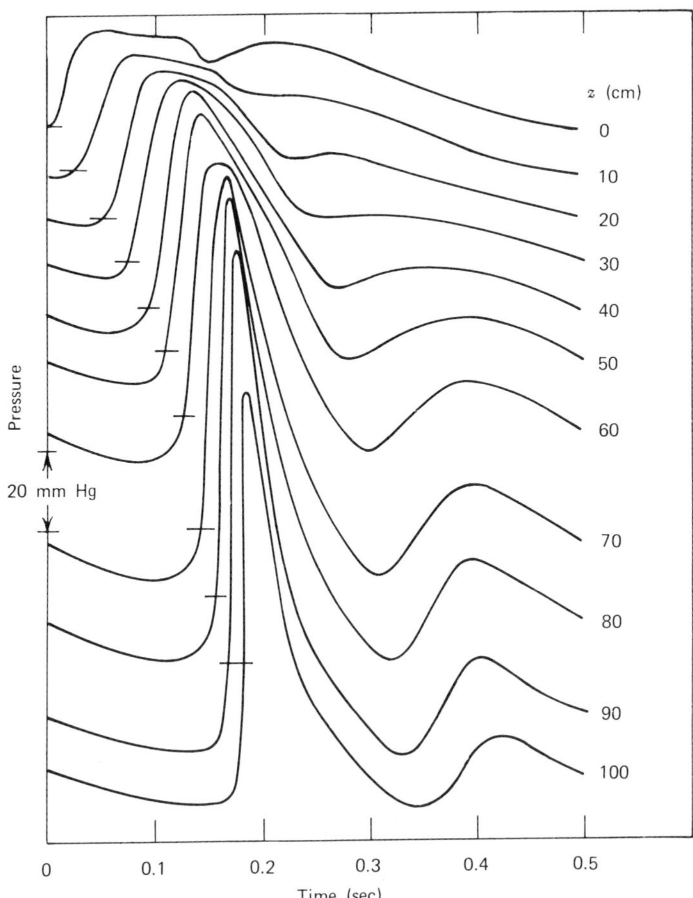


**Fig. 2.2.7.** Formation of an Arterial Shock Wave. Pressure-time relations are shown for different distances ($z$) from the heart. For each curve 80mmHg pressure is indicated by a short horizontal dash. From Max Anliker, in Fung, Perrone, and Anliker (1972).

degree of improvement in possibilities for accurate description. Each is best suited to different purposes, however, as indicated in Table 2.2.1, taken from Skalak's review.

### (b) *MASS-TRANSFER MODELS OF MAMMALIAN SYSTEMS*

Interest in physiological mass transfer dates back at least to the mass-balance studies of Sanctorius, but most significant contributions have been comparatively recent. Thus the importance of maintaining a relatively

**Table 2.2.1.** Levels of Cardio Vascular Modeling

| Concept | Introduction as a Research Model | Current Utility |
|---|---|---|
| 1. Pure resistance | Young (1809)<br>Poiseuille (1840) | Slow variations of mean blood pressure. Elasticity of microcirculation. Rheology of blood. |
| 2. Lumped-parameter model (*Windkessel*) | Otto Frank (1899) | Stroke volume from aortic pressure. Analog models. Analysis of cardiac assist devices. |
| 3. Distributed, linear model (transmission line) | Womersley (1957)<br>McDonald (1960) | Computation of flow from pressure gradient. Space and time distribution of pressure and flow. Input-impedance studies. |
| 4. Distributed, nonlinear model | Euler (1755)<br>Lambert (1958) | Accurate pressure and flow wave forms. Explanation of pistol-shot sounds. |

stable chemical environment was first recognized by Bernard in 1878, and the modern study of energy production in muscles begins with the work of Fletcher in 1898. Quantitative studies of transient mass transfer and mass transfer with chemical reaction are still more recent (Bassingthwaighte, 1970 and Bischoff, 1967).

Mass transfer occurs by both convection and diffusion, and it is normally accompanied by chemical reaction. In addition, physiological diffusional processes, particularly mass transport across cell membranes, can be exceedingly complex. As a result it is normally necessary to use gross anatomical simplifications as well as crude approximations for the mass-transfer processes being analyzed.

As in the case of blood flow, different levels of sophistication can be used in mass-transfer modeling, and the choice among them depends on both the system complexity and the type of information desired. It is usually convenient to categorize them as

(1) Stochastic descriptions (Aris, Bassingthwaighte, Levenspiel, Lightfoot et al.).
(2) Lumped-parameter models (Bischoff).
(3) Distributed-parameter models (see Section III.6).

All three approaches have proven useful, and all three are discussed in Section III.6. However, greatest emphasis is given there to lumped- and distributed-parameter modeling, as these follow directly from the conservation equations of Transport Phenomena. Lightfoot, Safford, and Stone (1971) provide a relatively extensive introduction to stochastic methods in applied biology as well as references to the pertinent literature.

## BIBLIOGRAPHY


Aris, Rutherford, *Introduction to the Analysis of Chemical Reactors*, Prentice-Hall (1965).

Bassingthwaighte, J. B., "Blood Flow and Diffusion in Mammalian Organs," *Science*, **167**, 1347–1353 (March 6, 1970).

Bird, R. B., W. E. Stewart, and E. N. Lightfoot, *Transport Phenomena*, Wiley (1960) Abbreviated *Tr. Ph.* in this book.

Bischoff, K. B., *Chemical Engineering in Medicine and Biology*, Daniel Hershey, Ed., Saunders (1967).

Bugliarello, George, "Engineering Implications of Biological Flow Processes," in *Chemical Engineering in Medicine*, E. F. Leonard, Ed., *Chem. Eng. Prog. Symp. Ser.*, **62**, 66 (1966).

Crick, Francis, lecture, University of Washington, 1966.

Fishman, A. P., and D. W. Richards, *Circulation of the Blood: Men and Ideas*, Oxford (1964).

Fung, Y. C., *Biomechanics*, ASME (1966).

Guyton, A. C., *Textbook of Medical Physiology*, Saunders (1967 printing).

Hershey, Daniel, Ed., *Chemical Engineering in Medicine and Biology*, Plenum (1967).

Lehninger, Albert, *Bio-energetics*, Benjamin (1971).

Levenspiel, Octave, and K. B. Bischoff, "Patterns of Flow in Chemical Process Vessels," *Adv. in Chem. Eng.*, **4**, Academic (1963).

Lightfoot, E. N., R. E. Safford, and D. R. Stone, "Biological Applications of Mass Transport Phenomena," *CRC Crit. Rev. Bioeng.*, **1**, 1:69–138 (1971).

Mahler, H. R., and E. H. Cordes, *Biological Chemistry*, Harper and Row (1966).

Rideout, V. C., and R. L. Schaeffer, "Hybrid Computer Simulation of the Transport of Chemicals in the Circulation," 6th Int. Hybrid Comput. Meet., Munich, 1970.

Salkowitz, E., L. Gerende, and L. Wingard, *Dimensions of Bio-engineering*, San Francisco Press, 1968.

Skalak, Richard, "Synthesis of a Complete Circulation," in *Cardiovascular Dynamics*, D. Bergel, Ed., Academic Press (1971).

White, Abraham, Philip Handler, and E. L. Smith, *Principles of Biochemistry*, 4th ed., McGraw-Hill (1968).

# II MOMENTUM TRANSFER

We begin our detailed discussion of biological transport phenomena with a study of fluid flow and rheology, and we put primary emphasis on the flow of blood. This is a logical place to start because of the vital importance of convective transport in all larger animals. We also, however, consider other flow problems of importance, such as ultrafiltration, the lubrication of joints, and the propulsion of motile organisms.

Many of these flow problems are complicated by the non-Newtonian nature of the fluids involved, and it is therefore necessary to start with a discussion of the rheology of biological fluids.

Following the discussion of rheology we consider flow systems of particular biological importance. We begin by reviewing the fluid-mechanical behavior of particulate suspensions, important to understanding the rheology of blood, the diffusion of proteins, and many other important aspects of biotechnology. We also consider briefly in this section the swimming of motile microorganisms.

We next turn our attention to duct flows and give primary emphasis to idealizations of the arterial system. However, we also consider potential flows and flow in porous-walled ducts, which are quite important both within the body and in extracorporeal circuits.

We conclude our discussion of fluid mechanics with approximate descriptions of the circulatory system as a whole. The treatment here is in many ways analogous to the discussion of friction factors in nonbiological texts.

# 1. DESCRIPTIVE BIORHEOLOGY

A few biological fluids, such as pulmonary gases and urine, are simple Newtonian fluids. Most, however, are not. Blood consists of a mixed suspension of collodial proteins. As a result its flow behavior shows considerable dependence on protein concentration and composition. Blood is nearly Newtonian at sufficiently high shear rates, and it can usually be considered so in arterial flow. Its viscosity increases markedly at low shear rates, however, and its flow behavior is strongly influenced by the volume percentage of red cells ("hematocrit") and by chemical composition. The tendency of red cells to aggregate is also important, both in flow and settling characteristics. Aggregation increases during most illnesses and can become very severe. Other fluids such as the lubricants in joints (synovial fluid) show very marked departures from Newtonian behavior and considerable viscoelasticity. The membranes of red cells are also viscoelastic, and it is possible that their rheological properties have an important bearing on their fragility in extracorporeal circuits. Chemical factors are very important here, however. Tendons also exhibit marked viscoelasticity.

## 1.1. Rheology of Blood and Plasma (Cokelet, 1971; Merrill, 1969)

Blood is a concentrated suspension of formed elements (primarily red cells, white cells, and platelets) in plasma. The plasma in turn is a colloidal suspension of the plasma proteins (of which serum albumin, the serum globulins, and fibrinogen will be of most importance to us) in an about

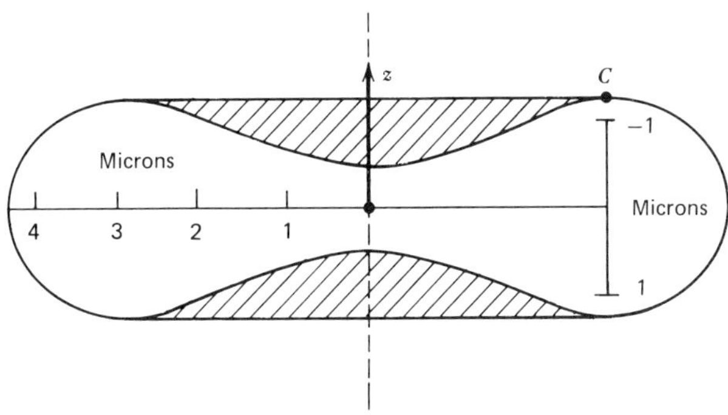


**Fig. 1.1.1.** An idealized human red cell. Diameter = $8.5 \pm 0.41\ \mu$; maximum thickness = $2.4 \pm 0.13\ \mu$; minimum thickness = $1.0 \pm 0.08\ \mu$; mean volume = $87\ \mu^3$; $= 8.7 \times 10^{-11}\text{cm}^3$. From E. Ponder, *Hemolysis and Related Phenomena*, Grune and Stratton (1948).

0.15 $N$ electrolyte of the composition shown previously. The predominant formed elements are the red cells, which are typically in the form of biconcave discs about 8 $\mu$ in diameter (see Fig. 1.1.1) when relaxed, but which can undergo very severe deformations. They typically make up about 93% by number of the formed elements ($5 \times 10^6$ cells/cm$^3$) or about 40 to 45% of the blood volume, and they have a very pronounced effect on blood rheology. The platelets are considerably smaller and are present to the extent of only about 5% of the red-cell volume. They have little direct effect on rheology, but play an important role in clotting. Clotting behavior in turn is strongly affected by flow conditions. The white cells or leucocytes occur in relatively small numbers, and are of little direct importance in rheology.

The major plasma proteins are

| Protein | Concentration (wt %) | Molecular Weight |
|---|---|---|
| Albumin | 4.5 | 65,200 |
| Globulins | 2.5 | 160,000 (gamma) |
| | | (35,000 to 1,000,000) |
| Fibrinogen | 0.3 | 330,000 |

They have a significant direct effect on the viscosities of both blood and plasma, and also an important indirect effect on the low-shear-rate rheology of blood, as is discussed further below. Others, such as the lipoproteins, which tend to denature during extracorporeal processing, may also be of engineering interest. They do not affect blood rheology, however.

From a hydrodynamic point of view the most important characteristics of blood are its apparent viscosity and the sedimentation rates of the red cells. It is also important to note that blood quickly coagulates on removal from the vascular system unless an anticoagulant such as heparin is added. Even then, deposits of the protein fibrin tend to form on container walls, and red cells tend to aggregate. The clotting of blood in extracorporeal circuits is a very serious practical problem.

### (a) *QUALITATIVE SUMMARY OF BLOOD RHEOLOGY*

Normal plasma (7% protein) has a viscosity about 1.8 times that of water at the same temperature; that is, it has a relative viscosity of 1.8. This relative viscosity is nearly independent of both temperature and shear rate. Blood rheology, however, exhibits several anomalous characteristics:

(i) EFFECT OF HEMATOCRIT (PERCENTAGE OF WHOLE-BLOOD VOLUME OCCUPIED BY CELLS)

Blood viscosity, however measured, increases rapidly with hematocrit, as indicated in Fig. 1.1.2. It can be seen that normal red cells cause a much smaller increase in viscosity than oil-in-water emulsions, suspended rigid spheres or discs, or sickle cells. This is undoubtedly due to the flexibility and deformability of the normal red cell and is discussed in the next chapter in connection with the viscosity of concentrated suspensions. It should be noted that because of its shape a red cell can deform without increase of membrane surface area. The extremely high viscosity of sickled-cell suspensions is due primarily to cell rigidity but in part to shape. Normal American hematocrits are about 42% for women and 45% for men, but they vary widely with the degree of bodily activity and the altitude at which one lives. Persons living at an altitude of 15,000 feet typically have blood counts one-third higher than normal, and people suffering from the disease polycythemia vera may have hematocrits of 70–80%.

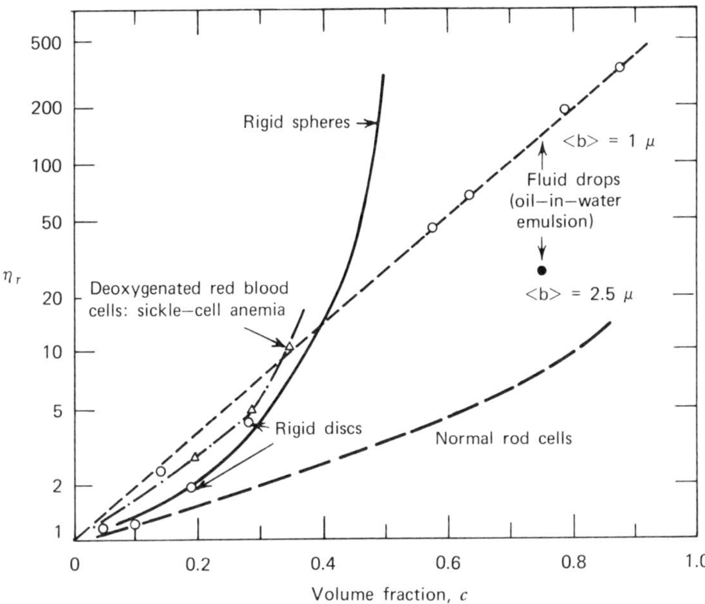


**Fig. 1.1.2** The apparent relative viscosities $\eta_r$ of a normal red-cell suspension as a function of concentration compared with those of suspensions of small rigid spheres, an oil-in-water emulsion (closed circles), and rigid discs (open circles). Also shown are the points obtained with sickled red blood cells (open triangles), demonstrating the effect of shape and rigidity on $\eta_r$. Taken from Goldsmith and Mason.

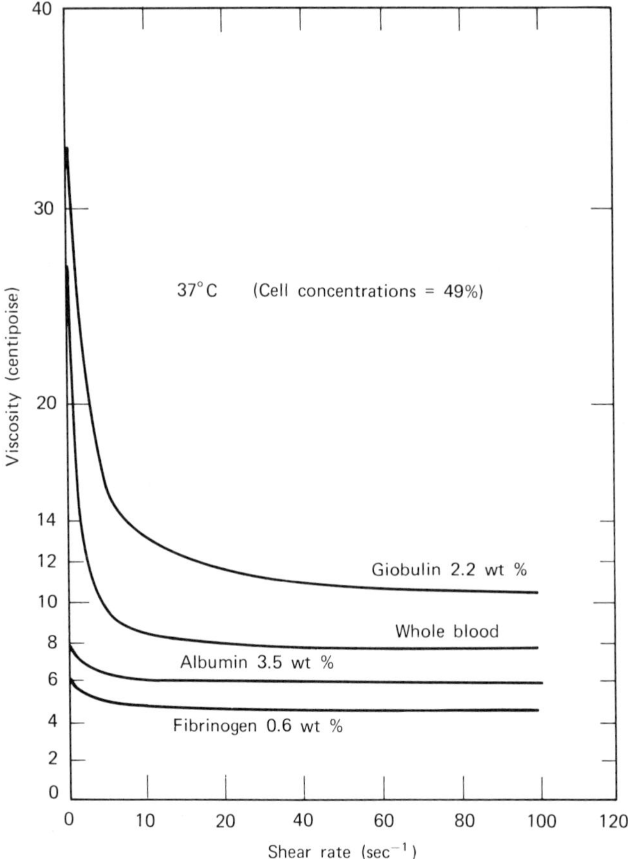


**Fig. 1.1.3.** Comparison of effects of protein fractions upon viscosity of red cell suspensions. From A. L. Copley, Ed., *Symposium on Bio-rheology*, Interscience (1965).

(ii) EFFECT OF TEMPERATURE

The relative viscosity of blood increases by about 10% as the temperature is lowered from 37 to 17°C.

(iii) EFFECT OF SHEAR RATE

The apparent viscosity of blood tends to be very high at low shear rates and to drop toward a lower asymptotic value as the shear rate is increased. Typical behavior is shown in Figure 1.1.3. The effect of higher shear is believed to be primarily in the breakup of aggregates (rouleaux) shown in

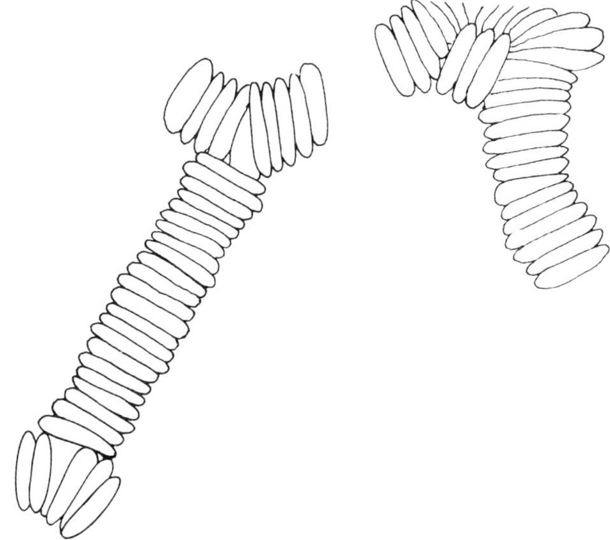

**Fig. 1.1.4.** Photomicrograph of rouleaux of human erythrocytes, showing two separate aggregates. Drawn from a photograph in Goldsmith and Mason. The variation in cell diameter shown is found in blood.

Fig. 1.1.4. Rouleaux formation is also strongly dependent on protein content of the blood, particularly fibrinogen, as discussed below. Rouleaux do not, for example, form when washed cells are suspended in physiological saline.

(iv) EFFECT OF CAPILLARY DIAMETER

When blood viscosity is measured in a capillary viscometer at a high shear rate, where Newtonian behavior is expected (and observed), the measured viscosity decreases with decrease in capillary diameter as shown in Figure 1.1.5. The reasons for this behavior are briefly discussed in the next subsection.

(v) EFFECT OF SERUM PROTEINS (R. E. Wells, Jr., in *Symp. on Biorheol., op. cit.*, pp. 431–435)

All three major serum proteins have an effect on the viscosity of blood, as shown in Figure 1.1.3. Plotted here are apparent viscosity–shear-rate relations as obtained from a Couette viscometer for various red-cell suspensions at a uniform hematocrit of 49%. Curves are shown for whole blood and for solutions of the individual serum proteins at the indicated

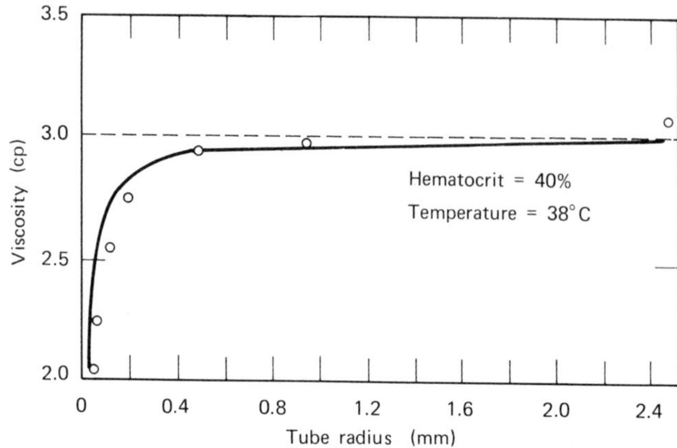


**Fig. 1.1.5.** Viscosity (in centipoise) of erythrocyte suspensions at high shear (asymptotic viscosity), as a function of tube radius. From P. Abbrecht, unpublished notes, Univ. of Michigan, Ann Arbor.

concentrations. Note that the globulin solution has a higher viscosity than the whole blood even though the total protein concentration is lower. The albumin solution shows the lowest viscosity even though it is the most concentrated of the three protein solutions. Fibrinogen results in a high viscosity even at the low concentration of 0.6% and, weight for weight, has slightly more effect than globulin. It is also noteworthy that the globulin has a disproportionately strong effect at very low shear rates and seems primarily responsible for the anomalously high viscosities exhibited at low shear. Similar effects occur *in vivo*, although here quantitative evaluation is difficult. These effects are thought to be related to the effects of the serum proteins on red-cell aggregation, or "rouleau" formation. The effects of low-shear viscosity increases are of most clinical importance in the microcirculation where shear rates tend to be low.

Low-shear-rate viscosity increases are observed in illness and following severe bruising of muscles, presumably because of a change in the serum protein balance (L. E. Gelin, *Symp. on Biorheol., op. cit.,* pp. 299–315; Bengt Zederfeldt, *ibid.,* pp. 397–408). This situation retards the rate of healing substantially and can be ameliorated by reducing blood viscosity—as through the use of plasma expanders.

### (b) *QUANTITATIVE DESCRIPTIONS OF BLOOD RHEOLOGY*

In this section we describe briefly the experimental methods used to characterize the flow behavior of blood and attempt to summarize the results that have been obtained. It should be emphasized at the outset that this is a complex subject, and that considerable care and judgement are required to obtain meaningful data. We shall not have time here to consider all of the practical problems encountered in hemorheology; rather we refer the reader to the authoritative review of G. R. Cokelet (1971).

(i) MEASUREMENT TECHNIQUES

Only three types of viscometers have been used to any appreciable extent for measuring the flow properties of blood: capillary-tube, concentric-cylinder, and cone-and-plate devices. We consider each of these briefly in turn.

*Capillary Viscometers* These are the simplest and cheapest of the three types, and they are of interest here for two reasons: they most nearly correspond to the geometries of actual blood vessels, and they provide simple examples of the anomalous behavior observed in all viscometers.

Their fundamental drawback is their nonuniform stress distribution (*Tr. Ph.*, Fig. 2.3-2):

$$\tau_{rz} = \tau_w (r/R), \tag{1.1.1}$$

where the nomenclature is that of the reference text *Transport Phenomena*,* and $\tau_w$ is the wall shear stress. One can, however, interpret capillary-tube viscometric data in such a way as to relate wall stress and rate of strain, if:

1. The test fluid is homogeneous and does not slip at the wall.
2. The fluid is in steady incompressible laminar flow.
3. The shear stress depends only on shear rate.
4. End effects are negligible.

The desired relation is obtained from the defining flow rate and Eq. 1.1.1 in the following way.

The volumetric flow rate through the tube is given by

$$Q = 2\pi \int_0^R r v_z \, dr \tag{1.1.2}$$

$$= \pi \left\{ \underline{\int_0^R d(r^2 v_z)} - \int_0^R r^2 \left( \frac{dv_z}{dr} \right) dr \right\} \tag{1.1.3}$$

The underlined integral in Eq. 1.1 − 3 is equal to zero, and the remaining

* Primary references, marked with an asterisk, are listed in the bibliography to Section I.

term can be rewritten with the aid of Eq. 1.1.1 as

$$\tau_w^3 Q = -\pi R^3 \int_0^{\tau_w} \dot{\gamma}_{rz} \tau_{rz}^2 \, d\tau_{rz} \tag{1.1.4}$$

where $\dot{\gamma}_{rz} = \partial v_z / \partial r$ is a function only of $\tau_{rz}$ according to assumption 3.

We now take the derivative of Eq. 1.1.4 with respect to $\tau_w$ to obtain

$$3Q\tau_w^2 + \tau_w^3 \frac{dQ}{d\tau_w} = -\pi R^3 \dot{\gamma}_w \tau_w^2 \tag{1.1.5}$$

or

$$-\dot{\gamma}_w = \left[ 3Q + \tau_w \frac{dQ}{d\tau_w} \right] / \pi R^3 \tag{1.1.6}$$

This is the desired relation, and it should be noted that $Q$ and $\tau_w$ are measurable. Thus

$$\tau_w = -\Delta \mathcal{P} \frac{R}{2L} \tag{1.1.7}$$

where $-\Delta \mathcal{P}$ represents the combined drop in pressure and fluid head across the tube (see *Tr. Ph.*, Eq. 2.3.10).

Since $\dot{\gamma}_w$ is a function only of $\tau_w$, it follows that $Q/R^3$ is a function only of $\tau_w = (-\Delta \mathcal{P})(R/2L)$. Then for a given fluid all capillary-tube data should fall on a single curve when $\tau_w$ is plotted against $Q/R^3$. A partial test of Eq. 1.1.7, taken from Merrill, Benis, Gilliland, Sherwood, and Salzman, is shown in Fig. 1.1.6. The prediction is confirmed over a wide range of operating conditions.

Significant dispersion of data is, however, observed at very low shear rates for all tube diameters, and dispersion of the type sketched in Fig. 1.1.5 would have been observed at high shear rates for smaller tubes. These two types of anomalous behavior are not thoroughly understood, but the basic reasons for their occurrence appear reasonably well established. The low-shear anomaly, which produces higher than expected apparent viscosities, appears to result from breakdown of the first assumption made—that the fluid is homogeneous. At low shear rates rouleaux and other less symmetrical aggregates form in whole blood. When their linear dimensions $\lambda$ are an appreciable fraction of the tube diameter $D$, the blood cannot be considered homogeneous, and a new parameter, $\lambda/D$, must be added to the correlation. Sticking of cells or aggregates to the tube wall has a similar effect. The high-shear anomaly, usually known as the Fåhreus-Lindqvist

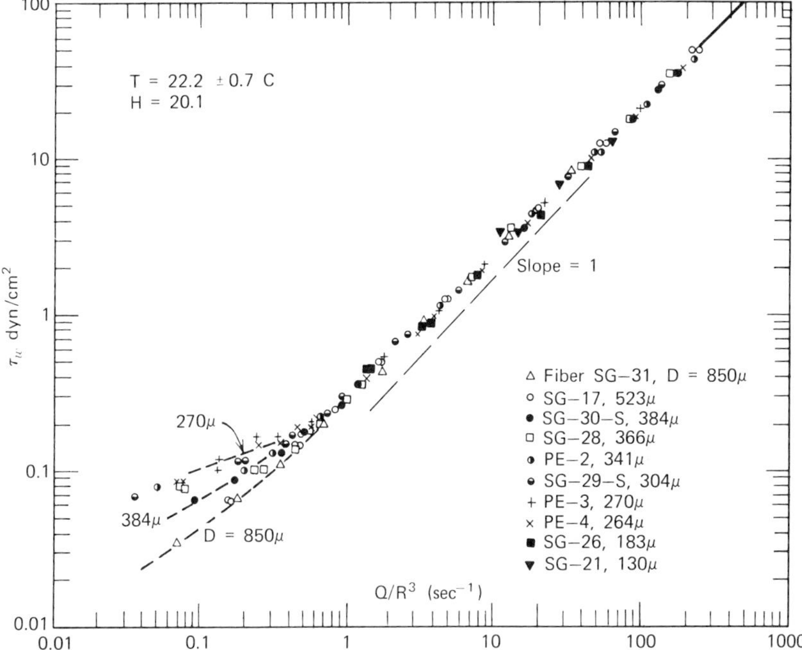


**Fig. 1.1.6.** Test of capillary viscometric measurements on human blood. Taken from Merrill, Benis, Gilliland, Sherwood, and Salzman.

effect, also appears to result from nonuniform blood distribution, and is apparently the result of three separate effects:

1. Red blood cells are partially excluded from the immediate neighborhood of the capillary wall, thereby producing a low-viscosity zone which does not decrease proportionately with diameter for smaller tubes. There is much argument over the origin of this cell-free layer, but its thickness is generally agreed to be of the order of 2 to 4 $\mu$ (see Copley, 1965). It is certainly due in part to the impossibility of a red-cell center's reaching the wall, but there is also a tendency for deformable particles to move toward the central regions of the tube.

2. Because red cells are concentrated in the central, faster-moving portions of the tube, their residence time is less and their mean concentration lower than in either the feed or outflowing blood.

3. There is considerable evidence that red cells are partially blocked from entering small tubes, so that the cell density in the actual feed to the tube is less than that in the feed reservoir.

There appears to be no doubt about the reality of the Fåhreus-Lindqvist effect, but there is still considerable uncertainty about the magnitude of the contributing causes and its biological significance. This effect has been critically discussed by Cokelet (1971) and is touched on in most of the general references listed at the end of this section.

The practical impact of the Fåhreus-Lindqvist effect is shown in Fig. 1.1.7 taken from Barbee and Cokelet (1970). Shown here is a Rabinowitsch plot for flow of blood through a 29-$\mu$-diameter tube, compared with corresponding results for an 811-$\mu$ tube. The circles represent data for flow from a feed reservoir with hematocrit $H_f = 0.559$, and from our foregoing discussion these should fall on the 811-$\mu$ curve—if the blood were truly homogeneous. It is found, however, that the actual hematocrit $H_t$ in the smaller tube is only 0.358. It is also found that a plot for $H_f = 0.358$ in the larger tube coincides with the small-tube data for $H_f = 0.559$, as indicated by the solid line through the circles. The departure from the Rabinowitsch prediction is thus entirely due to the discrepancy between $H_t$ and $H_f$ in this case. The same is true for the other hematocrits shown. The degree of

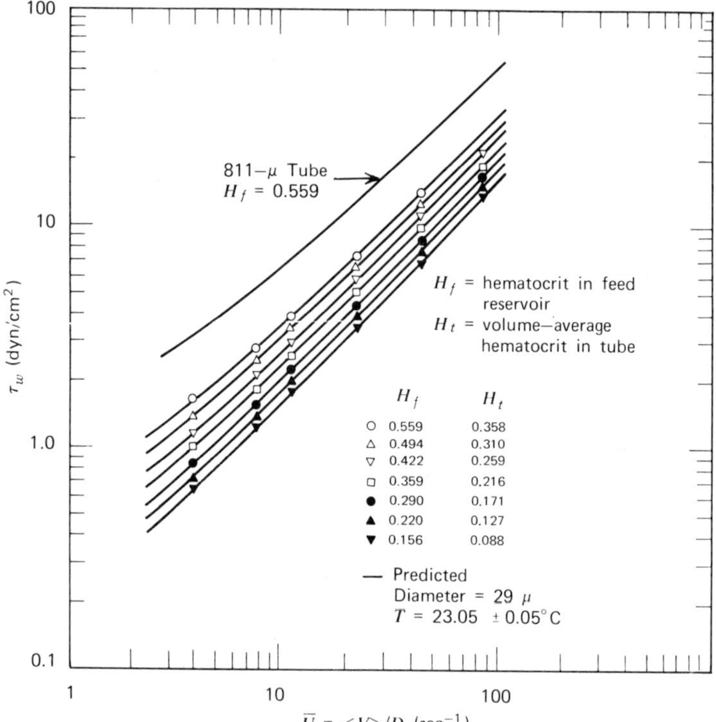


**Fig. 1.1.7.** Importance of the Fåhrens-Lindqvist effect on blood flow in small tubes. From G. R. Cokelet, in Fung, Perrone, and Anliker (1972).

red-cell exclusion and its effect on flow behavior are discussed in the two papers of Barbee and Cokelet.

In practice the interpretation of capillary-tube data is further complicated by settling of red cells, end effects, and other sources of experimental error. Species and individual variations are also substantial, and it is difficult to make general statements about such matters as the effect of red-cell aggregation. These factors are discussed at length by Cokelet (1971).

*Coaxial-Cylinder Viscometers* The instrument most widely used for precise hemorheologic measurements is a highly specialized coaxial-cylinder (couette) viscometer developed by Cokelet and Merrill. The basic geometry and fluid mechanics are very simple, and they are described in standard texts (see, for example, *Tr. Ph.*, Ex. 3.5.1). For these instruments the unidirectional flow field required for viscometric measurements is very closely approximated, and variation of shear stress is very small. The interpretation of the data is therefore straightforward for homogeneous fluids and need not be discussed here. These instruments do, however, display the same types of anomalous behavior resulting from the inhomogeneity of blood as capillary viscometers do. Of particular importance is the migration of red cells away from the two cylindrical faces, which results in a decrease of apparent viscosity with time at low shear rates. This once rather perplexing effect is explained very lucidly by Cokelet (1971).

*Cone-and-Plate Viscometers* These instruments have been used less extensively than the above, and their suitability for hemorheologic measurements has yet to be fully determined. Both the small cone angles and small vertical clearances, otherwise so advantageous (see *Tr. Ph.*, pp. 98, 119), are suspect here: Settling of red cells and wall effects of all types would appear to be particularly serious in this geometry.

(ii) NORMAL OBSERVED BEHAVIOR

Where a close approach to continuum behavior occurs good agreement has been achieved between different measurement techniques, and the following statements are widely accepted:

1. Whereas blood is generally found to be non-Newtonian, it does not appear to exhibit viscoelastic behavior.

2. Most available data can be correlated reasonably well by the Casson equation

$$\tau_{ij}^{1/2} = \tau_y^{1/2} + s\left[-\dot{\gamma}_{ij}\right]^{1/2} \tag{1.1.8}$$

for unidirectional flow. Here

$\tau_{ij}$ = $ij$th shear-stress component,

$\dot{\gamma}_{ij} = \partial v_i/\partial x_j$ = corresponding rate of strain,

$\tau_y^{1/2}$ = a constant (the "yield stress") characteristic of the blood,

$s$ = a second characteristic constant, corresponding to the square root of the apparent Newtonian viscosity approached at very high shear rates.

The first of these statements requires some amplification, and the second considerable qualification; both are provided below. In addition the rheological parameters in Eq. 1.1.8 can be correlated at least to a degree with the protein content of the blood.

All attempts to detect normal-stress effects in blood have failed, and transient effects observed experimentally appear to be due either to (1) damping characteristics of the viscometer, or (2) migration of red cells away from the active surfaces of the viscometer. The rheological behavior of blood is thus considered to be time independent whenever it may be considered homogeneous (in large ducts and at high shear rates).

The flow behavior of representative human blood of various hematocrits at low shear rates is shown in Fig. 1.1.8 (taken from Cokelet, 1971). Careful examination of these data shows that the Casson relation, Eq. 1.1.8,

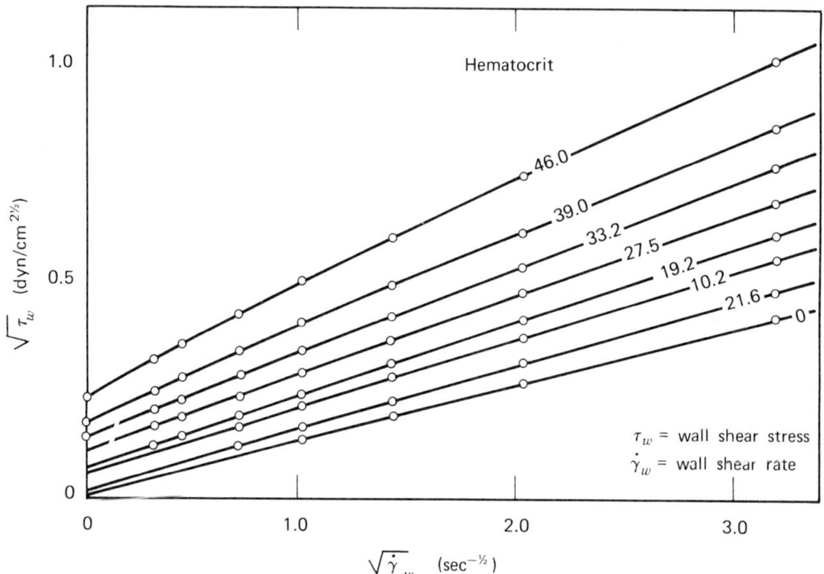


**Fig. 1.1.8.** Casson plot of data for blood of various hematocrits. From G. R. Cokelet, in Fung, Perrone, and Anliker (1972).

provides a good description for low shear rates at all but the higher hematocrits, and for this reason it is widely used for the characterization of arterial blood flows.* Arterial shear rates are, however, relatively high, as we shall see, and blood approaches Newtonian, rather than Casson, behavior under these conditions. Even though $\tau_w$ is much higher here than the yield stress $\tau_y$, the difference between these two models is substantial.

The Casson equation is less reliable in the microcirculation where shear rates are low, and departures from the Rabinowitsch equation are important. Whether there is a true yield stress is controversial; some evidence favors each side of this question. Cokelet (1971) makes a convincing case for the existence of such a limiting stress, partly on the basis of dynamic measurements, and he seems to have the best side of the argument insofar as blood acts as a continuum. The applicability of presently available viscometric measurements in the smaller vessels is, however, open to doubt. This is particularly true in the capillaries, which are barely big enough to accomodate individual red cells.

The chief present goal of hemorheology research is to relate the macroscopic flow behavior of blood to its chemical composition, and in particular its protein content. Such relations are useful both to the engineer (to provide reliable design equations) and to the pathologist and diagnostician. An example of the first type has been provided by Merrill et al. for banked type-O blood (containing ethylenediamine tetraacetic acid as anticoagulant:

$$\tau_y^{1/2} = (H - 0.017)^{1/2}(1.55c_F + 0.76) \tag{1.1.9}$$

$$s = \left[\frac{\mu_0}{(1-H)^{2k}}\right]^{1/2} \tag{1.1.10}$$

where $\tau_y$ = yield stress in dyn/cm$^2$
$H$ = hematocrit (or volume *percent* of red cells in whole blood)
$c_F$ = fibrinogen concentration, g/100 ml
$\mu_0$ = viscosity of the suspending plasma
$k$ = a dimensionless constant of the order of unity, varying with the concentrations of the other plasma proteins.

This is an empirical equation and provides little insight into the mechanism of momentum transfer. The physical significance of rheological

* Although this equation has some theoretical justification it must be considered essentially empirical. It is criticized by some rheologists for not meeting the requirements of material objectivity, but this does not inconvenience us. We are interested only in viscometric flows where the one-dimensional form of Eq. 1.18 is sufficient.

parameters is discussed briefly in Chapter 2 from the standpoint of suspension rheology.

Systematic rheological study of pathological blood is just beginning, but would appear to be quite promising. [See, e.g., Dintenfass in Hershey and Dintenfass (1971).]

## 1.2 Rheology of Hyaluronates: Synovial Fluid

Living organisms produce a very large number of highly viscoelastic and "slimy" fluids with excellent lubricating properties. This is evident, for example, to anyone who has tried to take hold of a living fish with his bare hands, or has successfully swallowed a mouthful of solid food. These lubricants are very dilute sols of hydrophilic polymers: proteins, polysaccharides, or protein-polysaccharide complexes. Typical of these, and of particular medical and engineering interest, is the synovial fluid that lubricates skeletal joints. We consider this material briefly by way of example.

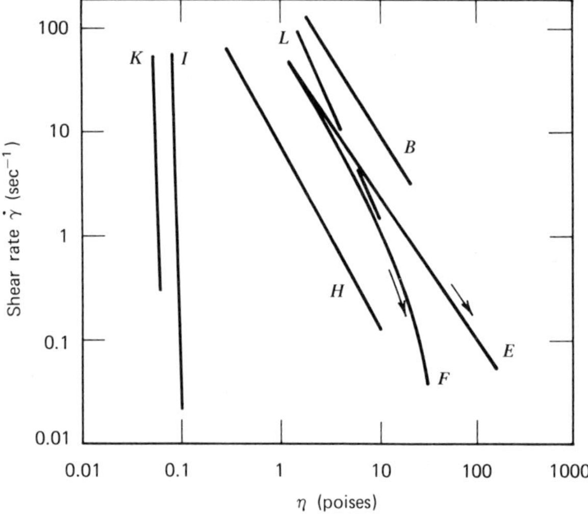


**Fig. 1.2.1.** Flow curves of human and animal synovial fluid. Viscosity $\eta$ is plotted against the rate of shear, $D$, on a log-log scale. $K$ and $I$ represent synovial fluid from rheumatoid arthritis cases; $B$, $E$, and $F$ represent synovial fluid from traumatic arthritis cases. Note that the viscosity of the fluid depends on the rheological history, as illustrated by curves $E$ and $F$, determined consecutively and in opposing directions of rate of shear. Curve $H$ represents ox synovial fluid. Curve $L$ represents a 0.24% solution of hyaluronic acid. Taken from Dintenfass (1966).

Synovial fluid (see bibliography) contains both proteins, for example serum albumin and globulins, and the polysaccharide known as hyaluronic acid, but its unusual rheological properties are due chiefly to the hyaluronic acid. It is known to be viscoelastic, and there is reason to believe that its elasticity is of importance in the lubrication of the joints.

Representative steady-flow measurements are shown in Fig. 1.2.1 taken from Dintenfass (1966). It can be seen that synovial fluid is strongly pseudoplastic (i.e., the apparent viscosity decreases with increasing rate of shear) and that it tends to follow the power-law model. However, it is clear from curves E and F of this figure that it is also elastic; that is, its flow behavior is dependent upon past history.

It is important to note that aqueous solutions of hyaluronic acid behave very much like normal synovial fluid and that they are distinctly non-Newtonian. The fluid of rheumatoid arthritics, on the other hand, is very nearly Newtonian. The frictional properties of normal synovial joints are 10- to 100-fold better than those of commercial friction bearings; frictional coefficients are reported to be less than $10^{-2}$ and as low as $2\times10^{-3}$.

The viscoelastic properties of purified hyaluronates have been measured by Gibbs, Merrill, Smith, and Balasz (1968) in an oscillatory couette viscometer. The inner cylinder of this device is oscillated sinusoidally through a small angle (on the order of 1 degree), and the rheological behavior of the test fluid is observed by measuring the torque exerted on the outer cylinder. The results of these measurements are expressed in terms of a *complex shear modulus* $G^*$ defined by

$$\tau_{r\theta} = -G^*\gamma_{r\theta} \tag{1.2.1}$$

where $\gamma_{r\theta}$ is the rate at which the tangential strain increases with radial distance.

In general the shear stress $\tau_{r\theta}$ and the resultant strain $\gamma_{r\theta}$ are not in phase, and it is therefore convenient to write

$$\tau_R + i\tau_I = (G' + iG'')\gamma_{r\theta}, \tag{1.2.2}$$

where $\tau_R, \tau_I$ = the real and imaginary parts of $\tau_{r\theta}$, respectively
$G', G''$ = the real and imaginary parts of the shear modulus.

The real part $G'$ of the shear modulus relates the in-phase components of stress and strain and thus corresponds to the shear modulus of an elastic solid; it describes the elasticity of the test fluid.

The imaginary part $G''$ of the shear modulus describes the viscous behavior of the fluid, and is closely related to the real part of the complex viscosity defined by

$$\tau_{r\theta} = -\eta^*\dot{\gamma}_{r\theta} \tag{1.2.3}$$

with

$$\eta^* = \eta' - i\eta''. \tag{1.2.4}$$

Thus for a sinusoidal time variation of strain,

$$\dot{\gamma}_{r\theta} = i\omega\gamma_{r\theta}, \tag{1.2.5}$$

where $\omega$ is the frequency of the oscillation and

$$\eta' = \frac{G''}{\omega} \tag{1.2.6}$$

It may be noted that for a Newtonian fluid $\eta' = \mu$, the shear viscosity, and $\eta'' = 0$.

The results of Gibbs, Merrill, Smith, and Balasz are shown in Fig. 1.2.1, taken from their paper. Both ordinate and abscissa are scaled here to adjust for variations of temperature, hyaluronic acid concentration, and pH. The ordinate is defined by

$$G_P = G\frac{T_0 c_0}{Tc}, \tag{1.2.7}$$

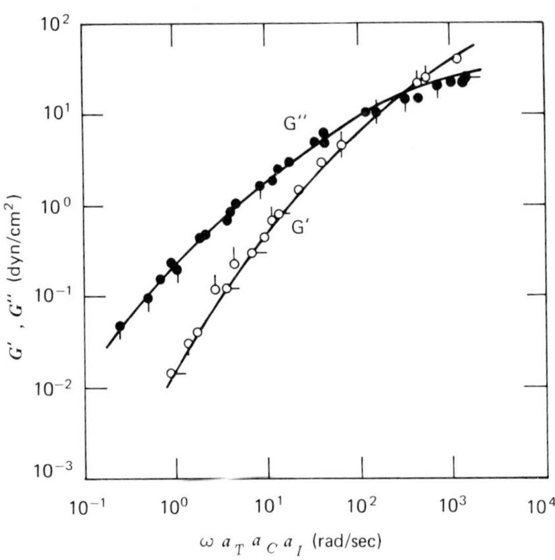


**Fig. 1.2.2.** Master curves for hyaluronic acid, pH7. The parameters are $c_0 = 2.1$ mg/ml, $T_0 = 25°\text{C}$, pH = 1.5, $I_0 = 0.1$.

where $G$ = measured shear modulus at $T$ and $c$
$T$ = absolute temperature of the measurement
$T_0$ = an arbitrary reference temperature
$c$ = hyaluronic concentration in the test fluid
$c_0$ = an arbitrary reference concentration
$\omega$ = oscillation frequency, radians per unit time
$I$ = ionic strength of the test solution
$I_0$ = arbitrary reference ionic strength

The factors $a_T$, $a_c$, and $a_I$ are shift factors to adjust the time scale for variations in temperature, hyaluronate concentration, and ionic strength; each is unity at the reference conditions. The pH and ionic strength were adjusted with HCl and NaCl, and the hyaluronic acid was obtained from umbilical cords stored in acetone.

The adjustments of ordinate and abscissa are essentially empirical, but quite successful. Data are well represented by the two curves shown over the ranges $\omega = 0.125$ to 10.5 rad/sec
$T = 276.66$ to 298.16°K
$c = 2.00$ to 4.09 mg/ml
$I = 0$ to 0.2
pH = 1.5 to 7.

It can be seen that the scatter of the data is not large.

From a practical standpoint the most impressive aspect of these results is the considerably more rapid increase of $G'$ with $\omega$ than that of $G''$. Because of this there is a relatively rapid transition from the elastic region at high frequency ($G > G''$) to the *terminal*, or viscous, region ($G' < G''$). In the elastic region viscous effects, (e.g., energy dissipation) are small. In the terminal region the solution exhibits little elasticity, as the relaxation time is short compared with the period of oscillation. This behavior contrasts strongly with the very prolonged transition normally observed for synthetic polymers.

Gibbs et al. take their success in superimposing data for a wide variety of conditions as an indication that the mechanism of stress relaxation is the same for all.

Unfortunately these investigators did not have the facilities for measuring normal stresses, and I am not aware of any such data from other sources. Normal stresses are to be expected, however, and should be important to the lubrication process.

## 1.3 Rheology of Red-Cell Membranes

It is important to characterize the rheology of red-blood-cell membranes in order to understand their behavior both under normal physiological conditions and in extracorporeal circuits. Red cells are continually exposed to high fluid shear rates and repeatedly pass through capillaries with diameters approximately equal to their own. They undergo frequent, very severe deformations, and yet the human red cell has a "life expectancy" of about 100 days. Furthermore, membrane rupture (hemolysis) is normally the result of inadequate metabolism associated with aging, rather than unusually large mechanical stresses. This is quite remarkable when one considers that the red-cell membrane is only about 70 to 100 Å thick.

Both the physiology and mechanics of red cells and their membranes have been studied extensively, and excellent reviews of these studies are available (see bibliography). We content ourselves here with a brief review of the state of stress in the normal human red cell (following Fung) and an approximate description of membrane rheology (following Rand).

### (a) *THE STATE OF A STRESS IN A BICONCAVE DISC AT EQUILIBRIUM*

Fung assumes the interior of the membrane to be liquid, and bending moments in the membrane to be zero. The first of these assumptions is supported by a large number of data, and the second is reasonable for so thin a membrane. It may now be seen that the membrane at point $C$ in Fig. 1.1.1 cannot restrain the inner and outer portions of the membrane from motion in the $z$ direction, and hence cannot counteract the effects of a nonzero pressure difference. It then follows that the membrane is unstressed in its biconcave form. This may also be seen from the fact that parts of the surface are convex and parts concave [see (b) below].

Fung also points out that the shape of the membrane cannot be determined on the basis of thin-shell analysis and suggests that the biconcave shape is "natural," in the same sense as the relaxed shape of a rubber glove. No convincing explanation has in fact been offered for the observed shape of the red cell, although numerous hypotheses have been put forth. Fung has shown experimentally that thin-walled rubber balloons with relaxed shapes similar to a red cell show swelling behavior similar to that of actual cells.

### (b) *CELL-MEMBRANE RHEOLOGY*

In practice red-cell membranes arc frequently stressed, either because of finite shear rates in the surrounding fluid or because of swelling ("sphering") due to the absorption of electrolytes from the surrounding plasma.

The response of red-cell membranes to tension has been studied carefully by Rand, using a somewhat oversimplified analysis, but a very elegant experimental technique. This method also neglects bending moments in the membrane. It is illustrated in Fig. 1.3.1. Portions of partially swelled red cells are drawn into the tips of capillary pipettes so that both the pipetted and the exposed membrane surfaces are portions of spheres. One may then write

$$p_3 - p_2 = \frac{2s_c}{R_c} \tag{1.3.1}$$

$$p_3 - p_1 = \frac{2s_p}{R_p} \tag{1.3.2}$$

where $R_c, R_p$ = cell and pipette radii,
$s_c, s_p$ = membrane tensions in the exposed and pipetted portions of the membrane.

Equations 1.3.1 and 1.3.2 are justified by the visual observation that the curved surfaces of the membrane shown are truly spherical sections.*

Rand and Burton now further assume that the membrane tension is uniform, so that $s_p = s_c = s$. It is then possible to eliminate the unknown cell interior pressure $P_3$ between Eqs. 1.3.1 and 1.3.2 and to write

$$s = \frac{p_2 - p_1}{2\left(\dfrac{1}{R_p} - \dfrac{1}{R_c}\right)} \tag{1.3.3}$$

It is thus possible to calculate the membrane tension $s$ in terms of the

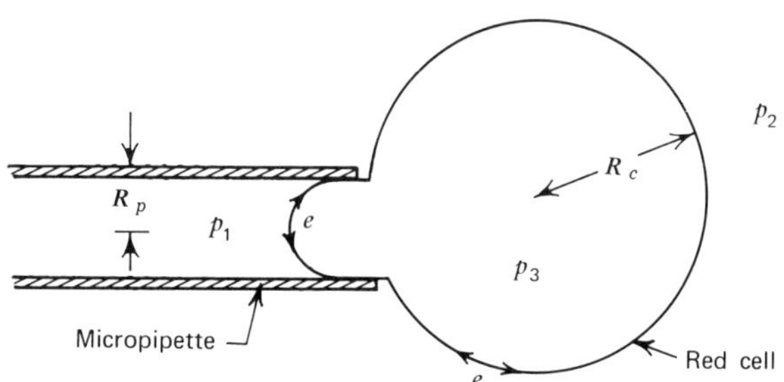


**Fig. 1.3.1** Schematic diagram of a red cell in a hanging drop that has been drawn into a micropipette of radius $R_p$ until the outer part is the portion of a sphere of radius $R_c$. The quantities $p_1$, $p_2$, and $p_3$ are the pressures inside the pipette, inside the hanging drop, and inside the cell, respectively. $p_3 - p_1 = 2s/R_p$; $p_3 - p_2 = 2s/R_c$. Therefore $p_2 - p_1 = 2s(1/R_p - 1/R_c)$.

* For a discussion of surface tension and surface curvature see, for example, Landau and Lifshitz, *Fluid Mechanics*.

measurable pressure difference $p_2 - p_1$. To complete the analysis of membrane rheology, we must still relate membrane tension to strain. This is difficult to do for such small cells, and Rand and Burton use the time to hemolysis following application of stress as an indirect measure of the strain. This must be done on a statistical basis, since there is a substantial dispersion of hemolysis times. Representative smoothed data covering a wide range of imposed tensions and pipette diameters from about 2 to 3 $\mu$ are shown in Fig. 1.3.2. In using these data to develop a rheological model, the following general observations must be considered:

1. The cell volume does not increase appreciably between the application of stress and rupture (hemolysis).
2. The response of the membrane to stress is similar to that of such viscoelastic fluids as polymer melts, and is not that of an elastic solid.
3. In Rand and Burton's experiments any applied stress, no matter how small, ultimately resulted in rupture. There is thus no evidence of a yield stress.

Rand and Burton suggest on the basis of these observations that rupture occurs at some rather small critical increase in surface area, and that the simplest realistic model for this strain is the linear viscoelastic model shown in Fig. 1.3.3. For it the strain $\delta$ is given for a step application of a constant stress $s$ by

$$\delta = s\left\{\frac{1}{Y_2} + \frac{1}{Y_1}\left[1 - \exp\left(-\frac{Y_1}{\eta_1}t\right)\right] + \frac{1}{\eta_2}t\right\} \tag{1.3.4}$$

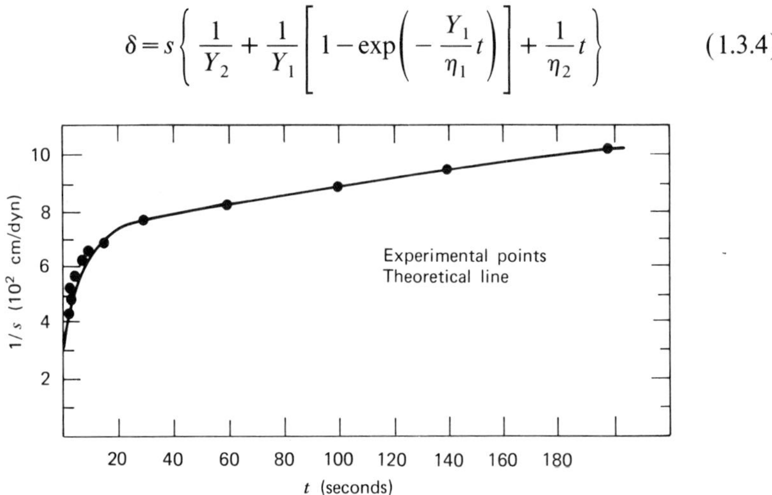


**Fig. 1.3.2.** Plot of the reciprocal of the membrane tension versus time for hemolysis. The experimental points were obtained by the technique outlined in the text. The theoretical line was obtained from the postulated viscoclastic model of Fig. 1.3.3 using the elastic and viscous parameters given in the text.

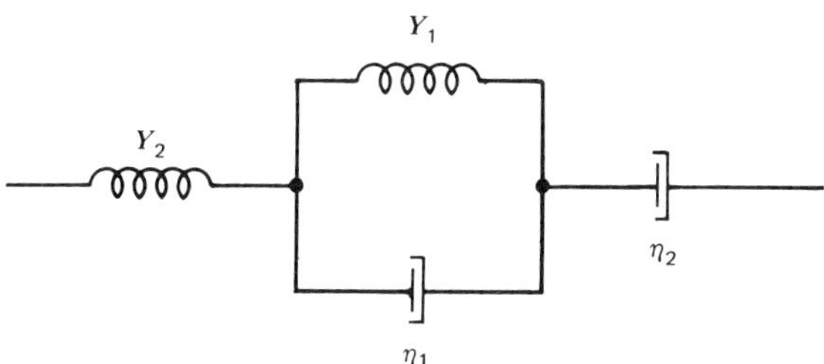


**Fig. 1.3.3.** Mechanical model of the cell membrane used to describe the kinetics of the membrane breakdown.

where $Y_2$ = the elastic (Young's) modulus of an elastic element ("spring") permitting a finite rapid response to imposed strain (this elastic element is needed to account for the apparently instantaneous rupture following application at a sufficiently large stress),

$\eta_2$ = viscosity of the viscous element ("dashpot") responsible for the observed ultimate rupture at even very low stress,

$Y_1$ and $\eta_1$ = corresponding characteristics for the viscoelastic element (spring and dashpot in parallel) required to account for the intermediate shape of the stress-hemolysis time relation.

We may then write

$$\frac{1}{s} = \frac{1}{\delta_c}\left\{\frac{1}{Y_2} + \frac{1}{Y_1}\left[1 - \exp\left(-\frac{Y_1}{\eta_1}t\right)\right] + \frac{1}{\eta_2}t\right\} \qquad (1.3.5)$$

where $\delta_c$ is the critical strain producing hemolysis. The curve of Fig. 1.3.3. is a plot of Eq. 1.3.5 for

$$1/Y_2\delta_c = 3.5 \times 10^{-2}\ \text{cm/dyn}$$

$$1/Y_1\delta_c = 3.9 \times 10^{-2}\ \text{cm/dyn}$$

$$1/\eta_2\delta_c = 1.5 \times 10^{-4}\ \text{cm/dyn sec}$$

$$Y_1/\eta_1 = (8\ \text{sec})^{-1}.$$

For a membrane thickness $d$ of 100 Å and a critical fraction strain $\delta/d = 0.1$,

$$Y_1 \sim Y_2 \sim 3 \times 10^8 \,\text{dyn/cm}^2$$

$$\eta_1 \sim 2.4 \times 10^8 \,\text{P}$$

$$\eta_2 \sim 6 \times 10^7 \,\text{P}$$

These estimates, *which are speculative*, suggest that the membrane viscosity corresponds to that of wax or pitch and that the elastic modulus is between that of elastin ($3 \times 10^6$) and collagen ($10^9$). They do indicate that the cell membrane is a very tough viscoelastic material.

Rand's results differ from earlier work by Katchalsky both in technique and analysis, and most particularly in the inclusion of series elements in his model. For many purposes the results of Rand and Katchalsky are comparable, however, especially if one considers the dispersion of the data.

It may be seen that Rand's analysis requires the red-cell membrane to undergo considerable distortion without damage or effect on the membrane stresses. This basic assumption is untested and may well prove unjustified in more refined experiments. This analysis and the associated experimental results do, however, appear to provide useful insight into the behavior of red cells under stress. It may, for example, explain the increased sensitivity of red cells to hemolysis under long-continued stress. Application of Rand's model to the study of hemolysis in extracorporeal circuits is discussed by Hochmuth, Das, Seshadri, and Sutera (1970).

Cell-membrane rheology is also discussed from a rather fundamental standpoint by Fung (1966, 1968—see specific references on blood rheology), who also provides a detailed criticism of Rand's analysis (1966).

A quite different approach to characterizing red-cell rheology has been taken by Hochmuth and his associates. This group and that of Fung have also made extensive studies of red-cell deformation.

## 1.4. Viscoelastic Solids

The rheological properties of many solid and semisolid portions of the body have also been investigated, in some cases in great detail. These include fibrin clots, the soft tissues of the body, arterial walls, tendons, muscles, and bones. An excellent review of these and other aspects of biomechanics has been provided by Fung.

# BIBLIOGRAPHY

## General

Copley, A. L., Ed., "Symposium on Biorheology," *Proceedings of the Fourth International Conference on Rheology*, Interscience (1965).

Copley, A. L., and G. Stainsby, Eds., *Flow Properties of Blood and Other Biological Systems*, Pergamon (1960).

Dintenfass, Leopold, *Blood Micro-rheology*, Appleton-Century-Crofts (1971).

Eirich, F. R., Ed., *Rheology*, Academic Press (continuing series).

Fung, Yuan-Cheng B., "Biomechanics," *Appl. Mech. Rev.*, **21**, 1: 1–20 (January 1968).

Fung, Yuan-Cheng B., et al., *Biomechanics: Its Foundations and Objectives*, Prentice-Hall (1972).

Wyssling, A. F., Ed., *Deformation and Flow in Biological Systems*, North Holland (1952).

## Blood Rheology

### *REVIEWS*

(See also, above, more general references, especially Fung.)

Attinger, O. A., *Pulsatile Blood Flow*, McGraw-Hill (1964).

Cokelet, G. R., "The Rheology of Human Blood," in Fung (1971)—see above.

Goldsmith, H. L., and S. G. Mason, "The Micro-rheology of Dispersions" in *Rheology*, Vol. Four, F. R. Eirich, Ed., Academic Press (1967).

Hershey, Daniel, Ed., *Blood Oxygenation*, Plenum (1971).

Merrill, E. W., *Physiol. Rev.*, **49**, No. 4 (October 1969).

Wayland, H., "Rheology and the Micro-circulation," *Gastroenterology*, **52**, 2: 342–355 (1967).

### *SPECIAL REFERENCES*

Barbee, J. H., and G. R. Cokelet, *Microvar. Res.*, **3**, 6–16 (1971); ibid 17–21 (1971).

Canham, P. B., and A. C. Burton, "Distribution of Size and Shape in Populations of Normal Human Red Cells," *Circ. Res.*, **22**, 405–422 (1968).

Cokelet, G. R., *Biorheology*, **4**, 123 (1967).

Dintenfass, L., "Blood Viscosity, Internal Fluidity of the Red Cell, Dynamic Coagulation, and the Critical Capillary Radius as Factors in the Physiology and Pathology of the Circulation and Micro-circulation," *Med. J. Aust.*, **1**, 688–696 (1968).

Fung, Y. C., "Theoretical Considerations of the Elasticity of the Red Cells and Small Blood Vessels," *Fed. Proc.*, **25**, 1761–1772 (1966).

Fung, Y. C., "Theory of the Sphering of Red Blood Cells," *Biophys. J.*, **8**, 175–198 (1968).

Goldsmith, H. L., "Red Cells and Rouleaux in Shear Flow," *Science*, **153**, 1406–1407 (1966).

Goldsmith, H. L., "Some Flow Properties of Erythrocytes and Rouleaux," *Bibl. Anat.*, **9**, 259–265 (1967).

Goldsmith, H. L., "Microscopic Flow Properties of Red Cells," *Fed. Proc.*, **26**, 1813–1820 (1967).

Harris, J., "Some Flow Properties of Whole Blood," *Nature*, **209**, 610–611 (1966).

Meiselman, H. J., and E. W. Merrill, "Hemo-rheology: the Effect of Hemodilution and Low Molecular Weight Dextran," *Bibl. Anat.*, **9**, 288–294 (1967).

Meiselman, H. J., "Influence of Plasma Osmolarity on the Rheology of Human Blood," *J. Appl. Physiol.*, **22**, 772–787 (1967).

Merrill, E. W., et al., "Viscosity of Human Blood: Transition from Newtonian to non-Newtonian," *J. Appl. Physiol.*, **23**, 178–182 (1967).

Merrill, E. W., A M. Benis, E. R. Gilliland, T. K. Sherwood, and E. W. Salzman, "Pressure-flow Relations of Human Blood in Hollow Fibers at Low Shear Rates," *J. Appl. Physiol.*, **20**, 954–967 (1965).

Merrill, E. W., W. G. Margetts, G. R. Cokelet, and E. R. Gilliland, "The Casson Equation and Rheology of Blood Near Zero Shear," *Symp. Biorheol., op. cit.*, 135–143.

Merrill, E. W., W. G. Margets, G. R. Cokelet, et al., *ibid.*, 601–611.

Replogle, R. L., H. J. Meiselman, and E. W. Merrill, "Clinical Implications of Blood Rheology Studies," *Circulation*, **36**, 148–160 (1967).

Replogle, R. L., et al., "Hemodilution: rheologic, hemodynamic, and metabolic consequences in shock," *Surg. Forum*, **18**, 157–162 (1957).

Wells, R., "Blood Flow in the Micro-circulation of Man and the Flow Properties of Blood: A Correlative Study," *Bibl. Anat.*, **9**, 520–524 (1967).

Whitmore, R. L., "The Flow Behavior of Blood in the Micro-circulation," *Nature (Lond.)*, **215**, 123–126 (1967).

## The Rheology of Synovial Fluid

Block, B., and Dintenfass, L., "Rheological Study of Human Synovial Fluid," *Aust. N. Z. J. Surg.*, **33**, 108–113 (1963).

Davies, D. V., "Synovial Fluid as a Lubricant," *Fed. Proc.*, **25**, 1069–1076 (1966).

Dintenfass, L., "Lubrication in Synovial Joints: A Theoretical Analysis," *J. Bone Joint Surg. (Amer.)*, **45**, 1241–1256 (1963).

Dintenfass, L., "Rheology in Medicine and Surgery," *Med. J. Aust.*, **2**, 926–930 (1964).

Dintenfass, L., "Rheology of Complex Fluids and Some Observations on Joint Lubrication," *Fed. Proc.*, **25**, 1054–1060 (1966).

Faber, J. J., et al., "Lubrication of Joints," *J. Appl. Physiol.*, **22**, 793–799.

Gibbs, D. A., E. W. Merrill, et al., "Rheology of Hyaluronic Acid," *Biopolymers*, **6**, 777–791 (1968).

Hamerman, D., et al., "Structure and Function of the Synovial Membrane," *Bull. Rheum. Dis.*, **16**, 396–399 (1966).

Lindstrom, J., "Rheological Analysis of Synovial Microcirculation," *Bibl. Anat.*, **7**, 404–409 (1965).

McCutchen, C. W., "Boundary Lubrication by Synovial Fluid: Demonstration and Possible Osmotic Explanation," *Fed. Proc.*, **25**, 1061–1068 (1966).

Ogston, A. G., and Stanier, J. E., "The Physiological Function of Hyaluronic Acid in Synovial Fluid; Viscous, Elastic and Lubricant Properties," *J. Physiol.*, **119**, 244–252 (1953).

### Red-Cell Stresses and Membrane Rheology

Fung, Y. C., "Theoretical Considerations of the Elasticity of Red Cells and Small Blood Vessels," *Fed. Proc.*, **25**, 6: 1761–1772 (1966).

Hochmuth, R. M., N. Mohan Das, V. Seshadri, and S. P. Sutera, "Deformation of Red Cells in Shear Flow," Paper No. 71-104, AIAA 9th Aerospace Sciences Meeting, New York City, January 25, 1971.

Hochmuth, R. M., R. N. Marple, and S. P. Sutera, "Capillary Blood Flow: I. Erythrocyte Deformation in Glass Capillaries," *Microvasc. Res.*, **2**, 409–419 (1970).

Lee, J. S., and Y. C. Fung, "Modeling Experiments of a Single Red Blood Cell Moving in a Capillary Blood Vessel," *Microvasc. Res.*, **1**, 221–243 (1969).

Lighthill, M. J., "Motion in Narrow Capillaries from the Standpoint of Lubrication Theory," in G. E. Wolstenholme and J. Knight, Eds., *Circulatory and Respiratory Mass Transport*, Little Brown (1969).

Rand, R. P., "Mechanical Properties of the Red Cell Membrane. II. Visco-elastic Breakdown of the Membrane," *Biophys. J.*, **4**, 303–316 (1964).

Rand, R. P., "Some Biophysical Considerations of the Red Cell Membrane," *Fed. Proc.*, **26**, No. 6, 1780–1784 (1967).

Seshadri, V., R. M. Hochmuth, P. A.Croce, and S. P. Sutera, "Capillary Blood Flow: III Deformable Model Cells Compared to Erythrocytes in Vitro," *Micro-vasc. Res*, **2**, 434–442 (1970).

Sutera, S. P., and R. M. Hochmuth, "Large-scale Modeling of Blood Flow in the Capillaries," *Biorheology*, **5**, 45–73 (1968).

## 2. THE FLOW BEHAVIOR OF PARTICULATE SUSPENSIONS

We now turn our attention to a quantitative description of the flow behavior of particulate systems at low Reynolds numbers. This is a biologically important subject because it provides improved understanding of such diverse subjects as the diffusion of proteins, the rheology of blood, and the swimming of microorganisms.

We begin by briefly reviewing the underlying hydrodynamics, and then consider a series of successively more complex situations: the translation of particles in a quiescent medium, rotation in a uniform shear field, translation in a nonuniform shear field (very briefly), and, finally, noninertial swimming.

### 2.1. The Quantitative Description of Creeping Flows

The description of any flow problem requires the determination of the three component equations of motion and pressure as functions of position and time. This must be done by integration of the equation of continuity and the three components equations of motion, for appropriate boundary conditions and equations of state. For the incompressible Newtonian fluids of interest to us here the governing equations take the form:

*Continuity*:
$$(\nabla \cdot \mathbf{v}) = 0 \tag{2.1.1}$$

*Motion*:
$$\rho\left[\frac{\partial \mathbf{v}}{\partial t} + \mathbf{v}\cdot\nabla\mathbf{v}\right] = \mu\nabla^2\mathbf{v} - \nabla p + \rho\mathbf{g} \tag{2.1.2}$$

The integration of these equations is hampered by the large number of dependent variables and the nonlinearity of the convective term $\mathbf{v}\cdot\nabla\mathbf{v}$. The first step in their solution is thus a systematic simplification.

We begin by taking the curl of the equation of motion to eliminate* pressure and the gravitational body force $\mathbf{g}$. For two-dimensional and axisymmetric flows one can go still further and express the two nonzero velocity components in terms of a single scalar, the *stream function* $\psi$, as shown in Table 2.1.1 for the three commonest coordinate systems. Since both the two-dimensional and the axisymmetric stream function automatically satisfy the equation of continuity (see Ex. 2.1.1), the three scalar equations describing such flows are reduced to one.

Here we shall be most interested in bodies of finite size and small Reynolds number, and for which the velocity changes slowly with time. Under these conditions, which are characteristic of very many biological situations, Eq. 2.1.2 takes the form

$$0 = \mu\nabla^2\mathbf{v} - \nabla\mathbf{p} + \rho\mathbf{g} \tag{2.1.3}$$

and the left sides of Eqs. A through D of Table 2.1.2 are zero. In this *pseudosteady creeping-flow* limit the equations are thus linear in velocity. The relative unimportance of particle acceleration for conditions of interest to us is shown in Ex. 2.1.2; the effect of neglecting the convective momentum transport $[\mathbf{v}\cdot\nabla\mathbf{v}]$ is discussed authoritatively by Happel and Brenner (1965), and illustrated for spheres in Section 6.3 of *Tr. Ph.*

In principle it now remains only to integrate the appropriate equations to determine velocity profiles. Differential equations for the pressure distribution can then be obtained by putting the velocities so obtained into Eq. 2.1.2 or 2.1.3, and drag forces can be obtained by generalizing the approach used in Section 6.3 of *Tr. Ph.*

In practice, however, one should be aware of the large number of creeping-flow solutions already available and of some very powerful theorems. These are discussed in the general references cited at the end of the chapter, particularly that of Happel and Brenner (1965). Among the

* Here we take advantage of the fact that the curl of the gradient of any scalar is zero. See, for example, Chapter 4 and Appendix A of *Tr. Ph.*

**Table 2.1.1.** Equations for the Stream Function

| Type of Motion | Coordinate System | Velocity Components | Differential Equations for $\psi$ which are equivalent to the Navier-Stokes equation[a] | Expression for Operators |
|---|---|---|---|---|
| Two-dimensional (planar) | Rectangular with $v_z=0$ and no $z$ dependence | $v_x = -\dfrac{\partial\psi}{\partial y}$ <br> $v_y = +\dfrac{\partial\psi}{\partial x}$ | $\dfrac{\partial}{\partial t}(\nabla^2\psi) + \dfrac{\partial(\psi,\nabla^2\psi)}{\partial(x,y)} = \nu\nabla^4\psi$ (A) | $\nabla^2 \equiv \dfrac{\partial^2}{\partial x^2} + \dfrac{\partial^2}{\partial y^2}$ <br> $\nabla^4\psi \equiv \nabla^2(\nabla^2\psi) \equiv \left(\dfrac{\partial^4}{\partial x^4} + 2\dfrac{\partial^4}{\partial x^2\partial y^2} + \dfrac{\partial^4}{\partial y^4}\right)\psi$ |
| | Cylindrical with $v_z=0$ and no $z$ dependence | $v_r = -\dfrac{1}{r}\dfrac{\partial\psi}{\partial\theta}$ <br> $v_\theta = +\dfrac{\partial\psi}{\partial r}$ | $\dfrac{\partial}{\partial t}(\nabla^2\psi) + \dfrac{1}{r}\dfrac{\partial(\psi,\nabla^2\psi)}{\partial(r,\theta)} = \nu\nabla^4\psi$ (B) | $\nabla^2 \equiv \dfrac{\partial^2}{\partial r^2} + \dfrac{1}{r}\dfrac{\partial}{\partial r} + \dfrac{1}{r^2}\dfrac{\partial^2}{\partial\theta^2}$ |
| Axisymmetric | Cylindrical with $v_\theta=0$ and no $\theta$ dependence | $v_z = -\dfrac{1}{r}\dfrac{\partial\psi}{\partial r}$ <br> $v_r = +\dfrac{1}{r}\dfrac{\partial\psi}{\partial z}$ | $\dfrac{\partial}{\partial t}(E^2\psi) - \dfrac{1}{r}\dfrac{\partial(\psi,E^2\psi)}{\partial(r,z)} - \dfrac{2}{r^2}\dfrac{\partial\psi}{\partial z}E^2\psi = \nu E^4\psi$ (C) | $E^2 \equiv \dfrac{\partial^2}{\partial r^2} - \dfrac{1}{r}\dfrac{\partial}{\partial r} + \dfrac{\partial^2}{\partial z^2}$ <br> $E^4\psi \equiv E^2(E^2\psi)$ |
| | Spherical with $v_\phi=0$ and no $\phi$ dependence | $v_r = -\dfrac{1}{r^2\sin\theta}\dfrac{\partial\psi}{\partial\theta}$ <br> $v_\theta = +\dfrac{1}{r\sin\theta}\dfrac{\partial\psi}{\partial r}$ | $\dfrac{\partial}{\partial t}(E^2\psi) + \dfrac{1}{r^2\sin\theta}\dfrac{\partial(\psi,E^2\psi)}{\partial(r,\theta)} - \dfrac{2E^2\psi}{r^2\sin^2\theta}\left(\dfrac{\partial\psi}{\partial r}\cos\theta - \dfrac{1}{r}\dfrac{\partial\psi}{\partial\theta}\sin\theta\right) = \nu E^4\psi$ (D) | $E^2 \equiv \dfrac{\partial^2}{\partial r^2} + \dfrac{\sin\theta}{r^2}\dfrac{\partial}{\partial\theta}\left(\dfrac{1}{\sin\theta}\dfrac{\partial}{\partial\theta}\right)$ |

[a] The Jacobians are defined by $\dfrac{\partial(f,g)}{\partial(x,y)} = \begin{vmatrix} \partial f/\partial x & \partial f/\partial y \\ \partial g/\partial x & \partial g/\partial y \end{vmatrix}$ and $\nu$ is kinematic viscosity.

most important of these solutions are the following, discussed in the indicated sections of Happel and Brenner:

1. A general solution of Eq. 2.1.3 in spherical harmonics, due to Lamb (see Section 3-2).
2. Integral representations, including the point-force approximations of Burgers (see Section 3-4).
3. A description of the rotation of ellipsoids in a shear field, due to Jeffery (see Section 2-3).

We refer to these solutions and to specific theorems below.

### *EX. 2.1.1. THE AXISYMMETRIC STREAM FUNCTION IN SPHERICAL COORDINATES*

Develop the relations between $\mathbf{v}$ and $\psi$ given in connection with Eq. D of Table 2.1.1, and show that any velocity distribution obtained from $\psi$ automatically satisfies Eq. 2.1.1. Use the notation of the Fig. 2.1.1 and the definition

$$\psi(\mathbf{R},t)=\frac{Q}{2\pi} \tag{1}$$

Here $Q$ is the instantaneous volumetric flow rate through the surface produced by rotating the curve $\mathbf{R}$–$\mathbf{R}_0$ about the axis of symmetry ($z$ axis), taken as positive in the negative $z$ direction. Note that for the axisymmetric flow of any incompressible fluid, $Q$ depends only upon $\mathbf{R}$ (and the nature of the flow), not on $\mathbf{R}_0$ or the shape of the curve $\mathbf{R}$–$\mathbf{R}_0$.

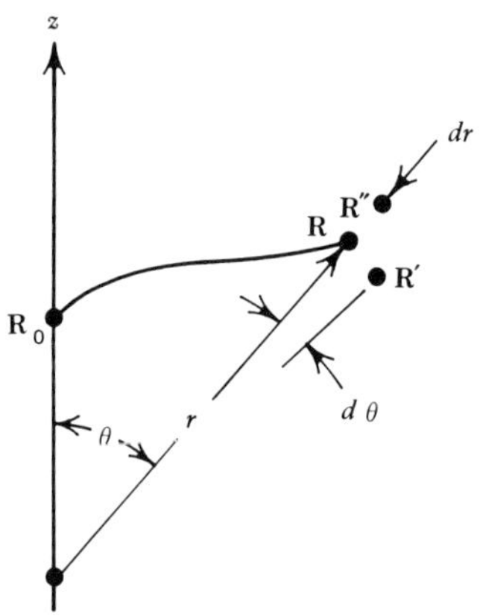


**Fig. 2.1.1.** Coordinate system for Ex. 2.1.1.

SOLUTION

We begin by considering two points $\mathbf{R}'$ and $\mathbf{R}''$ which are close to $\mathbf{R}$ with $\mathbf{R}'$ at the same radial distance from the origin and $\mathbf{R}''$ on the same straight line through the origin.

The flow between $\mathbf{R}'$–$\mathbf{R}_0$ and $\mathbf{R}$–$\mathbf{R}_0$ is

$$(dQ)_r = 2\pi(d\psi)_r \tag{1}$$

$$= -(2\pi r \sin\theta)(r\,d\theta)v_r \tag{2}$$

where

$$dr = |\mathbf{R}' - \mathbf{R}|.$$

Then

$$v_r = -\frac{1}{r^2 \sin\theta}\frac{\partial \psi}{\partial \theta} \tag{3}$$

Similarly

$$(dQ)_\theta = 2\pi(d\psi)_\theta \tag{4}$$

$$= (2\pi R \sin\theta)(dr)v_\theta \tag{5}$$

where

$$d\theta = \frac{|\mathbf{R}'' - \mathbf{R}|}{r}$$

and

$$v_\theta = \frac{1}{r\sin\theta}\frac{\partial \psi}{\partial r} \tag{6}$$

These are the entries in the table.

To complete our solution we note that Eq. 2.1.1 takes the form

$$\frac{1}{r^2}\frac{\partial}{\partial r}r^2 v_r + \frac{1}{r\sin\theta}\frac{\partial}{\partial \theta}v_\theta \sin\theta = 0 \tag{7}$$

in spherical coordinates for axisymmetric flow. Inserting the expressions for $v_r$ and $v_\theta$ (Eqs. 3 and 6 above) into Eq. 7 shows that it is indeed satisfied for any $\psi$.

For a more general and authoritative discussion of axisymmetric flows, see Chapter 4 of Happel and Brenner.

## *EX. 2.1.2. PSEUDOSTEADY FLOWS*

Investigate the importance of particle acceleration by considering the unsteady creeping motion of a sphere.

SOLUTION

It has been shown by Villat (see, e.g., Berker, 1963) that the force required to move a sphere of radius $R$ along a one-dimensional trajectory at arbitrarily varying speed $V(t)$ is

$$F = 6\pi\mu R V(t) + \frac{M}{2} V'(t) + 6\sqrt{\pi\mu\rho}\, R^2 \int_0^t \frac{V'(\tau)}{\sqrt{t-\tau}}\, d\tau \tag{1}$$

Here $M$ = mass of displaced fluid,
$V'(t) = dV/dt$, the instantaneous acceleration,
$t$ = time.

Note that force-velocity relations for three-dimensional trajectories can easily be obtained by superposition because of the linearity of the system being considered.

To illustrate the characteristics of Eq. 1 let us consider a period of constant acceleration followed by a period of constant velocity:

$$\text{For} \quad 0 < t < t_0 \qquad V = at$$

$$\text{For} \quad t > t_0 \qquad V = V_0 = at_0$$

where the acceleration $a$ is a constant. For times later than $t_0$ the integral in Eq. 1 takes the form

$$I = \int_0^t \frac{V'(\tau)\, d\tau}{\sqrt{t-\tau}} = a \int_o^{t_0} \frac{d\tau}{\sqrt{t-\tau}} \tag{2}$$

since the acceleration is zero for $t > t_0$. Changing the variable of integration from $\tau$ to $t-\tau$, we obtain

$$I = -a \int_t^{t-t_0} \frac{d(t-\tau)}{\sqrt{t-\tau}} = 2a\sqrt{t}\left(1 - \sqrt{1 - \frac{t_0}{t}}\right) \tag{3}$$

For $t \gg t_0$ we may write

$$\sqrt{1-\frac{t_0}{t}} \doteq 1-\frac{1}{2}\frac{t_0}{t} \tag{4}$$

or

$$I \doteq \frac{at_0}{\sqrt{t}} = \frac{V_0}{\sqrt{t}} \tag{5}$$

This final result is independent of the acceleration $a$ and is exact for the limiting case of an impulsive (infinitely fast) acceleration to $V_0$. For such a situation Stokes's law should be replaced by

$$F = 6\pi\mu R V_0\left[1+\frac{1}{\sqrt{\pi}}\sqrt{\frac{R^2\rho}{\mu t}}\right] \tag{6}$$

It remains then only to determine the time constant $R^2\rho/\mu$ for any given particle.

For a red blood cell, $R$ is of the order of $4\mu$, and $\mu/\rho$ is about 0.04 $cm^2/sec$ (4cS). Then

$$R^2\rho/\mu \sim 4\times 10^{-6}\,\text{sec}$$

The drag force drops to within 10% of its Stokes's-law value when

$$\left(\frac{R^2\rho}{\pi\mu t}\right)^{1/2} = \tfrac{1}{10}$$

and we may therefore define

$$t_{1/10} = \frac{100R^2\rho}{\pi\mu} \doteq 10^{-4}\,\text{sec}$$

or about one-tenth of a millisecond. Since we are normally interested in times much longer than a millisecond, we can usually assume pseudo-steady-state behavior with little loss in accuracy. Very similar results are obtained for the approach to terminal velocity for a solid particle initially at rest (Villat, 1943) and for fluid spheres (Sy, Taunton, and Lightfoot, 1973). Response times clearly decrease very rapidly with particle radius. As a result it is possible to neglect hydrodynamic transients (though not

Brownian motion) in describing the behavior of protein molecules—which do behave as hydrodynamic particles. We return to this topic in the next section.

## 2.2 Translation of Spheroids in Creeping Flow

We begin here with quantitative descriptions of simple model systems and then discuss the biological significance of these models. Justification of the expressions presented is given only for steady Stokes flow of a sphere. A treatment of steady flow about nonspherical particles and an introduction to the use of the appropriate curvilinear coordinate systems are provided by Happel and Brenner (1965). The works of Jeffery (1915, 1922) and of Burgers (1938) are also of interest.

We consider first the system of Section 2.6, *Tr. Ph.*: a stationary sphere of radius $R$ with fluid approaching from the negative $z$ direction at a uniform approach velocity $v_\infty$. (See Fig. 2.2.1.) We assume the Reynolds number to be sufficiently low that the convective terms in $\mathbf{v} \cdot \nabla \mathbf{v}$ of the

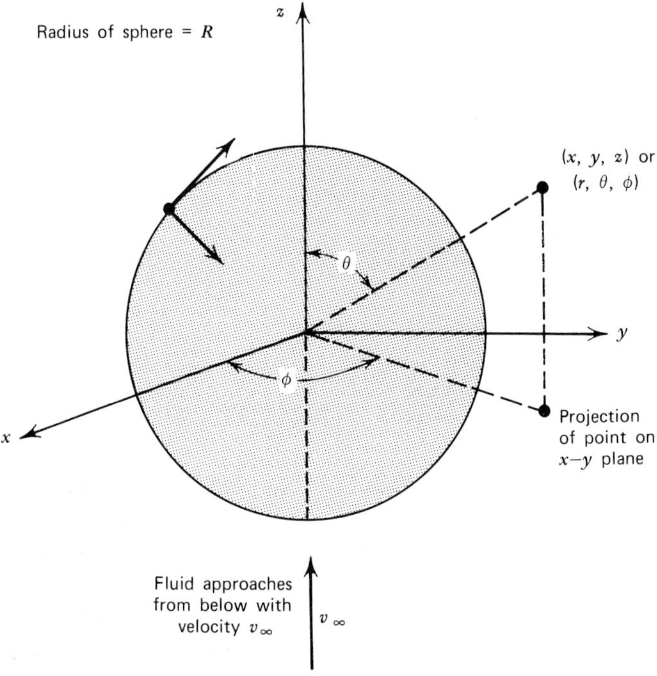


**Fig. 2.2.1.** Coordinate system used in describing the flow of a fluid about a rigid sphere.

equation of motion can be neglected. Then from Eq. D of Table 2.1.1,

$$\left[\frac{\partial^2}{\partial r^2}+\frac{\sin\theta}{r^2}\frac{\partial}{\partial\theta}\left(\frac{1}{\sin\theta}\frac{\partial}{\partial\theta}\right)\right]^2\psi=0 \tag{2.2.1}$$

Equation 2.2.1 is to be solved with the aid of the boundary conditions

1. $v_r=-\dfrac{1}{r^2\sin\theta}\dfrac{\partial\psi}{\partial\theta}=0 \quad \text{at} \quad r=R$

2. $v_\theta=\dfrac{1}{r\sin\theta}\dfrac{\partial\psi}{\partial r}=0 \quad \text{at} \quad r=R$

3. $\psi\to-\dfrac{1}{2}v_\infty r^2\sin^2\theta \quad \text{as} \quad r\to\infty$

The first two boundary conditions describe the clinging of the fluid to the solid sphere surface, and the third implies that $\mathbf{v}=\hat{\delta}_z v_\infty$ far from the sphere, where $\hat{\delta}_z$ is the unit vector in the $z$ direction. This last boundary condition suggests that $\psi(r,\theta)$ is of the form

$$\psi=f(r)\sin^2\theta \tag{2.2.2}$$

When this function is substituted into Eq. 2.2.1, we get*

$$\psi(r,\theta)=\left(\frac{A}{r}+Br-\tfrac{1}{2}v_\infty r^2\right)\sin^2\theta \tag{2.2.3}$$

Application of the remaining two boundary conditions gives

$$\psi(r,\theta)=-\tfrac{1}{2}v_\infty R^2\sin^2\theta\left[\eta^2-\tfrac{3}{2}\eta+\tfrac{1}{2}\eta^{-1}\right] \tag{2.2.4}$$

where $\eta=r/R$. We then find from Table 2.1.1 that

$$\frac{v_r}{v_\infty}=\left[1-\tfrac{3}{2}\eta^{-1}+\tfrac{1}{2}\eta^{-3}\right]\cos\theta \tag{2.2.5}$$

$$\frac{v_\theta}{v_\infty}=-\left[1-\tfrac{3}{4}\eta^{-1}-\tfrac{1}{4}\eta^{-3}\right]\sin\theta \tag{2.2.6}$$

*Note that the third B.C. requires terms in $r^4$ to be zero. See Section 4.2 of *Tr. Ph.* or L. M. Milne-Thomsen, *Theoretical Hydrodynamics*, MacMillan (1955) for a more detailed discussion.

It now only remains to determine the pressure distribution to complete the description of the system.

The pressure can be determined by putting Eqs. 2.2.5 and 2.2.6 into 2.1.3 and integrating the result. However, we are normally interested in only the drag force $F_z$ exerted by the fluid on the particle, and this can be obtained directly from Eq. 2.2.4 and the expression

$$F_z = 8\pi\mu \lim_{r\to\infty} \left\{ \frac{r(\psi - \psi_\infty)}{(r\sin\theta)^2} \right\} \tag{2.2.7}$$

Equation 2.2.7 is applicable[†] for any axisymmetric flow about an arbitrary (but axisymmetric) particle in an unbounded fluid with stream function $\psi_\infty$ at large distances from the particle. Here Eqs. 2.2.4 and 2.2.7 yield

$$F_z = 6\pi\mu R v_\infty \tag{2.2.8}$$

which is the familiar result known as Stokes's law.

We shall return to this result shortly, but first we note that many of the particles we wish to consider later, for example, red blood cells and protein molecules, are nonspherical. It is therefore important to consider the effect of particle shape on velocity profiles and drag forces. Oblate and prolate ellipsoids of revolution are particularly important, as these are frequently used to approximate both red cells and protein molecules.

For nonspherical particles the drag force is dependent on orientation, but for very small particles such as protein molecules Brownian motion is very vigorous. Hence we are interested only in drag coefficients properly averaged for random orientation. We designate the longer and shorter half axes as $a$ and $b$, respectively, and describe the drag force as*

$$F = 6\pi\mu R_0 V \frac{f}{f_0} \tag{2.2.9}$$

where $R_0$ is the radius of a *sphere of equal volume*. For *prolate ellipsoids* (semiaxes $a > b = c$ or football-shaped):

$$\frac{f}{f_0} = \frac{\sqrt{1 - b^2/a^2}}{\left(\frac{b}{a}\right)^{2/3} \ln\left(\frac{1 + \sqrt{1 - b^2/a^2}}{b/a}\right)} \tag{2.2.10}$$

[†] See Happel and Brenner (1965), Section 4-14. An even more powerful result for spheres is given in Section 3-2 of that reference.

* See Tanford (1961) for a lucid discussion.

and for *oblate ellipsoids* (semiaxes $c=a<b$)

$$\frac{f}{f_0}=\frac{\sqrt{a^2/b^2-1}}{(a/b)^{2/3}\tan^{-1}\sqrt{a^2/b^2-1}} \tag{2.2.11}$$

Note that

$$R_0^3=ab^2 \qquad \text{(prolate)} \tag{2.2.12a}$$

$$=a^2b \qquad \text{(oblate)} \tag{2.2.12b}$$

and that $f/f_0$ is always greater than unity.

The above results can be used to determine sedimentation rates of red cells and other particles, and this is important in centrifugal separations. However, they are most useful for describing the diffusional behavior of proteins and the rheology of suspensions. This is indicated in the following examples.

### *EX. 2.2.1. THE HYDRODYNAMIC THEORY OF DIFFUSION*

Show how the above results can be used to predict the diffusion coefficients for dilute solutions of large spheroidal molecules such as proteins. Look ahead to Chapter IV, Section 1 to obtain the generally applicable relation.

$$\mathbf{d}_A=\frac{x_A x_B}{Ð_{AB}}(\mathbf{v}_B-\mathbf{v}_A) \tag{1}$$

where $cRT\mathbf{d}_A$ is the force per unit volume of solution acting to move "solute" $A$ relative to "solvent" $B$, $x_i$ is the mole fraction of species $i$, $\mathbf{v}_i$ is the velocity of species $i$, $Ð_{AB}$ is the diffusivity, $R$ is the international gas constant, $T$ is the absolute temperature, and $c$ is the total molar concentration.

SOLUTION

For our hydrodynamic model, Eq. 2.2.9, we may set

$$cRT\mathbf{d}_A=6\pi\mu R_A(\mathbf{v}_A-\mathbf{v}_B)\frac{f}{f_0}\left(\tilde{N}c_A\right) \tag{2}$$

where $\tilde{N}$ is Avogadro's number and $c_A=cx_A$ is the molar concentration of "solute."

For a dilute solution of large molecules, $x_B \approx 1$, and substitution of Eq. 2 into Eq. 1 gives

$$Đ_{AB} = \frac{RT}{6\pi\mu R_A (f/f_0)\tilde{N}} \tag{3}$$

which is our desired result. It was first obtained for spheres ($f = f_0$) by Einstein, using a very different approach, and is disucssed in standard references (e.g., Tanford, *Tr. Ph.*). It has been widely used for estimating the shapes of protein molecules* as well as in the prediction of diffusion coefficients. The nature of the forces producing relative motion at a molecular level are discussed in Chapter IV, Section 1.

*EX. 2.2.2. ROTATION OF A RIGID DUMBBELL*

The above analysis of pseudosteady translation of spheres can be extended to a rotating dumbbell as first suggested by Kuhn. Begin by considering two spheres of radius $R$ connected by a rigid thin rod of length $2l$ in a velocity field

$$\mathbf{v} = \hat{\delta}_x \dot{\gamma} y$$

where $\dot{\gamma}$ is a constant (see Fig. 2.2.2). Describe the rate of rotation of the dumbbell and the force on the connecting shaft. To do this assume that (i)

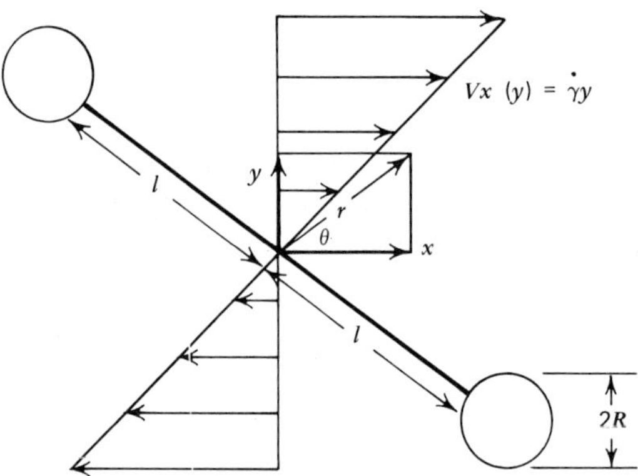


**Fig. 2.2.2.** A rigid dumbbell rotating in a shear field.

* Although much more powerful methods are now available, these are expensive and require highly specialized equipment. Measurements of mobility and viscosity are therefore still useful.

The spheres are too far apart to interact hydrodynamically, (ii) the pseudosteady form of stokes's law holds for each sphere, and (iii) perturbation of the shear field due to the presence of each sphere is negligible.

SOLUTION

We begin by comparing the motion of one sphere with that of the fluid:

*Sphere velocity*:

$$\mathbf{v}_s = \hat{\delta}_\theta l\dot{\theta}$$

$$v_{xs} = -l\dot{\theta}\sin\theta \tag{1}$$

$$v_{ys} = l\dot{\theta}\cos\theta$$

*Fluid velocity*:

$$\mathbf{v}_f = \hat{\delta}_x \dot{\gamma} l \sin\theta \tag{2}$$

where $\dot{\theta} = d\theta/dt$.

We next calculate the forces in the $x$ and $y$ directions exerted on the sphere because of its motion relative to the fluid:

$$\mathbf{F} = 6\pi\mu R(\mathbf{v}_f - \mathbf{v}_s) \tag{3}$$

$$= 6\pi\mu R\left[\hat{\delta}_x\left(\dot{\gamma}l\sin\theta + \dot{\theta}l\sin\theta\right) + \hat{\delta}_y\left(-\dot{\theta}l\cos\theta\right)\right] \tag{4}$$

from Eq. 2.2.8.

We now note that the torque exerted on the connecting shaft is

$$\mathcal{T} = \mathbf{F}\times\mathbf{r} = \hat{\delta}_z(F_x y - F_y x)$$

$$= 6\pi\mu R l^2 \hat{\delta}_z\left[\left(\dot{\gamma} + \dot{\theta}\right)\sin^2\theta + \dot{\theta}\cos^2\theta\right] \tag{5}$$

Since the torque must be zero within the framework of our assumptions,

$$\dot{\theta} = -\dot{\gamma}\sin^2\theta \tag{6}$$

and the dumbbell ultimately spends all of its time aligned with the shear field. This result, however, ignores Brownian* motion, and torques exerted directly on the spheres (if they are of finite size).

* Perturbations resulting from Brownian motion will tend to keep the sphere rotating, and also have other profound effects, as described, for example, by Bird, Warner, and Evans (1971).

Even though Eq. 6 is unreliable for predicting rotation rate at very small $\theta$, it is useful for other orientations and can be used for calculating the force on the rod. This force must be radial and is given by

$$F_r = F_y \sin\theta + F_x \cos\theta$$

$$= 6\pi\mu\dot{\gamma} l \mathrm{R} \sin\theta \cos\theta \qquad (7)$$

It is alternately compressive and tensile and has a maximum magnitude given by

$$|F_{\max}| = 3\pi\mu R\dot{\gamma} l \qquad (8)$$

when $\theta = \pi/4 + n\pi/2$, where $n$ is any integer.

This simple treatment is sufficient to explain qualitatively the breakup of rouleaux in a shear field and the effect of shear rate on the apparent viscosity of blood. The force tending to separate neighboring cells in an aggregate increases linearly with shear rate, as predicted by Eq. 8, and in a more complex way with aggregate length. The maximum stable rouleau length thus decreases with increasing shear rate, and, as we shall see in Section 2.3, the apparent viscosity of a suspension decreases as the length-to-diameter ratio of the suspending particles becomes smaller. A surprising amount of quantitative information can also be obtained by extensions of this treatment which are, *in principle*, straightforward.

The most obvious extensions are to consider three-dimensional flows and more complex shapes. It has been shown by Burgers (1938) (see also Happel and Brenner, 1965, expecially pp. 82–85) that interactions between neighboring particles distributed along the rigid rod of our example are only of secondary importance, and can be neglected for many practical purposes (the *point-force approximation*). One can in this way obtain useful descriptions of the motion of a wide variety of particles in shear fields. Among the most important shapes that have been investigated are ellipsoids of revolution and chains of spheres. Purely hydrodynamic analyses —either such approximate treatments or the more rigorous analyses by Jeffery—are normally sufficient to describe the flow behavior of particles the size of red cells or larger. We return to them in Section 2.3.

For smaller particles, such as protein molecules, Brownian motion must be considered. Such motion tends to randomize the particle orientation and can have a profound effect on the rheology of suspensions. The interaction of hydrodynamic and Brownian forces on suspended particles has been intensively investigated, and is discussed in several of the references in the bibliography. The work of R. B. Bird and his associates, summarized in Bird, Warner, and Evans (1971), is of particular interest for the insight it provides into polymer rheology. These investigators have shown that suspensions of rigid dumbbells exhibit all of the viscoelastic behavior associated with polymer suspensions and thus provide a very useful rheological model.

## 2.3 The Motion of Spheroids in a Uniform Shear Field

In this section we wish to provide an elementary introduction to suspension rheology and a brief description of the rotation of such particles as red blood cells in a shear field. We do this by considering the behavior of ellipsoids of revolution suspended in the simple velocity field

$$\mathbf{v} = \hat{\delta}_x \dot{\gamma}_0 y \tag{2.3.1}$$

for which the shear rate $\dot{\gamma}_0$ is a constant (see Fig. 2.3.1). The results we obtain are useful for describing many much more complex flow situations, in particular the effective viscosity of suspensions. They are not applicable to all flow situations by any means, however, and the range of their utility is not always known. The very comprehensive and authoritative review of suspension rheology by Howard Brenner is highly recommended to those seeking either deeper insight or more detailed information.

We begin by considering the effect of an isolated solid sphere on the system of Fig. 2.3.1, where the shear rate $\dot{\gamma}_0$ far from the sphere is $V/Y$. The perturbation of this velocity field resulting from the presence of the sphere can be shown to be

$$v_x' = -C\frac{x^2y}{r^5} + \underline{B\left(\frac{3y}{r^5} - \frac{15x^2y}{r^7}\right)} \tag{2.3.2}$$

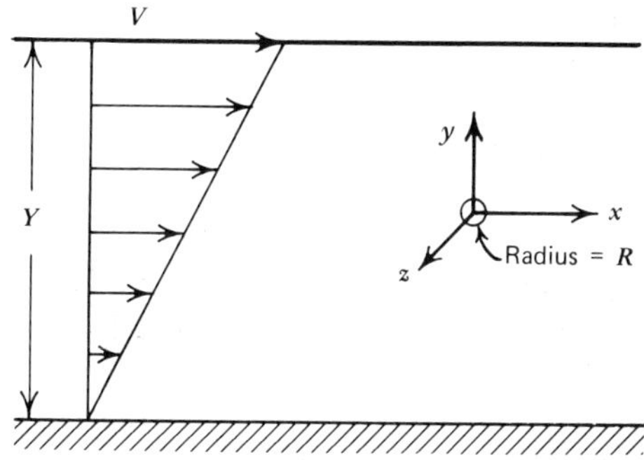


**Fig. 2.3.1.** A solid sphere in a shear field.

$$v_y' = -C\frac{xy^2}{r^5} + \underline{B\left(\frac{3x}{r^5} - \frac{15xy^2}{r^7}\right)} \tag{2.3.3}$$

$$v_z' = -C\frac{xyz}{r^5} - \underline{B\left(\frac{15xyz}{r^7}\right)} \tag{2.3.4}$$

where

$$r = \sqrt{x^2+y^2+z^2}\ ; \tag{2.3.5}$$

$$B = -\tfrac{1}{6}\dot{\gamma}_0 R^5; \tag{2.3.6}$$

$$C = \tfrac{5}{2}\dot{\gamma}_0 R^3 \tag{2.3.7}$$

The total velocity field $\mathbf{v} = \mathbf{v}^0 + \mathbf{v}'$ approaches $\mathbf{v}^0$ at large $r$, but the effect of $\mathbf{v}'$ is very important, as we shall see. The underlined terms, however, can often be neglected. Inspection of these equations shows that the presence of the sphere

(1) slows the $x$ component of the fluid velocity above the sphere ($y>0$) and increases it below;

(2) causes a downward motion "before" the sphere ($x>0$) and an upward motion "behind" it ($x<0$).

The sphere itself is rotated by the fluid at an angular velocity $\Omega = \frac{1}{2}\dot{\gamma}_0$ about the $z$ axis. These motions have three effects of particular interest to us:

1. The circulation of fluid about the sphere enhances mass transfer, particularly in the $y$ direction, an effect that appears to have first been noted by Blackshear (1965). It follows that the apparent diffusion coefficient in a particulate suspension undergoing shear is in general anisotropic, and its magintude is dependent on the rate of shear. Suspensions cannot, therefore, always be treated as continua. This effect is more important for larger particles and is noticeable, but of minor significance, for red cells.

2. The surface of the sphere is subjected to shear stresses which tend to deform it, and which are greatest at the equator. Similar stresses occur for fluid spheroids and are discussed by Goldsmith and Mason (1967). The shear-induced distortion of red cells is discussed by Hochmuth, Marple, and Sutera (1970), by Nevaril, Hellums, Alfrey, and Lynch (1969), and by Shapiro and Williams (1970), among others.

3. The presence of the suspended solids increases the effectiveness of momentum transfer and hence the *apparent* or *effective viscosity* of the suspension.

The remainder of this section is devoted primarily to a description of the effective suspension viscosity and is based on an earlier development due to Burgers.

First, however, we review very briefly the behavior of nonspherical particles, which is much more complex than that of spheres and can result in such varied phenomena as flow birefringence and non-Newtonian rheological behavior. The orientation of the particle is now important, and it in turn is affected both by flow conditions and by Brownian motion. Both flow birefringence and rheological behavior have been much studied in attempts to determine the structure of macromolecules such as proteins. Important results obtained up to about 1960 are summarized by Tanford (1961). The results of hydrodynamic analyses have been reviewed by Happel and Brenner (1965) and more recently by Brenner (1970).

The creeping-flow motion of ellipsoids of revolution in simple shear fields of the type considered above was treated in detail by Jeffery in 1922. Jeffery's analysis has since been extended to particles of arbitrary shape by Brenner (1964), but we shall consider only the simpler geometries of Jeffery.

The orientation of an ellipsoid of revolution can be specified in terms of the spherical coordinates $\phi$ and $\theta$ of any point on its axis, as shown in Fig. 2.3.2. For the velocity field of Eq. 2.3.1 one may write (in the absence of Brownian motion):

$$\frac{\tan\theta}{\tan\theta_0} = \left(\frac{a^2\cos^2\phi_0 + b^2\sin^2\phi_0}{a^2\cos^2\phi + b^2\sin^2\phi}\right) \tag{2.3.8}$$

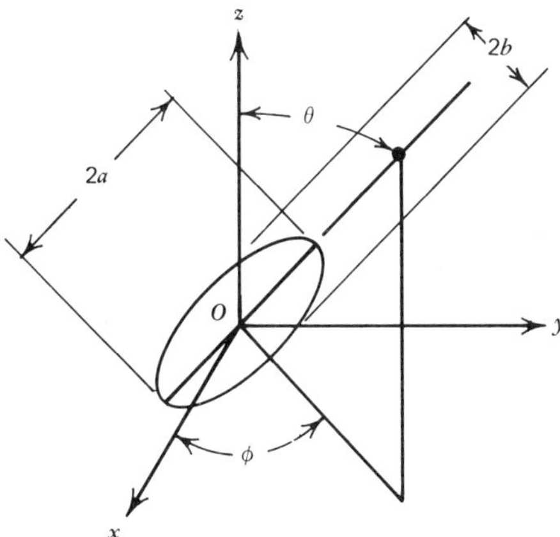


**Fig. 2.3.2.** Orientation of an ellipsoid.

where $(\theta_0, \phi_0)$ represents a reference position, and the half axes of the ellipsoid are $(a,b,b)$ as indicated in the figure. The rate of rotation is given by

$$\tan\phi = \frac{a}{b}\tan\left(\frac{\dot{\gamma}abt}{a^2+b^2}\right) \tag{2.3.9}$$

and the period of rotation is given by

$$T = 2\pi\frac{a^2+b^2}{ab\dot{\gamma}} \tag{2.3.10}$$

The sphere corresponds to the special case of $a=b$, so that for it $\theta=\theta_0, \phi=\frac{1}{2}\dot{\gamma}t$, and $d\phi/dt=\Omega=\frac{1}{2}\dot{\gamma}$, as previously described. For $a\neq b$ the ends of the ellipsoid describe elliptical paths, and the rate of rotation is nonuniform. As for the dumbbell described earlier in an approximate way, the orientation changes most slowly when the ellipsoid is most nearly aligned with the shear field (its axis in the $x$-$z$ plane).

The fluid stresses acting on the ellipsoid surface can be expressed as the sum of two couples, one tending to put the axis in the $x$-$y$ plane and the other tending to align the ellipsoid with the $z$ axis. The velocity fields about rotating ellipsoids have been calculated by Jeffery, but are not given here.*

We are now ready to indicate the significance of the above results by way of representative examples.

### *EX. 2.3.1. THE EFFECTIVE VISCOSITY OF SUSPENSIONS*

Show that the macroscopic flow behavior of a particulate suspension can be approximated as that of a homogeneous fluid of greater viscosity than the suspending medium. Calculate the effective viscosity for dilute suspensions of spheres, and discuss the rheology of blood in the light of the rheology of suspensions. Assume that the suspension contains a very large number of particles and that both particle diameters and particle spacing are small relative to the characteristic lengths† of the system in which they flow.

* See also Burgers for useful approximations to these velocity fields and corresponding surface stresses.

† This implies that fractional changes in velocity between adjacent particles are small.

SOLUTION

We define the effective viscosity* in a manner analogous to that in Section 1.1 of *Tr. Ph.* for the couette system of Fig. 2.3.1:

$$\mu_{\text{eff}} = \frac{-\tau_{yx}}{V_x/Y} \tag{1}$$

$$= -\frac{\left(\tau_{yx}^0 + \tau_{yx}'\right)Y}{V_x^0 + V_x'} \tag{2}$$

where $V_x$ = the velocity of the upper surface of our figure relative to that of the lower.

$\tau_{yx}$ = the magnitude of the shear stress at either surface.

Once again the superscript 0 refers to the unperturbed conditions existing in he absence of the particles, and the prime refers to the perturbations resulting from the presence of particles.

Since we are interested only in such large systems that $R \ll Y$ the underlined terms in Eqs. 2.3.2 to 2.3.4, which are of the order $r^{-4}$, may be neglected.† We next note that for a sufficiently large *number concentration* of particles, perturbations in the neighborhood of the bounding surfaces $s$ are uniform over the surfaces. Then, if the *volume fraction* of particles is small enough that we may ignore particle interactions, we need only consider surface-average perturbations from representative particles. Furthermore if we consider the dimensions of the bounding surfaces to be large relative to $Y$ the surface-average perturbation velocities are

$$(\mathbf{V}')_m = \frac{1}{s}\int_{-\infty}^{\infty}\int_{-\infty}^{\infty} \mathbf{v}'\,dx\,dz \tag{3}$$

We next note that since both $v_y'$ and $v_z'$ are odd functions of $x$,

$$(v_y')_m = (v_z')_m = 0 \tag{4}$$

and we need only consider $(v_x')_m$ further.

* Our approach is similar to that of Burgers (1938). For alternate treatments see Happel and Brenner (1965) or Landau and Lifshitz (1959).

† It is worth while to note that the remaining terms can be obtained rather easily using the point-force technique of Burgers (1938). This technique can be very useful for particles of complex shape.

We may also see that $(v_x')_m$ is independent of $y$-position:

$$(v_x')_m = -yC\left\{\frac{1}{s}\int_{-\infty}^{\infty}\int_{-\infty}^{\infty}\frac{x^2\,dx\,dz}{(x^2+y^2+z^2)^{5/2}}\right\} \tag{5}$$

$$= \mp\tfrac{5}{4}\dot{\gamma}\frac{\frac{4}{3}\pi R^3}{s} \tag{6}$$

Here the minus and plus signs refer to conditions at the upper and lower surfaces, respectively. This is the key to the argument, and it follows both that any one particle is representative and that $\tau_{yx}'$ is zero.

If we now put Eq. 5 into Eq. 1 we obtain

$$\mu_{\text{eff}} = \frac{\mu\dot{\gamma}_0}{\dot{\gamma}_0\left[1-\frac{5}{2}\left(\frac{4}{3}n\pi R^3 Ys\right)\right]}$$

$$= \frac{\mu}{1-\frac{5}{2}\varphi} \tag{7}$$

where $n$ is the total number of particles, and

$$\varphi = \frac{\frac{4}{3}n\pi R^3}{Ys}$$

is the volume fraction of particles in the suspension. This result is independent of particle size and can be extended to any three-dimensional flow consistent with our basic assumptions. The specific geometry and flow conditions are thus unimportant, but the limitations of Eq. 7 should always be kept in mind. For example, its validity in turbulent flows is open to doubt unless the smallest eddies are much bigger than the interparticle distances.

Somewhat surprisingly, Eq. 7 is in good agreement with available data for finite $\varphi$. Thus the expansion of this equation

$$\frac{\mu_{\text{eff}}}{\mu} \doteq 1 + \tfrac{5}{2}\varphi + \left(\tfrac{5}{2}\varphi\right)^2 + \cdots \tag{8}$$

agrees reasonably well with data on glass spheres, milk fat, latex, and asphalt suspensions to order $\varphi^2$ (see Happel and Brenner, 1965, p. 464). It also agrees reasonably well with an analytical extension due to Kynch:

$$\frac{\mu_{\text{eff}}}{\mu} \doteq 1 + \tfrac{5}{2}\varphi + \tfrac{30}{4}\varphi^2 + \cdots \tag{9}$$

and the empirical expression due to Mooney:

$$\ln\left(\frac{\mu_{\text{eff}}}{\mu}\right) \doteq \frac{\frac{5}{2}\varphi}{1-c\varphi} \tag{10}$$

where $c$ is a constant between 1.35 and 1.91 specific to the particles used [see Happel and Brenner (1965); Goldsmith and Mason (1967)]. This agreement is, however, accidental.

The primary source of viscous dissipation in concentrated suspensions of spheres is due to particle-particle interactions, and most probably to the high velocity gradients produced when two neighboring particles are forced toward each other (see Frankel and Acrivos, 1967). This effect was not even considered in the preceding argument.

We now turn our attention to ellipsoids and note that the effective viscosity can be calculated for any instantaneous orientation using the pseudosteady-state approximation discussed above. The results may be expressed in the form

$$\Lambda^{(\theta,\phi)} = \left(\frac{\mu_{\text{eff}}^{(\theta,\phi)}}{\mu} - 1\right)\frac{1}{\varphi} \tag{11}$$

so that $\Lambda = \frac{5}{2}$ for a sphere. The superscripts $(\theta, \phi)$ denote limitation to a particular orientation.

Since all possible orientations may be expected to exist in a real suspension, it remains to take a properly weighted average of these various orientations. This is in general difficult to do, but two simple limiting situations may be readily handled:

1. Probabilities calculated on a purely hydrodynamic basis from Eqs. 2.3.8 and 2.3.9, with all $\theta_0$ and $\phi_0$ equally probable.
2. All $\theta$ and $\phi$ considered equally probable.

Since hydrodynamic probabilities favor orientations producing minimum specific viscosities, the first of these limiting calculations predicts the lower viscosities. It is usually considered to represent a limit for large particles and high shear rates, where the effect of Brownian motion is small. The second set of calculations is thought to be reasonable for small particles and low shear rates, where the randomizing effect of Brownian motion is dominant. The differences between $\mu$ calculated in these two ways increases with asymmetry, as shown in the Table 2.3.1 (taken from Burgers, 1938) for prolate (football-like) ellipsoids.

For particles of molecular dimensions the high limiting values (or *low-shear-rate* limit) are of particular interest. These have been calculated by

**Table 2.3.1.** Calculated Specific Viscosities for Prolate Ellipsoids.

| a/b | $\Lambda^a$ | | a/b | $\Lambda^a$ | |
|---|---|---|---|---|---|
| | (1) | (2) | | (1) | (2) |
| 1 | 2.5 | 2.5 | 20 | 4.45 | 13.7 |
| 2 | 2.54 | 2.58 | 25 | 4.91 | 18.8 |
| 4 | 2.84 | 3.08 | 50 | 6.83 | 55.3 |
| 6 | 3.07 | 3.84 | 100 | 10.1 | 177. |
| 8 | 3.29 | 4.80 | 200 | 15.9 | 595. |
| 10 | 3.50 | 5.93 | 300 | 21.3 | 1227. |
| 15 | 3.98 | 9.38 | | | |

[a] (1) Hydrodynamic probabilities used; (2) all orientations weighted equally.

Simha for ellipsoids of revolution. The results of such calculations, and also the drag factors $f/f_0$ of Section 2.2, have been used to estimate the shapes of protein molecules (see, e.g., Tanford, 1961). This is useful from a qualitative standpoint, but can be misleading. Viscosity measurements have, for example, been used to suggest that the fibrinogen molecule is a prolate ellipsoid with $a/b=20$. The much more interesting structure indicated by electron microscopy is shown in Fig. 2.3.3.

An additional warning is in order. Ellipsoids tend to increase the effective viscosity of the suspension in two ways:

1. Via a rotatory motion of the fluid similar to that discussed for a sphere
2. Via an axial drag caused by the difference in velocity between the ellipsoid and the fluid. This is the dominant effect for a long thin rod and is the source of the compressive and tensile forces in the dumbbell, that were described previously.

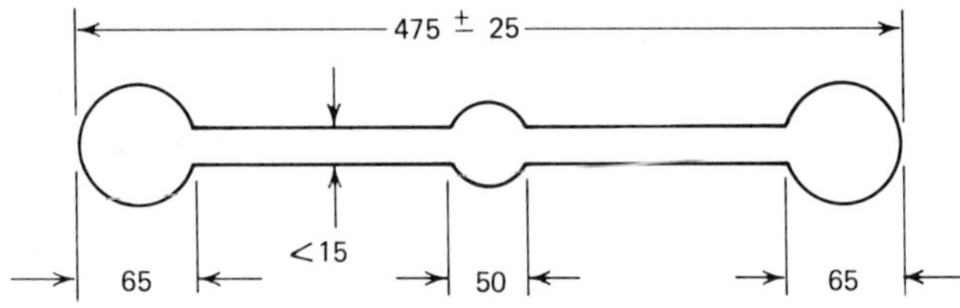


**Fig. 2.3.3.** Sketch of a fibrinogen molecule (see Tanford, 1961). All dimensions are in angstroms.

The effects of these two factors on apparent viscosity are different for different flow fields (see, e.g., Brenner, 1970). It follows that the apparent viscosity of a suspension of nonspherical particles is not an intrinsic property of the suspensions, except perhaps in the low-shear-rate limit where Brownian motion is dominant. Much remains to be learned about suspension rheology.

We now turn our attention to hemorheology and begin with plasma, which is effectively an aqueous suspension of globular proteins. It should thus behave like an isotropic suspension with $\Lambda$ on the order of 3, and this is just what is observed.*

Blood is much more complex, in part because of the large size and asymmetry of red cells, their tendency to aggregate (e.g., into rouleaux), and the high volume fraction of red cells in normal blood. In addition, their flexibility has a profound effect on apparent viscosity. Dilute suspensions of isolated red cells act much like oblate spheroids and are not much affected by Brownian motion at physiologically interesting shear rates. They are distorted, and may rupture, at very high rates, as indicated earlier. They are remarkably tough, however, and shear-induced hemolysis is not important in well-designed equipment.†

The aggregation of red cells into rouleaux increases the asymmetry of the suspended particles and hence their apparent viscosity. As indicated in Ex. 2.2.2, aggregate length decreases with increasing shear rate; this effect is in fact the starting point for the derivation of the Casson equation, and is the primary reason for the decrease in blood viscosity with increasing shear that has been measured by rheologists. The situation in the body is, however, much more complex because the bulk of the pressure drop occurs in the microcirculation. Here the diameters of the vessels are of the same order as those of the red cells, and blood does not behave as a homogeneous fluid. The rheology of microcirculation is a currently active research area (see the reviews by Skalak (1966) and by Aroesty and Gross (1971)).

Perhaps the most striking aspect of blood rheology is the very small increase of apparent viscosity with hematocrit, shown for example in Fig. 1.1.3. The fundamental reason for this is the ability of the red cell to deform and thus prevent the formation of such high-shear areas as are shown in Fig. 2.3.1. Moreover, this deformation can take place without increase of surface, and hence without work against surface forces. The

* Serum albumin for example acts like a prolate ellipsoid with axis ratio ($a/b$) of about 3.5 to 4.0 (depending on the assumed degree of hydration—see Tanford, (1961)).

†High rates of shear can, however, increase the importance of other factors and are often accompanied by damaging turbulence. See Blackshear and Watts (1971) and Forstrom, Blackshear, Keshaviah, and Dorman (1971) for references to recent work.

disclike red cell thus produces an even smaller effect on viscous dissipation than spherical droplets do.*

## *EX. 2.3.2. THE SIZE DISTRIBUTION OF RED CELLS*

It is common practice to estimate the size distribution of red blood cells from measurements of the electrical resistance of small orifices through which highly diluted blood is pumped (see Fig. 2.3.4). The resistance increases when red cells move into the orifice, and at one time the increase in resistance was assumed proportional to cell volume.

It can be seen from Fig. 2.3.5 that the size distribution so calculated is strongly skewed, and it was at one time proposed that this skewness was due to the presence of two types of cells: a major population of smaller cells plus a smaller population of large ones.

Discuss this proposal in the light of Eqs. 2.3.8 and 2.3.9

### SOLUTION

It is almost certain that the apparent skewness results from the variability of red-cell resistance with orientation: most cells move through the orifice with their axes of rotation nearly parallel to the plane of the orifice, but some cells move through with their axes nearly perpendicular to that plane. The former produce a small resistance increase and correspond to the hypothetical dominant proportion of small cells; the latter produce a large resistance increase and correspond to the hypothetical secondary population of large cells.

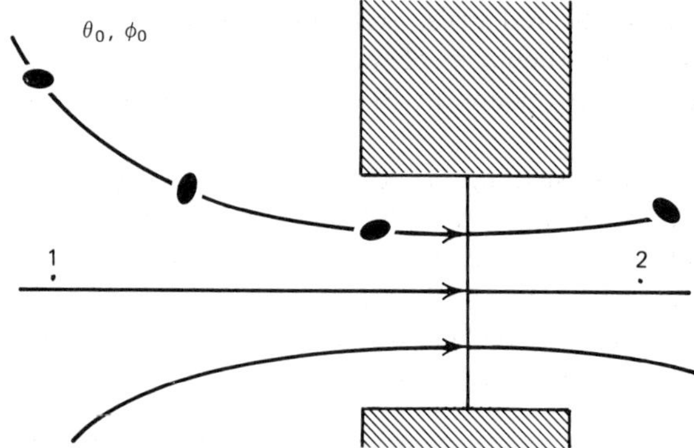


**Fig. 2.3.4.** Flow of red cells through an orifice. The presence of cells in the orifice increases the electrical resistance across it, for example, between points 1 and 2.

* For the latter, $\Lambda = [\lambda + \frac{2}{5})/(\lambda + 1)](5/2)$ with $\lambda$ the ratio of drop to continuum viscosity.

This hypothesis was tested by Breitmeyer, Lightfoot, and Dennis, by approximating red cells as oblate ellipsoids of revolution with $a/b=4$. With the aid of Eqs. 2.3.8 and 2.3.9, and expressions for the relative resistance as a function of orientation,* it is possible to calculate the probability of any resistance for a given cell; the results of this calculation are shown in Fig. 2.3.6. It is now possible to use these results to recalculate the size distribution, and it is found that the bulk of the initial skewness is due to orientation effects. The true volume distribution of red cells is thereby seen to follow a normal probability curve very closely.

Further confirmation of this hypothesis is obtained by "sphering" the red cells through the addition of saponin. It can be seen from Fig. 2.3.5(*b*) that such sphered cells show a much less strongly skewed distribution.

## 2.4 Note on the Motion of Spheroids in Non-uniform Shear Fields

Nonuniform shear rates are the rule both in the body and in extracorporeal circuits, and the relatively simple case of tube flow provides an interesting as well as important example. In such nonuniform fields, particles tend to migrate perpendicularly to the direction of flow; this is thought to produce nonuniform distributions of red cells in blood vessels. In steady tube flow the shear stress varies linearly with radial position, and

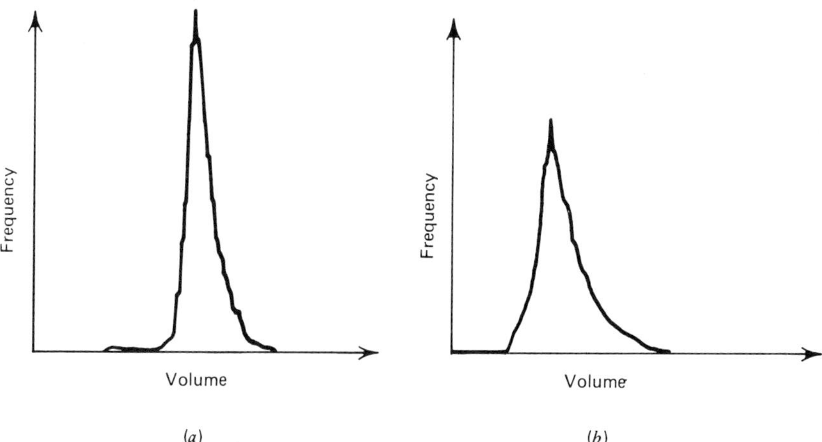


**Fig. 2.3.5.** Apparent size distributions of red cells. The volume is calculated as proportional to the increase in electrical resistance produced by the cell in the orifice. The effect of saponin is to make the red cells more nearly spherical. (*a*) Normal; (*b*) 7 min after addition of saponin.

* See, for example, Carslaw and Jaeger (1959), p. 39.

the shear rate approaches a linear relation with the wall stress at large shears. On the other hand in pulsatile flows such as are characteristic of the arterial system, both shear stress and shear rate vary in a complex way with position and time. Because of this complexity, and that added by the high concentration and deformability of red cells, it is presently feasible to provide a detailed description of red-cell migration. We therefore content ourselves here with listing what are believed to be the major effects of nonuniform shear, and giving references to the published literature.

There appears to be at least two mechanisms tending to produce radial particle migration in duct flows:

1. "Bernoulli" forces tending to drive rigid hydrodynamic particles across shear fields.
2. Additional but poorly understood forces acting on deformable particles.

The Bernoulli forces have been calculated by Rubinow and Keller (1949) on the basis of the Stokes and Oseen expansions, and the behavior of

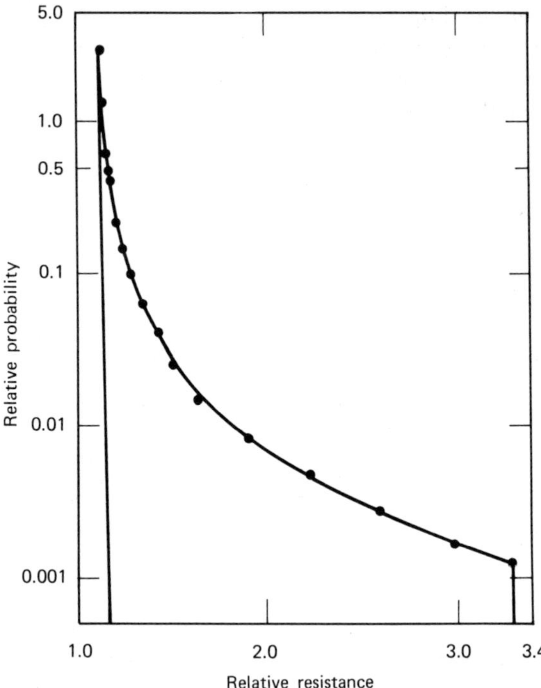


**Fig. 2.3.6.** The probability distribution of electrical resistance for a red cell moving through an orifice.

deformable particles has been studied* experimentally by Mason and his associates.

As discussed by Happel and Brenner (1965, Section 7-3), the Bernoulli force on a particle in a tube always tends to move it toward the tube axis. The magnitude of this force is, however, very small for particles the size of red cells, and predictions based on its dominance are not in qualitative agreement with experiment. For example, Segre and Silberberg observed that the equilibrium position of a very small neutrally buoyant sphere is about 0.6 tube radius from the tube axis.

It has also been found by S. G. Mason and his associates, that solid spheres the size of red cells do not migrate appreciably, whereas deformable particles do. The nature of the forces acting on deformable particles does not seem to have been explained even qualitatively as yet, but they may be closely related to those producing swimming motions, as discussed in the next section.

## 2.5 Propulsion at Very Low Reynolds Numbers

We close our discussion of particulate systems by a brief review of the swimming motions of microorganisms, which is a special case of low-Reynolds-number propulsion. We base our discussion on an early paper by G. I. Taylor (1951), who was the first to demonstrate the possibility of propulsion in the absence of inertial forces. The later literature in this field has been reviewed by Wu, who also considers the swimming of large animals, such as fish, for which inertial effects are dominant.

Following Taylor, we shall consider the propulsion of a sheetlike "tail" subject to sinusoidal lateral displacements. This motion is similar to that of a fish, but the hydrodynamic conditions are far different for such large objects as fish and such small ones as microorganisms. This is because the governing parameter, the Reynolds number, is very different in these two cases. Thus we may define

$$\mathrm{Re} = \frac{LV}{\nu} \tag{2.5.1}$$

where $L$ is the length of the organism and $V$ its velocity relative to the water. We then find, according to Taylor, that Re is many thousands for most fish, of the order of $10^2$ for tadpoles, and of the order of $10^{-3}$ or less for such small objects as spermatozoa which have lengths of the order of 50 $\mu$. We may also define an oscillatory Reynolds number $\mathrm{Re}_\omega = \omega D^2/\nu$,

* See Goldsmith and Mason (1967).

where $\omega$ is the frequency of displacement and $D$ is the diameter of the "tail." For spermatozoa,

$$\omega \sim 10^2/\text{sec}$$

$$D \sim 10^{-5}\ \text{cm}$$

and for water $\nu \doteq 10^{-2}\ \text{cm}^2/\text{sec}$. Then

$$\text{Re}_\omega \sim 10^{-6}$$

It is clear from this low value value of $\text{Re}_\omega$ and our earlier discussion of transient effects in creeping-flow systems that flow about spermatozoa can be described in terms of the pseudosteady creeping-flow equation of motion

$$0 = \mu \nabla^2 \mathbf{v} - \nabla \mathcal{P} \tag{2.5.2}$$

This is the equation we use in our analysis below.

Before proceeding to this analysis, however, we should note that creeping-flow propulsions can be produced by systems of greatly different geometry using quite different types of motion. These include cilial action (e.g., in mucous transport in the lungs) and wavelike deformation of spheroidal organisms. All of these, however, appear to depend upon viscous transport of momentum and hence are closely related to the example we now consider in detail.

### *EX. 2.5.1. MOTION OF AN INEXTENSIBLE SHEET*

We consider as our system a large *inextensible* sheet oscillating in a large body of quiescent fluid about the plane $y=0$. We assume the oscillation to take the form

$$y_0 = b \sin(kx - \omega t) \tag{1}$$

in coordinates fixed relative to the mean position of the particles in the sheet. Then the disturbance has an amplitude $b$ and a wavelength $\lambda = 2\pi/k$. It is propagated in the positive $x$ direction at a celerity

$$c = \frac{\omega}{k}$$

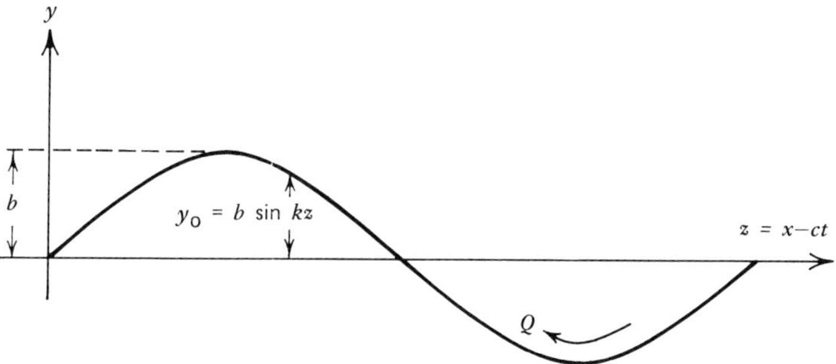


and appears stationary in the *wave-fixed coordinates* $(y, z)$, where $z = x - ct$, as shown in the accompanying diagram. In this wave-fixed coordinate system, points on the sheet are moving in the negative $z$ direction at a constant speed $Q$ along the sinusoidal path $y_0 = b \sin kz$.

We postulate that some unspecified energetic process is producing the lateral displacement, and our purpose is to determine whether or not the sheet is moving relative to the fluid at large $y$. To do this we must determine the $x$ component of velocity in the surrounding fluid.

This in turn must be determined from the equation of motion and continuity, which can be expressed (see Table 2.1.1, Eq. A) in the form

$$\frac{\partial^2 \psi}{\partial x^2} + \frac{\partial^2 \psi}{\partial y^2} = 0 \tag{2}$$

with

$$V_x = -\frac{\partial \psi}{\partial y}; \quad v_y = \frac{\partial \psi}{\partial x} \tag{3}$$

The boundary conditions at the sheet surface, at

$$y = b \sin(kx - \omega t),$$

are obtained from the requirement of inextensibility:

$$\frac{1}{c} v_x = -\frac{1}{c}\frac{\partial \psi}{\partial y} = \tfrac{1}{4}(bk)^2 \cos(2kz) + O\left((bk)^4\right) + v \tag{4}$$

$$\frac{1}{c} v_y = \frac{1}{c}\frac{\partial \psi}{\partial x} = -\left[bk - \tfrac{1}{8}(bk)^3\right]\cos(kz)$$

$$+ \tfrac{1}{8}(bk)^3 \cos(3kz) + O((bk^5)) \tag{5}$$

Specification of the problem is completed by requiring that $v_x$ be finite and $v_y$ be zero at large $y$. Solutions of the form

$$\frac{1}{c}\psi=\sum_{n=0}^{\infty}\{(A_n y+B_n)e^{-2ny}\sin(2nkz)$$

$$+(C_n y+D_n)e^{-(2n+1)y}\cos[(2n+1)kz]\}$$

$$-Vy/c \tag{6}$$

meet these requirements, and there will be a translation of the sheet (relative to the fluid at infinity) with velocity $-V$ in the $x$ direction, if this quantity is nonzero.

Taylor determined the following constants by series expansion of Eq. 6 to the powers of $bk$ indicated:

$$A_1=-bk\left[1-\tfrac{1}{2}(bk)^2\right],\qquad B_1=-bk\left[1-\tfrac{1}{4}(bk)^2\right]$$

$$C_2=\tfrac{1}{4}(bk)^2-\tfrac{1}{6}(bk)^4,\qquad D_2=\tfrac{1}{12}(bk)^4$$

$$A_3=O\left((bk)^5\right),\qquad B_3=\tfrac{1}{12}(bk)^3$$

$$C_4=\tfrac{99}{192}(bk)^4,\qquad D_4=\tfrac{1}{24}(bk)^4$$

It follows from these that

$$\frac{V}{c}=\tfrac{1}{2}(bk)^2\left[1-\tfrac{19}{16}(bk)^2\right] \tag{7}$$

It has therefore been shown that when small but finite waves travel down a sheet immersed in a viscous fluid, they propel the sheet relative to the fluid in the opposite direction.

In a later paper (1952) Taylor extended the above analysis to a rodlike rather than sheetlike tail. He then confirmed his calculations qualitatively by experiments with a scaled-up mechanical model of a spermatozoon propelled through glycerine by a rubber-band motor.

## PARTICULATE SYSTEMS BIBLIOGRAPHY

### Books and Reviews

Berker, R., *Handbuch der Physik*, Vol. VIII-2, 1–384 (1963).

Bird, R. B., H. R. Warner, Jr., and D. C. Evans, "The Kinetic Theory and Rheology of Rigid Dumbell Suspensions with Brownian Motion," *Adv. Poly. Phys.* (1971).

Brenner, Howard, "Rheology of Two-phase Systems," *Ann. Rev. Fluid Mech.*, **2**, 137–176 (1970).

Brenner, Howard, "Suspension Rheology," 1970 International Seminar on Heat and Mass Transfer in Rheologically Complex Fluids, Herceg Novi, Yugoslavia, Sept. 8–12 (1970).

Burgers, J. M., "Second Report on Viscosity and Plasticity," *Kon. Ned. Akad. Wet., Ver.* (Eerste Sectie), D1. XVI, No. 4 (1938), pp. 1–287. Amsterdam: N.V. Noord-Hollandsche Vitgeversmaatschappi (1938).

Frisch, H. L., and Robert Simha, "The Viscosity of Colloidal Suspensions" in *Rheology*, Vol. 1, F. R. Eirich, Ed., Academic Press (1967).

Goldsmith, H. L., and S. G. Mason, "The Microrheology of Dispersions" in *Rheology*, Vol. 4, F. R. Eirich, Ed., Academic Press (1967).

Gross, J. F., and J. Aroesty, "Mathematical Models of Capillary Flows: a Critical Review," Rand Corp., Santa Monica, Calif. (1971).

Happel, John, and Howard Brenner, *Low Reynolds Number Hydrodynamics*, Prentice-Hall (1965).

Skalak, R., "Mechanics of the Micro-circulation,:" in *Biomechanics*, Y. C. Fung, Ed., Am. Soc. Mech. Eng. (1966).

Tanford, Charles, *Physical Chemistry of Macromolecules*, Wiley (1961).

Wu, T. Y., "The Mechanics of Swimming" in *Biomechanics*, Y. C. Fung, Ed., Am. Soc. Mech. Eng. (1966).

## Research Papers and Minor References

Basset, A. B., *A Treatise on Hydrodynamics*, Dover (1961).

Blackshear, P. L., Jr., "On Transport of Heat, Mass, and Momentum in Blood Due to Particulate Motion," Bioeng. Memo No. One, July 1965, Department of Mechanical Engineering, University of Minnesota, Minneapolis.

Blackshear, P. L. Jr., and Christopher Watts, "Observation of Red Blood Cells Hitting Solid Walls," *Adv. Bio-eng.*, R. G. Buckles, Ed., A.I.Ch.E., C.E.P. Symposium Series, No. 114, **67**, 60 (1971).

Breitmeyer, M. O., E. N. Lightfoot, and W. H. Dennis, "Model of Red Cell Rotation, etc.," *Biophys. J.*, **11**, 146 (1971).

Brenner, Howard, "The Slow Motion of a Sphere through a Viscous Fluid Towards a Plane Surface," *Chem. Eng. Sci.*, **16**, 242–251 (1961).

Brenner, Howard, "The Stokes Resistance of an Arbitrary Particle–III. Shear Fields," *Chem. Eng. Sci.*, **19**, 631–651 (1964).

Carslaw, H. S., and J. C. Jaeger, *Conduction of Heat in Solids*, 2nd ed., Oxford (1959).

Einstein, Albert, *Ann. Phys. (Leipz.)*, **19**, 289–306 (1906); erratum **24**, 591–592 (1911).

Forstrom, R. J., P. L. Blackshear, Jr., Prakash Keshaviah, and F. D. Dorman, "Fluid Dynamic Lysis of Red Cells," *Adv. Bio-eng.*, R. G. Buckles, Ed., A.I.Ch.E., C.E.P. Symposium Series, No. 114, **67**, 69 (1971).

Frankel, N. A., and Andreas Acrivos, *Chen. Eng. Sci.*, **22**, 847–853 (1967).

Hochmuth, R. M., R. N. Marple, and S. P. Sutera, *Microvas. Res.*, **2**, 212–220 (1970).

Jeffery, G. B., *Proc. Roy. Soc.*, **A102**, 161 (1922).

Jeffery, G. B., *Proc. London Math Soc.*, **14**, 327 (1915).

Kuhn, W., *Z. Physik. Chem.*, **A161**, 1 (1932).
Landau, L. D., and E. M. Lifshitz, *Fluid Mechanics*, Addison-Wesley (1959).
Nevaril, C. G., E. C. Lynch, C. P. Alfrey, Jr., and J. D. Hellums, *A.I.Ch.E. J.*, **15**, 707 (1969).
Rubinow, S. I., and J. B. Keller, *J. Fluid MH.*, **11**, 447 (1949).
Segré, G., and A. Silberberg, *J. Fluid Mech,*, **4**, 115, 136 (1966).
Shapiro, S. I., and M. C. Williams, A.I.Ch,E. J., **16**, 575 (1970).
Sy, Francisco, J. W. Taunton, and E. N. Lightfoot, *A.I.Ch.E. Journal*, **16**, 386–391 (1970).
Taylor, G. I., Proc. Roy Soc., **A209**, 447–461 (1951).
Taylor, G. I., *ibid.*, **A211**, 225–239 (1952).
Taylor, G. I., *ibid.*, **A214**, 158–183
Villat, H., *Lecons sur les Fluides Visqueux*, Gauthier-Villars, Paris (1943).

# 3. FLOW IN DUCTS

In this chapter we review biologically important aspects of duct flows, with particular emphasis on flow in the mammalian body. We begin with a brief summary of anatomic and physiological considerations and then discuss a series of representative flow problems, starting with the fundamental cases of steady flows in rigid and elastic ducts. We then consider the effect of wall porosity, which is important in all vascular beds, most particularly the kidney, and proceed from there to the very important case of pulsatile flows. These latter are characteristic of the entire circulatory system, and it appears at the time of writing that they are beneficial in extracorporeal circuits. We conclude the chapter with a brief discussion of the particulate flows in the microcirculation.

## 3.1 Anatomic and Physiological Considerations

The circulatory system must provide convective transport between the major organs of the body and also diffusional transport to the tissues within these organs. In addition, many specialized operations must be performed, such as glomerular filtration in the kidney.

The general nature of the circulatory system as a complex series-parallel network is shown in Fig. 3.1.1, and the anatomy of specific body regions is discussed later at several different points. The sizes and general characteristics of representative blood vessels are shown in Fig. 3.1.2 and Table 3.1.1.

Note that the Reynolds numbers in Table 3.1.1 are below 2100 except in the aorta. Because of this, laminar flow has normally been assumed in the

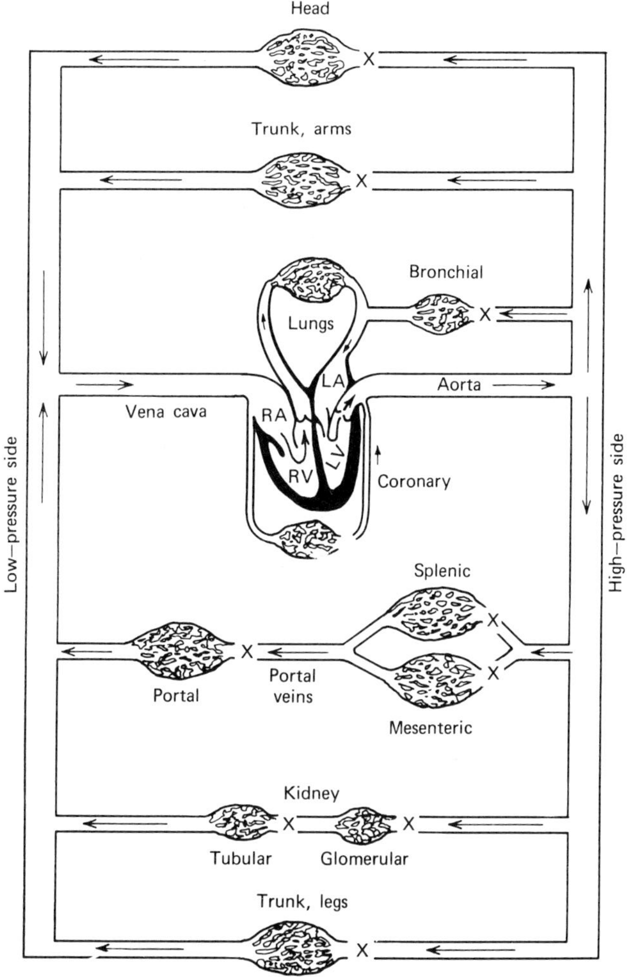


**Figure 3.1.1.** Schematic diagram of circulation showing multiple parallel routes from the arterial to the venous side. Reprinted from H. D. Green, "Circulation: Physical Principles" in *Medical Physics*, Glassman, Ed. Vol. I, Year Book Publisher, Inc., Chicago, 1944. Used by permission of the author and publisher.

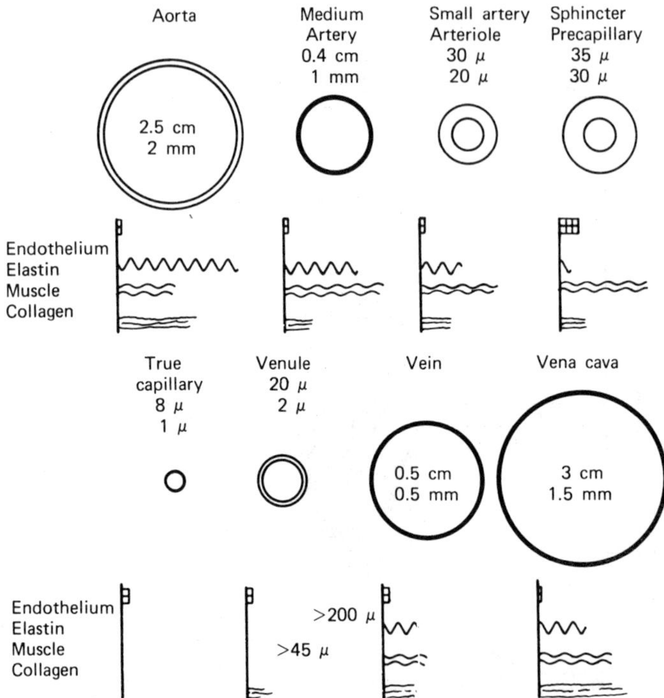


**Figure 3.1.2.** Schematic of relative sizes, wall-to-lumen ratios, and proportions of components. From Alan C. Burton, *Physiology and Biophysics of the Circulation* Year Book Medical Publishers (1944). Used by permission of Year Book Medical Publishers.

analysis of blood flow. Inertial effects are probably important in the aorta, however, and turbulence is possible here. Very complex flow patterns are also observed at branches, and there is evidence that inertial forces, here only those due to convective momentum transport ($\mathbf{v} \cdot \nabla \mathbf{v}$), may be important even in the microcirculation.

The walls of blood vessels are complicated structures, and contain four major structural components, as described briefly in Table 3.1.2. The approximate proportions of these components in typical vessels are indicated in Fig. 3.1.2.

The larger vessels leading from the heart and down through the first few branches are known as elastic arteries. Their walls are made up of as many as fifty superimposed networks of elastic tissue mesh interwoven with collagen. Muscle is sparse and entirely of the tension type. A representative static stress-strain curve for such vessels is shown in Fig. 3.1.3. The

**Table 3.1.1.** Approximate Dimensions and Blood Velocities in Various Segments of the Cardiovascular System *for a 13-kg Dog*. Assumed Cardiac Output: 2.4 l/min.[a]

| Segment | Number | Diameter, mm | Cross section, $cm^2$ | Length, cm | Volume, ml | Blood velocity, cm/sec | Reynolds number |
|---|---|---|---|---|---|---|---|
| Left atrium | — | — | — | — | 25 | | |
| Left ventricle | — | — | — | — | 25 | | |
| Aorta | 1 | 10. | 0.8 | 40. | 30 | 50. | 2,500. |
| Large arteries | 40 | 3. | 3.0 | 20. | 60 | 13.4 | 201. |
| Main arterial branches | 600 | 1. | 5. | 10. | 50 | 8. | 40. |
| Terminal arteries | 1,800 | 0.6 | 7.0 | 1.0 | 5 | 6. | 9. |
| Arterioles | $40 \times 10^6$ | 0.02 | 125. | 0.2 | 25 | 0.32 | 0.03 |
| Capillaries | $12 \times 10^8$ | 0.008 | 600. | 0.1 | 60 | 0.07 | 0.003 |
| Venules | $80 \times 10^6$ | 0.03 | 570. | 0.2 | 114 | 0.07 | 0.01 |
| Terminal veins | 1,800 | 1.5 | 30. | 1.0 | 30 | 1.3 | 9.8 |
| Main veins | 600 | 2.4 | 27. | 10. | 270 | 1.48 | 18. |
| Large veins | 40 | 6.0 | 11. | 20. | 220 | 3.6 | 108. |
| Venae cavae | 1 | 12.5 | 1.2 | 40. | 50 | 33.4 | 2,090. |
| Right atrium | — | — | — | — | 25 | | |
| Right ventricle | — | — | — | — | 25 | | |
| Main pulmonary artery | 1 | $12^b$ | 1.1 | 2.4 | 24 (combined with next row) | 36.4 | 2,090. |
| Lobar pulmonary artery branches | 9 | $4^b$ | 1.19 | 17.9 | | 33.6 | 670. |
| Smaller arteries and arterioles | — | — | — | — | 18 | | |
| Pulmonary capillaries | — | 0.008 | 300. | 0.05 | 16 | 0.14 | 0.006 |
| Pulmonary veins | $6 \times 10^2$ | — | — | — | 52 (combined with next row) | | |
| Large pulmonary veins | 4 | — | — | — | | | |

[a] Source: After H. Green, in O. Glasser (ed.), *Medical Physics*, Vol. I, p. 210, The Year Book Medical Publishers, Inc., Chicago, 1944, and modified by the editor.

[b] Mean of major and minor semiaxes of the elliptic cross section.

**Table 3.1.2.** Constituents of Blood-Vessel Walls

1. *Endothelium:* The innermost layer, continuous through the circulatory system. In its most general form it comprises a pavement-like layer of single cells. Its integrity is essential to prevent clotting of the contained blood, and to maintain normal permeability to solutes and water.

2. *Collagen:* This is a protein of high tensile strength and relatively low extensibility, with Young's modulus probably of the order of $10^8$ dyn/cm$^2$. In the vessel wall, collagen fibers assume a serpentine shape, so that some wall extension is possible without fiber stretching. They contribute primarily in wall stiffness (see Fig. 3.1.3), and thus protect the wall against failure.

3. *Elastin:* A second protein substance of high extensibility, similar in many respects to rubber. It has been investigated as single fibers and shows a Young's modulus increasing with extension to about $6\times10^6$ dyn/cm$^2$. The elasticity of this material is important to moderate the pressure pulses produced by the flow pulsations of the heart. If, however, Young's modulus is too low to protect the vessels from failure at peak pressures, such protection is supplied by the collagen.

4. *Smooth Muscle:* The significant characteristic of smooth muscle is its ability to contract on suitable stimulation–chemical, or as a result of strain. Chemical stimulation may be initiated by nerve impulses.

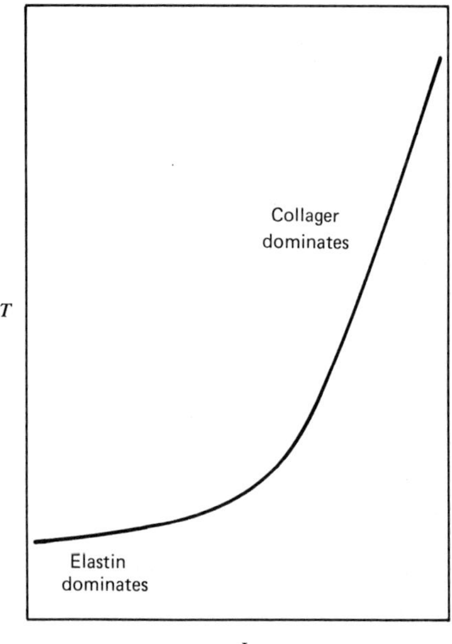


**Figure 3.1.3.** Generalized tension-length diagram for vessel walls (redrawn after Burton).

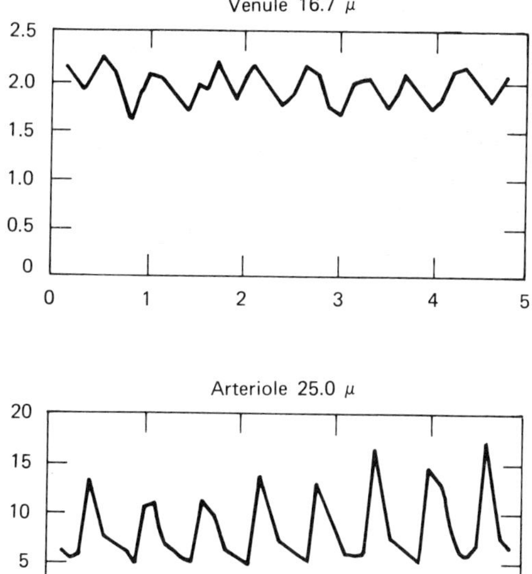


**Figure 3.1.4.** Characteristic pressure pulsations in microcirculation.

viscoelastic properties of arteries are briefly considered later. These vessels are of primary interest during our analysis of pulsatile arterial flow. The smaller vessels contain higher ratios of muscle to elastic layers, and the latter ultimately take the form of two thin membranes at the inner and outer margins of an increasingly prominent muscular coat. In addition all but the smallest vessels are enveloped by a net of vessels in their walls, the vasa vasorum. Veins are particularly dependent on this blood supply and deteriorate rapidly if it is interrupted.

The term microcirculation is a recent one. It refers to all blood vessels not visible to the naked eye and thus includes quite a variety of sizes, anatomical and mechanical properties, and flow conditions. Excellent reviews of microcirculatory flow have been provided by Gross and Aroesty (1971) and by Skalak (1966). The structure and function of the small vessels, and the nomenclature used to describe them, are summarized by Sobin (1966).

The first direct observation of capillary vessels was by Malphigi in 1661, and only seven years later Leeuwenhoek described individual red cells

**Figure 3.1.5.** Red-cell distortion in capillary flow. Flow is from left to right.

flowing in the tail capillaries of a live tadpole.* In 1930 Krogh had summarized many of the important characteristics of capillary flow. For example, he recognized that capillary flow was pulsatile, but that the pulsations were not accompanied by appreciable changes in capillary diameter. Characteristic pulsations are shown in Fig. 3.1.4; the inextensibility of capillaries was explained mathematically by Fung in 1966. Krogh also observed "plasma skimming" and the distortion† of red cells in the smallest vessels, where they must proceed in single file (see Fig. 3.1.5, drawn from a photograph in Lew and Fung, 1969).

The permeability of the capillaries to macromolecules as well as to electrolytes is thought to be due to spaces between adjoining endothelial cells. There is general agreement that these cells are separated over most of their circumference by distances of the order of 200 Å. We shall find in our later discussions of mass transfer that capillary permeability is quite important physiologically.

## 3.2 Steady Flows in Ducts with Impermeable Walls

We begin here by considering flow in long rigid ducts of constant cross section and then discuss briefly the problem of entrance flows. We then generalize our discussion to description of Hele-Shaw flows and steady flow in distensible ducts.

* See Fishman (1966) and Richards (1964); a brief historical summary is available in Skalak (1966).

† See Gross and Aroesty, and Skalak, for references on distortion.

(a) *DEVELOPED FLOWS IN STRAIGHT DUCTS OF CONSTANT CROSS SECTION*

For a cylindrical duct of arbitrary cross section, as shown in Fig. 3.2.1a, the equations of motion and continuity reduce to

$$\frac{\partial^2 v_z}{\partial x^2} + \frac{\partial^2 v_z}{\partial y^2} = \frac{1}{\mu}\frac{d\mathcal{P}}{dz} \tag{3.2.1}$$

with

$$\frac{d\mathcal{P}}{dz} = \frac{dp}{dz} - \rho g_z = \frac{\Delta\mathcal{P}}{L} \tag{3.2.2}$$

for a duct of length L, with $\Delta\mathcal{P} = \mathcal{P}(0) - \mathcal{P}(L)$. These equations have the general solution*

$$v_z = \chi + \frac{1}{4\mu}\frac{d\mathcal{P}}{dz}(x^2+y^2) \tag{3.2.3}$$

where $\chi$ is defined by the relations

$$\frac{\partial^2 \chi}{\partial x^2} + \frac{\partial^2 \chi}{\partial y^2} = 0 \tag{3.2.4}$$

with

$$\chi = -\frac{\Delta\mathcal{P}}{4\mu L}(x^2+y^2) \tag{3.2.5}$$

on the duct boundary. The duct-flow problem is thus reduced to solution of Laplace's equation for $\chi$, and such solutions are readily available.

*EX. 3.2.1. POISEUILLE'S LAW*

Use Eqs. 3.2.3 and 3.2.4 to develop the pressure-flow relation for a duct of circular cross section and radius R. Compare with the solution for an ellipsoidal cross section.

SOLUTION

It is simplest to use cylindrical coordinates here, so that $x^2+y^2$ is replaced

* For the limiting case of flow between close-spaced parallel sheets, one may set $\partial^2 v_z/\partial x^2 = 0$ if $y$ is measured perpendicular to the sheets. The 4 in Eqs. 3.2.3 and 3.2.5 must then be replaced by a 2.

by $r^2$ and Eq. 3.2.3 and 3.2.5 take the form

$$\frac{1}{r}\frac{d}{dr}r\frac{d\chi}{dr}=0 \tag{1}$$

and

$$\chi(R)=-\frac{\Delta\mathcal{P}}{4\mu L}R^2 \tag{2}$$

Equation 1 may be integrated to

$$\chi=c_1\ln r+c_2 \tag{3}$$

and $c_1$ must be zero to provide a finite velocity at the tube axis. Then

$$v_z=\left[\frac{(-\Delta\mathcal{P})R^2}{4\mu L}\right]\left[1-\left(\frac{r}{R}\right)^2\right] \tag{4}$$

This is a well-known result which may be readily obtained by direct integration of Eq. 3.2.1 (see, e.g. *Tr.Ph.*, Section 2.3); this is, however, a particularly simple example.

Equation 4 may now be integrated over the tube cross section to give Poiseuille's law for the volumetric flow rate $Q$:

$$Q=\frac{\pi}{8\mu L}(\mathcal{P}_0-\mathcal{P}_L)R^4 \tag{5}$$

This expression was first developed to describe blood flow, and it is reasonable for the smaller arteries, as we shall see in our discussion of pulsatile flow.

Blood vessels normally have more nearly an elliptical cross section, however, and their surface can be better approximated as

$$\frac{x^2}{a^2}+\frac{y^2}{b^2}=1 \tag{6}$$

than by $r=R$. The flow for this shape is given by

$$v_z=\frac{(-\Delta\mathcal{P})a^2b^2}{2\mu L(a^2+b^2)}\left[1-\left(\frac{x}{a}\right)^2-\left(\frac{y}{b}\right)^2\right] \tag{7}$$

and

$$Q = \frac{\pi(-\Delta\mathcal{P})}{4\mu L} \frac{a^3 b^3}{a^2 + b^2} \tag{8}$$

which is of the same general form as Eq. 5.

Much more complex shapes are encountered in extracorporeal processing equipment, and the nature of the velocity profile can have a marked effect on equipment performance.

### (b) *HELE-SHAW AND RELATED FLOWS*

Flow between closed-spaced parallel sheets can easily be shown (see footnote to Eq. 3.2.3) to be described by

$$\mathrm{v} = -\frac{B^2}{2\mu}\left[1 - \left(\frac{z}{B}\right)^2\right]\nabla\mathcal{P} \tag{3.2.6}$$

where $z$ = distance measured perpendicularly from a plane midway between the two sheets, and a distance $B$ from each of them

$\nabla\mathcal{P} = \delta_x \partial\mathcal{P}/\partial x + \delta_y \partial\mathcal{P}/\partial y$

The idealized blood-flow path of Fig. 3.2.1($b$) is a system of this type. We may now integrate this velocity over the distance $2B$ between the boundaries to obtain

$$\langle \mathbf{v} \rangle = \hat{\delta}_x \langle v_x \rangle + \hat{\delta}_y \langle v_y \rangle = -\left(\frac{1}{3}\frac{B^2}{\mu}\right)\nabla\mathcal{P} \tag{3.2.7}$$

where

$$\langle \mathbf{v} \rangle = \frac{1}{2B}\int_{-B}^{B} \mathbf{v}\, dz \tag{3.2.8}$$

is the flow-average velocity. If we next integrate the equation of continuity with respect to $z$ we find

$$\nabla \cdot \langle \mathrm{v} \rangle = \frac{\partial \langle v_x \rangle}{\partial x} + \frac{\partial \langle v_y \rangle}{\partial y} = 0 \tag{3.2.9}$$

so that

$$\nabla^2 \mathcal{P} = \frac{\partial^2 \mathcal{P}}{\partial x^2} + \frac{\partial^2 \mathcal{P}}{\partial y^2} = 0 \tag{3.2.10}$$

Equations 3.2.7 and 3.2.10 are the equations describing two-dimensional potential flows and reduce the problem of determing $\langle v \rangle$ as a function of $x$ and $y$ to the solution of Laplace's equation.

Two-dimensional flows have been very thoroughly investigated, and solutions are available, for example in Milne-Thomson (1967), for a very wide variety of boundary conditions. For this situation it is convenient to use the complex plane and represent the fluid velocity and the pressure distribution in terms of a complex potential

$$w(z) = \phi(x,y) + i\psi(x,y) \tag{3.2.11}$$

where $z$ is now $x + iy$
(i.e., the position in the complex plane),
$\phi = (B^2/3\mu)$
and $\psi$ = the two-dimensional stream function (see Table 2.1.1). The streamlines are lines of constant $\psi$ (the imaginary part of $w$), and the velocity is obtained from Eq. 3.2.7. To describe any flow problem it is then necessary only to determine the proper complex potential $w$. Means for doing this are discussed at length in standard references, and illustrated in the first example below.

Equation 3.2.7 can also be used for the variable $B$, provided the changes are not too abrupt. For this situation Eq. 3.2.9 takes the form

$$\nabla \cdot B\langle \mathbf{v} \rangle = 0 \tag{3.2.12}$$

In at least some situations of biological importance, the spacing is pressure dependent, and one may write as a first approximation

$$B = B_0(1 + \alpha \mathcal{P}) \tag{3.2.13}$$

Then for small strains Eqs. 3.2.7 and 3.2.12 can be combined with Eq. 3.2.13 to give the unexpectedly simple result:

$$\nabla^4 B = 0 \tag{3.2.14}$$

An application of this equation is given in the second example.

Finally, one must keep in mind that in real situations the flat-sheet geometry may represent an unacceptable idealization. In Fig. 3.2.1($b$) and ($c$), for example, this is clearly the case for the dialysate and quite possibly for the blood. In addition, the permeability is different in the two directions, so that we must write*

$$\langle \mathbf{v} \rangle = -\kappa \cdot \nabla \mathcal{P} \tag{3.2.15}$$

* Equation 3.2.7 can also be generalized to three-dimensional flows, as in vascular beds. Here, however, a very complex coordinate system may be required.

with

$$\kappa = \begin{bmatrix} \hat{\delta}_x \hat{\delta}_x \kappa_{xx} & 0 \\ 0 & \hat{\delta}_y \hat{\delta}_y \kappa_{yy} \end{bmatrix} \tag{3.2.16}$$

The permeabilities may be calculated from Eqs. 3.2.3 to 3.2.5.

## *EX. 3.2.2. FLOW IN A HEMODIALYZER*

Here we shall consider, as an example of some medical importance, blood flow in the simple, and very commonly used, hemodialyzer ("artificial kidney") of Fig. 3.2.1. The blood is enclosed by two nearly flat and parallel

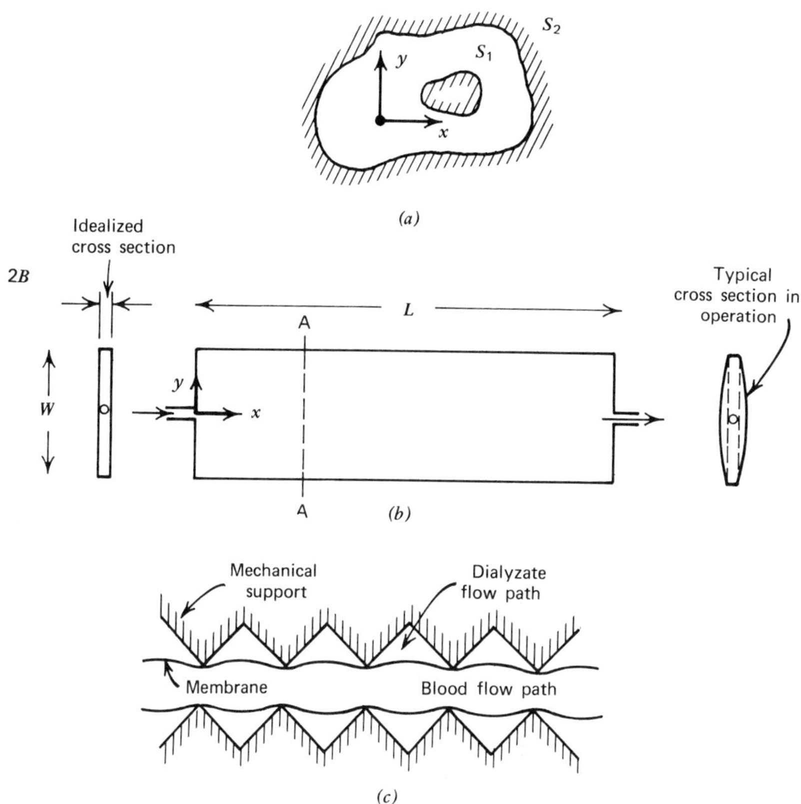


**Figure 3.2.1.** Representative duct flows. (*a*) A generalized duct cross section. Any number of cores can be present. (*b*) Schematic representation of a commonly used hemodialyzer. Normally $B \sim 2 \times 10^{-1}$ mm, $W \sim 1/3$ m, $L \sim 1$ m. (*c*) Cross section $AA$ for the above dialyzer.

membranes and by a rectangular frame. It enters at the center of one short side of this frame and leaves at the center of the other. Describe the streamlines and flow distribution.

SOLUTION

We begin by considering the membranes to be flat and parallel and the diameter of the blood ports to be very small. If we assume in addition that $L \gg W$ we may write for small $x$ (see Milne-Thomson, 1967, p. 268):

$$w = -\frac{Q}{\pi} \ln(\sinh Z) \tag{1}$$

where

$$Z = X + iY = \frac{\pi z}{W} \tag{2}$$

and $Q$ is the volumetric rate of blood input to the dialyzer. It remains to determine the velocity and pressure profiles.

We begin by writing sinh $Z$ in the polar form

$$\sinh Z = Re^{i\theta} \tag{3}$$

where

$$R = \sqrt{\sinh^2 X \cos^2 Y + \cosh^2 X \sin^2 Y} \tag{4}$$

$$\theta = \arctan\left(\frac{\cosh x \sin Y}{\sinh x \cos Y}\right) \tag{5}$$

Then

$$W = -\frac{Q}{\pi} \ln(\sinh Z) = \ln R + i\theta \tag{6}$$

It follows that

$$\phi = -\frac{1}{2}\frac{Q}{\pi} \ln(\sinh^2 x \cos^2 Y + \cosh^2 x \sin^2 Y) \tag{7}$$

$$\psi = \arctan\left(\frac{\cosh x \sin Y}{\sinh x \cos Y}\right) \tag{8}$$

Sample streamlines, which are lines of constant $\psi$, are shown in Fig. 3.2.2. Note that for our system the extreme values of $\psi$ are $\pm\pi/2$. The flow contained between any two streamlines $\psi = \psi_1$ and $\psi = \psi_2$ is proportional to $\psi_2 - \psi_1$ (see, for example, Batchelor); and for our case Eq. 2 has been normalized so that the constant of proportionality is unity. It follows that the streamlines for $\psi = \pm 0.9\pi/2$ on Fig. 3.2.2 contain 90% of the flowing

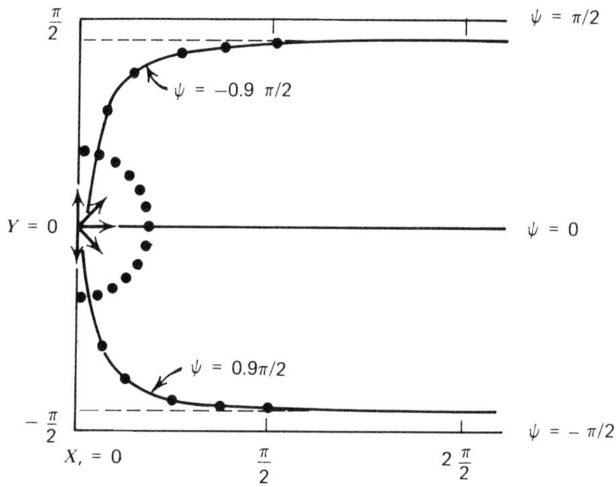


**Figure 3.2.2.** Flow in a "long" hemodialyzer.

blood. It can be seen from this figure that the streamlines are essentially fully developed for $x^* > \pi/2$, so that end effects due to the localized blood input are unimportant beyond this point. Then for $L > W$, which is the usual case, Eq. 2 can be used to describe the whole upstream half of the dialyzer without significant error.

In practice it proves very difficult to maintain constant spacing between the membranes, and as a result, there is usually a very appreciable variation in both $B$ and $K$ over any flow cross section. The result is a maldistribution of flow and a decrease in mass-transfer effectiveness. It is not unusual for a dialyzer to operate at 50% or less of the efficiency predicted for uniform flow.

### *EX. 3.2.3. SHEET FLOW IN LUNG ALVEOLI*

The capillary network providing blood flow around the alveolar sacs of the lung can be approximated as two elastic endothelial membranes or sheets held together by frequent connecting columns of septal tissue (see Fig. 3.2.3). Blood flows between these sheets in a close approximation of the Hele-Shaw flow just described (see Eq. 3.2.14), except that the permeability is reduced by the presence of the columns.

Discuss the effect of distensibility.

SOLUTION

The geometry of the alveolar capillary net is indicated in Fig. 3.2.3, taken from Sobin and Fung (see Fung, 1968). The septal columns show clearly in

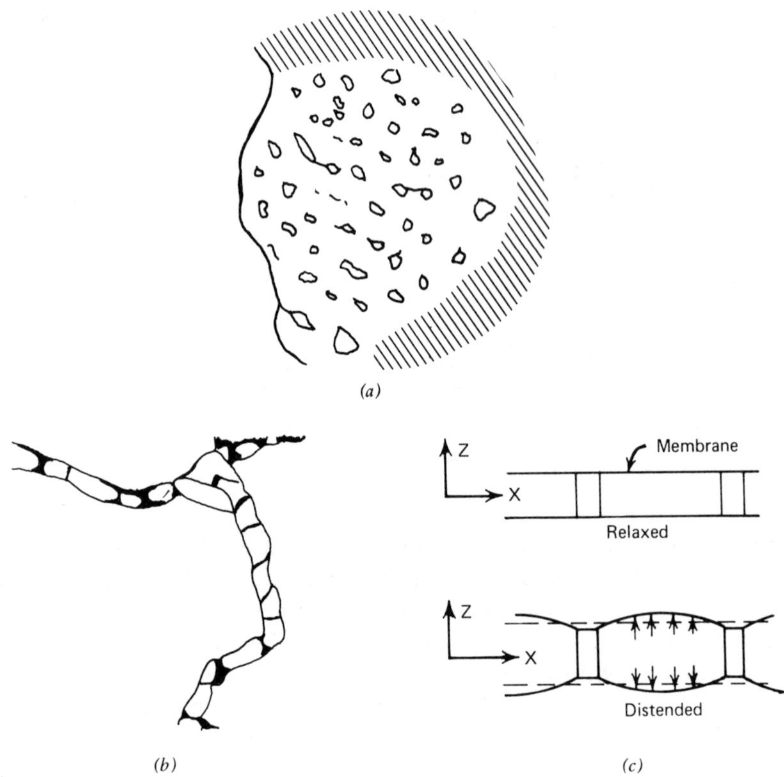


**Figure 3.2.3.** Anatomy of the alveolar capillary net. Drawings from photographs by Fung and Sobin (see Fung, 1968). (*a*) Plan view of alveolar wall. (*b*) Cross section of alveolar sheet. (*c*) Effect of internal pressure.

the plan view (*a*), and the endothelial sheets can be seen in the cross section of (*b*). The distending effect of hydrostatic pressure is shown schematically in (*c*).

The blood flow paths in the alveoli are not known in detail, but two rather plausible possibilities are shown in Fig. 3.2.4, also taken from Sobin and Fung. The alveolar nets are indicated by the roughly rectangular elements in this figure, and in each case blood flow is down the efferent venule from the afferent arterioles. Figure 3.2.4*a* is drawn for rigid sheets, and for it the streamlines and equipotential lines can be calculated much as above. In fact, the flow situation is essentially that of the previous example.

The situation is rather different for the elastic models shown in Fig. 3.2.4. Here the greatest pressure gradients occur at the downstream ends of

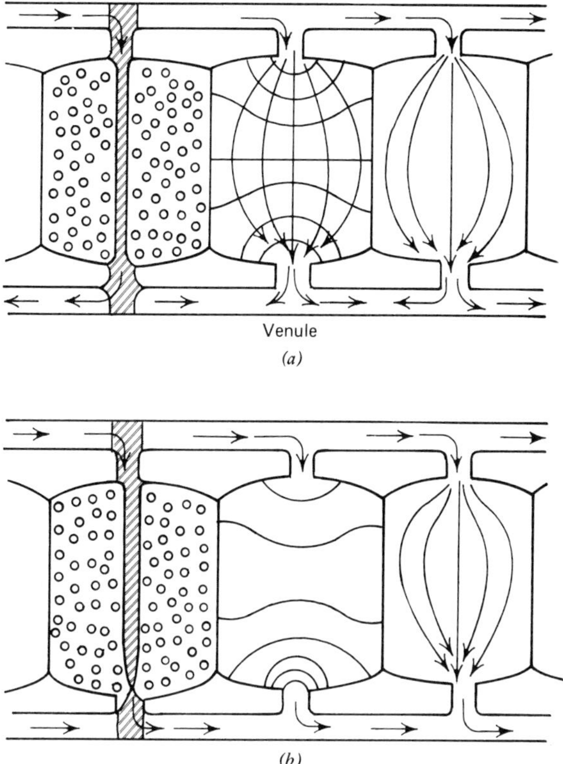


**Figure 3.2.4.** Effect of elasticity on alveolar flow distribution for one geometry. From Sobin and Fung (see Fung, 1968). (*a*) Rigid membrane; (*b*) linearly elastic membrane.

the flow paths where the hydrostatic pressure, and hence the permeability, are lowest.

A quantitative description of an elastic system can be easily obtained and is provided by Sobin and Fung. It is shown in the section on pulsatile flow that the effects of transients produced by the cyclic operation of the heart and lungs are unimportant for such small systems. The particulate nature of the blood and the deformability of red cells probably have only a secondary effect.

## 3.3 Seepage Flows

Up to this point we have considered all duct walls impermeable to the contained fluid. This is a good approximation in the major blood vessels, but not in the microcirculation.

**Table 3.3.1** Velocity Distributions in Tube and Slit Flow at Low Seepage Rates[a]

(1) $\alpha r \ll 1$: Tube

$$v_z = v_z(0,z)\left[1-\left(\frac{r}{R}\right)^2\right]$$

$$v_r(r,z) = v_r(R,z)\left[2\frac{r}{R}-\left(\frac{r}{R}\right)^3\right]$$

(2) $\alpha y \ll 1$: Slit

$$v_x(y,x) = v_x(0,x)\left(1-\frac{y^2}{B^2}\right)$$

$$v_y(y,x) = \frac{v_y(b,x)}{2}\left[3\frac{y}{b}-\left(\frac{y}{B}\right)^3\right]$$

[a] See Ex. 3.3.1 for the Significance of $\alpha$.

Seepage from the arterial ends of the capillary beds, and corresponding return flow at the venous end, is an important aid to mass transfer and is critical to the functioning of the kidneys. About 180 l of water per day pass through the glomeruli of the kidneys by ultrafiltration, and 99% of this water is reabsorbed in the tubules. Water removal by ultrafiltration is also normally necessary to maintain water balance in patients being treated for renal insufficiency by hemodialysis. On the order of 1 to 2 l of water is ordinarily removed during the course of each dialysis, over an 8- to 12-hour period.

Finally, membrane processes for selectively removing water from solutions of proteins and other macromolecules (ultrafiltration) or of salts and other small solutes (reverse osmosis) show considerable promise. Many potential applications are biological and are closely related to the practice of medicine.

In most situations of current interest, the flow is either between parallel sheets or in round tubes, and is laminar. In addition the boundary conditions normally result in two-dimensional or axisymmetric flow, respectively, and Reynolds numbers are low enough that fluid inertia may be safely neglected (Berman, 1953, 1958; Macey, 1963, 1965). The equations of continuity and motion then reduce to Eq. A or C of Table 2.1.1 with the left side negligibly small. These equations have been solved by Kozinski, Schmidt, and Lightfoot (1970) for arbitrary dependence of seepage rate on axial position. For the (mathematically) low filtration rates characteristic

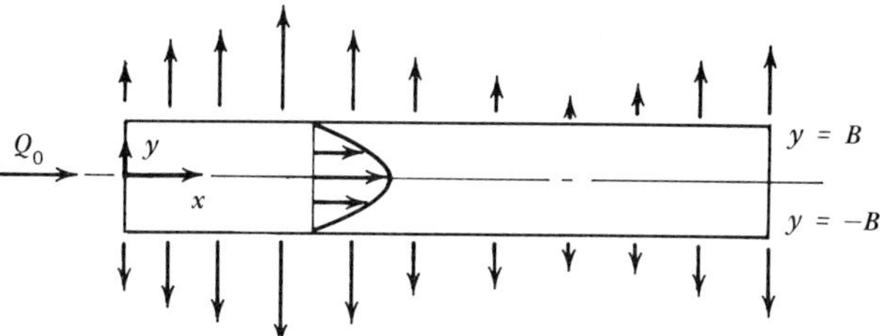


**Figure 3.3.1.** Slit Flow with arbitarary axial variation of seepage rate. For tube flow the axial and normal coordinates are $z$ and $r$, respectively.

of all existing or currently envisaged systems, the solutions reduce to the very simple expressions shown in Table 3.3.1 (see also Fig. 3.3.1).

The axial-velocity profiles are seen to be the same as for no seepage at the same local axial flow rate, and the distribution of velocity normal to the axis has the same form for all seepage rates. Furthermore, since the pressure gradients depend only on the axial-velocity gradient at the wall (see, e.g., Section 1.1), these are independent of seepage rate as well. In practice seepage rates are normally determined by membrane characteristics and diffusional considerations, and must in principle be evaluated by simultaneous solution of the diffusion equations and the equations of motion. The simplifications represented in Table 3.3.1 are thus quite important. Their origin and their range of validity are indicated in the following example for the mathematically somewhat simpler case of flow between parallel sheets.

### *EX. 3.3.1. ULTRAFILTRATION IN SLIT FLOW*

Integrate Eq. $A$ of Table 2.1.1 for slit flow and an exponential variation of seepage rate with position. Discuss the significance of your result and its relation to Table 3.3.1.

SOLUTION

Suitable boundary conditions for this problem are:

1. At $y=B$, $\quad v_x=0 \quad$ or $\quad \dfrac{\partial\psi}{\partial y}=0 \qquad (1)$

2. At $y=B$, $\quad v_y=v_0e^{-\alpha x} \quad$ or $\quad \dfrac{\partial\psi}{\partial x}=v_0e^{-\alpha x} \qquad (2)$

3. At $y=0$, $\quad \dfrac{\partial v_x}{\partial y}=0 \quad$ or $\quad \dfrac{\partial^2\psi}{\partial y^2}=0 \qquad (3)$

4. At $y=0$, $\quad v_y=0 \quad$ or $\quad \frac{\partial\psi}{\partial x}=0 \qquad (4)$

5. At $x=0$, $\quad \int_{-B}^{B} v_x dy=\frac{Q_0}{2W} \quad$ or $\quad \psi(B)-\psi(-B)=\frac{Q_0}{2W} \qquad (5)$

Here $v_0$ and $\alpha$ are constants characterizing the seepage rate, $Q_0$ is the volumetric flow rate to the system, and $W$ is the half width (in the $z$ direction).

Equations 1 through 4 follow from the problem statement and system symmetry, but Eq. 5 deserves some discussion. It was chosen deliberately to avoid mathematical difficulties in describing flow development at the inlet and will give a velocity profile roughly equivalent to fully developed steady flow.

We now note that Eq. A has the form

$$\left(\frac{\partial^2}{\partial x^2}+\frac{\partial^2}{\partial y^2}\right)\left(\frac{\partial^2}{\partial x^2}+\frac{\partial^2}{\partial y^2}\right)\psi=0 \qquad (6)$$

which suggests seeking solutions of the form

$$\psi_n=f_n(x)\cdot g_n(y) \qquad (7)$$

Furthermore inspection of the boundary conditions shows that there are only two possible solutions for $f$:

$$f_1=f_0, \quad \text{a constant}$$

$$f_2=e^{-\alpha x}$$

It follows that the defining equations for $g_n$ are

$$\frac{d^4 g_1}{dy^4}=0 \qquad (8)$$

and

$$\left(\alpha^2+\frac{\partial^2}{\partial y^2}\right)\left(\alpha^2+\frac{\partial^2}{\partial y^2}\right)g_2=0 \qquad (9)$$

These are well-known equations with solutions

$$g_1=a+by+cy^2+dy^3 \qquad (10)$$

$$g_2=A\sin\alpha y+B\cos\alpha y$$

$$+Cy\sin\alpha y+D\cos\alpha y \qquad (11)$$

**Table 3.3.2.** Velocity and Pressure Distributions in Slit Flow

$$v_y = \left[ \frac{(\alpha \cos \alpha b)(y \cos \alpha y) - (\cos \alpha b)(\sin \alpha y) + (\alpha b \sin \alpha b)\sin \alpha y}{\alpha b - \cos \alpha b \sin \alpha b} \right] v_0 e^{-\alpha x}$$

$$v_x = \left[ \frac{(\alpha b \sin \alpha b)(\cos \alpha y) - (\alpha y \cos \alpha b)(\sin \alpha y)}{\alpha b - \cos \alpha b \sin \alpha b} \right] v_0 e^{-\alpha x}$$

$$+ \left( \frac{3}{4} \frac{Q_0}{Wb} - \frac{3}{2} \frac{v_0}{\alpha b} \right) \left( 1 - \frac{y^2}{b^2} \right)$$

$$P - P(0,0) = 2\alpha \mu v_0 \cos \alpha b \left[ \frac{\cos \alpha y - e^{-\alpha x}}{\alpha b - \sin \alpha b \cos \alpha b} \right] + \left[ \frac{3 v_0 \mu}{\alpha b^3} - \frac{3}{2} \frac{Q_0 \mu}{W b^3} \right] x$$

Application of the boundary conditions now yields the velocity profiles and pressure distributions shown in Table 3.3.2.

If we now consider only the $x$-dependent portion of $v_x$, we see that

$$\alpha = \frac{d}{dx} \ln[v_x(y,x) - v_x(y,0)] \tag{12}$$

Then $\alpha^{-1}$ is the length of duct over which this velocity changes by a factor $e \doteq 2.718$. This is large relative to $B$ in any rational design, and it follows that

$$|\alpha y| \ll 1 \quad \text{for} \quad |y| \leqslant B \tag{13}$$

If the terms in Table 3.3.2 are now expanded in powers of $y$, and quadratic and higher powers neglected, the entries in Table 3.3.1 are obtained. This is a very good approximation from a practical standpoint, and a similar situation exists for tube flow.

To analyze the effects of arbitrary variation in seepage rate we merely allow $\alpha$ to be imaginary. We thus obtain solutions for oscillatory variation of seepage with position, and can extend these to arbitrary variation by the use of a Fourier series approximation. Once again, however, Eq. 13 normally applies, and the entries of Table 3.3.1 should still be valid—a remarkably simple result.

## 3.4 Entrance Effects and the Response to Transients

One of the most striking features of the circulatory system is its frequent bending and branching. The longest reasonably straight and uniform arterial section is the abdominal aorta between the mesenteric and renal arteries, and it has a length-to-diameter ratio of only about 15 to 20. Probably much more typical is the ratio of about 4.5 used by Skalak and Stathis (see Fung, 1966) in developing their tapered-tube model of arterial flows. Total arterial lengths are, however, much longer (see e.g., Table 3.1.1). A second characteristic feature is the unsteadiness of blood flow, which results both from the pulsatile action of the heart and the effect of body activity on the heart rate. In this section we take a look at both entrance effects and transient-response characteristics in duct flows of biological interest.

The entrance length $L_e$ is normally defined as the distance required for the center-line velocity to approach within 1% of its asymptotic (long-tube) value. For steady flow it can be shown by boundary-layer arguments (see Ex. 3.4.1) that to a first approximation

$$\frac{L_e}{D} \doteq c\,\mathrm{Re} \tag{3.4.1}$$

where $c$ is a constant characteristic of the duct geometry and entrance flow distribution, and $\mathrm{Re}=D\langle v\rangle/\nu$ is a Reynolds number for the flow. Here $D$ is a characteristic length, and $\langle v\rangle$ is the flow-average velocity.

Perhaps the best known analysis is that of Schiller for ducts of circular cross section and a flat entrance velocity profile. Schiller found $c$ to be about 0.0288 (with $D$ the tube diameter) and calculated the excess pressure drop resulting from flow development to be a bit more than $\rho\langle v\rangle^2$. (This is in addition to that required to produce the kinetic energy at the entrance.) More refined calculations by Lew and Fung (1969) show $L_e/D$ to approach a finite limit slightly over unity as the Reynolds number goes to zero. At high Reynolds numbers their analysis agrees with Eq. 3.4.1 and yields a value of 0.16 for $c$. Analyses of this type indicate that the entrance region is relatively long in arteries, but they are developed for a quite different situation from that actually occurring in the body. We return to this point after looking briefly at transient responses.

We consider here as representative the sudden imposition of a pressure difference across a long circular duct filled with a Newtonian liquid (see Ex. 4.1.2 in *Tr. Ph.*). If this pressure difference is maintained constant once it has been applied, the system response depends only on the dimensionless time $\tau=\nu t/R^2$, where $R$ is duct radius. Furthermore, over 90% of the

**Table 3.4.1.** Arterial Response Times

| Artery | $R$, (cm) | $t$(90%), (sec) |
|---|---|---|
| Aorta | 0.5 | 5.0 |
| Main | 0.15 | 0.45 |
| Terminal | 0.05 | 0.05 |
| Arterioles | 0.03 | 0.018 |

response will have occurred when $\tau = \frac{1}{2}$; representative values for the entries of Table 3.1.1, a 13-kg dog, are shown in Table 3.4.1. These times are generally short compared to those over which the heart rate changes, but are of the same order as the cardiac cycle (about 1 sec), except in the terminal arteries and arterioles. It is therefore desirable to determine the effect of pulsations on entrance length. Available analyses (see, e.g., Atabek in Attinger, 1964) indicate that the entrance length oscillates about that for a steady flow of the same time-average value, and that the time-average energy loss is greater than for the corresponding steady flow.

Kuchar and Ostrach (in Berman, 1953, 1958) have investigated entrance effects in distensible tubes for both steady and periodic flows. Effects of wall distensibility were found to be of secondary importance within the physiologically normal range. However, significant harmonic interaction and harmonic generation were noted. In addition, it is pointed out by these authors that the high local shear stresses near the inlet may be of interest as a source of intimal injury. Endothelial injury has been linked to mechanical stresses imposed by the blood, on the one hand, and to the development of atherosclerosis, on the other (see e.g., Texon, 1968; Fry, 1968).

Unfortunately the geometry of blood vessels is very complex. It is clear, for example, that the flow at branches cannot be axisymmetric, and it has been shown by Attinger (1964) that the cross sections of the major arteries are elliptical rather than circular. Investigation of these geometric complications is far from complete, but it appears that they do add appreciably to the energy losses. In addition, since they introduce nonlinear terms into the equations of motion, they can be expected to cause harmonic interaction and the generation of harmonics much like those predicted by Kuchar and Ostrach.

Nonlinear secondary flows can also be expected from bends in the larger vessels, and particularly in the arch of the aorta. It is difficult to reference observations on nonlinearities effectively, as they tend to be widely scattered. The reader is therefore referred to the general reviews cited and advised to watch for further developments. These secondary flows do appear to be of secondary importance, but their aggregate effect is far from negligible.

## *EX. 3.4.1. ENTRANCE LENGTHS IN SLIT FLOW*

Develop an approximate description of the entrance region for flow between parallel sheets, with a flat initial velocity profile, by using a von Kármán boundary-layer approximation.

### SOLUTION

We begin by considering all velocity gradients to be concentrated in boundary layers adjacent to the solid surfaces, and we assume that within these layers,

$$v_x = v_m\left[2\frac{y}{\delta} - \left(\frac{y}{\delta}\right)^2\right] \tag{1}$$

where $v_m(x)$ is the velocity outside the boundary layer, $\delta(x)$ is the local boundary layer thickness, and $y$ is the distance measured from the adjacent wall into the fluid (see Fig. 3.4.1). Finally, we assume hydrostatic equilibrium across the boundary layer. The equations of motion and continuity for this system then take the form*

$0<y<\delta(x)$:

$$v_x\frac{\partial v_x}{\partial x} - \left(\int_0^y \frac{\partial y_x}{\partial x}dy\right)\frac{\partial v_x}{\partial y} = \nu\frac{\partial^2 v_x}{\partial y^2} - \frac{1}{\rho}\frac{d\mathcal{P}}{dx} \tag{2}$$

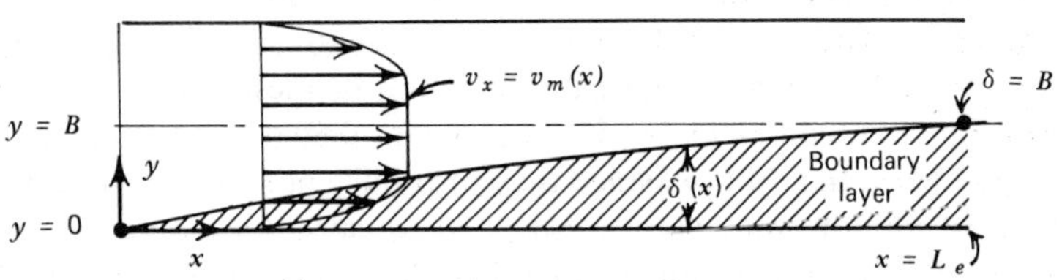


**Figure 3.4.1.** Development of slit flow.

* It should be noted that Eq. 1 is only a plausible approximation to the true velocity profile. This description is therefore neither exact nor entirely self-consistent. It is, however, useful. The nature and utility of the von Kármán approximation are discussed at length in Schlichting. See also Ex. 4.4.2 of *Tr. Ph.*

$\delta < y < B$:

$$v_m \frac{dv_m}{dx} = -\frac{1}{\rho}\frac{d\mathcal{P}}{dx} \tag{3}$$

with $v_m$, $\delta$, and $\mathcal{P}$ to be determined with the aid of Eq. 1 and a macroscopic mass balance.

We begin by substituting Eq. 3 into Eq. 2 to eliminate the pressure and then substituting Eq. 1 to eliminate $v_x$. We may then integrate with respect to $y$ over the boundary-layer thickness $\delta$ to obtain*

$$\int_0^\delta v_x \frac{\partial v_x}{\partial x} dy \quad = \tfrac{8}{15}\delta v_m v_m{}' - \tfrac{7}{30}\delta' v_m^2 \tag{4}$$

$$+\int_0^\delta \left(\int_0^\delta \frac{\partial v_x}{\partial x} dy\right)\frac{\partial v_x}{\partial y} dy = \tfrac{2}{15}\delta v_m v_m{}' - \tfrac{1}{10}\delta' v_m^2 \tag{5}$$

$$\nu \int_0^\delta \frac{\partial^2 v_x}{\partial y^2} dy = -2\nu v_m/\delta \tag{6}$$

The integrated form of Eq. 2 may then be written as

$$\tfrac{9}{15}\delta^2 v_m{}' + \tfrac{4}{30} v_m \delta\delta' = 2\nu \tag{7}$$

To complete the solution it is now only necessary to relate $v_m$ and $\delta$ via the macroscopic mass balance:

$$B\langle v\rangle = \int_0^\delta v_x dy + v_m(B-\delta)$$

where $\langle v\rangle$ is the flow-average velocity. Then

$$\delta = 3B\left(1 - \frac{\langle v\rangle}{v_m}\right); \qquad \delta' = 3B\langle v\rangle \frac{v_m{}'}{v_m^2} \tag{8,9}$$

and it follows from Eq. 8 that

$$v_m = \langle v\rangle, \qquad x = 0 \tag{10a}$$

$$= \tfrac{3}{2}\langle v\rangle, \qquad x = L_e \tag{10b}$$

since $x = L_e$ when $\delta = B$.

* The prime indicates differentiation with respect to $x$.

If Eq. 7 is now integrated over the entrance length we find

$$\frac{L_e}{B} = \left(\tfrac{123}{60} - \tfrac{24}{5}\ln\tfrac{3}{2}\right)\frac{B\langle v\rangle\rho}{\mu} \doteq 0.104\frac{B\langle v\rangle\rho}{\mu} \tag{11}$$

which is of the expected form. If we integrate Eq. 3 over the velocity range indicated by Eq. 10 we find

$$\mathcal{P}(L_e) - \mathcal{P}(0) = -\tfrac{5}{8}\rho\langle v\rangle^2 \tag{12}$$

which is of the right order of magnitude.

Although Eqs. 11 and 12 are only approximations, they tell us all we need to know about flow in typical extra-corporeal circuits of this type. Since Reynolds numbers, velocities, and half-widths tend to be small (of the order of $10^2$, $10^2$ cm/sec, and $10^{-2}$ cm, respectively, for a common hemodialyzer) both entrance lengths and the associated pressure drops are normally quite small.

## 3.5 Pulsatile Flow in Elastic-Walled Ducts: Models of the Arterial System

We consider in this section simple linearized models that have been proposed to describe the flow behavior of the individual ducts making up the arterial system. These are sufficiently realistic to provide a general understanding of the nature of arterial flow and in particular the transmission of velocity and pressure pulses. In addition they give a useful background for the discussions of the circulatory system as a whole in the next chapter. We begin with formal mathematical descriptions, first of rigid-walled and then of elastic ducts, and then discuss briefly the numerical aspects of duct flows and representative examples. In all cases we can limit ourselves to periodic flows without serious error; this is because of the unimportance of transients shown by Table 3.4.1. We also neglect end effects—with somewhat less justification.

### (a) *PERIODIC FLOWS IN RIGID-WALLED DUCTS*

We begin our analysis by considering the system of Fig. 3.5.1 where $L \gg D$, so that $v_r$ and $v_\theta$ are zero. We focus our attention on a long-continued oscillatory flow produced by the pressure gradient

$$\frac{\mathcal{P}_0 - \mathcal{P}_L}{L} = A\sin\omega t + B\cos\omega t \tag{3.5.1}$$

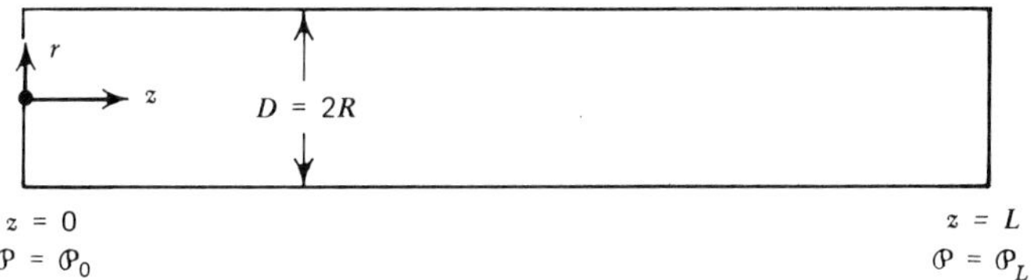


**Figure 3.5.1.** A rigid cylindrical duct.

For the linear system under consideration we may build up a solution for more complex flows by the superposition of individual solutions for various frequencies that is, by Fourier analysis.

The equations for this system take the form

(continuity)

$$\frac{\partial v_z}{\partial z} = 0; \qquad v_z = v_z(r,t) \tag{3.5.2}$$

(motion)

$$r \text{ component:} \quad 0 = -\frac{\partial \mathcal{P}}{\partial r} \tag{3.5.3}$$

$$\theta \text{ component:} \quad 0 = -\frac{1}{r}\frac{\partial \mathcal{P}}{\partial \theta} \tag{3.5.4}$$

$$\mathcal{P} = \mathcal{P}(z,t)$$

$$z \text{ component:} \quad \rho\left(\frac{\partial v_z}{\partial t} + \left[v_z \frac{\partial v_z}{\partial z}\right]\right) = \mu \frac{1}{r}\frac{\partial}{\partial r}\left(r\frac{\partial v_z}{\partial r}\right) - \frac{\partial \mathcal{P}}{\partial z} \tag{3.5.5}$$

where the term in [ ] is identically zero, from Eq. 3.5.2.

The $z$ component of the equation of motion may be rearranged to

$$\rho\frac{\partial v_z}{\partial t} - \mu\frac{1}{r}\frac{\partial}{\partial r}\left(r\frac{\partial v_z}{\partial r}\right) = -\frac{\partial \mathcal{P}}{\partial z} \tag{3.5.6}$$

Equation 3.5.2 implies that the left side of Eq. 3.5.6. is a function only of $r$ and $t$, and Eqs. 3.5.3 and 3.5.4 imply that the right side is a function only of $z$ and $t$. It follows that each side must be a function only of time:

$$\rho\frac{\partial v_z}{\partial t} - \mu\frac{1}{r}\frac{\partial}{\partial r}\left(r\frac{\partial v_z}{\partial r}\right) = c(t) \tag{3.5.7}$$

$$\frac{\partial \mathcal{P}}{\partial z} = c(t) \tag{3.5.8}$$

Equation 8 may be readily integrated with respect to $z$ to obtain

$$\frac{\mathcal{P}_L - \mathcal{P}_0}{L-0} = c(t) \tag{3.5.9}$$

To complete our analysis then it is only necessary to know the pressure difference across the tube as a function of time and to integrate Eq. 3.5.7 for this function.

Before proceeding to a detailed solution it is desirable to rewrite the system description in dimensionless variables:

$$\phi_\omega = \frac{v_z(r,t)}{R\omega}$$

$$\eta = \frac{r}{R} \qquad \zeta = \frac{z}{R}$$

$$P = \frac{\mathcal{P}}{\omega\mu} \qquad \frac{\partial P}{\partial \zeta} = -Ge^{i\omega t} \qquad \text{(real part)}$$

$$G = (A - iB)\frac{R}{\mu\omega}$$

$$\tau = \omega t$$

$$\alpha^2 = \frac{R^2\omega}{\nu} = \text{a characteristic Reynolds number for a given frequency, } \omega.$$

The equation of motion then takes the form

$$\alpha^2 \frac{\partial \phi_\omega}{\partial \tau} - \frac{1}{\eta}\frac{\partial}{\partial \eta}\left(\eta \frac{\partial \phi_\omega}{\partial \eta}\right) = -\frac{\partial P_\omega}{\partial \zeta} = Ge^{i\tau} \tag{3.5.10}$$

As boundary conditions on $\eta$ we require

$$\text{At } \eta = 1, \qquad \phi_\omega = 0 \tag{3.5.11}$$

$$\text{At } \eta = 0, \qquad \phi_\omega \text{ finite} \tag{3.5.12}$$

Since we are interested only in a periodic solution, we do not need an initial condition. Furthermore, the periodic solution must oscillate at frequency $\omega$, and we therefore assume a trial solution of the form

$$\phi_\omega = u(\eta)Ge^{i\tau} \tag{3.5.13}$$

Putting this expression into Eq. 3.5.10, we obtain

$$\alpha^2 iue^{i\tau} - \frac{1}{\eta}\frac{d}{d\eta}\left(\eta\frac{du}{d\eta}\right)e^{i\tau} = e^{i\tau} \tag{3.5.14}$$

and

$$\alpha^2 iu - \frac{1}{\eta}\frac{du}{d\eta} - \frac{d^2u}{d\eta^2} = 1 \tag{3.5.15}$$

This result can be converted to a Bessel equation by writing

$$w = u - \frac{1}{i\alpha^2} \tag{3.5.16}$$

to obtain

$$\eta^2\frac{d^2s}{d\eta^2} + \eta\frac{dw}{d\eta} + (-i\alpha^2\eta^2 - 0)w = 0 \tag{3.5.17}$$

and

$$w = c_1 J_0(i^{3/2}\alpha\eta) + c_2 Y_0(i^{3/2}\alpha\eta) \tag{3.5.18}$$

When $\eta = 0$, $u$ and hence $w$ must be finite, and $c_2$ must be zero. Since $u$ must equal zero at $\eta = 1$,

$$-\frac{1}{i\alpha^2} = c_1 J_0(i^{3/2}\alpha) \tag{3.5.19}$$

so that

$$u - \frac{1}{i\alpha^2} = -\frac{1}{i\alpha^2}\frac{J_0(i^{3/2}\alpha\eta)}{J_0(i^{3/2}\alpha)} \tag{3.5.20}$$

and

$$\frac{1}{G}\phi_\omega = ue^{i\tau} = \left[1 - \frac{J_0(i^{3/2}\alpha\eta)}{J_0(i^{3/2}\alpha)}\right]\frac{e^{i\tau}}{i\alpha^2} \tag{3.5.21}$$

This is the fundamental solution from which all others can be built up in Fourier series. It is important to note that it does not contribute to the time-average flow.

The numerical calculation of velocity profiles is complex, except in the limits of very low or very high $\alpha$, and is normally of little interest. However, the reader may find it of interest to refer to Womersley (1958) to see how complicated velocity profiles in the arteries are under physiological conditions.

We are usually interested only in the dimensionless flow-average velocity

$$V \equiv \frac{2\pi\int_0^R \phi r\,dr}{2\pi\int_0^R r\,dr} = 2\int_0^1 \phi\eta\,d\eta \tag{3.5.22}$$

For the profile of Eq. 3.5.21,

$$V = \frac{1}{i\alpha^2}\left[1 - \frac{2}{i^{3/2}\alpha}\frac{J_1(i^{3/2}\alpha)}{J_0(i^{3/2}\alpha)}\right]Ge^{i\tau} \tag{3.5.23}$$

The nature of this expression is shown for the limiting cases of small and large $\alpha^2$ in Ex. 3.5.1.

First, however, we note that this expression may be written in the form

$$V = -\frac{\partial P}{\partial \zeta}\frac{1}{Z} \tag{3.5.24}$$

where $Z$, the effective (dimensionless) *impedance* per unit (dimensionless) length of duct, is defined by

$$Z = \left\{\frac{1}{i\alpha^2}\left[1 - \frac{2}{i^{3/2}\alpha}\frac{J_1(i^{3/2}\alpha)}{J_0(i^{3/2}\alpha)}\right]\right\}^{-1}$$

$$\equiv \mathcal{R} + i\mathcal{L} \tag{3.5.25}$$

Here $\mathcal{R}$ and $\mathcal{L}$ are the real and imaginary parts of $Z$, which is in general complex. Since

$$V = \tilde{V}e^{i\tau}; \quad \frac{\partial V}{\partial \tau} = iV \tag{3.5.26}$$

(where $V$ is a complex constant) we may also write

$$\frac{\partial P}{\partial \zeta} = -\left(\mathcal{R}V + \mathcal{L}\frac{\partial V}{\partial \tau}\right) \tag{3.5.27}$$

These equations correspond to those for a series electrical circuit contain-

**Table 3.5.1.** Useful Analogies

| | Hydrodynamic | Electric |
|---|---|---|
| Potential | $P$<br>(Pressure) | $E$<br>(Electric potential) |
| Flux | $V$<br>(Mass flow) | $L$<br>(Electric current) |
| Resistance | $\mathcal{R}$<br>(Viscous) | $R$<br>(Ohmic) |
| Inductance | $\mathcal{L}$<br>(Inertial) | $L$<br>(Electromagnetic) |

ing both a resistance $R$ and an inductance $L$, and operating at frequency $\omega$:

$$\frac{\partial E}{\partial z} = -(R + i\omega L)I \tag{3.5.28}$$

$$= -\left(RI + L\frac{\partial I}{\partial t}\right) \tag{3.5.29}$$

The frequency $\omega$ (or $\alpha^2$) does not appear in Eq. 25 or 27 because of the choice of dimensionless variables used. The quantities appearing in these two sets of equations are, however, clearly analogous (see Table 3.5.1). We shall find these analogies very useful and will add an effective capacitance in the next section. We do, however, have to be careful with them:

1. The quantities $\mathcal{R}$ and more particularly $\mathcal{L}$ vary with the dimensionless frequency (or Reynolds number) $\alpha^2$.
2. The boundary conditions at branches are much more complex for hydrodynamic than for electric systems, and end effects are very much larger.

Much more doubtful approximations must therefore be made in analysis of hydrodynamic networks than for their fundamentally much simpler electric analogs.

We now consider briefly the practical problem of obtaining numerical values for the dimensionless hydrodynamic impedance

$$Z = \mathcal{R} + i\mathcal{L} = \frac{i\alpha^2}{J} \tag{3.5.30}$$

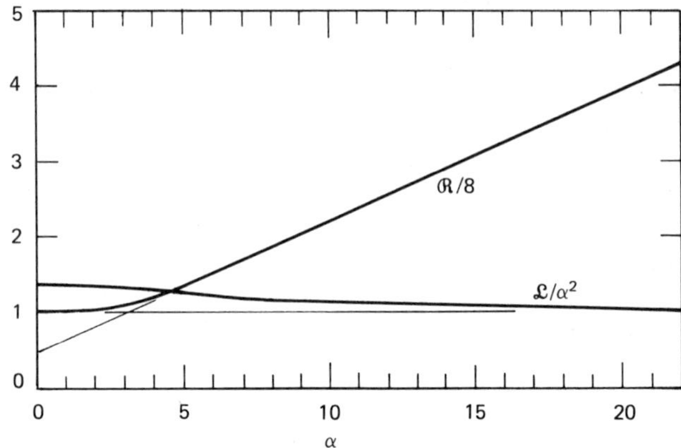


**Figure 3.5.2.** Dimensionless resistance ℛ and inductance ℒ for viscous oscillating flow in a rigid pipe as a function of $\alpha = R(\omega/\nu)^{1/2}$.

where*

$$J = 1 - \frac{2}{i^{3/2}\alpha}\frac{J_1(i^{3/2}\alpha)}{J_0(i^{3/2}\alpha)} \tag{3.5.31}$$

The quantities ℛ $(\alpha)$ and ℒ $(\alpha)$ are shown graphically in Fig. 3.5.2. It may be seen from this figure that

1. The combination $\mathcal{L}/\alpha^2$ varies only from $\frac{4}{3}$ at $\alpha = 0$ to unity in the limit of large $\alpha$. The large-$\alpha$ limit is approached within about 3% at $\alpha = 20$.
2. The quantity ℛ becomes very nearly linear in $\alpha$ for large $\alpha$ and very nearly independent of $\alpha$ as $\alpha$ approaches zero.

We may approximate the system behavior over much of the range of interest by the following approximations:

$\alpha \to 0$:

$$R \sim 8 + O(\alpha^4) \tag{3.5.32}$$

$$L \sim \tfrac{4}{3}\alpha^2 + O(\alpha^4) \tag{3.5.33}$$

* We write the impedance in terms of $J$ because this function is described in detail by Womersley (1957) in the polar form

$$J = M_{10}' e^{i\epsilon_{10}'}.$$

$\alpha \to \infty$:

$$R \sim \sqrt{2}\alpha + 3 + \frac{2\sqrt{2}}{\alpha} + O(\alpha^{-2}) \tag{3.5.34}$$

$$L \sim \alpha^2 \left[ 1 - \frac{\sqrt{2}}{\alpha} + \frac{4\sqrt{2}}{\alpha^2} + O(\alpha^{-3}) \right] \tag{3.5.35}$$

It can thus be seen that the impedance is largely resistive at low $\alpha$ and inductive as $\alpha$ becomes large. These limiting expressions are normally as reliable as the physiological measurements from which they must ultimately be derived. As a rough rule of thumb, Eqs. 3.5.32 and 3.5.33 should be used for $\alpha \ll 3$, and Eqs. 3.5.34 and 3.5.35 for $\alpha \gg 3$. The limiting behavior is further considered in the example immediately below.

### *EX. 3.5.1. LIMITING BEHAVIOR FOR OSCILLATORY FLOWS IN RIGID DUCTS*

Obtain first approximations for $Z$ in the limits of very large and very low Reynolds number from Eqs. 3.5.30 and 3.5.31 and the definitions of the Bessel functions.

SOLUTION

We begin with the low-Reynolds-number limit, for which the following convergent series are convenient:
Convergent series:

$$J_0(z) = 1 - (z/2)^2 + \frac{(z/2)^4}{1^2 2^2} - \frac{(z/2)^6}{1^2 2^2 3^2} + \cdots \tag{1}$$

$$J_1(z) = (z/2) - \frac{(z/2)^3}{1^2 2} + \frac{(z/2)^5}{1^2 2^2 3} + \cdots \tag{2}$$

It then follows by direct substitution of these equations into Eqs. 3.5.31 and 3.5.32 that

$$Z = \frac{8}{1 - \frac{1}{6}\alpha^2 + O(\alpha^4)}$$

$$= 8 + i\tfrac{4}{3}\alpha^2 \qquad (\alpha \to 0) \tag{3}$$

which gives the leading terms in Eq. 3.5.32 and 3.5.33.

For large $\alpha$, Eqs. 1 and 2 are very inconvenient, and it is preferable to use an asymptotic series, the leading terms of which are

$$J_0(z) \sim \left(\frac{2}{\pi z}\right)^{1/2}\left[\cos\left(z-\frac{\pi}{4}\right)\right] \qquad (z \to \infty) \tag{4}$$

$$J_1(z) \sim \left(\frac{2}{\pi z}\right)^{1/2}\left[\cos\left(z-\frac{3\pi}{4}\right)\right] \tag{5}$$

It follows that to a first approximation

$$\frac{J_1(z)}{J_2(z)} = \tan\left(z-\frac{\pi}{4}\right) \tag{6}$$

$$= \tan(i^{3/2}\alpha) - \frac{\pi}{4} \qquad (\alpha \to \infty) \tag{7}$$

$$= \tan\left(\sqrt{2}\,\alpha(1-i) - \frac{\pi}{4}\right) \tag{8}$$

This in turn may be rewritten as

$$\frac{J_1(i^{3/2}\alpha)}{J_0(i^{3/2}\alpha)} \doteq -\frac{\sinh(2\sqrt{2}\,\alpha)}{\cosh(2\sqrt{2}\,\alpha)}i - i \tag{9}$$

It follows that

$$Z = \frac{i\alpha^2}{1+\dfrac{2}{\alpha\sqrt{i}}} \doteq i\alpha^2\left[1-\frac{2}{\alpha}\left(\frac{1-i}{\sqrt{2}}\right)\right] \tag{10}$$

and that

$$R \sim \sqrt{2}\,\alpha; \qquad L \sim \alpha^2\left(1-\frac{\sqrt{2}}{\alpha}\right) \tag{11}$$

which are the leading terms of Eqs. 3.5.34 and 3.5.35.

We may now see from Eq. 3 that for $\alpha^2$ less than $\frac{2}{3}$ more than 90% of the impedance will be viscous, and hence dissipative. For the fundamental frequency of the human heart

$$\omega_0 \doteq 2\pi \sec^{-1}$$

and

$$\nu \doteq 0.04\,\text{cm}^2/\text{sec}$$

the corresponding internal diameter is

$$D_{\text{viscous}} = 2\sqrt{\tfrac{2}{3}(0.04\,\text{cm}^2/\text{sec})(2\pi/\text{sec})} \doteq 1.3\,\text{mm}.$$

For vessels smaller than this (from terminal arteries down to the microcirculation), inertial effects are predicted to be small*, and the pseudo-steady Poiseuille's law should be valid. This is also the region accounting for the bulk of energy dissipation in the circulation.

It follows similarly from Eq. 10 that for $\alpha^2$ greater than 162 more than 90% of the impedance will be inertial. The corresponding diameter is

$$D_{\text{inertial}} = 2(9\sqrt{2}\,)\sqrt{\frac{0.04}{2\pi}}\ \text{cm} \doteq 2\ \text{cm}$$

This suggests that viscous effects are barely appreciable for the fundamental frequency in the great vessels but that inertial effects dominate the behavior there. This is in accord with the observation that viscous dissipation is of secondary importance in the major arteries. Secondary flows in the aortic arch and at major branches can, however, be expected to be important, and they should be investigated carefully.

### (b) *FLOW IN EXPANSIBLE DUCTS*

Healthy arterial walls are viscoelastic, as has already been pointed out, and they undergo a significant change in diameter as a result of the normal physiological pressure differences occurring during one heartbeat. The effects of such changes on the velocity and pressure distribution in the arterial system have been extensively studied. The works of Womersley (1957), Skalak (in Bergel, 1971; and Fung, 1966 and 1968) and M. G. Taylor (see Skalak, 1971), and their associates, are particularly important. Attinger's and Fung's texts and Fung's review of biomechanics provide a comprehensive introduction to the literature.

The primary effects of wall viscoelasticity are to lower the peak systolic pressure and to produce pressure pulses which travel down the arterial

* There is, however, reason to believe that inertial effects are appreciable here as a result of vessel taper, end effects at branches, and other factors resulting in secondary flows. This is discussed briefly in the next chapter.

system at speeds much greater than the mean liquid velocity. The velocity of propagation in the human body is of the order of 6 to 8 m/sec, and the pulse reaches the whole body before the end of systole. For most mammals the wavelength for the fundamental frequency (the heart rate) is about 5 times the distance from the heart to the furthest extremity. The wall elasticity does not have an appreciable effect on the local velocity profile within any one long duct. It has, for example, been shown by Womersley (1957) that Eq. 3.5.23 applies to elastic arteries within the degree of approximation normally used.

We now wish to extend the previous rigid-tube analysis to elastic-walled ducts. To do this we shall have to reconsider the equation of continuity and write a description of stress-strain relations in the confining wall. Once again we assume symmetry about the tube axis.

The continuity equation may be written in the form

$$-\frac{1}{\rho}\left(\frac{\partial \rho}{\partial t}+v_r\frac{1}{r}\frac{\partial \pi}{\partial r}+v_z\frac{\partial \rho}{\partial z}\right)=\frac{1}{r}\frac{\partial}{\partial r}rv_r+\frac{\partial v_z}{\partial z}$$

The left side of Eq. 3.5.32, which describes fluid compression, is concerned only with acoustic waves and if of no present interest. We therefore set the right side of this equation equal to zero and write

$$\frac{\partial v_z}{\partial z}=-\frac{1}{r}\frac{\partial}{\partial r}rv_r$$

We now integrate this expression over the cross-sectional area of the duct:

$$2\pi\int_0^R\left(\frac{\partial v_z}{\partial z}\right)r\,dr=-2\pi\int_0^R\left(\frac{\partial}{\partial r}rv_r\right)dr$$

Interchanging the order of integration and differentiation on the left, and then carrying out the indicated integrations, we obtain

$$\frac{\partial\langle v_z\rangle}{\partial z}=-\frac{2}{R}v_r\Big|_{r=R} \tag{3.5.36}$$

where

$$\langle v_z\rangle=\frac{1}{\pi R^2}\int^R 2\pi r v_z\,dr=VR\omega$$

is the flow-average velocity defined earlier.

We next write an equation of motion (conservation of momentum) for the membrane, assuming elastic behavior, a small ratio of wall thickness to diameter, and complete tethering* of the artery to the surrounding tissue (so that the membrane is incapable of axial motion).

For these assumed conditions we need only the relation†

$$\rho_a h \frac{\partial^2 \xi}{\partial t^2} = \mathcal{P} - \mathcal{P}_e - \frac{Eh\xi}{R^2} \tag{3.5.37}$$

where $\rho_a$ is the density of the arterial wall, $h$ is the wall thickness, $\xi$ is the radial strain $\Delta R$ relative to the radius at rest, $\mathcal{P}_e$ is the pressure in the tissue surrounding the artery, and $E$ is Young's modulus. The left side of this equation represents the inertia of the tube wall, but should also contain contributions from the surrounding tissue if $\mathcal{P}_e$ is assumed time independent. These contributions are normally considered negligible, and we will neglect them here. Since gravitational effects are small over an arterial cross section, we can assume $(\mathcal{P} - \mathcal{P}_e)$ to be independent of position and write

$$\mathcal{P} = \frac{Eh\xi}{R^z} + \mathcal{P}e \tag{3.5.38}$$

Differentiating this expression with respect to time, we obtain

$$\frac{\partial \mathcal{P}}{\partial t} = \frac{Eh}{R^z} v_r \bigg|_{r = R + \xi} \tag{3.5.39}$$

where $v_r = \partial \xi / \partial t$ at the wall. If strains are small,

$$v_r|_{R+\xi} \doteq v_r|_R \tag{3.5.40}$$

and we may combine Eqs. 3.5.36 and 3.5.39 to obtain

$$\frac{\partial \langle v_z \rangle}{\partial z} = -\frac{2R}{Eh} \frac{\partial \mathcal{P}}{\partial t} \tag{3.5.41}$$

* It should be noted in connection with the third assumption that the larger vessels in their natural state are stretched to about twice their relaxed length.

† In Eq. 3.5.37 and all subsequent relations containing Young's modulus

$$E = \frac{\mathcal{E}}{1 - \sigma^2}$$

where $\mathcal{E}$ is the modulus as normally defined in the engineering mechanics literature and $\sigma$ is Poisson's ratio. Inclusion of $(1 - \sigma^2)^{-1}$ is necessary to allow for the tethering of blood vessels, which prevents their axial motion in the natural state. In practice, however, the difference between $E$ and $\mathcal{E}$ is small compared to the uncertainty with which either is known.

We may then write the defining equations for our system, Eqs. 3.5.24 and 3.5.41, as

$$\frac{\partial P}{\partial \zeta} = -ZV = -\left(\mathcal{R}V + \mathcal{L}\frac{\partial v}{\partial \tau}\right) \tag{3.5.42}$$

and

$$\frac{\partial V}{\partial \zeta} = -C\frac{\partial P}{\partial \tau} \tag{3.5.43}$$

where

$$C = \frac{2R\omega\mu}{Eh}$$

is the effective "capacitance" of the system, and the remaining variables are as previously defined. The definition of our system is now complete except for the boundary conditions needed for specific problems.

We can obtain a general solution to Eqs. 3.5.42 and 3.5.43 very simply by noting that

$$P = \tilde{P}(\zeta)e^{i\tau} \tag{3.5.44}$$

$$V = \tilde{V}(\zeta)e^{i\tau} \tag{3.5.45}$$

where $\tilde{P}$ and $\tilde{V}$ are complex functions of position to be determined. We begin by substituting Eqs. 3.5.44 and 3.5.45 into 3.5.42 and 3.5.43 to obtain

$$\tilde{P}' = -Z\tilde{V} \tag{3.5.46}$$

$$\tilde{V}' = -iC\tilde{P} \tag{3.5.47}$$

where primes denote differentiation with respect to $\zeta$. The functions $\tilde{V}$ and $\tilde{P}$ can each be eliminated between these two equations to obtain

$$\tilde{P}'' = iCZ\tilde{P} \tag{3.5.48}$$

$$\tilde{V}'' = iCZ\tilde{V} \tag{3.5.49}$$

These are a very well-known pair of equations mathematically analogous to current-voltage relations describing the behavior of electrical transmission lines. Their general solution is of the form:

$$\tilde{V} = A_1 \sinh\left(\sqrt{iCZ}\,\zeta\right) + A_2 \cosh\left(\sqrt{iCZ}\,\zeta\right) \tag{3.5.50}$$

$$\tilde{P} = B_1 \sinh\left(\sqrt{iCZ}\,\zeta\right) + B_2 \cosh\left(\sqrt{iCZ}\,\zeta\right) \tag{3.5.51}$$

where $A_1$, $A_2$, $B_1$, and $B_2$ are (generally complex) constants, to be determined from the system boundary conditions.

We have now accomplished a great deal. First, we reduced the entire description of pulsatile flow in long tubes to a pair of very simple ordinary differential equations, Eqs. 3.5.48 and 3.5.49. Since the parameters $C$ and $Z$ appearing in these equations are independent of position, we were able to integrate these equations to obtain their general solution, Eqs. 3.5.50 and 3.5.51. To obtain solutions for arbitrary flow or pressure pulses we turn to harmonic analysis. It then will remain only to consider boundary conditions to extend our formal description to networks. This apparent simplicity is, however, to a degree deceptive, because it was achieved only by the neglect of end effects and nonlinearities in the defining equations. These complications are discussed briefly in the next chapter.

It now only remains to rewrite our description of velocity and pressure distributions. We begin by recalling that the fundamental solution for either $P$ or $V$ is

$$Ae^{\pm\sqrt{iCZ}\,\zeta + i\tau} \tag{3.5.52}$$

where $A$ is a complex constant. It now proves convenient to rewrite the coefficient of $\zeta$ as

$$\sqrt{iCZ} = \kappa + i\beta \tag{3.5.53}$$

and thus to separate it into its real and imaginary parts, $\kappa$ and $\beta$,

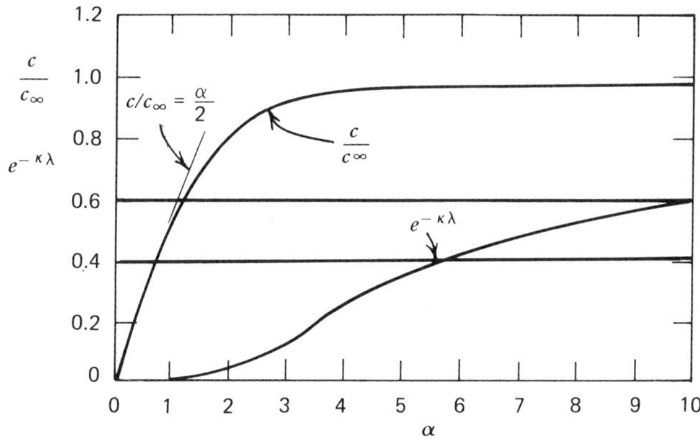


**Figure 3.5.3.** Dimensionless phase velocity $(c/c_\infty)$ and inverse attenuation per wave length $(e^{-\kappa\lambda})$ as a function of $\alpha = R(\omega/\nu)^{1/2}$. Based on Womersley (1957) taken from Skalak.

respectively. The expression 3.5.52 may then be written in the alternative form

$$Ae^{\pm \kappa\zeta + i(\tau \pm \beta\zeta)} \tag{3.5.54}$$

Here $-\kappa$ = the attenuation of the wave per unit dimensionless length
$-1/\beta$ = the dimensionless celerity of the wave; that is, the dimensionless rate of propagation in the $z$ direction.

The Reynolds-number dependence of $\kappa$ and $\beta$ is shown in Fig. 3.5.3. In this figure $c$ is the celerity or phase velocity, and the reference celerity

$$c_\infty = \sqrt{\frac{hE}{2R\rho}} \tag{3.5.55}$$

is known as Young's velocity. We return shortly to a discussion of these quantities, but we note here that:

1. The celerity of a wave is

$$c = \left.\frac{\partial z}{\partial t}\right|_{(\tau - \beta\zeta)} = \frac{\omega R}{\beta} \tag{3.5.56}$$

2. The fractional rate of reduction in amplitude with distance is

$$\frac{\partial \ln P}{\partial z} = -\frac{\kappa}{R} \tag{3.5.57}$$

3. The wavelength $l$ of the pulse under consideration is the distance over which $\beta h$ changes by $2\pi$. Hence

$$\frac{\beta l}{R} = 2\pi$$

$$l = \frac{2\pi R}{\beta} \tag{3.5.58}$$

It follows from statement (1) that Eqs. 3.5.50 and 3.5.51 each represent two waves, one traveling forward (in the positive $z$ direction) and one backward. Furthermore, since it can be shown (see Examples) that $\beta$ and $\kappa$ always have the same sign, the amplitude of each wave decreases in the direction of propagation. Such a wave is shown in Fig. 3.5.4.

It remains to develop a convenient means for numerical calculation of $\kappa$ and $\beta$, that is, to separate $iCZ$ into its real and imaginary parts. We begin

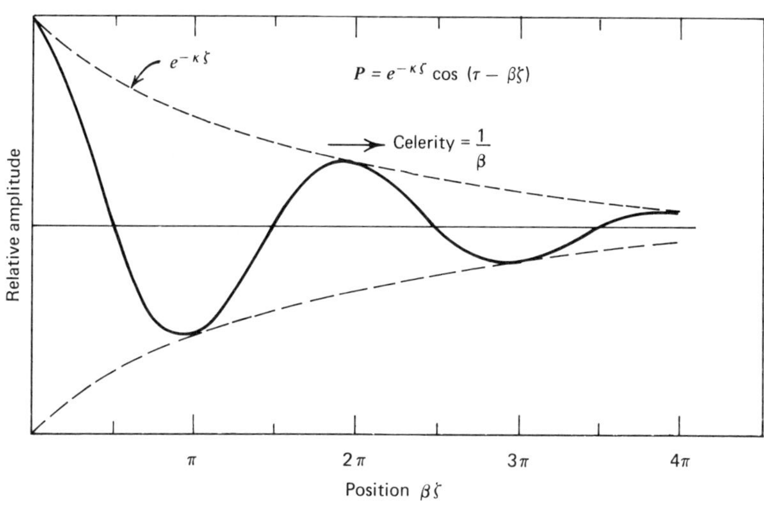


**Figure 3.5.4.** A damped traveling wave.

by noting that the dimensionless hydrodynamic capacitance

$$C = 2R\omega\mu / Eh \tag{3.5.59}$$

is real, and therefore that

$$iCZ = C(-\mathcal{L} + i\mathcal{R}) \tag{3.5.60}$$

$$= re^{i\theta} \tag{3.5.61}$$

where

$$r = C\sqrt{\mathcal{R}^2 + \mathcal{L}^2}$$

$$\theta = -\tan^{-1}(\mathcal{R}/\mathcal{L})$$

or

$$\cos\theta = -\mathcal{L}/\sqrt{\mathcal{R}^2 + \mathcal{L}^2}$$

It may now readily be shown that

$$\kappa = \pm\alpha\sqrt{C_0/2}\,(\mathcal{R}^2 + \mathcal{L}^2)^{1/4}\left(1 - \frac{\mathcal{L}}{\sqrt{\mathcal{R}^2 + \mathcal{L}^2}}\right)^{1/2} \tag{3.5.62}$$

$$\beta = \pm \alpha \sqrt{C_0/2}\,(\mathcal{R}^2 + \mathcal{L}^2)^{1/4}\left(1 + \frac{\mathcal{L}}{\sqrt{\mathcal{R}^2 + \mathcal{L}^2}}\right)^{1/2} \tag{3.5.63}$$

Here

$$C_0 = \alpha^2 C = 2\mu\nu / E L h$$

is used in place of $C$ because it has no frequency dependence.

*For small* $\alpha$, where viscous forces predominate, we may write, to order $\alpha^2$,

$$\left.\begin{aligned} \kappa &\sim 2\alpha C_0\left(1 - \frac{\alpha^2}{12}\right) && (35.64) \\ \beta &\sim 2\alpha C_0\left(1 + \frac{\alpha^2}{12}\right) && (3.5.65) \end{aligned}\right\} \quad (\alpha \to 0)$$

$$\left.\begin{aligned} c &= \alpha c_\infty / 2 && (3.5.66) \\ l &= \frac{\pi R}{\alpha}\left(\frac{ERh}{2\nu^2\rho}\right)^{1/2} && (3.5.67) \end{aligned}\right\} \quad (\alpha \to 0)$$

where $c_\infty = \sqrt{Eh/2R\rho}$. The low-frequency waves are thus seen to be very long and to travel very slowly.

*For large* $\alpha$, where inertial forces predominate, the corresponding expressions are

$$\left.\begin{aligned} \kappa &\sim \sqrt{C_0}\left(\frac{\alpha}{\sqrt{2}} + \frac{3}{2} + \frac{1}{2\sqrt{2}\alpha} + \frac{3}{4\alpha^2}\right) && (3.5.68) \\ \beta &\sim \sqrt{C_0}\left(\alpha^2 + \frac{1}{4} - \frac{1}{8\alpha^2}\right) && (3.5.69) \end{aligned}\right\} \quad (\alpha \to \infty)$$

$$\left.\begin{aligned} c &\sim c_\infty && (3.5.70) \\ l &\sim \left(\frac{2\pi R}{\alpha^2}\right)\sqrt{\frac{ERh}{2\nu^2\rho}} && (3.5.71) \end{aligned}\right\} \quad (\alpha \to \infty)$$

Celerities are now much higher, and nearly frequency independent; wavelengths are much shorter and decrease rapidly with frequency.

We are now ready to consider some representative examples.

## *EX. 3.5.2 PULSATILE FLOW IN A STRAIGHT ELASTIC DUCT*

Describe the velocity and pressure distribution for a straight duct of dimensionless length $\lambda$ with a pressure $\tilde{P}_0 e^{i\tau}$ at $\zeta=0$ and zero at $\zeta=\lambda$.

SOLUTION

We may immediately determine $B_1$ and $B_2$ in Eq. 3.5.51 by application of the boundary conditions:

$$\tilde{P}_0 = 0 + B_2 \tag{1}$$

$$0 = B_1 \sinh(\sqrt{iCZ}\,\lambda) + B_2 \cosh(\sqrt{iCZ}\,\lambda) \tag{2}$$

It then follows directly that

$$P = \tilde{P}_0 e^{i\tau}\left(\cosh\sqrt{iCZ}\,\zeta - \sinh\sqrt{iCZ}\,\zeta/\tanh\sqrt{iCZ}\,\lambda\right)$$

$$= \tilde{P}_0 e^{i\tau}\frac{\sinh\sqrt{iCZ}\,(\lambda-\zeta)}{\sinh\sqrt{iCZ}\,\lambda} \tag{3}$$

The velocity distribution is found from Eq. 3.5.46 and the result above:

$$\tilde{V} = -\frac{1}{Z}\tilde{P}' \tag{4}$$

Then

$$V = \tilde{V} e^{i\tau} \tag{5}$$

$$= \tilde{P}_0 e^{i\tau}\sqrt{\frac{iC}{Z}}\;\frac{\cosh\sqrt{iCZ}\,(\lambda-\zeta)}{\sinh\sqrt{iCZ}\,\lambda} \tag{6}$$

Note once again that we are interested only in the real parts of $P$ and $V$.

The rigid-duct limit is obtained by taking the limit of the expressions above as $C$ approaches zero. We thus obtain from Eqs. 3 and 6

$$P = \tilde{P}_0 e^{i\tau}\left(\frac{\lambda-\zeta}{\lambda}\right) \tag{7}$$

and

$$V = \frac{\tilde{P}_0 e^{i\tau}}{\lambda Z} \tag{8}$$

which are the results obtained previously.

For very long tubes, Eqs. 3 and 6 take the limiting forms

$$P=\tilde{P}_0 e^{i\tau} e^{-\sqrt{iCZ}\,\zeta} \tag{9}$$

$$V=\tilde{P}_0 e^{i\tau}\sqrt{\frac{iC}{Z}}\; e^{-\sqrt{iCZ}\,\zeta} \tag{10}$$

We thus find that both the pressure and the velocity pulses are damped exponentially with distance. For an infinitely long rigid tube with a finite driving pressure the velocity is identically zero. Since this system is linear, we can easily obtain solutions for finite pressure at $\lambda$ by superposition.

*EX. 3.5.3. PULSATILE FLOW IN A SIMPLE BRANCHED CIRCUIT*

We consider here an extension of the previous problem in which a bifurcation occurs:

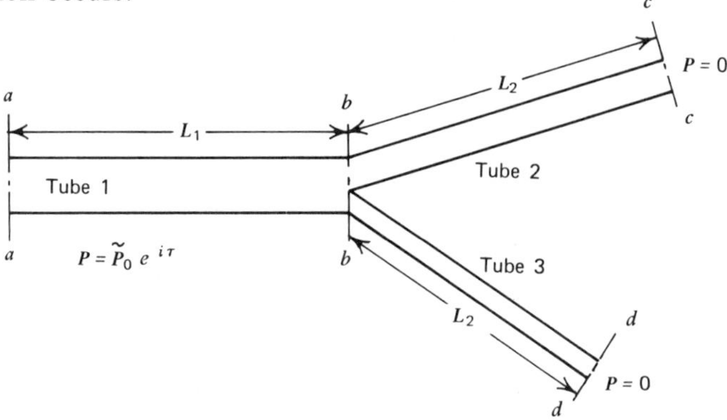


The diameters of tubes 1 to 3 are $D_1$, $D_2$, and $D_3$, respectively and tubes 2 and 3 are connected to a reservoir at zero pressure (sections *c–c* and *d–d*). The pressure at *a–a* is specified. It is desired to obtain the pressure and flow distributions in this system.

SOLUTION

We may write immediately that the pressure *b–b* is

$$P_b=\tilde{P}_b e^{i\tau} \tag{1}$$

where $\tilde{P}_b$ is a complex constant to be determined. Equation 1 follows directly from the fact that all pressures and the velocities must be oscillating at the same frequency for the Fourier element under consideration.

The pressure and flow relations for tubes 2 and 3 can now be obtained directly from Eqs. 6 and 8 of the previous example*:

$$P_2=\tilde{P}_b e^{i\tau}\left[\frac{\sinh\sqrt{iC_2Z_2}\,(\lambda_2-\zeta_2)}{\sinh\sqrt{iC_2Z_2}\,\lambda_2}\right] \tag{2}$$

$$P_3=\tilde{P}_b e^{i\tau}\left[\frac{\sinh\sqrt{iC_3Z_3}\,(\lambda_3-\zeta_2)}{\sinh\sqrt{iC_3Z_3}\,\lambda_3}\right] \tag{3}$$

$$V_2=\tilde{P}_b e^{i\tau}\sqrt{\frac{iC_2}{Z_2}}\left[\frac{\cosh\sqrt{iC_2Z_2}\,(\lambda_2-\zeta_2)}{\sinh\sqrt{iC_2Z_2}\,\lambda_2}\right] \tag{4}$$

$$V_3=\tilde{P}_b e^{i\tau}\sqrt{\frac{iC_3}{Z_3}}\left[\frac{\cosh\sqrt{iC_3Z_3}\,(\lambda_3-\zeta_3)}{\sinh\sqrt{iC_3Z_3}\,\lambda_3}\right] \tag{5}$$

The pressure and flow relations for tube 1 can be found from the results of the previous example by superposition. That is, to the previous results we add the corresponding pressures and flows for the boundary conditions

$$\text{at } \zeta_1=0, \qquad P=0 \tag{6}$$

$$\text{at } \zeta_2=\lambda_1, \qquad P=P_b \tag{7}$$

Therefore,

$$P_1=\tilde{P}_0 e^{i\tau}\left[\frac{\sinh\sqrt{iC_1Z_1}\,(\lambda_1-\zeta_1)}{\sinh\sqrt{iC_1Z_1}\,\lambda_1}\right]$$

$$+\tilde{P}_b e^{i\tau}\left[\frac{\sinh\sqrt{iC_1Z_1}\,\zeta_1}{\sinh\sqrt{iC_1Z_1}\,\lambda_1}\right] \tag{8}$$

* We assume here that pressure is continuous across the junction, as is usually done (Womersley, Skalak, Martin). This assumption is, however, an approximation which does not in general meet the requirement of conservation of energy. This point is further discussed in connection with the macrospic balances.

and

$$V_1=\tilde{P}_0 e^{i\tau}\sqrt{\frac{iC_1}{Z_1}}\left[\frac{\cosh\sqrt{iC_1Z_1}\,(\lambda_1-\zeta_1)}{\sinh\sqrt{iC_1Z_1}\,\lambda_1}\right]$$

$$+\tilde{P}_b e^{i\tau}\sqrt{\frac{iC_1}{Z_1}}\left[\frac{\cosh\sqrt{iC_1Z_1}\,\zeta_1}{\sinh\sqrt{iC_1Z_1}\,\lambda_1}\right] \tag{9}$$

Equations 2 through 9 contain only $\tilde{P}_b$ as unknown. This quantity can be eliminated through the requirement of conservation of mass at the intersection:

$$S_1V_1=S_2V_2+S_3V_3 \tag{10}$$

at section $b$–$b$. Here $S$ is the duct cross section. It follows then from Eqs. 4, 5, and 9 that:

$$\tilde{P}_b=\tilde{P}_0(\sinh\Lambda_1)\left[\frac{\sqrt{C_2S_1Z_1/C_1S_2Z_2}}{\tanh\Lambda_2}+\frac{\sqrt{C_3S_1Z_1/C_1S_3Z_3}}{\tanh\Lambda_3}-\frac{1}{\tanh\Lambda_1}\right] \tag{11}$$

where $\Lambda=\sqrt{iCZ}\,\lambda$.

More complicated problems can be handled by similar techniques.*

### *EX. 3.5.4. HARMONIC ANALYSIS OF PULSATILE FLOWS*

To this point we have considered only pure sinusoidal oscillations at a single frequency. To describe the much more complicated pulses characteristic of the circulatory system, we make use of harmonic analysis. That is, we decompose the complicated physiological wave forms into sinusoidal components or Fourier elements by use of Fourier series. We can then follow each component, or harmonic, as it travels through the system, and we can add them at any point to determine the waveform.

Illustrate this procedure by making a Fourier analysis of the pressure represented by the following data (points in Fig. 3.5.5).

* See, for example, work by Skalak, Taylor, and their associates referenced in the bibliography.

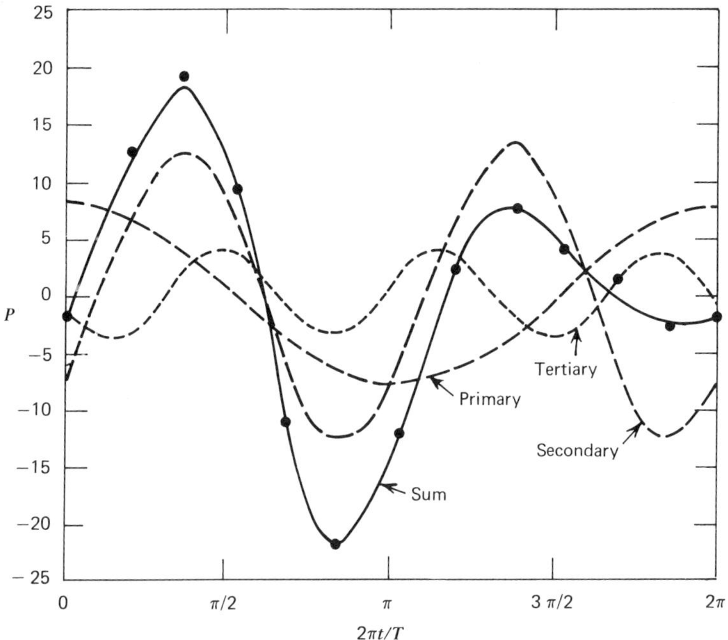


**Figure 3.5.5.** Harmonic analysis of a pressure pulse. $a_0 = 1.256$; $a_1 = 7.47$, $b_1 = -0.349$; $a_2 = -6.27$, $b_2 = 11.51$; $a_3 = -1.74$, $b_3 = -3.66$.

SOLUTION

We begin here by simply stating that a periodic function $f(x)$ can be represented by a Fourier series, i.e. that

$$f(x) = \sum_{n=0}^{\infty} (a_n \cos nx + b_n \sin nx) \tag{1}$$

with

$$a_n = \frac{1}{\pi} \int_{-\pi}^{\pi} f(x) \cos nx \, dx \tag{2}$$

$$b_n = \frac{1}{\pi} \int_{-\pi}^{\pi} f(x) \sin nx \, dx \tag{3}$$

and refer the reader to standard references for detailed discussions of such series representations.

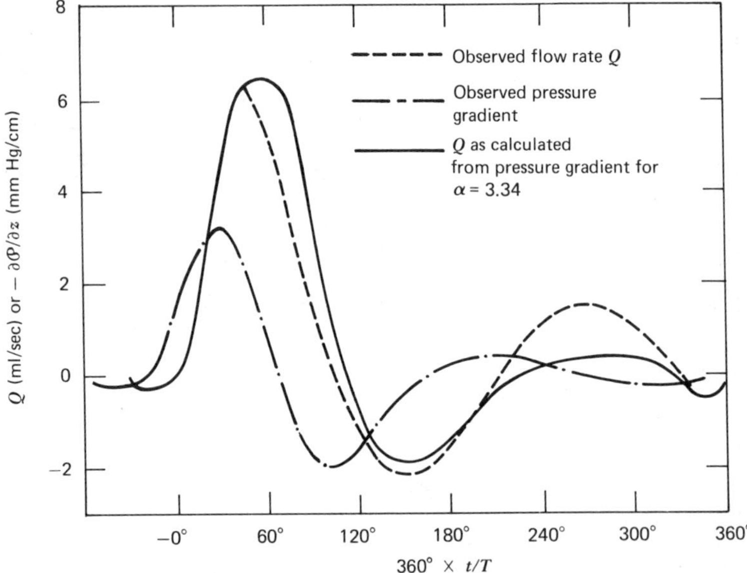


**Figure 3.5.6.** Pressure and flow pulses in the femoral artery of a dog (Womersley, 1958).

In Fig. 3.5.5, Fourier components are shown for $n=0$ to 3, with the integrals of Eqs. 2 and 3 approximated by Simpson's rule:

$$\int_{x_0}^{x_n} f(x)\,dx \doteq \frac{\Delta x}{3}\left[f_0+f_n+4(f_1+f_3+\cdots+f_{n-1})+2(f_2+f_4+\cdots+f_{n-2})\right] \tag{4}$$

where $\Delta x$ is the (uniform) interval between the $n+1$ values of $f(x)$ used. Note that $n$, here 12, must be even in the application of Eq. 4.

In this figure the dashed curves are the individual harmonics, and the solid curve is their sum. The circles represent the experimental data used to determine the pulse shape. It can be seen that three harmonics give a reasonably good but not perfectly accurate fit.

A very large number of harmonics are normally needed near the heart, where waveforms are quite complex. As can be seen from the previous section, specifically Eqs. 3.5.64 and 3.5.68, however, the higher harmonics (large $\omega$) are more rapidly damped out, and the waveforms become progressively simpler.

Harmonic analysis is also useful for calculating velocity profiles from pressure pulses as shown in Fig. 3.5.6 (taken from Womersley, 1958). This can be done, for example, by applying Eq. 3.5.42 if the physical constants of the arterial wall are known.

*EX. 3.5.5 SIGNIFICANCE AND UTILITY OF THE LINEAR THEORY*

Among the most significant of recent experiments are the perturbation studies performed by Anliker and his associates on large arteries and the vena cava. In this work, short sinusoidal pressure pulses (on the order of three to four cycles) of very small amplitude are superimposed on the heartbeat in selected regions of the circulatory network. The celerity and attenuation of these pulses are then measured over a wide range of frequencies for various carefully controlled physiological conditions. Discuss this work, described below, in the light of the linear theory of wave propagation.

SOLUTION

Since the amplitude of the superimposed pulses is quite small relative to those produced by the heart, they can be analyzed independently of the natural pulse in much the same way as described above for any one harmonic. In addition, it has been established that the short pulses used can be analyzed as if a cyclic steady state had been achieved. Finally, it should be noted that since only large blood vessels were investigated, the dimensionless Reynolds numbers $\alpha^2$ were always quite large. Thus, in the aorta of a typical test dog with an internal diameter of about 1 cm,

$$\alpha^2 \doteq \frac{(0.25\ \text{cm}^2)(2\pi N)}{0.04\ \text{cm}^2/\text{sec}} \tag{1}$$

where $N$ is the number of pulse cycles per unit time. The lowest frequencies used were of the order of 50 Hz (cycles/sec), so that the smallest values of $\alpha^2$ were

$$\alpha^2_{\text{min}} \sim 2 \times 10^3$$

It can be seen from our above development that for these high Reynolds numbers, inertial effects predominate over viscous resistance. Then from

Eq. 3.5.70 the celerity of the superimposed pressure pulse should be very nearly*

$$c \doteq c_{\infty} = \sqrt{\frac{Eh}{2R\rho}} \tag{2}$$

Similarly from Eq. 3.5.68 the attenuation may be expressed as

$$\kappa = \frac{\partial \ln \tilde{P}}{\partial (z/l)} \doteq -\sqrt{2\pi^2 \frac{\nu}{\omega R^2}} \tag{3}$$

where $l$ is the wavelength of the disturbance. Measurements of celerity then give two dynamic means of estimating the elastic modulus of the wall, and, as we shall see, measurements of attenuation give a measure of the wall viscoelasticity.

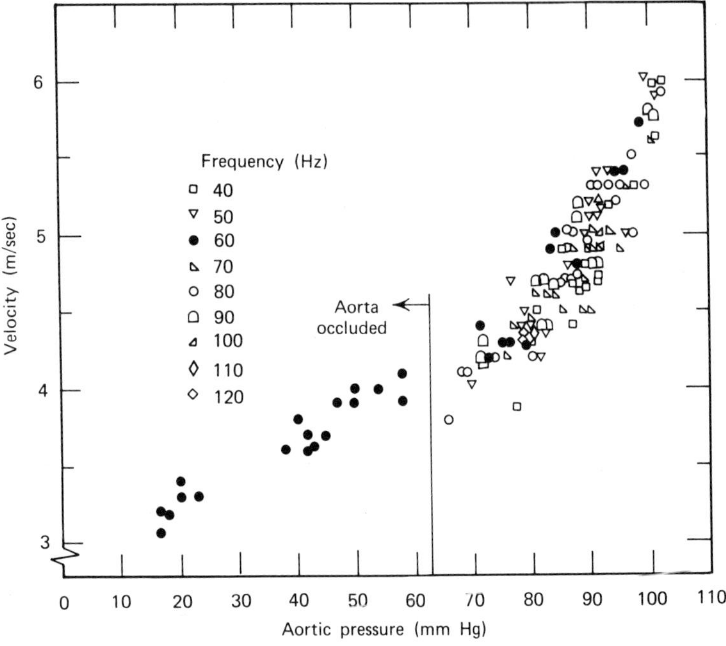


**Figure 3.5.7.** Celerities of high-frequency perturbations in the canine aorta as a function of pressure (Anliker, Histand, and Ogden).

* Again note that $E$ should be written as $E/(1-\sigma^2)$ for a tethered artery.

In Fig. 3.5.7, taken from Anliker, Histand, and Ogden, are shown wave celerities for propagation downstream in the canine aorta. The prediction of a frequency-independent wave velocity is borne out by the data shown, but the dependence of wave velocity on pressure level is not in agreement with the linear theory. This pressure dependence is quite marked over the physiological pressure range, and also below it. It shows that the apparent Young's modulus increases with stress, as indicated qualitatively in Fig. 3.1.3 for static tests.

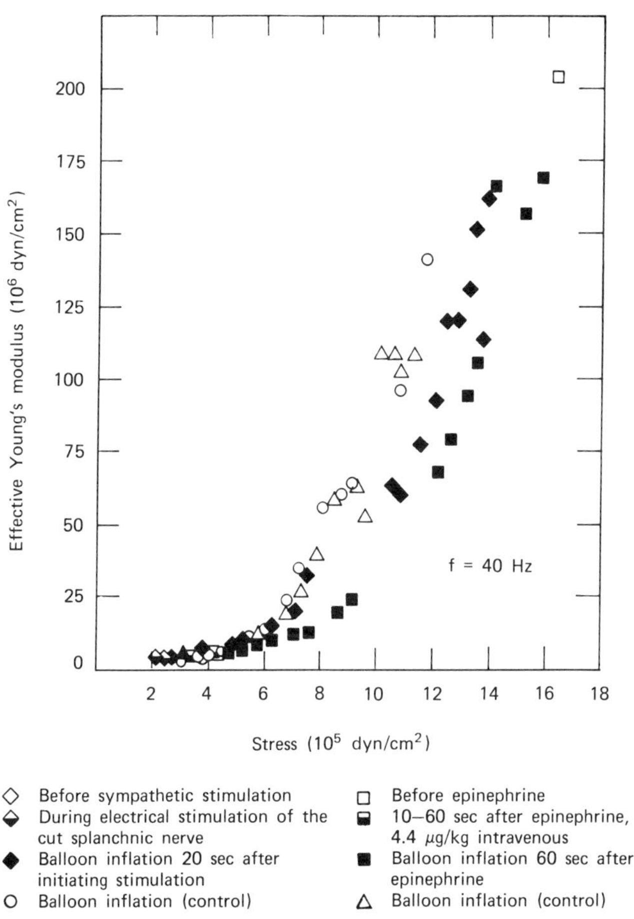


**Figure 3.5.8.** Effects of transmural pressure, sympathetic stimulation, and epinephrine on the effective Young's modulus of the abdominal vena cava (from Yates and Anliker, in press). The relaxed vena cava exhibits a noticeably higher Young's modulus than the stimulated wall at the same wall stress.

A study of the vena cava by Yates and Anliker (see Fig. 3.5.8) also shows an increase in the apparent Young's modulus as well as a marked effect of physiological control mechanisms. Once again the wave speed increases with pressure, but the effect is now much larger: the effective Young's modulus is found to vary from about $1 \times 10^6$ dyn/cm$^2$ at circumferential stresses of $2 \times 10^5$ dyn/cm$^2$, to $200 \times 10^6$ at $12 \times 10^5$. In its fully distended state then the vena cava is quite rigid. The cross section of the vena cava is physiologically controlled and can be decreased by a factor of more than 2 from its relaxed state by epinephrine injection or stimulation of the splanchnic nerve. The effective Young's modulus is substantially less in the contracted state.

Anomalies are also found in the pulse attenuation, as shown in Fig. 3.5.9 (from Histand and Anliker). Here the same attenuation constant $\kappa \doteq 0.9$

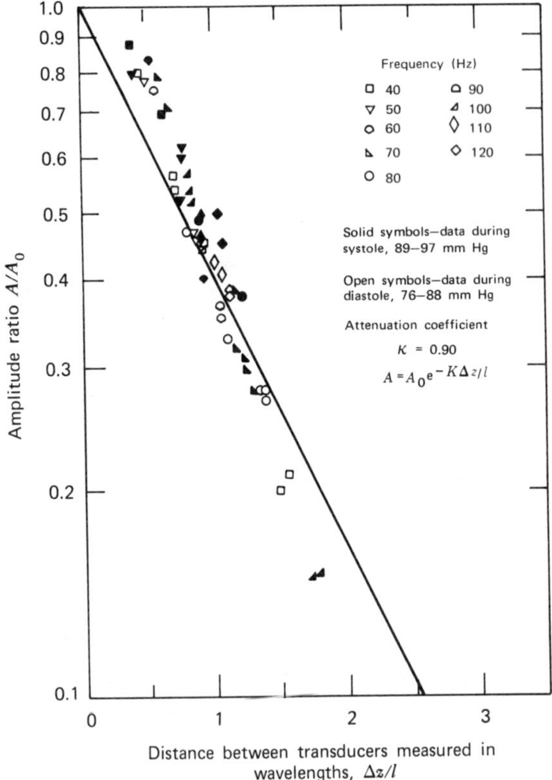


**Figure 3.5.9.** Attenuation of sinusoidal pressure waves propagating in the downstream direction and induced at different aortic pressures.

was found to within experimental error for all frequencies. The corresponding values of $\kappa$ from the linear theory (Eq. 3) vary from 0.113 at 40 Hz to 0.0645 at 120 Hz. Note that this result is for waves traveling downstream. For waves traveling upstream, the attenuation constant is even larger (1.2 to 1.6) and again frequency independent. The discrepancy between upstream and downstream constants is attributed to vessel taper. The unexpectedly higher values of the attentuation constant cannot be due to viscous dissipation in the blood as required by our theory. Rather they must arise from viscoelastic elements in the arterial wall and surrounding tissue.

Another useful result of Histand and Anliker's work is the finding that pressure waves travel faster downstream than upstream and that the difference between upstream and downstream celerities is twice the instantaneous flow-average velocity. The waves are thus convected with the fluid velocity.

It is worth noting that more general linear analyses are available for evaluating such effects as axial and torsional motion of the arterial wall and axial prestressing. It is, for example, found by Moritz and Anliker that the arterial wall is strongly anisotropic. These points are reviewed by Skalak, but they are omitted from this introductory discussion.

## 3.6 Flow in the Microcirculation

This is an important area with a rapidly growing literature, to which we can provide only a brief and somewhat speculative introduction. General references and a summary of salient features were given in the introduction to this chapter. We limit discussion here primarily to the geometry of capillaries and their flow behavior.

Two important characteristics of the microcirculation are the effective rigidity of capillaries and the extreme deformability of red cells. The lack of distensibility in a capillary, once it is fully opened, has been shown by Fung (1966) to result primarily from the effectively infinite thickness of surrounding tissue and not from the mechanical properties of the capillary wall. Red-cell deformation is described at some length in Brånemark's monograph (1969) and has been the subject of much recent research (see, for example, the work of Hochmuth and his associates). The axisymmetric "parachute" shape suggested by Fig. 3.1.5 does occur, but it is only one of many possibilities. Hochmuth et al. found, for example, that red cells adopt an "edge-on" configuration when flowing in glass tubes below about 8 $\mu$ in diameter. Here the cell axis is perpendicular to that of the tube, the cell rim is pushed inward at the tube wall, and the originally concave regions are pushed out. A finite clearance between the red cell and the vessel wall is necessary to provide slip between these solid surfaces, and large fluid pressures tend to form in regions of small clearance. These pressures act to deform the red cell and reduce the viscous dissipation produced by the flow.

Attempts to predict the shapes of red cells in capillary flow mathematically, also reviewed by Skalak, (1966) have not been completely successful, but they have provided useful insight. The lubrication-theory analyses of flow past deformable bodies by Lighthill (1969) and Fitz-Gerald (1969) are of particular interest. Hochmuth et al. find experimentally that the effective thickness of the peripheral plasma layer, between the red cell and the tube wall, increases with velocity and becomes essentially constant for velocities above about 1 mm/sec.

Measurements of capillary pressure drops in whole blood and red-cell suspensions made by Prothero and Burton (1961) and by Braasch and Jennett (1969) show apparent viscosities about 2 to 2.5 times that of plasma for normal hematocrits. Unfortunately the effect of flow rate was not systematically investigated. Experiments with cell models, referred to above, show a decrease of apparent viscosity with increasing velocity, resulting from increasing red-cell deformation. The data of Lee and Fung (1969) can be fitted fairly well by assuming the excess pressure drop (i.e., that produced by the cells) to be proportional to the square root of the velocity, as predicted by Lighthill (in Wolstenholme and Knight, 1969) on the basis of lubrication theory. However, tests by Sutera and Hochmuth (1968) show the excess pressure drop increasing toward an asymptotic value. Seshardi et al. (1970) show reasonably good agreement between tests on rubber models and the results of Braasch and Jennett when the experiments are properly scaled; this is a nice exercise in dimensional analysis.

The relative mean velocities of red cells and plasma have been measured many times *in vivo*, as by residence-time measurements, and the red cells are found consistently to move faster. This effect is due primarily to the axial segregation of red cells; it has been investigated quantitatively by Lee and Fung using liquid-filled ballons in tubes about 1 to $1\frac{3}{4}$ in. in diameter. These results cannot be safely scaled down to describe capillary flows, but they do give both an indication of the magnitude of velocity differences and a partial explanation for low organ hematocrits.

The circulation of plasma between red cells in the capillaries has received a great deal of attention because of its possible importance to mass transfer, particularly of oxygen, in flowing blood. The general nature of this *bolus* flow, shown in Fig. 3.6.2 was first described experimentally by

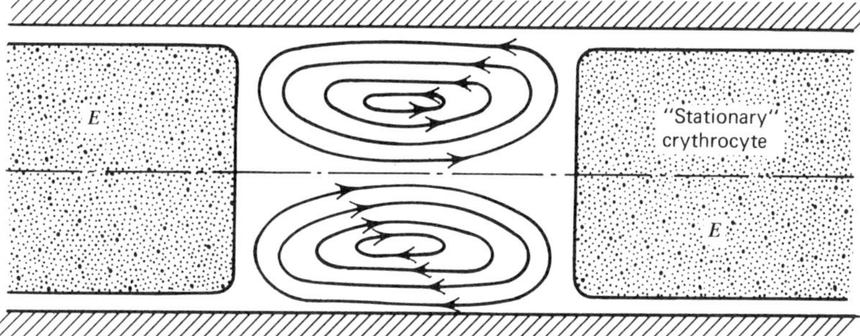


**Fig. 3.6.2.** Bolus flow: An idealized flow pattern produced by cylindrical erythrocytes very nearly filling the capillary. The diagram is drawn for an observer moving with the red cells.

Prothero and Burton (1961).* Bolus flows were first analyzed quantitatively by Bugliarello and his associates, (1967, 1970) and in more detail by Aroesty and Gross (1973) and by Lew and Fung (1969). As we shall see later, the effect of circulation on mass transfer is surprisingly small and of little practical importance.

It has already been mentioned several times that flow across capillary walls is important, and we consider this topic in some depth in the mass-transfer section of these notes. Such flows result from a combination of hydrodynamic and diffusional forces and were treated quantitatively by Starling in 1896. In this now classic paper he proposed that

$$v_r|_{r=R} = K(p_i - p_0 - \pi_i + \pi_0) \tag{3.6.1}$$

where $p_i$, $p_0$ = hydrostatic pressures inside the vessel and out in the surrounding tissue, respectively

$\pi_i$ and $\pi_0$ = the corresponding *colloid* osmotic pressures

and $K$ = a permeability coefficient.

It appears that the changes in hydrostatic and osmotic pressures in typical vascular beds are such that fluid passes into the surrounding tissue at the arterial end and returns to the blood at the venous end. Water transport occurs by both diffusion and convection, and detailed discussion must be left till later. It is, however, worth noting here that values of the permeability coefficient $K$ have been measured by Zweifach (1961) and by Intaglietta et al. (1970). Their measurement techniques show promise for

* These experiments were not, however, carried out at the extremely low Reynolds numbers characteristic of capillary flows.

hydrodynamic investigations of the microcirculation.

The question of pressure-flow relations in the microcirculation as a whole is particularly complex and cannot be discussed very authoritatively at present.

It can be seen from Fig. 2.2.3 of Chapter I that most of the drop in time-average pressure takes place in the microcirculation. However, the measured flow resistance includes contributions from a wide variety of vessels, whose geometry and interrelations are not completely understood. The flow conditions in these vessels vary from continuum flow in the largest through a very poorly understood transition to the "two-phase" flow in the capillaries. It is thus clear that not much depth of understanding can be gained from overall measurements.

In addition, available measurements are to some degree contradictory. It is shown in Fig. 2.2.4 of Chapter I.2 that total peripheral resistance decreases with increase in flow rate. The probable causes of such a decrease are the distensibility of the larger vessels of the microcirculation, the increase in the number of functioning capillaries, and red-cell deformability. All this seems reasonable. However, Benis et al. (1973) found that the pressure-flow relation included an inertial term and was described by the expression

$$-\Delta p = A\mu Q + BQ^2 \tag{3.6.2}$$

in perfusing isolated hind paws of dogs. Here $Q$ is the total flow rate through a paw produced by the pressure drop $-\Delta p$, and $A$ and $B$ are functions of the vascular anatomy. The ratio of inertial losses $BQ^2$ to viscous losses $A\mu Q$ varied from about 0.005 to 1 for perfusion with an albumin Ringer solution as the pressure drop was increased from about 6 to 200 mm Hg. Since the oscillatory Reynolds numbers $\alpha^2$ were always very low in these experiments, the inertial losses must have been due to convective momentum transport $(\mathbf{v}\cdot\nabla\mathbf{v})$.

It remains for future workers to reconcile these two sets of observations and to clarify the roles of the various factors affecting flow resistance.

## BIBLIOGRAPHY

### Books and Major Reviews

American Society of Mechanical Engineers, *Biomedical Fluid Mechanics Symposium* (Fluids Engineering Conference, Denver, Col., April 25-27, 1966), ASME (1966).

Attinger, E. O., Ed., *Pulsatile Blood Flow*, Blakiston Div., McGraw-Hill (1964).

Bergel, D., Ed., *Cardiovascular Dynamics*, Academic Press (1971).

Brånemark, P. I., *Intra-vascular Anatomy of Blood Cells in Man*, Karger, Basel (1969).

Copley, A. L., and G. Stainsky, Eds., *Flow Properties of Blood*, Pergamon (1960).

Fishman, A. P., and D. W. Richards, Eds., *Circulation of the Blood: Men and Ideas*, Oxford (1964).

Fung, Y. C., N. Perrone, and Max Anliker, Eds., *Foundations of Biomechanics*, Prentice-Hall (1972).

Fung, Y. C., Ed., *Biomechanics*, ASME (1966).

Fung, Y. C., "Biomechanics," *Appl. Mech. Rev.*, **21**, No. 1, 1—20 (1968).

Gross, J. F., and J. Aroesty, "Mathematical Models of Capillary Flows: a Critical Review," Rand Corp., Santa Monica (1971).

Krogh, A., *The Anatomy and Physiology of Capillaries*, Hafner (1930).

McDonald, D. A., *Blood Flow in Arteries*, Williams and Wilkins (1960).

Skalak, R., "Synthesis of a Complete Circulation," in Bergel (1971).

Skalak, R., "Mechanics of the Micro-circulation," in Fung (1966).

Skalak, R., "Wave Propagation in Blood Flow," in Fung (1968).

Sobin, S. S., "The Architecture and Function of the Microvasculature," in Fung, (1968).

Wolstenholme, G. E., and J. Knight, Eds., *Circulatory and Respiratory Mass Transport*, Little Brown (1969).

Womersley, J. R., *An Elastic-tube Theory of Pulse Transmission and Oscillatory Flow in Mammalian Arteries*, Wright Air Development Center, WADC Rep. TR56-614 (1957).

Zweifach, B. W., *Functional Behavior of the Micro-circulation*, Thomas (1961).

## Research Papers and Minor References

### *PULSATILE TUBE FLOW*

Anliker, Max, M. B. Histand, and Eric Ogden, "Disbursion and Attenuation of Small Artificial Pressure Waves in the Canine Aorta," *Circ. Res.*, **23**, 539–551 (1968).

Anliker, Max, W. G. Yates and Eric Ogden, "Transmission of Small Pressure Waves in the Canine Vena Cava,", *Am. J. Physiol.*, **221**, 644–651 (1971).

Dwight, H. B., *Tables of Integrals and Other Mathematical Data*, 4th ed., MacMillan, (1961).

Gow, B. S., and M. G. Taylor, *Circ. Res.*, **23**, 111—122 (1968).

Skalak, R., and T. Stathis, "A Porous Elastic Tube Model of a Vascular Bed," in Fung, (1968).

Taylor, M. G., "Input Impedance of Randomly Branching Elastic Tubes," *Biophys. J.*, **6**, 29–51 (1966).

Taylor, M. G., "Wave Transmission through an Assembly of Randomly Branched Tubes," *ibid.*, **6**, 697–716 (1966).

Taylor, M. G., "An Introduction to Some Recent Developments in Haemodynamics," *Aust. Ann. Med.*, **15**, 71–86 (1966).

Wiley, E. B., "Flow through Tapered Tubes with Non-linear Wall Properties," in Fung (1968) above.

Womersley, J. R., "Oscillatory Flow in Arteries II: The Reflection of the Pulse Wave at Junctions and Rigid Inserts in the Arterial System," *Phys. in Med. and Biol.*, **2**, 313–323 (1958).

Yates, W. G., and Max Anliker, "Active and Passive Behavior of the Canine Vena Cava in Vivo," *Am. J. Physiol.*, in press.

## *SEEPAGE FLOWS*

Berman, A. S., *J. Appl. Phys.*, **24**, 1232 (1953); **29**, 71 (1958).

Kelman, R. B., *Bull. Math. Biophys.*, **24**, 303 (1962).

Kozinski, A. A., F. P. Schmidt, and E. N. Lightfoot, "Velocity Profiles in Porous-walled Ducts," *Ind. Eng. Chem. Fund.*, **9**, 502–505 (1970).

Macey, R. I., *Bull. Math. Biophys.*, **25**, 1–9 (1963); **27**, 117–24 (1965).

## *POTENTIAL FLOWS*

Fung, Y. C., and S. S. Sobin, "Theory of Sheet Flow in Lung Alveoli," *J. Appl. Physiol.*, **26**, No. 4, 472-488 (1969).

Milne-Thomson, L. M., *Theoretical Hydronamics*, 5th ed., MacMillan, (1967).

## *ENTRANCE EFFECTS AND SECONDARY FLOWS*

Atabek, H. B., "End Effects," in Attinger (1964) above.

Attinger, E. O., in Attinger, (1964) above.

Fry, D. L., "Acute Vascular Endothelial Changes Associated with Increased Blood Velocity Gradients," *Circ. Res.*, **22**, 165–197 (1968).

Kuchar, N. R., and S. Ostrach, "Flows in the Entrance Regions of Elastic Tubes," American Society of Mechanical Engineers (1966) above.

Kuchar, N. R., and S. Ostrach, "Unsteady Entrance Flows in Elastic Tubes with Application to the Vascular System," Paper No. 70-786, 3rd Fluid and Plasma Dynamics Conference, Los Angeles (June 29-July 1, 1970).

Lee, H. S., and Y. C. Fung, "The Motion of the Plasma between the Red Cells in the Bolus Flow," *Biorheol.*, **6**, 109–119 (1969).

Lee, H. S., "Entry Flow into Blood Vessels at Arbitrary Reynolds Number," *J. Biomech.*, **3**, 23–38 (1970).

Schiller, L., *Angew. Math. und Mech.*, **2**, 96–106 (1922).

Skalak, R., and T. Stathis, in Fung (1968) above.

Texon, M., "The Role of Fluid Mechanics in the Development of Atherosclerosis," *Proc. Ann. Conf. Med. Biol.*, **10**, 49B.2 (1968).

## *FLOW IN THE MICROCIRCULATION*

Aroesty, J., and J. F. Gross, "Convection and Diffusion in the Micro-circulation," *Microvascular Research*, in press.

Baez, S., H. Lampert, and A. Baez, "Pressure Effects in Living Microscopic Vessels," Copley and Stainsky (1960) above.

Benis, A. M., S. Usami, and S. Chien, "Role of Viscosity and Internal Losses in Pressure-flow Relations Studied on Perfused Canine Hind Paw," *Microvascular Research*, in press.

Braasch,D., and W. Jennett, "Erythrocyte Flexibility, etc," in *Fifth European Conference on the Microcirculation*, H. Harders, Ed., Karger (1969) 109—112.

Bugliarello, G., and G. C. C. Hsiao, "Numerical Simulation of Three-dimensional Flow in the Axial Plasmatic Gaps of Capillaries," *Proc. 7th Int. Cong. Med. Biol. Eng.*, Stockholm, 1967.

Bugliarello, G., and G. C. Hsiao, "A Mathematical Model of the Flow in the Axial Plasmatic Gaps of the Smaller Vessels," *Biorheology*, **7**, No. 1, 5 (1970).

Fitz-Gerald, J. M., "Mechanics of Red-cell Motion through Very Narrow Capillaries," *Proc. Roy. Soc. (Lond.)*, **B174**, 193–227 (1969).

Fung, Y. C., "Theoretical Considerations of the Elasticity of Red Cells and Small Blood Vessels," *Fed. Proc.*, **25**, 1761–1722 (1966).

Fung, Y. C., "Blood Flow in the Capillary Bed," *J. Biomech.*, **2**, 353-372 (1969).

Fung, Y. C., B. W. Zweifach, and M. Intaglietta, "Elastic Environment of the Capillary Bed," *Circ. Res.*, **19**, 441–461 (1966).

Gaehtgens, P., H. J. Meiselman, and H. Wayland, "Erythrocyte Flow Velocities in Mesenteric Micro-vessels of the Cat," *Microvasc. Res.*, **2**, 151–162 (1970).

Hochmuth, R. M., R. N. Marple, and S. P. Sutera, "Capillary Blood Flow: I Erythrocyte Deformation in Glass Capillaries," *Microvasc. Res.*, **2**, 409–419 (1970).

Intaglietta, M., R. F. Pawula, and W. R. Tompkins, "Pressure Measurements in the Mammalian Microvasculature," *Microvasc. Res.*, **2**, 212–220 (1970).

Lee, J. S., and Y. C. Fung, "Modeling Experiments of a Single Red Blood Cell Moving in a Capillary Blood Vessel," *Microvasc. Res.*, **1**, 221–243 (1969).

Lee, H. S., and Y. C. Fung: see "Entrance Effects and Secondary Flows," above.

Lighthill, M. J., "Motion in Narrow Capillaries from the Standpoint of Lubrication Theory," Wolstenholme and Knight, (1969).

Prothero, J., and A. C. Burton, "The Physics of Blood Flow in the Capillaries. I–Nature of the Motion," *Biophys. J.*, **1**, 565 (1961).

Seshadri, V., R. M. Hochmuth, P. A. Croce, and S. P. Sutera, "Capillary Blood Flow: III. Deformable Model Cells Compared to Erythrocytes in Vitro," *Micro-vasc. Res.*, **2**, 434–442 (1970).

Sutera, S. P., and R. M. Hochmuth, "Large-scale Modeling of Blood Flow in the Capillaries," *Biorheol.*, **5**, 45–73 (1968).

## 4. THE MACROSCOPIC MASS, MOMENTUM, AND ENERGY BALANCES

The macroscopic balances treated in Chapter 7, of *Tr. Ph.*, can be applied in all the situations discussed above, and for the particulate systems this is generally straightforward and uninteresting. However, the oscillatory nature of arterial blood flow, and the harmonic analysis we have employed for its description, make further discussion of the macroscopic balances worth while. We shall therefore summarize these balances and consider briefly their application to periodic flows.

We begin by rewriting the mass, momentum, and mechanical-energy balances in a form convenient for branches in the circulation and show how they have been used to obtain simple approximate boundary conditions for matching single-tube analyses. We then specialize these equations for periodic flows and reexamine the boundary conditions at branches. Finally, we give a brief review of progress in the analysis of flow networks in the light of our previous discussions.

## 1. Presentation of the Macroscopic Balances

The macroscopic balances useful for our purposes are the mass, momentum, angular-momentum, and mechanical-energy balances. These are essentially bookkeeping relations and are useless for *a priori* description of any but the simplest systems. They have, however, proven very useful for applying detailed information, obtained experimentally or by integrating the microscopic equations of change, to the engineering analysis of complex systems. We are particularly interested in calculating mass-flow and pressure distributions, as well as the dissipation of mechanical energy in systems with multiple connections to their surroundings. Examples are the

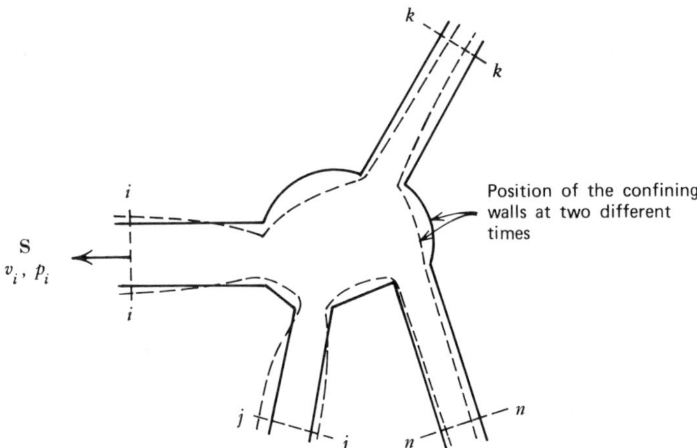


**Figure 4.1.1.** The system over which the macroscopic balances are made. The system is a mass of fluid separated from its surroundings by an impermeable wall and mathematical "control surfaces" across each of $n$ ducts permitting direct convective exchange of mass. The system walls are distensible, so that the contained fluid can do work on them. The control surface across any duct $i$ is represented by a vector $\mathbf{S}_i$ of magnitude equal to the flow cross section and direction out of the system along the duct axis. Other assumptions are as in Chapter 7 of *Tr. Ph.*

pressure and velocity changes at arterial branches and the forces on and energy expenditure by the heart. We therefore write our balances for the rather complex representative system of Fig. 4.1.1. This system, described briefly in the figure caption, is essentially a generalization of Fig. 7.0-1 in *Tr. Ph.* to multiple inlets (or outlets). For it the macroscopic balances take the form

*Mass*:
$$\frac{dm_{\text{tot}}}{dt} = -\sum_{i=1}^{n} \rho \int_{Si} (\mathbf{v}_i \cdot d\mathbf{S}_i) = -\sum_{i=1}^{n} (\mathbf{n}_1 \cdot \mathbf{w}_i) \tag{4.1.1}$$

Here $m_{\text{tot}}$ = instantaneous total contained fluid mass,
$\mathbf{w}_i$ = rate of mass flow at any control surface, assumed directed along the duct axis, as in *Tr. Ph.*, Chapter 7.
$\mathbf{n}_i$ = a unit vector in the direction of $\mathbf{S}_i$.

*Momentum*:
$$\frac{d}{dt}\mathbf{P}_{\text{tot}} = -\sum_{i=1}^{n}\left[\int_{\mathbf{S}_i} \rho \mathbf{v}_i \mathbf{v}_i \cdot d\mathbf{S}_i + p_i \mathbf{S}_i\right] - \mathbf{F} + m_{\text{tot}}\,\mathbf{g}$$
$$= -\sum_{i=1}^{n}\left(\mathbf{w}_i \frac{\langle v_i^2\rangle}{\langle v_i\rangle} + p_i \mathbf{S}_i\right) - \mathbf{F} + m_{\text{tot}}\,\mathbf{g} \tag{4.1.2}$$

Here $\mathbf{P}_{\text{tot}}$ = instantaneous total contained fluid momentum
$\mathbf{F}$ = total instantaneous force exerted by the contained fluid on the confining walls
$\mathbf{g}$ = gravitational acceleration
$\langle v_i^n\rangle = \dfrac{1}{S_i}\displaystyle\int_{S_i} (\mathbf{n}_i \cdot \mathbf{v}_i)^n dS_i$
$\mathbf{n}_i$ = unit normal in the direction of $\mathbf{S}_i$.

The summation terms in this equation are a straightforward generalization of the corresponding terms in Eq. 7.2-2 of *Tr. Ph.*

*Angular Momentum*
$$\frac{d}{dt}\mathbf{L}_{\text{tot}} = -\sum_{i=1}^{n}\left[\mathbf{r}_i \times \left(\mathbf{w}_i \frac{\langle v_i^2\rangle}{\langle v_i\rangle} + p\mathbf{S}_i\right)\right] - \mathbf{T} + \mathbf{T}_{\text{ext}} \tag{4.1.3}$$

where $\mathbf{L}_{tot}$ = instantaneous total contained angular momentum

$\mathbf{r}_i$ = position vector of the centroid of the $i$th control surface (whose diameter is assumed small relative to $|\mathbf{r}_i|$).

$\mathbf{T}$ = total instantaneous torque exerted by contained fluid on surroundings

$\mathrm{T}_{ext}$ = total instantaneous torque exerted on fluid via body forces

It should be kept in mind that all fluid is assumed to flow parallel to the duct axes. (See *Tr. Ph.*, footnote 3, p. 212). The angular-momentum balance is included here because of its potential in such applications as ballistocardiography. It does not seem to have been applied yet in hemodynamics.

*Mechanical Energy**:

$$\frac{d}{dt}(K_{tot}+\Phi_{tot}) = -\sum_{i=1}^{n} \rho \int_{S_i}\left(\mathbf{v}_i\left[\frac{1}{2}v_i^2+\hat{\Phi}_i+\frac{pi}{\rho}\right]\cdot d\mathbf{S}_i\right) - W - E_v$$

$$= -\sum_{i=1}^{n}(\mathbf{n}_i\cdot\mathbf{w}_i)\left(\frac{1}{2}\frac{\langle v_i^3\rangle}{\langle v_i\rangle}+\tilde{\Phi}_i+\frac{p_i}{e}\right) - W - E_v \qquad (4.1.4)$$

Here $K_{tot}$ = total instantaneous kinetic energy of contained fluid

$\Phi_{tot}$ = total instantaneous potential energy of contained fluid

$\tilde{\Phi}$ = potential energy per unit mass

$W$ = instantaneous total rate at which work is done by contained fluid on confining walls.

$E_v$ = instantaneous total rate of viscous dissipation of mechanical energy in contained fluid.

The presentation of the balances is now complete.

## *EX. 4.1.1. AN APPROXIMATE MATCHING CONDITION AT BRANCHES*

In most analyses of arterial networks the long-tube pressure-flow relationships are assumed to hold over the entire length of each individual tubular segment of the network. The descriptions for adjacent sections are then matched at each branch point by the application of the macroscopic balances. The system is then bounded by the confining walls and control surfaces across the ends of each of the blood vessels meeting at the branch (typically three). Since end effects are neglected, the volume of fluid contained within the control surfaces and arterial walls of the system is negligible. Discuss the matching process.

* It is assumed here that flow is isothermal and incompressible.

SOLUTION

It follows from the lack of a finite volume that the time derivatives of Eqs. 4.1.1 to 4.1.4 can be neglected. It also follows that the work $W$ done by the contained fluid against the distensible arterial walls is negligible because no appreciable wall area has been included within the system. Similarly $E_v$ must be zero, because there is no appreciable volume within the system, and the potential energy per unit mass, $\Phi$, must be the same for all streams. When these simplifications have been made Eqs. 4.1.1 to 4.1.4 reduce to

*Mass:*

$$\sum_{i=1}^{n} (\mathbf{n}_i \cdot \mathbf{w}_i) = 0 \qquad (4.1.5)$$

*Momentum:*

$$\sum_{i=1}^{n} \left( \mathbf{w}_i \frac{\langle v_i^2 \rangle}{\langle v_i \rangle} + p_i \mathbf{S}_i \right) + \mathbf{F} = 0 \qquad (4.1.6)$$

*Angular momentum:*

$$\sum_{i=1}^{n} \left[ \mathbf{r}_i \times \left( \mathbf{w}_i \frac{\langle v_i^2 \rangle}{\langle v_i \rangle} \right) + p_i S_i \right] - \mathbf{T} = 0 \qquad (4.1.7)$$

*Mechanical energy:*

$$\sum_{i=1}^{n} (\mathbf{n}_i \cdot \mathbf{w}_i) \left( \frac{1}{2} \frac{\langle v_i^3 \rangle}{\langle v_i \rangle} + \frac{p_i}{\rho} \right) = 0 \qquad (4.1.8)$$

Note that **F** and **T** cannot be neglected. To date, attention has been centered on the mass and mechanical-energy balances, and in essentially all published analyses of networks, the kinetic energy has been neglected. That is, it has been assumed that

$$\frac{\langle v_i^3 \rangle}{\langle v_i \rangle} \ll \frac{p_i}{\rho} \qquad (4.1.9)$$

Equation 4.1.8 then reduces to

$$0 = \sum_{i=1}^{n} (\mathbf{n}_i \cdot \mathbf{w}_i) p_i \qquad \text{(low-velocity limit)} \qquad (4.1.10)$$

Equations 4.1.5 and 4.1.10 are clearly satisfied by

$$p_i = p_j \qquad (i, j = 1, 2, \ldots, n) \qquad (4.1.11)$$

that is, by the continuity of pressure. This is the solution normally assumed. Thus for the system of Ex. 2 of Section 3.2 we may write

$$w_1 = w_2 + w_3 \tag{4.1.12}$$

$$p_1 = p_2 = p_3 \tag{4.1.13}$$

These are the matching conditions used in Ex. 3.5.3. It also follows from the orthogonality relationships between sines and cosines introduced in Eqs. 3.2.138 to 3.2.140 that Eqs. 4.1.12 and 4.1.13 must hold separately for each harmonic of a more complicated pulse. Equations 4.1.12 and 4.1.13 can then be used in the harmonic analysis of network flows. Unfortunately the simplicity of Eqs. 4.1.12 and 4.1.13 is bought at the price of neglecting nonzero terms in the equations of motion as well as the macroscopic balances.

Under resting conditions in the human body the kinetic-energy flux $w_i\langle v_i^2\rangle/\langle v_i\rangle$ is never more than about 5% of the work term (Martin, 1966); here neglect of kinetic-energy changes at branches may not lead to serious errors. However, under conditions of vigorous exercise the two terms are of the same order, and neglect of the kinetic-energy cannot be justified.

A more fundamental, and quite possibly more serious, source of error is the neglect of end effects, discussed in Section 3.5. These include both the localized viscous dissipation resulting from abrupt changes in the direction of flow at branches, and boundary-layer buildup at tube entrances. Evidence of flow disturbances at branches is shown in Fig. 4.1.2, taken from McDonald (1960).

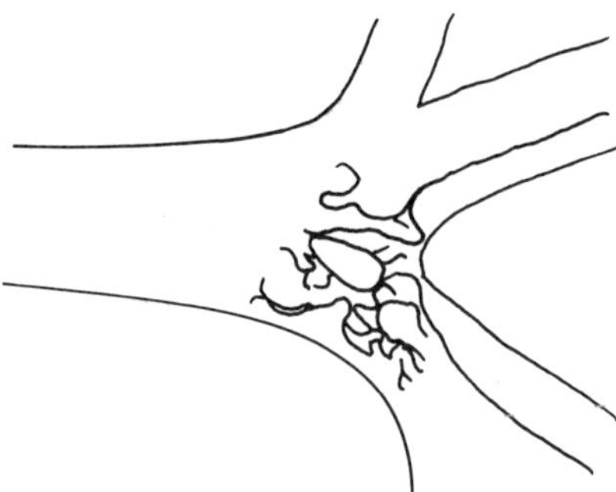

**Figure 4.1.2.** Flow disturbances in the neighborhood of a branch: flow patterns with injected dye at the distal end (the "bifurcation") of the rabbit aorta. These are tracings of single frames of a high-speed film (16-mm) record, during diastole.

## 4.2. Time-Averaging the Macroscopic Balances for Periodic Flows*

It is very frequently impossible to evaluate the macroscopic balances in the form presented above, because of their transient nature. It is therefore once again convenient to take advantage of the near periodicity of physiological flows and to time-average the macroscopic balances over one cardiac cycle. The procedure we use is identical to that described in Chapter 5 of *Tr. Ph.* for turbulent flows, except that there is now no ambiquity concerning the time period over which the average is taken. We also find that the periodicity of our motion permits us to eliminate some of the oscillatory terms from the time-averaged balances, because of the orthogonality relations between sines and cosines.

Before considering the macroscopic balances in their entirety, we look briefly at representative terms. We begin by noting that the time average of any quantity $q$ is defined as

$$\bar{q}=\frac{1}{T}\int_0^T q\,dt \tag{4.2.1}$$

with the overbar denoting a time-averaged quantity. We note then that we may write

$$p=\bar{p}+\sum_{\nu=1}^{\infty} p_\nu \tag{4.2.2}$$

$$v_z=\bar{v}_z+\sum_{\nu=1}^{\infty}(v_z)_\nu \tag{4.2.3}$$

where the subscript $\nu$ now represents a given harmonic. We now are ready to look at the individual balances, and note that the time derivatives vanish on time averaging.

The *mass balance* now takes the form

$$0=-\rho\sum_{i=1}^{n}\frac{1}{T}\int_0^T\int_{S_i}(\mathbf{v}_i\cdot d\mathbf{S}_i)\,dt \tag{4.2.4}$$

* To the author's knowledge this approach was first suggested by Martin.

All the fluctuating components vanish in this simple case, and we are left with

$$0 = -\rho \sum_{i=1}^{n} \int_{S_i} (\bar{\mathbf{v}}_i \cdot d\mathbf{S}_i)$$

$$= -\sum_{i=1}^{n} (\mathbf{n}_i \cdot \bar{\mathbf{w}}_i) \tag{4.2.5}$$

Note that the overline represents the steady component.

The *momentum balance* is a bit more complicated in that some of the oscillatory terms must now be retained. We now write*

$$0 = \frac{1}{T} \int_0^T \left[ -\sum_{i=1}^{n} \left\{ \rho \int_{S_i} \mathbf{v}_i \mathbf{v}_i \cdot d\mathbf{S}_i + p\mathbf{S}_i \right\} - \mathbf{F} + m_{\text{tot}} \mathbf{g} \right] dt \tag{4.2.6}$$

$$= -\sum_{i=1}^{n} \left\{ \rho \int_{S_i} \bar{\mathbf{v}}_i \bar{\mathbf{v}}_i \cdot d\mathbf{S}_i + \bar{p}\mathbf{S}_i \right\} - \bar{\mathbf{F}} + \bar{m}_{\text{tot}} \mathbf{g}$$

$$-\sum_{i=1}^{n} \rho \int_{S_i} \left( \sum_{\nu=1}^{\infty} v_{\nu i}^2 \right) d\mathbf{S}_i \tag{4.2.7}$$

This equation may also be written in the alternative form

$$0 = -\sum_{i=1}^{n} \mathbf{S}_i [\, \rho \langle \bar{v}^2 \rangle_i + \bar{p}_i] - \bar{\mathbf{F}} + \bar{m}_{\text{tot}} \mathbf{g}$$

$$-\sum_{i=1}^{n} \mathbf{S}_i \rho \left( \sum_{\nu=1}^{\infty} \langle v_\nu^2 \rangle_i \right) \tag{4.2.8}$$

Here the fluctuating convective momentum flux components contribute because the time average of their square, $\langle v_\nu^2 \rangle$, is not zero. They will thus make a contribution to the time-average reaction force $\bar{\mathbf{F}}$.

The *mechanical-energy balance* also retains oscillatory terms in its time-averaged form. Here we may write:

$$0 = -\rho \sum_{i=1}^{n} \frac{1}{T} \int_0^T \left\{ \int_{S_i} \left( \mathbf{v}_i \left[ \tfrac{1}{2} v_i^2 + \hat{\Phi}_i + \frac{p_i}{\rho} \right] \cdot d\mathbf{S}_i \right) - W - E_v \right\} dt \tag{4.2.9}$$

* Note that the order of integration can be simply changed because each term is a product of a function of position and a function of time.

or

$$0 = -\sum_{i=1}^{n} (\mathbf{n}_i \cdot \overline{\mathbf{w}}_i)\left(\frac{1}{2}\frac{\langle \bar{v}_i^3 \rangle}{\langle \bar{v}_i \rangle} + \hat{\Phi}_i + \frac{\bar{p}_i}{\rho}\right)$$

$$-\overline{W} - \bar{E}_v$$

$$-\sum_{i=1}^{n}\left(\sum_{\nu=1}^{\infty}\left[\tfrac{3}{2}\rho\langle \bar{\mathbf{v}}\, \overline{v_\nu^2} \rangle_i + \overline{p_{\nu_i}\langle \mathbf{v}_\nu \rangle_i}\right]\cdot \mathbf{S}_i\right)$$

$$\times \sum_{i=1}^{n}\sum_{\nu=1}^{\infty}\sum_{\kappa=1}^{\infty}\sum_{\lambda=1}^{\infty}\left(\tfrac{1}{2}\rho\langle \overline{\mathbf{v}_\nu \cdot \mathbf{v}_\kappa \mathbf{v}_\lambda} \rangle_i \cdot \mathbf{S}_i\right) \tag{4.2.10}$$

It remains* to determine the relative magnitudes of these terms.

For these order-of-magnitude purposes, consider the limiting velocity and pressure distributions for a long tube given as Eqs. 9 and 10 of Ex. 3.5.2

$$\frac{V}{P} = \frac{\sqrt{iCZ}}{Z} \tag{4.2.11}$$

The velocity and pressure pulses will in general be out of phase, but we can write the absolute value of their ratio as

$$\left|\frac{V}{P}\right| = \frac{\sqrt{C}\,(\mathcal{L}^2 + \mathcal{R}^2)^{1/4}}{(\mathcal{L}^2 + \mathcal{R}^2)^{1/2}} = \sqrt{C}\,/(\mathcal{L}^2 + \mathcal{R}^2)^{1/4} \tag{4.2.12}$$

For the high Reynolds numbers of primary interest in the major arteries,

$$(\mathcal{L}^2 + \mathcal{R}^2)^{1/4} \sim \alpha$$

and

$$\left|\frac{V}{P}\right| = \frac{\sqrt{C}}{\alpha} = \frac{\sqrt{2\mathcal{R}\mu\omega/Eh}}{\alpha} \tag{4.2.13}$$

We now recall that

$$V = \frac{v_z}{R\omega}$$

$$P = \frac{\mathcal{P}}{\mu\omega}$$

$$\alpha = \sqrt{\frac{R^2\omega\rho}{\mu}}$$

* Note that relatively few of the terms $\langle \overline{\mathbf{v}_\nu \cdot \mathbf{v}_\kappa \mathbf{v}_\lambda} \rangle$ are finite.

Then

$$\left|\frac{v_z}{\mathcal{P}}\right| \sim \sqrt{\frac{2R}{Eh\rho}} \tag{4.2.14}$$

or

$$\mathcal{P} \sim \rho \langle v_z \rangle c_\infty$$

where $c_\infty = \sqrt{hE/2\mathcal{R}\rho}$ is on the order of $10^3$ cm/sec. This relation is independent of $\alpha$ and, though it is hardly quantitative, it indicates that the work term $p\langle v \rangle$ will normally dominate over the kinetic energy term for all harmonics. In particular, it is suggested by Martin that the last set of terms in Eq. 4.2.10 (with a dashed underline) can be neglected, and, as we have seen, the small amount of available evidence indicates that they are small. These terms are of considerable theoretical interest, however, in that they show direct interaction between harmonics—and thus invalidate our harmonic analysis whenever they are important. Similarly, the terms $\langle \overline{\bar{\mathbf{v}} v_\nu^2} \rangle$ show an indirect interaction via the steady-flow component $\bar{\mathbf{v}}$. It is not yet known how to deal with either set of terms, but a plausible approach is suggested below.

The remaining terms in Eq. 4.2.10 correspond closely to their counterparts in steady flows, but the work term $\overline{W}$ and viscous dissipation $\overline{E}_v$ are worth discussing briefly.

All of the work $(-W)$ done on the blood is by definition via motion of the confining walls. The major, and most obvious, part of this is contributed by contraction of the heart. However, a considerable amount of blood pumping is done by virtue of muscle contraction and consequent motion of the blood vessel walls. This type of energy input is distributed throughout the body. Conversely, work is done by the blood on the vessel walls and surrounding tissue because of their imperfect elasticity; this effect is shown by the work of Anliker et al. but is not predicted by the linear flow theories of earlier sections.

The viscous dissipation is entirely within the flowing blood, and can be calculated from the velocity profiles and an appropriate rheological model as

$$\overline{E}_v = \frac{1}{T}\int_0^T \int_V (-\boldsymbol{\tau} : \nabla \mathbf{v})\, dV \tag{4.2.15}$$

where $V$ is system volume. For the high shear rates of the major arteries it is reasonable to consider the blood a Newtonian fluid; in this case the dissipation rate can be calculated from

$$(\boldsymbol{\tau} : \nabla \mathbf{v}) = -\tfrac{1}{2}\mu(\boldsymbol{\Delta} : \boldsymbol{\Delta}) \tag{4.2.16}$$

with

$$\Delta_{ij} = \frac{\partial v_i}{\partial x_j} + \frac{\partial v_j}{\partial x_i} \tag{4.2.17}$$

It follows for this limiting, but closely approximated, condition that

$$\bar{E}_v = \frac{\mu}{2} \int_V \overline{\boldsymbol{\Delta} : \boldsymbol{\Delta}}\, dV \tag{4.2.18}$$

Since $\boldsymbol{\Delta} : \boldsymbol{\Delta}$ contains only quadratic velocity terms it follows that

$$\bar{E}_v = \frac{\mu}{2} \left\{ \int_V \bar{\boldsymbol{\Delta}} : \bar{\boldsymbol{\Delta}}\, dV + \sum_{\nu=1}^{\infty} \overline{\boldsymbol{\Delta}_\nu : \boldsymbol{\Delta}_\nu}\, dV \right\} \tag{4.2.19}$$

It follows that the viscous dissipation for each harmonic can be calculated separately. This convenient, and at first glance surprising, result is valid for any flow regime, provided only that the blood behaves as a Newtonian fluid.

It suggests the definition of a friction loss factor analogous to that introduced for steady flows in Section 7.4 of *Tr. Ph.* We begin by defining

$$\left(\hat{E}_v\right)_\nu = \left(\bar{E}_v\right)_\nu / \rho S \sqrt{\overline{\langle v^2 \rangle}_\nu} \tag{4.2.20}$$

which reduces to the definition just below Eq. 7.4-4 on p. 215 of *Tr. Ph.*, for the steady component ($\nu = 0$). We next define a dimensionless dissipation function $\boldsymbol{\Delta}^* : \boldsymbol{\Delta}^*$ by

$$\left(\bar{E}_v\right)_\nu = \tfrac{1}{2}\rho \left( \overline{\langle v \rangle^2} \right)^{3/2} S \left( \frac{\mu}{R^2 \omega \rho} \right) \int_{V^*} \boldsymbol{\Delta}^* : \boldsymbol{\Delta}^*\, dV^* \tag{4.2.21}$$

and then a friction loss factor $(e_v)_\nu$ by

$$\left(\hat{E}_v\right)_\nu = \tfrac{1}{2}\rho\, \overline{\langle v_\nu \rangle^2}\, (e_v)_\nu \tag{4.2.22}$$

just as for the steady systems. For the special case of a section of a single straight duct one may write

$$(e_v)_\nu \equiv 4f_\nu \frac{L}{D} \tag{4.2.23}$$

again by analogy with steady flows. Unfortunately, reliable values of friction loss factors are not yet available for physiological systems.

To further illustrate the utility of the time-averaged equations, we now reconsider the problem of determining matching conditions at a branch. This analysis represents some improvement on Ex. 4.1.1 by taking into account both the kinetic energy of the fluid and the oscillatory nature of the flow. We do, however, have to neglect viscous dissipation and work done across vessel walls, for lack of available data, and we choose to neglect the third-order harmonic (underlined) terms in Eq. 4.2.10: these latter are most probably of little importance.

We begin by taking a rather general look at the bifurcation discussed in Ex. 3.5.3 to illustrate the approach suggested by Martin, and we concentrate our attention on Eq. 4.2.10. First, however, we note that the macroscopic mass balance requires that

$$S_a\langle v_a\rangle = S_b\langle v_b\rangle + S_d\langle v_d\rangle \tag{4.2.24}$$

This equation takes the same form on either an instantaneous or a time-averaged basis, and it must be satisfied separately for each harmonic.

We next follow Martin's suggestion (1966) and assume that the steady terms satisfy this equation independently of the oscillatory contributions. It then follows that

$$\tfrac{1}{2}\rho\langle\bar{v}_a^3\rangle + \bar{p}_a\langle\bar{v}_a\rangle = \frac{S_b}{S_a}\left(\tfrac{1}{2}\rho\langle\bar{v}_b^3\rangle + \bar{p}_b\langle\bar{v}_b\rangle\right) + \frac{S_d}{S_a}\left(\tfrac{1}{2}\rho\langle v_d^3\rangle + \bar{p}_d\langle\bar{v}_d\rangle\right) \tag{4.2.25}$$

The oscillatory components can now be separately calculated from the relations

$$\tfrac{3}{2}\rho\overline{\langle \bar{v}_a v_{va}^2 \rangle} + \overline{p_{va}\langle v_{va}\rangle}$$

$$= \frac{S_{2a}}{S_a}\left(\tfrac{3}{2}\rho\langle \bar{v}_b \overline{v_{vb}}\rangle + \overline{p_{vb}\langle v_{vb}\rangle}\right)$$

$$+ \frac{S_{2b}}{S_a}\left(\tfrac{3}{2}\rho\langle \bar{v}_d \overline{v_{vd}^2}\rangle + \overline{p_{vd}\langle v_{vd}\rangle}\right) \tag{4.2.26}$$

Equation 4.2.26 can now be considerably simplified by noting that all of the fluctuating quantities $q_\nu$ are of the form

$$q_\nu = \bar{q}_\nu(\zeta, \eta)e^{i\nu\tau} \tag{4.2.27}$$

Hence the time averages in this equation can be replaced by the corresponding instantaneous terms. One can then relate the fluctuating pressures and local velocities, just as in combining Eqs. 3.2.84 or 3.2.85 with 3.2.88 and 3.2.89 for the flow-averaged velocity. One is, however, still left with the problem of computing the flow-average values of $p_\nu v_\nu$ and $v_\nu^2$. This must be done numerically even for symmetrically branching systems, and the interested reader is referred to Martin for further discussion and specific examples.

One can, however, proceed with surprising ease for situations of large oscillatory Reynolds number $\alpha^2$, because here the fluctuating velocity profiles may be considered flat. This is shown in the following example.

### *EX. 4.2.1. PRESSURE CHANGES ACROSS AN ENLARGEMENT*

Estimate the pressure change across the enlargement shown in Fig. 4.2.2 assuming that:

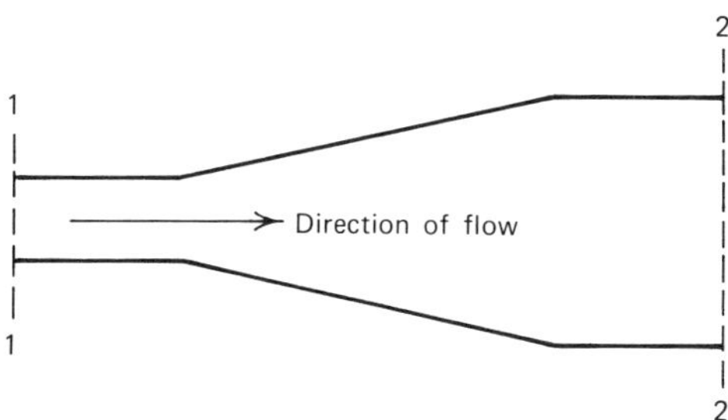


**Figure 4.2.2.** A gradual enlargement.

1. The fluid is Newtonian and the flow laminar.
2. The oscillatory Reynolds numbers are all large, so that the oscillatory profiles are essentially flat.
3. No net work is done on the contained fluid and the confining walls ($\overline{W}=0$).
4. Viscous dissipation is negligible ($\overline{E}_v=0$).

In addition, neglect the third-order harmonic terms in Eq. 4.2.10.

SOLUTION

For steady laminar flow the velocity profile is parabolic and

$$\langle \bar{v}^3 \rangle = 2\langle \bar{v} \rangle^2 \tag{1}$$

Therefore

$$\bar{p}_2 - \bar{p}_1 = \tfrac{1}{2}\rho v_1^2 \left(1 - \frac{S_1^2}{S_2^2}\right) \tag{2}$$

which is a well-known result.

The oscillatory terms also take a very simple form:

$$\tfrac{3}{2}\rho\langle \bar{v}_1 \rangle v_{\nu 1}^2 + p_{\nu 1} v_{\nu 1} = \frac{S_2}{S_1}\left(\tfrac{3}{2}\rho\langle \bar{v}_2 \rangle v_{\nu 2}^2 + p_{\nu 2} v_{\nu 2}\right) \tag{3}$$

Using the macroscopic mass balance in the form of Eq. 4.2.24, we may simplify this result to

$$p_{\nu 2} - p_{\nu 1} = \tfrac{3}{2}\rho\langle \bar{v}_1 \rangle v_{\nu 1}\left(1 - \frac{S_1}{S_2}\right) \tag{4}$$

Thus the fluctuating pressure change across the enlargement is in phase with the fluctuating velocity, for this limiting case of flat profiles.

The total pressure change is then

$$\begin{aligned} p_2 - p_1 = {} & \tfrac{1}{2}\rho\langle \bar{v}_1 \rangle^2 \left[1 - \left(\frac{S_1}{S_2}\right)^2\right] \\ & + \tfrac{3}{2}\rho\langle \bar{v}_1 \rangle \left(1 - \frac{S_1}{S_2}\right) \sum_{\nu=1}^{\infty} v_{\nu 1} \\ & + \rho g (h_1 - h_2) \end{aligned} \tag{5}$$

where $h_1$ and $h_2$ are the mean distances of control surfaces 1 and 2 above an arbitrary gravitational reference point. The last term in Eq. 5, which represents the effect of the hydrostatic head, should clearly contain no fluctuating term. It is normally very small.

The above development is useful, at least qualitatively, for explaining the pressure changes occurring at *aneurysms* (local enlargements in blood vessel diameter) and *stenoses* (local contractions). In the case of aneurysms, for example, pressures tend to be higher than either upstream or downstream because of the lower blood velocities in the enlarged region. These pulsating higher pressures act on the already weakened wall of the enlargement to expand it further.

Thus, according to Eq. 5, if the oscillatory profiles may be considered flat,

$$p_2 = \bar{p}_1 + \rho \bar{v}_1^2 \left[ 1 - \left( \frac{S_1}{S_2} \right)^2 \right] + \sum_{\nu=1}^{\infty} \left[ p_{1\nu} + \tfrac{3}{2} \rho \langle \bar{v}_1 \rangle \left( 1 - \frac{S_1}{S_2} \right) v_{\nu 1} \right] \tag{6}$$

Since the upstream flow pulses $v_{\nu 1}$ lag the pressure pulses $p_{\nu 1}$ by less than 90°, they tend to reinforce each other. As a result there are both time-average and oscillatory contributions to the pressure rise on enlargement.

Equation 5 must not be taken too literally, because the real situation is very complex. From a hydrodynamic standpoint it must be remembered that viscous dissipation, known to be important in steady flows (see, e.g., *Tr. Ph.*, Chapter 7), tends to reduce the pressure rise. Second, from a physiological standpoint, it is known that a spongy deposit forms in the enlarged region and reduces the drop in velocity. It also seems likely that the biochemistry of the wall is important.

On balance, however, the hydrodynamic aspects of aneurism formation are important, and analyses like the above do provide at least a partial explanation for their occurrence. Such analyses also show that their growth is aggravated by large pulses, both in pressure and in flow, and suggest that the moderation of these pulses is highly desirable.

## 4.3. Modeling the Circulatory System

Modeling large circulatory networks has always been an important goal of hemodynamicists, and it is fitting that we return briefly to this subject as we conclude our discussion of momentum transfer. It is clear from the above discussions that *a priori* modeling is exceedingly difficult and that many types of complications are encountered.

From a practical medical standpoint the least sophisticated models are at present the most useful, in large part because they require only a small number of simple measurements. As has already been pointed out, the total peripheral resistance has long been used routinely as a measure of physiological condition. More recently the *Windkessel* model has proven useful for the determination of the heart stroke volume from pressure measurements. As a specific example, Warner, Gardner, and Toronto have combined the *Windkessel* model with an empirical characterization of arterial compliance to obtain a one-parameter relation between cardiac output and aortic pressure. Once this parameter has been determined (as by calibration using the dye-dilution techniques discussed in Chapter I.2), the stroke volume can be determined from the time variation of the aortic pulse pressure. Even without such calibration, relative changes in cardiac output can be determined.

More sophisticated models, requiring more detailed information, have so far proven useful primarily for providing added insight into the nature of arterial flow. Linear models of arterial flow, for example, require pressure measurements at two points along an artery to permit the calculation of flow rates.

The first extensive linear models of the circulation were by Weiner, Morkin, and Skalak (1966) and by Taylor (1966). These were done numerically by an extension of Ex. 3.5.3 to multiply branching systems. Wiener, Morkin, Skalak, and Fishman (1966) modeled the canine lung and obtained good agreement between the linear theory and experimental observations. Taylor considered randomly branching systems. It appears that a certain degree of randomness is necessary to achieve realistic results, in particular to reproduce the remarkably flat impedance-frequency relation (see Bergel and Milnor, 1965).

More recently Skalak and Stathis (1966) used the linear solutions in a tapered-porous-tube model of the circulation which corresponds closely to symmetrical binary branching with uniform length-to-diameter ratios in each generation. As a boundary condition, the branched elastic system was connected in series to a pure resistance, representing the microcirculation, which comprised about two-thirds of the total pressure drop. This model shows the typical observed strong downstream pressure rise at low frequencies, which cannot be predicted on the basis of a linear model. As discussed by Skalak (1971), this behavior is primarily produced by nonlinearities in the arterial system, which are outside the scope of the present text.

## BIBLIOGRAPHY

Attinger, E. O., Ed., *Pulsatile Blood Flow*, Blakiston Div., McGraw-Hill (1963).

Bergel, D. H., and W. R. Milnor, "Pulmonary Vascular Impedance in the Dog," *Circ. Res.*, **16**, 401–415 (1965).

McDonald, D. A., *Blood Flow in Arteries*, Williams and Wilkins (1960).

Martin, J. D., "An Extension to the Theory of Wave Reflections," *Biomedical Fluid Mechanics Symposium*, ASME, Fluids Eng. Div., 70–77 (1966).

Skalak, Richard, "Synthesis of a Complete Circulation," in *Cardiovascular Dynamics*, D. Bergel, Ed., Academic Press (1971).

Skalak, R., and T. Stathis, "A Porous Tapered Elastic Tube Model of a Vascular Bed," in *Biomechanics*, Y. C. Fung, Ed., ASME (1966).

Taylor, M. G., "Input Impedance of an Assembly of Randomly Branching Elastic Tubes," *Biophys. J.*, **6**, No. 1, 29–51 (1966).

Warner, H. R., R. M. Gardner, and A. F. Toronto, "Computer-based Monitoring of Cardiovascular Functions in Post-operative Patients," Supplement II to *Circulation*, Vols. XXXVII and XXXVIII, April 1968.

Wiener, F., E. Morkin, R. Skalak, and A. P. Fishman, "Wave Propagation in the Pulmonary Circulation," *Circ. Res.*, **19**, 834–50 (1966).

# III MASS TRANSFER

We now turn our attention to the very important and challenging area of mass transfer. There is no doubt that mass-transfer considerations play the dominant role in the organization and structure of living organisms, as well as in the design of many extracorporeal devices. On the one hand, the need to carry out many different types of chemical reactions separately from one another requires the erection of mass-transfer barriers, not only between individual organs, but also between adjacent regions within individual cells. On the other hand the slowness of *diffusional* transport of major metabolites, (such as $O_2$, $CO_2$, sugars, and amino acids), through body tissues requires development of a piping system for *convective* transport. Thus selectively permeable membranes are found in the mitochondria of cells, and at least some rudimentary convection occurs in organisms as small as daphnia, which are on the order of a half millimeter in diameter.

The problem of maintaining large specialized mass-transfer surfaces also becomes acute for large organisms such as ourselves. A primary difficulty here is to maintain effective mass transfer simultaneously with a suitable mechanical structure. In the lung, for example, special surfactants must be produced continually to prevent the collagse of the alveoli across whose walls $O_2$, $CO_2$, and $H_2O$ are transferred. In "artificial kidneys" the corresponding problem is to provide sufficiently rigid membrane support to maintain the desired geometry without unduly reducing diffusional access to the membrane.

Even the structure of the nervous system is strongly influenced by diffusional considerations, since nerve impulses are propagated chemically by a complex sequence of diffusional processes.

Both homogeneous and heterogeneous chemical reactions are also very important. Facilitated diffusion and active transport across cell membranes are thought to require heterogeneous reactions at the membrane boundaries, and homogeneous reactions are involved both in $O_2$ and $CO_2$ transport in blood. We shall, therefore, be interested in the whole gamut of mass transfer operations familiar to chemical engineers.

We begin in Section 1 by developing a reliable framework for the quantitative description of mass and energy transport in complex systems. This framework is used in Section 2 to describe diffusion in fluids for those situations in which convection is of secondary importance. In Section 3 we describe mass transport in macroscropic membranes and show how to account for the mechanical restraints on the membrane matrix. In Section 4 we describe the salient features of biological membranes and interpret their diffusional behavior in terms of the previous discussions.

In Section 5 we turn to convective mass transfer and its applications to both biological modeling and the design of extracorporeal processing equipment. We conclude in Section 6 with a discussion of macroscropic mass balances. Here we consider the body as a system of interacting components. We investigate the effects of the body and extracorporeal devices on each other and discuss the design of extracorporeal processing systems. This is a most important area of direct clinical importance, and we shall speculate on possible clinical applications. Here we impinge directly on the key problem of all living organisms: *homeostasis*, the maintenance of a viable steady state in a highly variable environment.

## 1. MASS AND ENERGY TRANSPORT IN MULTICOMPONENT SYSTEMS

Here we lay the foundation for the discussion of mass transfer in such varied applications as pulmonary gas exchange, active transport in kidney tubules, and hemodialysis in the presence of pressure gradients. Past experience indicates that the soundest approach is to begin with the necessary conservation relations and a rather general set of rate equations based on the principles of irreversible thermodynamics. These may then be specialized for individual applications as conditions require.

Accordingly the conservation laws will be presented in Section 1.1, and the flux equations needed to relate mass and energy fluxes to the "driving forces" producing them will be developed in Section 1.2. In the next three subsections the flux equations are specialized for the description of diffusion in free solution, for membrane transport, and for the specification of solute and potential distributions at phase boundaries. The remainder of this section is devoted to techniques for obtaining useful approximate descriptions of multicomponent mass transport when available thermodynamic and transport-property data are incomplete.

Chapters 16 and 18 of *Tr. Ph.* provide a useful introduction to our discussions, but here we go substantially beyond the coverage there, both in the formulation of the rate equations and their application to specific systems. Other references are provided as appropriate.

### 1.1 The Multicomponent Equations of Change in Terms of the Fluxes

The conservation relations for multicomponent systems are introduced in *Tr. Ph.,* Section 18.3, and the discussion there is generally adequate for our purposes. They may be summarized as follows for a system of $n$ components:

Continuity:
$$\frac{\partial c_i}{\partial t} = -(\nabla \cdot \mathbf{N}_i) + R_i \qquad (i=1,2,3,\ldots,n) \qquad (1.1.1)$$

Motion:
$$\frac{\partial}{\partial t}\rho\mathbf{v} = -[\nabla \cdot \boldsymbol{\phi}] + \sum_{i=1}^{n} \rho_i \mathbf{g}_i \qquad (1.1.2)$$

Energy:
$$\frac{\partial}{\partial t}\rho\left(\hat{U} + \frac{1}{2}v^2\right) = -(\nabla \cdot \mathbf{e}) + \sum_{i=1}^{n} (\mathbf{N}_i M_i \cdot \mathbf{g}_i) \qquad (1.1.3)$$

Here $c_i$ and $\rho_i$ are the molar and mass concentrations of species $i$ respectively, the $\mathbf{g}_i$ are body forces per unit mass acting on species $i, \hat{U}$ is the internal energy per unit mass, $M_i$ is the molecular weight of species $i$, and $\mathbf{v}$ is the mass-average velocity. The quantities $\mathbf{N}_i = c_i\mathbf{v}_i$, $\boldsymbol{\phi}$ , and $\mathbf{e}$ are the total fluxes of species $i$ (in moles), momentum, and energy, respectively, relative to any (arbitrarily chosen) fixed coordinates. The remaining terms are

$R_i$ = rate of formation of $i$ in moles per unit volume and time by chemical reaction

$\sum_{i=1}^{n} \rho_i \mathbf{g}_i$ = volumetric rate of momentum increase resulting from body forces $\mathbf{g}_i$ per unit mass of each species

$\sum_{i=1}^{n} (\mathbf{N}_i M_i \cdot \mathbf{g}_i)$ = rate at which work is done on the system per unit volume by virtue of the body forces

Any one of the $n$ species continuity equations may be replaced by their sum, the equation of continuity.

Each of the fluxes may be written as the sum of a *convective* flux, resulting from the motion of the fluid, and a *diffusion* flux relative to this motion. Thus

$$\mathbf{N}_i = c_i\mathbf{v} + \mathbf{J}_i \tag{1.1.4}$$

$$\boldsymbol{\phi} = \rho\mathbf{v}\mathbf{v} + \boldsymbol{\pi} \tag{1.1.5}$$

$$\mathbf{e} = \rho\left(\hat{U} + \frac{1}{i}v^2\right)\mathbf{v} + \mathbf{q} + \boldsymbol{\pi}\cdot\mathbf{v} \tag{1.1.6}$$

Here the mass-average velocity $\mathbf{v}$ has been used as a measure of the fluid motion, and the diffusion fluxes $\mathbf{J}_i$, $\boldsymbol{\pi}$, and $\mathbf{q}$ are fluxes with respect to $\mathbf{v}$. This choice is completely arbitrary, although it is the natural choice for $\boldsymbol{\phi}$, the momentum flux. The forms of the flux equations for other reference velocities are provided in rather general form by de Groot and Mazur (1962); a somewhat more accessible summary is given by Lightfoot, Cussler, and Rettig (1962). We are primarily concerned with the equations of continuity, and during discussions of convective mass transfer we normally use previously obtained solutions of the equations of motion. There are, however, important situations for which the equations of continuity and motion must be solved simultaneously.

Equations 1.1.1 through 1.1.3, are rather straightforward conservation statements, and Eqs. 1.1.4 through 1.1.6 are little more than definitions. For the solution of most problems it is necessary to relate the mass, momentum, and energy fluxes to concentration, pressure, electrostatic potential, velocity, and temperature gradients. This is done in the next section with the aid of a fourth "conservation" equation, the entropy balance, and a number of fundamental assumptions, or postulates.

## 1.2. Diffusion and the Mechanisms of Mass Transport: Nonequilibrium Thermodynamics and the Generalization of Fick's Law

Our purpose here is to present convenient but reliable relations between mass and energy diffusion fluxes and the forces producing mass and energy transport. We use for this purpose generalizations of the Stefan-Maxwell equations, which simply state that the total force required to move any species relative to the solution as a whole is linearly related to the velocity differences between it and each of the other species present. Our major problems are to obtain a precise and reliable expression for the force acting on any diffusing species and to make use of all the relations existing between the many diffusion coefficients required to characterize multicomponent mass transfer. For the complex chemical systems of interest in this course, the most effective framework for solving these problems is the nonequilibrium thermodynamic analysis of Lars Onsager. Just enough of this analysis will be repeated here to make the generalized Stefan-Maxwell equations plausible; those wishing a more complete discussion are referred to the texts cited below. The text by Fitts (1962) is probably the most accessible to the beginning reader, and that of de Groot and Mazur (1962) is the most exhaustive. The discussion below most closely parallels that of Hirschfelder, et al. (1954), however, and this reference is recommended as being authoritative, brief, and concerned primarily with the diffusion processes of interest to us.*

Diffusion, like the viscous transfer of momentum, heat conduction, and all other spontaneous processes, is irreversible. In all such natural processes there is a tendency toward randomization which makes a spontaneous return to an earlier state unlikely—and, for systems of very large numbers of molecules such as we ordinarily deal with, effectively impossible. The quantitative measure of the degree of randomness or disorder in a system is its entropy, and all natural processes are accompanied by an increase in entropy. It seems reasonable, therefore, that a complete description of entropy changes taking place in a system should also provide a description of all spontaneous processes which are possible in that system. This assumption is the starting point for nonequilibrium thermodynamic analyses.

The first step in these developments is to write a general conservation equation for entropy similar to those used earlier for mass, momentum, and energy:

$$\rho \frac{D\hat{S}}{Dt} = -(\nabla \cdot \mathbf{j}_s) + \sigma \tag{1.2.1}$$

where $\hat{S}$ = entropy per unit mass
$\mathbf{j}_s$ = entropy flux relative to the mass-average velocity
$\sigma$ = volumetric rate of entropy production

This equation is nothing but a definition, and the real problem to be solved is to write expressions for the individual terms in it, most particularly for the rate of entropy production. Once we have a complete description for the rate of entropy production, we also have a description of all the natural processes possible in the system being considered; as indicated above, these include heat and mass transfer.

Our result can be meaningful only if it is written in terms of measurable quantities such as temperature, pressure, and composition. Hence, we must

* The approach of Kedem and Katchalsky (1958) is discussed later in this chapter, and in Chapter 3.

relate the entropy to such quantities, and to do this we need the first postulate of nonequilibrium thermodynamics:

I. *Departures from local equilibrium are sufficiently small that all thermodynamic quantities may be defined locally by the corresponding relations for systems at equilibrium.*

This postulate is almost certainly reliable for transport through macroscopic fluids, or thick membranes such as the cellophane used in extracorporeal circuits. Its applicability to very thin membranes such as those surrounding individual cells is considerably more doubtful, but the results of thermodynamic analyses are used here as well, for lack of a suitable alternative.

From the second law of thermodynamics and the first postulate we may write

$$T\rho\frac{D\hat{S}}{Dt}=\rho\frac{D\hat{U}}{Dt}+p\rho\frac{D\hat{V}}{Dt}-\sum_{i=1}^{n}\frac{\mu_i}{M_i}\rho\frac{D\omega_i}{Dt} \tag{1.2.2}$$

where $T$ is absolute temperature, $\hat{V}=1/\rho$ is the specific volume, $\mu_i$ is partial molal free energy, and $\omega_i$ is the mass fraction of species $i$ in the mixture. We may now put the equations of energy and continuity into this expression and rearrange the result to obtain*

$$\begin{aligned}\rho\frac{D\hat{S}}{Dt}=&-\left(\nabla\cdot\left[\frac{\mathbf{q}-\mathbf{j}_i\mu_i/M_i}{T}\right]\right)\\&-\frac{1}{T}\left\{(\mathbf{q}\cdot\nabla\ln T)+\sum_{i=1}^{n}\left(\mathbf{j}_i\cdot\left[T\nabla\left(\frac{\mu_i}{TM_i}\right)-\mathbf{g}_i\right]\right)\right.\\&\left.+(\boldsymbol{\tau}:\nabla\mathbf{v})+\sum_{i=1}^{n}\frac{\mu_i r_i}{M_i}\right\}\end{aligned} \tag{1.2.3}$$

We now define the entropy flux as

$$\mathbf{j}_s=\frac{1}{T}\left(\mathbf{q}-\mathbf{j}_i\frac{\mu_i}{M_i}\right) \tag{1.2.4}$$

* See also the table of nomenclature in the Appendix. In most cases the nomenclature is that of the reference text *Transport Phenomena*.

and the remainder* of the right side of Eq. 1.2.3 as the volumetric rate of entropy production, $\sigma$.

This latter term can now be rearranged to the form

$$-T\sigma = \left(\left[\mathbf{q} - \sum_{i=1}^{n} \frac{\overline{H}_i}{M_i}\mathbf{j}_i\right]\cdot\nabla \ln T\right) + (\boldsymbol{\tau}:\nabla\mathbf{v})$$

$$+ \sum_{i=1}^{n} (\mathbf{j}_i\cdot\boldsymbol{\Lambda}_i) + \sum_{i=1}^{n} \mu_i r_i \qquad (1.2.5)$$

where

$$\boldsymbol{\Lambda}_i = \nabla_{T,p}\frac{\mu_i}{M_i} + \frac{\overline{V}_i}{M_i}\nabla p - \mathbf{g}_i \qquad (1.2.6)$$

and†

$$\nabla_{T,p}\mu_i = RT\nabla(\ln a_i)_{T,p} = \sum_{j=1}^{\nu}\left(\frac{\partial \mu_i}{\partial x_j}\right)_{T,p,x_k}\nabla x_j \qquad (1.2.7)$$

where $R$ is the international gas constant, and $a_i$ is thermodynamic activity of species $i$. It should now be noted that Eq. 1.2.5 is "complete," in the sense that it contains all terms contributing to entropy production, and that each term is the product of a flux and a "driving force." In addition to

*In addition to plausibility, this choice meets three reasonable criteria:

1. The entropy production is zero at equilibrium
2. The source term is invariant on transformation to the form

$$\frac{\partial}{\partial t}\rho\hat{S} = -\left(\nabla\cdot\left[\rho\hat{S}\mathbf{v} + \frac{q}{T} - \sum_{i=1}^{n}\mathbf{j}_i\frac{\mu_i}{T}\right]\right) + \sigma$$

3. The proper expression is obtained for a closed system, which can exchange only conduction heat with its surroundings:

$$\frac{d\hat{S}_{\text{tot}}}{dt} = \int_{V_{\text{tot}}}\left(\frac{\partial}{\partial t}\rho\hat{S}\right)dV = -\int_{A_{\text{tot}}}\frac{1}{T}(\mathbf{q}\cdot\mathbf{n})\,dA + \int_{V_{\text{tot}}}\sigma\,dV \geqslant \int_{A_{\text{tot}}}\frac{1}{T}(\mathbf{q}\cdot\mathbf{n})\,dA$$

† The use of $\nabla_{T,p}\mu_i$ rather than $\nabla\mu_i$ is convenient because it describes only that part of the chemical-potential gradient resulting from concentrate gradients. The contribution of pressure and temperature gradients is described separately.

species body forces, these latter contain gradients in temperature, pressure, velocity, and chemical potential (or concentration).

It follows that the entropy production rate should depend only upon the fluxes, the forces, and the chemical nature of the system. In addition, for any local temperature, pressure, and composition (i.e., for any thermodynamic state) the fluxes should depend only on the driving forces.

It remains to determine these dependences, and this cannot be done by the thermodynamic arguments used up to this point. Before going further it is, however, worth while to note that the mass fluxes with respect to the mass-average velocity sum to zero:

$$\sum_{i=1}^{n} \mathbf{j}_i = 0 \tag{1.2.8}$$

so that Eq. 1.2.5 is also satisfied if we add any arbitrary quantity to each of the $\mathbf{\Lambda}_i$. We find it convenient to replace the $\mathbf{\Lambda}_i$ by

$$\mathbf{\Lambda}_i - \frac{1}{\rho}\nabla p + \sum_{i=1}^{n} \omega_i \mathbf{g}_i \tag{1.2.9}$$

where the added quantity

$$-\frac{1}{\rho}\nabla p + \sum_{i=1}^{n} \omega_i \mathbf{g}_i = \text{total force per unit mass acting on the fluid} \tag{1.2.10}$$

We may then write that

$$\sum_{i=1}^{n} (\mathbf{j}_i \cdot \mathbf{\Lambda})_i = \sum_{i=1}^{n} \left( \mathbf{j}_i \cdot \left[ \mathbf{\Lambda}_i - \frac{1}{\rho}\nabla p + \sum_{k=1}^{n} \omega_k \mathbf{g}_k \right] \right)$$

$$= cRT \sum_{i=1}^{n} ([\mathbf{v}_i - \mathbf{v}] \cdot \mathbf{d}_i) \tag{1.2.11}$$

where

$$cRT\mathbf{d}_i \equiv c_i \nabla_{T,p} \mu_i + \left(c_i \overline{V}_i - \omega_i\right) \nabla p - \rho_i \left(\mathbf{g}_i - \sum_{k=1}^{n} \omega_k \mathbf{g}_k\right) \tag{1.2.12}$$

$c$ = total molar concentration

$c_i$ = molar concentration of species $i$

Here we have recognized that $\mathbf{j}_i = \rho_i(\mathbf{v}_i - \mathbf{v})$ and have chosen to associate the $\rho_i$ with the "driving force" $\mathbf{\Lambda}_i$ rather than the "flux" $(\mathbf{v}_i - \mathbf{v})$. We shall see shortly that Eq. 1.2.12 is very convenient. This is because the rather fearsom quantity $cRT\mathbf{d}_i$ has a very simple physical significance and because the $\mathbf{d}_i$, defined by Eq. 1.2.11, are simply related. Specifically

$$cRT\mathbf{d}_i = \text{force per unit volume of solution tending to move species } i \textit{ relative to the solution.} \tag{1.2.13}$$

It also follows either from Eq. 1.2.12 and the Gibbs-Duhem equation or from Eq. 1.2.13 that

$$\sum_{i=1}^{n} \mathbf{d}_i = 0 \tag{1.2.14}$$

This equation is valid for nonisothermal nonisobaric reacting mixtures and is used shortly to complete our development. This modification of the mass-transfer driving force is useful because it permits us to consider diffusion in systems undergoing acceleration: The flux relations given below are therefore not restricted to systems at "mechanical equilibrium."

To determine the flux-force dependence we must make the second postulate:

II. *The fluxes are linearly and homogeneously related to the driving forces*:

$$j_i = \sum_j \beta_{ij} x_j \tag{1.2.15}$$

*where* $j_i =$ *any flux appearing in the expression for* $\sigma$; *that is, mass, momentum, or energy* $(j_{ix}, q_x, \text{ or } \tau_{yx})$

$x_j =$ *any corresponding driving force*;

*and the* $\beta_{ij}$ *are the "phenomenological coefficients"—or transport properties.*

The flux-force relations must clearly be homogeneous, because there must be no entropy production at equilibrium. They are not necessarily linear, however, and the above linear relation must be considered as a limiting expression for very small gradients.

The second postulate is supported by kinetic theory for low-density gases and by a large mass of experimental data for more complex systems. There is very little reason to doubt its validity for diffusion or heat transfer in systems large compared to molecular dimensions—or more specifically, when fractional changes in concentration or temperature per mean free path are small. Significant departures do occur in the flow of non-Newtonian fluids, and they may occur in mass transport through bimolecular lipid membranes.

Fortunately one does not have to consider all of the interrelations indicated by Eq. 1.2.15. It can be shown on the basis of symmetry* that coupling can occur only between driving forces and fluxes of the same order or between those varying in rank by even multiples. This means, for example, that mass fluxes can be caused only by temperature gradients and by the sum of "mechanical" forces represented by the $\mathbf{d}_i$.

One final simplification is made possible by the third postulate, which simply states:

III. *The* $\beta_{ij}$ *are symmetric*†: $\beta_{ij}=\beta_{ji}$.

This postulate was developed by Onsager on the basis of statistical-mechanical arguments and is supported by available data, as well as by kinetic theory in the case of gases.

It follows from the second and third postulates that‡

$$cRT\mathbf{d}_i=\sum_{j=1}^{n}\beta_{ij}(\mathbf{v}_j-\mathbf{v})+\beta_{i0}\nabla T \tag{1.2.16}$$

and

$$\frac{1}{T}\nabla T=\beta_{00}\mathbf{q}+\sum_{j=1}^{n}\beta_{0j}(\mathbf{v}_j-\mathbf{v}) \tag{1.2.17}$$

with

$$\beta_{ij}=\beta_{ji}\qquad (i,j=0,1,2,\ldots,n) \tag{1.2.18}$$

To complete our development it is now only necessary to recognize the restriction on the $\beta_{ij}$ provided by Eq. 1.2.14. Carrying out this summation, we obtain

$$0=\sum_{j=1}^{n}\sum_{i=1}^{n}\left[\beta_{ij}(\mathbf{v}_j-\mathbf{v})+\beta_{i0}\nabla T\right] \tag{1.2.19}$$

* One must be careful in applying these symmetry arguments in flow systems. Both turbulent convection and the tumbling of suspended solids in suspensions undergoing shear can have a profound effect on mass transfer. This effect is the result of convection, however, and not caused directly by velocity gradients.

† This statement is correct for flux-force pairs appearing in this text. For restrictions on the third postulate, see the general references cited in the bibliography.

‡ The use of $\nabla T$ rather than $\mathbf{q}$ here is arbitrary but convenient. We will be primarily interested in isothermal rather than locally adiabatic systems.

Since the diffusion velocities $\mathbf{v}_j - \mathbf{v}$ are mathematically independent of each other and of the heat flux $\nabla T$, it must follow that

$$\sum_{i=1}^{n} [\beta_{ij}(\mathbf{v}_j - \mathbf{v}) + \beta_{i0}\nabla T] = 0$$

so that

$$\sum_{i=1}^{n} \beta_{ij} = 0 \qquad (j = 0, 1, 2, \ldots, n) \tag{1.2.20}$$

Equation 1.2.20 essentially completes our development, and it remains only to rewrite the key equations for later convenience. We do this by using Eq. 1.2.20 to replace the reference velocity $\mathbf{v}$ by any arbitrary species velocity $\mathbf{v}_k$, and by rewriting the phenomenological coefficients. Our entire development can then be summarized by a set of $n-1$ rate expressions, the *Stefan-Maxwell equations*, and three sets of restraints:*

$$\mathbf{d}_i = \sum_{\substack{j=1 \\ j \neq k}}^{n} \frac{x_i x_j}{Đ_{ij}} (\mathbf{v}_j - \mathbf{v}_k) + \underline{\beta_{i0}\nabla T} \tag{1.2.21}$$

$$Đ_{ij} = Đ_{ji}; \qquad \sum_{i=1}^{n} \frac{x_i}{Đ_{ij}} = 0 \tag{1.2.22; 23}$$

$$\sum_{i=1}^{n} \mathbf{d}_i = 0 \tag{1.2.24}$$

Here the multicomponent mass diffusivities are defined by

$$Đ_{ij} = \frac{x_i x_j}{\beta_{ij}} cRT \tag{1.2.25}$$

This choice is a convenient one both because the $Đ_{ij}$ usually show a smaller composition dependence than the $\beta_{ij}$ and because they reduce to the more familiar $\mathfrak{D}_{AB}$ of Fick's first law (see Chapter 16, *Tr. Ph.*) for ideal binary solutions.

It is most common to let $k = i$ in Eq. 1.2.21, and this is in many respects the most natural choice: For $k = i$ all the $Đ_{ij}$ are positive. Frequently, however, it is convenient to single out one species as "solvent," for

* The thermal diffusion term $\beta_{i0}\nabla T$ is normally unimportant in biological systems and is not further considered.

example water, and to use it as the reference species $k$; similarly, in membrane transport it is often convenient to select the membrane matrix for reference.

There are also many other ways the above diffusion relations can be written.* A particularly common approach is to "invert"the set of Eqs. 1.2.21 to give explicit relations for the fluxes as weighted sums of driving forces. This is particularly useful when only concentration diffusion need be taken into account, and we consider this formulation later.

We are now, however, ready to consider some characteristic examples.

### *EX. 1.2.1. FICK'S LAW AND THE STEFAN-MAXWELL EQUATIONS*

Show the relation between Eq. 1.2.21 and Eq. 16.2-1 of *Tr. Ph.*:

$$\mathbf{J}_A^{\star} = -c\,\mathfrak{D}_{AB}\nabla x_A$$

SOLUTION

Eq. 16.2-1 is written for a binary system, containing only the two species $A$ and $B$. Equation 1.2.21 for this special case can be written in the form

$$\mathbf{d}_A = \frac{x_A x_B}{\mathcal{D}_{AB}}(\mathbf{v}_B - \mathbf{v}_A) \tag{1}$$

with $k=i=A$. Since $x_B = 1 - x_A$, we may also write

$$x_B(\mathbf{v}_B - \mathbf{v}_A) = x_B\mathbf{v}_B + x_A\mathbf{v}_A - \mathbf{v}_A$$

$$= \mathbf{v}^{\star} - \mathbf{v}_A \tag{2}$$

and

$$-c\mathcal{D}_{AB}\mathbf{d}_A = c_A(\mathbf{v}_A - \mathbf{v}^{\star}) = \mathbf{J}_A^{\star} \tag{3}$$

Making use of the definition of $\mathbf{d}_i$, Eq. 1.2.12, we find that

$$\mathbf{J}_A^{\star} = -c\mathcal{D}_{AB}\Bigg[x_A\nabla\ln a_A + \frac{(\phi_A - \omega_A)}{cRT}\nabla p$$

$$- \frac{\rho_A}{cRT}\omega_B(\mathbf{g}_A - \mathbf{g}_B)\Bigg] \tag{4}$$

* By no means are all of these trustworthy, and it is only prudent to test any new set for consistency with the above-described nonequilibrium-thermodynamic framework.

Equation 4 is clearly more general than Eq. 16.2-1 in containing pressure and forced-diffusion terms. However, it also differs in the concentration driving force.

Inasmuch as

$$x_A \nabla \ln a_A = \nabla x_A + x_A \nabla \ln \gamma_A \tag{5}$$

where $\gamma_A$ is the activity coefficient, it follows that

$$x_A \nabla \ln a_A = \nabla x_A \left( 1 + \frac{\partial \ln \gamma_A}{\partial \ln x_A} \right) \tag{6}$$

and

$$\mathfrak{D}_{AB} = \left( 1 + \frac{\partial \ln \gamma_A}{\partial \ln x_A} \right) Ð_{AB} \tag{7}$$

The activity-based diffusion coefficient $Ð_{AB}$ is in general much less concentration dependent than $\mathfrak{D}_{AB}$. However, its use requires rather extensive and accurate activity data, and is therefore not widespread.

For multicomponent mixtures of *low-density gases* the activity coefficients $\gamma_i$ are all unity, and

$$\mathbf{d}_i = \nabla x_i \tag{8}$$

for concentration diffusion. Also from kinetic theory we find that

$$Ð_{ij} = \mathfrak{D}_{ij} \tag{9}$$

the corresponding binary diffusivity. Equation 1.2.21 thus permits the extension of Eq. 18.4-19 (*Tr. Ph.*) to nonisobaric systems.

## *EX. 1.2.2. PRESSURE DIFFUSION AND THE ULTRACENTRIFUGE*

We consider here the steady-state operation of a tube in a centrifuge containing an arbitrary number of uncharged solute species. Rotation is about the axis, and the centrifugal force is in the radial direction.

We begin by noting that for steady operation all species velocities are the same:

$$\mathbf{v}_i = \mathbf{v} = \boldsymbol{\delta}_\theta r \Omega \tag{1}$$

We may then set

$$\mathbf{d}_i = 0 \tag{2}$$

and

$$\mathbf{g}_i = \mathbf{g} \tag{3}$$

Equation 2 follows from Eq. 1 and Eq. 3 follows from the fact that no body forces other than that of gravitational attraction are being considered.

We may now solve the equation of motion to find that

$$\nabla p \sim \boldsymbol{\delta}_r \rho \Omega^2 r \tag{4}$$

Putting this result into Eq. 1.2.12, we find that

$$\left(1 + \frac{d \ln \gamma_i}{d \ln x_i}\right) \frac{dx_i}{dr} = (\omega_i - \phi_i) \frac{M}{RT} \Omega^2 r \tag{5}$$

where $M = \rho / c =$ the molar mean molecular weight of the mixture (see Table 16.1-1, *Tr. Ph.*). For dilute solutions in low-molecular-weight solvents such as water, $M$ is very nearly the molecular weight of the solvent.

Equation 5 is surprisingly simple and informative. No diffusivities appear, because there is no relative motion between molecules. Hence the chemical nature of the system appears only in the term $1 + d \ln \gamma_i / dx_i$, which is of the order of unity for most systems. More specifically,

$$\frac{d \ln \gamma_i}{d \ln x_i} \to 1 \qquad \text{as} \qquad x_i \to 0$$

for all nonreactive species. We thus find that the more dense materials, those for which the mass fraction $\omega_i$ is greater than the volume fraction $\phi_i = c_i \overline{V}_i$, tend to be displaced toward the periphery of the system. This is what we would expect intuitively. However, we also find that the ratio of the mole fraction gradient $dx_i / dr$ to the "driving force" $\omega_i - \phi_i$ is the same for all species. Thus at the same mass concentration sodium chloride, with a partial specific volume on the order of $\frac{1}{2}\ \text{cm}^3/\text{g}$ exhibits a larger mole-fraction gradient than proteins, with $\overline{V}_i / M_i \doteq 0.8\ \text{cm}^3/\text{g}$. However, the fractional increase of concentration with distance is much greater for proteins because of their extremely high molecular weights. For the same reason, the determination of protein molecular weights by the above-described steady-state ultracentrifugation is a routine analytical procedure even though the term $(M/RT)\Omega^2 R_0$ is normally rather small.*

* For aqueous solutions $cRT \doteq 1360$ atm. For gases, on the other hand, $cRT = p$, and large fractional concentration changes are easy to obtain. Unsteady ultracentrifugation is also of interest, and here diffusion coefficients do appear. These measurements, combined with molecular-weight determinations and the hydrodynamic theory of diffusion, give useful measures of the protein shape. (See, e.g. Tanford (1959).

Centrifugation techniques are also widely used in biology for isolating subcellular particles and whole cells as well as protein molecules; sedimentation coefficients for these various types of particles vary over about six orders of magnitude (Anderson), but the behavior of all is described by Eq. 1.2.21. However, equipment design considerations become very important in any real operation, and any one experiment can be used only for a very small fraction of this range.

## 1.3. Diffusion in Free Solution under the Influence of Concentration, Pressure, and Electrostatic Potential Gradients

We consider here the important special case in which the only body forces are those due to electrostatic potential gradients. We may then write

$$\mathbf{g}_i = \mathbf{g} + \mathbf{g}_i^{(\mathrm{el})} \tag{1.3.1}$$

with

$$\mathbf{g}_i^{(\mathrm{el})} = -\frac{\nu_i \mathfrak{F}}{M_i} \nabla \phi \tag{1.3.2}$$

where $\nu_i$ = the ionic charge of solute $i$ (e.g., $-2$ for $SO_4^{2-}$).
$\mathfrak{F}$ = Faraday's constant = $9.652 \times 10^4$ absolute coulombs per gram-equivalent.
$\phi$ = electrostatic potential.

The use of Faraday's constant, like that of the gravitational conversion factor $g_c$, is optional. It is, however, convenient because it is very nearly universal practice to express the electrostatic potential in volts. Equation 1.3.1 is sufficiently general for our purposes *in the absence of mechanical constraints* on the diffusing species. Such constraints occur only in membrane transport, which is discussed in the next section.

On insertion of Eqs. 1.3.1 and 1.3.2, Eq. 1.2.25 takes the form

$$cRT\mathbf{d}_i = c_i \nabla_{T,p} \mu_i + \left(c_i \overline{V}_i - \omega_i\right) \nabla p$$

$$+ \left(c_i \nu_i - \omega_i \underline{\sum_{k=1}^{n} c_k \nu_k}\right) \mathfrak{F} \nabla \phi \quad \text{(no mechanical constraints)} \tag{1.3.3}$$

The underlined term represents the net electrical body force on the fluid as a whole. This net force is identically zero for an electrically neutral solution (cations and anions in exact balance). Normally it is negligibly small (as discussed by Newman) in bulk solution, except very near solid surfaces where diffuse electrical double layers are frequently encountered. We may then normally write

$$cRT\mathbf{d}_i = c_i \nabla_{T,p}\mu_i + \underline{\left(c_i \bar{V}_i - \omega_i\right)\nabla p} + c_i \nu_i \mathfrak{F} \nabla \phi$$

(no mechanical constraints or net electric charge) (1.3.4)

As a practical matter the pressure-diffusion term, underlined in Eq. 1.3.4, is negligibly small for liquids in the absence of very strong centrifugal fields. It can be quite important for gases.

Equation 1.3.4 will be suitable for describing the diffusion of individual ions in free solution because for these, $\omega_i \ll 1$ and $\sum_{k=1}^{n} c_k \nu_k \leqslant c_i \nu_i$ under most circumstances of interest. However, the effect of the net charge on the solvent, and hence on the solution as a whole, can be large under the circumstances described in the following paragraph. It should be noted that this equation is consistent with the kinetic theory of electrolytes—for example, the Debye-Hückel theory and its extensions.

The only important exception to the existence of electroneutrality is found in the electrical double layers in electrolyte solutions immediately adjacent to solid surfaces and the surfaces of macromolecules such as proteins. Appreciable departures from neutrality occur in these, and the potential gradients are correspondingly very large. Double layers play a key role in many electrokinetic phenomena and are quite important biologically—for example, in the coagulation of proteins. The diffuse portion of the double layer thickens with decreasing electrolyte concentration and extends indefinitely into the solution. However, at physiological electrolyte concentrations the bulk of the electrical imbalance takes place in a few tens of angstroms. The double-layer thickness is appreciable in comparison with that of hydrodynamic and mass-transfer boundary layers for colloidal particles and very small pores. It is usually considered negligible for macroscopic equipment, but begins to become important in the tips of glass microelectrodes for tip lumens of the order of 10 $\mu$. Double layers in such small openings can apparently contribute to "tip potentials" and produce appreciable permselectivity.*

Outside the double layers, departures from electroneutrality are truly negligible. They can, for example, be neglected in the calculation of salt diffusion and junction potentials and in the description of convective mass transfer in physiological electrolytes. In all subsequent discussions of

* These effects are discussed by several contributors to *Glass Micro-electrodes*, Lavallée, Schanne, and Hébert, Eds., Wiley (1969).

diffusion in free solution, electrical neutrality is assumed except where the contrary is explicitly stated.

We now illustrate the characteristics of Eq. 1.3.4 by considering three simple examples of diffusion in solutions of a single 1, 1-electrolyte. In the first we show that the diffusion of a salt in aqueous solution can be treated as a simple case of binary diffusion—because the requirement of no current flow provides a mathematical constraint reducing the number of independent rate equations to one. It is nevertheless useful to treat even this relatively simple process as an example of ternary diffusion, because we can then calculate the diffusivity of the salt from the mobilities of the constituent ions. Later we extend this discussion to multicomponent systems, which unfortunately are much more complex. In the second example we describe diffusion potentials in binary salt solutions and show that these are a manifestation of their ternary nature. This example will also be generalized later to the description of diffusion in multicomponent solutions and across membranes. In the last example we show the effects of solute-solute interaction by contrasting salt conductivity with salt diffusion.

*EX. 1.3.1. SALT DIFFUSION*

Consider the diffusion of a salt $M^+X^-$ (for example, NaCl) from a region of high concentration to one of low concentration, as indicated schematically in Fig. 1.3.1, and develop an expression for the effective diffusivity of

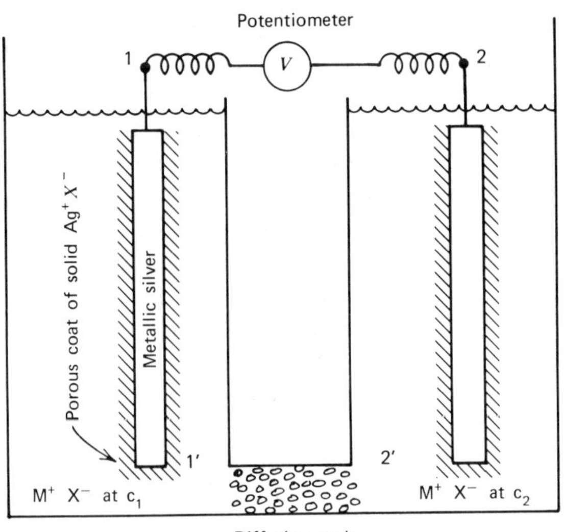


**Figure 1.3.1.** Salt diffusion and diffusion potentials.

the salt in terms of the $Ɖ_{ij}$ for the two constituent ions and water.

SOLUTION

We begin by noting that the pressure gradients are very small relative to $cRT$ in systems of this type, so that we may ignore pressure diffusion and write

$$\mathbf{d}_i \doteq x_i \nabla \ln a_i + x_i \nu_i \frac{\mathfrak{F}}{RT} \nabla \phi \tag{1}$$

As indicated in connection with Eq. 1.2.21, only $n-1$, or here two, such equations are independent. We choose those for the two ionic species and develop an expression for the salt diffusion by eliminating the electrostatic potential between them. In doing this we may take advantage of the requirements of electroneutrality and no electric current flow. We may then write

$$x_M = x_X = x_S = 1 - x_W \tag{2}$$

$$\mathbf{N}_M = \mathbf{N}_X = \mathbf{N}_S \tag{3}$$

where the mole fractions of the ionic species are conventionally but somewhat arbitrarily defined as $c_{MX}/(c_{MX}+c_W)$. In addition, the activities of the ions are determined solely by their concentrations at any given temperature and pressure. The system is thus binary from a thermodynamic standpoint, but a constrained ternary one for diffusional purposes.

The Stefan-Maxwell equations may now be written as

$$\frac{1}{cƉ_{MW}}(x_W \mathbf{N}_S - x_S \mathbf{N}_W) = -\frac{\partial \ln a_M}{\partial \ln x_S} \nabla x_S - \frac{x_S \mathfrak{F}}{RT} \nabla \Phi \tag{4}$$

$$\frac{1}{cƉ_{XW}}(x_W \mathbf{N}_S - x_S \mathbf{N}_W) = -\frac{\partial \ln a_X}{\partial \ln x_S} \nabla x_S + \frac{x_S \mathfrak{F}}{RT} \nabla \Phi \tag{5}$$

Note that the ion-ion interaction term $Ɖ_{MX}$ does not appear, because the relative velocity of the two ions is zero. The electrostatic potential may be eliminated by adding these two equations, and the result may be rearranged to give

$$\mathbf{N}_S = -\left(\frac{1}{cƉ_{MW}} + \frac{1}{cƉ_{XW}}\right)^{-1} \frac{\partial \ln a_M a_X}{\partial \ln x_S} \nabla x_S + x_S(\mathbf{N}_S + \mathbf{N}_W) \tag{6}$$

or

$$\mathbf{N}_S = -\mathfrak{D}_{SW} \nabla x_S + x_S(N_S + N_W) \tag{7}$$

where the binary salt-water diffusivity

$$\mathfrak{D}_{SW}=2\left(\frac{Đ_{MW}Đ_{XW}}{Đ_{MW}+Đ_{XW}}\right)\left(1+\frac{\partial \ln \gamma_S}{\partial \ln x_S}\right) \tag{8}$$

and $\gamma_S=\sqrt{\gamma_M\gamma_X}$ is the mean activity coefficient for the salt. The ion-water diffusivities in turn may be estimated from the limiting ionic conductances according to the expression

$$K_\infty=\lim_{x_i\to 0}\frac{\nu_i c_i Đ_{iW}\mathfrak{F}^2}{RT} \tag{9}$$

where $K_\infty$ is the limiting specific conductance of the ion. (Note that the equivalent conductance $\lambda_i=K_i/c_i$.) These diffusivities are less concentration dependent than the conductances, which are influenced by ion-ion interactions, and salt diffusivities for concentration on the order of 1 $N$ can be estimated with fair accuracy from limiting conductances. Note that this result is completely independent of flow conditions and geometry; that is not true of multicomponent systems.

### *EX. 1.3.2. BINARY JUNCTION POTENTIALS*

Extend the previous example to describe the *junction potential*, that is, the electrostatic potential difference that arises between compartments 1 and 2 as a result of the diffusion process.

SOLUTION

Junction potentials in a solution may be readily calculated by eliminating the mass fluxes from the Stefan-Maxwell equations. One thus obtains

$$\frac{d\Phi}{dx_S}=-\frac{RT}{\mathfrak{F}}\frac{1}{x_S}\left[\frac{\frac{\partial \ln a_M}{\partial \ln x_S}Đ_{MW}-\frac{\partial \ln a_X}{\partial \ln x_S}Đ_{XW}}{Đ_{MW}+Đ_{XW}}\right]$$

Such solution junction potentials are not, however, directly measurable, because the electrodes used to measure them also contribute to the measured voltage $V$. Not surprisingly, $V$ depends on the nature of the electrodes.

We shall assume here that we are using electrodes operating according to the reaction

$$A^+X^-(\text{saturated})+\epsilon\to A(\text{solid})+X^- \tag{1}$$

The commonest examples are Ag$-$AgCl and calomel electrodes, both of which are widely used in applied electrochemistry and described in standard references. Since both A and $A^+X^-$ are present as pure solids, they may be assigned unit activities. The free-energy change for the above reaction is then

$$\Delta G = \Delta G_0 + (RT \ln a_{X^-} - \mathfrak{F}\,\phi_{\text{solution}}) + \mathfrak{F}\,\phi_{\text{solid}} \tag{2}$$

and the *equilibrium* half-cell potential is

$$\phi_{\text{solid}} - \phi_{\text{solution}} = -\phi_0 - \frac{RT}{\mathfrak{F}} \ln a_{X^-} \tag{3}$$

The first term on the right side of Eq. 2 is the free energy of formation of $X^-$, and the second, that for the consumption of the electron. The reference free energy $G_0$ and the potential $\phi_0$, relative to the standard hydrogen electrode, are available in standard references. We do not need them here.

The voltage difference $V$ measured by the potentiometer in this system can be obtained by summing the voltages around the closed circuit shown. We assume here that no current is flowing and that the two electrodes are *reversible*, that is, that the potential across them is their reversible half-cell potential. Then

$$\begin{aligned} V &= (\phi_2 - \phi_1)_{\text{solid}} \\ &= \frac{RT}{\mathfrak{F}} \ln \frac{a_{X1}}{a_{X2}} + (\phi_2 - \phi_1)_{\text{solution}} \end{aligned} \tag{4}$$

where $(\phi_2 - \phi_1)_{\text{solution}}$ is the electrostatic potential difference in the solution. This latter term can be evaluated from Eq. 10 of Ex. 1.3.1 if we neglect* changes in the diffusivity ratio $\mathcal{D}_{MW}/\mathcal{D}_{XW}$. We thus obtain

$$\begin{aligned} \phi_2 - \phi_1 = -\frac{RT}{\mathfrak{F}} \Bigg( & \frac{\mathcal{D}_{MW}}{\mathcal{D}_{MW} + \mathcal{D}_{XW}} \ln \frac{a_{M2}}{a_{M1}} \\ & - \frac{\mathcal{D}_{XW}}{\mathcal{D}_{MW} + \mathcal{D}_{XW}} \ln \frac{a_{X2}}{a_{X1}} \Bigg) \end{aligned} \tag{5}$$

*This can be justified in a rough and ready way by noting that transport numbers are very nearly concentration independent. See Section 2.4 for a more complete and definitive treatment.

It follows that the measured potential difference is

$$V = \frac{RT}{\mathfrak{F}} \frac{\mathit{Ð}_{MW}}{\mathit{Ð}_{MW} + \mathit{Ð}_{XW}} \ln \frac{a_{S1}}{a_{S2}} \tag{6}$$

where $a_S = a_M a_X$. Our final result then contains only activities of the neutral salt S, which are directly measurable by standard thermodynamic techniques.Equation 6 provides an unambiguous means of determining the diffusivity ratio $\mathit{Ð}_{MW}/\mathit{Ð}_{XW}$. The individual ion-water diffusivities can then be determined from simultaneous measurements of junction potentials and salt diffusivities.

Unfortunately one cannot always find suitable reversible half-cells for biological systems. One must then resort to the use of salt bridges, as discussed later.

### *EX. 1.3.3. CONDUCTIVITY OF A 1, 1-ELECTROLYTE*

Develop an expression for the conductivity of a well-stirred solution of the salt $M^+X^-$ in water. Assume the stirring prevents the development of any concentration gradients, and use the complete Stefan-Maxwell equations.

SOLUTION

We begin by defining the *specific conductivity* $\kappa$ by

$$\mathbf{I} = -\kappa \nabla \phi \tag{1}$$

where **I** is the local current density. We next note that

$$\mathbf{I} = (\mathbf{N}_{M^+} - \mathbf{N}_{X^-})\mathfrak{F} \tag{2}$$

where $\mathfrak{F}$ is Faraday's constant. We assume that pressure diffusion is negligible.

This system is described by

$$M^+ : x_S\left(\frac{\mathfrak{F}}{RT}\right)\nabla\phi = \frac{x_S\mathbf{N}_W - x_W\mathbf{N}_{M^+}}{c\mathit{Ð}_{WM}} + \frac{x_S(\mathbf{N}_{X^-} - \mathbf{N}_{M^+})}{c\mathit{Ð}_{MX}} \tag{3}$$

$$X^- : -x_S\left(\frac{\mathfrak{F}}{RT}\right)\nabla\phi = \frac{x_S\mathbf{N}_W - x_W\mathbf{N}_{X^-}}{c\mathit{Ð}_{WX}} + \frac{x_S(\mathbf{N}_{M^+} - \mathbf{N}_{X^-})}{c\mathit{Ð}_{MX}} \tag{4}$$

We may eliminate $\mathbf{N}_W$ between these two expressions to obtain

$$cx_S \frac{\mathfrak{F}}{RT} \nabla\phi(\mathcal{D}_{WM} + \mathcal{D}_{WX}) = (\mathbf{N}_{X^-} - \mathbf{N}_{M^+})\left[x_W + x_S\left(\frac{\mathcal{D}_{WM} + \mathcal{D}_{WX}}{\mathcal{D}_{MX}}\right)\right] \tag{5}$$

This can be rearranged to give

$$\mathbf{I} = -\left(\frac{c_S \mathfrak{F}^2}{RT}\right)\left[\frac{\mathcal{D}_{WM} + \mathcal{D}_{WX}}{x_W + x_S(\mathcal{D}_{WM} + \mathcal{D}_{WX})/\mathcal{D}_{MX}}\right]\nabla\phi \tag{6}$$

so that

$$\kappa = \left(\frac{c_S \mathfrak{F}^2}{RT}\right)\left[\frac{\mathcal{D}_{WM} + \mathcal{D}_{WX}}{x_W + x_S(\mathcal{D}_{WM} + \mathcal{D}_{WX})/\mathcal{D}_{MX}}\right] \tag{7}$$

We note that here the ionic interaction remains in the expression for the conductivity except in the limit as $x_S$ approaches zero. This is because there is in general a relative motion of the two ions in the presence of an electric current. As a result, the conductivity is much more concentration dependent than the diffusivity.

### 1.4. Diffusion in Mechanically Restrained Membranes

It is customary to treat membranes as homogeneous phases, much as is done in Section 17.6 of *Tr. Ph.* for porous catalysts, except that we now must take into account all of the diffusional interactions required by Eq. 1.2.21. We do this by the method of Scattergood and Lightfoot (1968). The membrane matrix is treated as one of the diffusing species, except that it is constrained to zero velocity, and normally to near-zero concentration gradients. A similar treatment has been used by Evans, Watson, and Mason (1962) for the diffusion of gases in porous solids, and our discussion below parallels theirs closely.

In membrane transport processes the membrane itself will tend to move, just like any other diffusing species, unless it is mechanically constrained. The force required to prevent membrane movement is not accounted for in Eq. 1.3.1 and must now be introduced. We begin by considering a supported membrane opposing a "hydraulic" flow as pictured in Fig. 1.4.1. Clearly the drag of the moving species on the membrane matrix $m$ must be

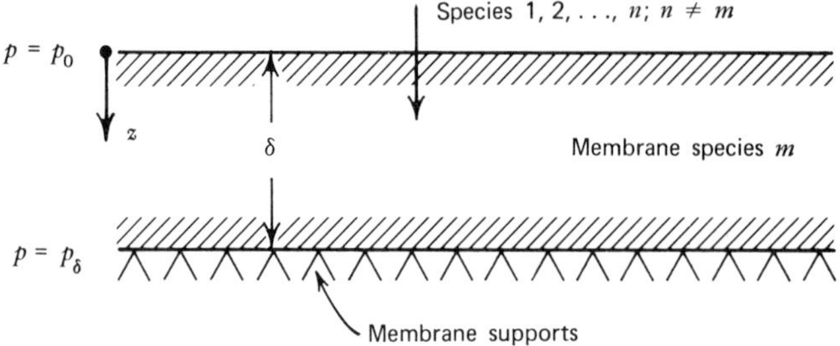


**Figure 1.4.1.** Membrane restraints.

counteracted by the supports. The force exerted by the supports on the membrane is

$$\mathbf{F}_s = -\boldsymbol{\delta}_z(p_0 - p_s)A \tag{1.4.1}$$

where $A$ is the area of the membrane surface. Since the pressure gradient is

$$\nabla p = \boldsymbol{\delta}_z \frac{\partial p}{\partial z}$$

it follows that

$$\frac{\mathbf{F}_s}{A\delta} = (\nabla p)_{av} \tag{1.4.2}$$

All of this force is transmitted to the membrane matrix itself and can be considered as a body force $\mathbf{g}_m^{(s)}$ acting on the membrane as a result of the support. It may be expressed as

$$(\mathbf{g}_m^{(s)})_{av} = (\nabla p)_{av} \frac{1}{\rho_m} \tag{1.4.3}$$

That is,

$$\frac{\text{force}}{\text{mass of matrix}} = \frac{(\text{force})/(\text{total volume})}{(\text{mass of matrix})/(\text{total volume})}$$

In the case of compression of the membrane (which does occur, for example, in ultrafiltration), $\nabla p$ and $\mathbf{g}_m^{(s)}$ are not uniform. However, the part of $\mathbf{F}_s$ required to restrain any layer of the membrane is transmitted through the "downstream" portions of the membrane, which act as the support for the layer in question. Hence we may write locally

$$\mathbf{g}_m^{(s)} = \nabla p \frac{1}{\rho_m} \tag{1.4.4}$$

and

$$\mathbf{g}_m = \mathbf{g} + \mathbf{g}_m^{(\mathrm{el})} + \mathbf{g}_m^{(\mathrm{s})} \tag{1.4.5}$$

where the superscripts (el) and (s) refer to the electrical body force and to that supplied by the support, respectively.

One must therefore single the membrane out for special treatment, since its behavior is different from that of the mobile species. However, we may describe diffusional behavior completely in terms of the $n-1$ equations for the mobile species. For these,

$$\mathbf{g}_i = \mathbf{g}_i^{(\mathrm{el})} + \mathbf{g} \qquad (i \neq m) \tag{1.4.6}$$

and in the absence of a net electric charge in the membrane phase as a whole,

$$\mathbf{g}_i - \sum_{k=1}^{n} \omega_k \mathbf{g}_k = \mathbf{g}_i^{(\mathrm{el})} - \frac{1}{\rho} \nabla p \qquad (i \neq m) \tag{1.4.7}$$

Since

$$\mathbf{g}_i^{(\mathrm{el})} = -\frac{\nu_i \mathfrak{F}}{M_i} \nabla \phi \tag{1.4.8}$$

we may write

$$cRT\mathbf{d}_i = c_i \nabla_{T,p} \mu_i + c_i \bar{V}_i \nabla p + c_i \nu_i \mathfrak{F} \nabla \phi \tag{1.4.9}$$

($i = 1, 2, \ldots, n, \neq m$;
$m$ constrained to zero velocity;
system as a whole electrically neutral)

This expression differs from Eq. 1.3.4, its free-solution analog, only in the omission of the term $\omega_i \nabla p$. Partly for this reason and partly because pressure drops across membranes are often large, the pressure-diffusion term is frequently important in membrane transport.

We return to the subject of membrane transport after the discussion of interphase equilibria in Section 1.5.

## 1.5. Interphase Equilibria

Our purpose here is to use the results of previous sections to describe the abrupt changes in concentration, pressure, and electrostatic potential that occur across the boundary between a membrane and the adjoining free solution. The interfacial region is a very small one, so that concentration, pressure, and electrostatic-potential gradients in the two bulk phases are negligible by comparison: the interfacial gradients are *very* large. Therefore, in the absence of slow surface reactions accompanying the transfer of solute, the resistance to transfer here is negligibly small. It follows from Eq. 1.2.34 that in this very thin interfacial region we may write

$$\mathbf{d}_i = 0 \tag{1.5.1}$$

which is equivalent to assuming interfacial equilibrium. Since pressure gradients can develop only as a result of mechanical restraints on the matrix, the proper expression for the $\mathbf{d}_i$ is Eq. 1.4.9:

$$\nabla_{T,p} \ln a_i + \frac{\nu_i \mathfrak{F}}{RT} \nabla \phi + \frac{\overline{V}_i}{RT} \nabla p = 0 \tag{1.5.2}$$

It now remains only to integrate this expression from the external phase $e$ to the membrane phase $m$ to obtain:

$$\ln\left(\frac{a_{ie}}{a_{im}}\right)_{T,p} + \frac{\nu_i \mathfrak{F}}{RT}(\phi_e - \phi_m) + \frac{\left(\overline{V}_i\right)_{\text{av}}}{RT}(p_e - p_m) = 0 \tag{1.5.3}$$

where

$$\left(\overline{V}_i\right)_{\text{av}} = \frac{1}{p_e - p_m} \int_{p_m}^{p_e} \overline{V}_i \, dp$$

Note that the chemical activity of $i$ in any phase, $(a_i)_{T,p}$, is defined for existing local chemical composition, but that the (arbitrary) reference temperature and pressure are *the same for both phases*. It is by no means always possible to separate the effects of composition and pressure within the membrane phase. Some authors therefore prefer to define activity *relative to the pressure in the external phase*. Defining this activity by $(a)_{T,p_e}$, we may put Eq. 1.5.3 in the equivalent alternative form

$$\ln\left(\frac{a_{ie}}{a_{im}}\right)_{T,p_e} + \frac{\nu_i \mathfrak{F}}{RT}(\phi_e - \phi_m) = 0 \tag{1.5.4}$$

This point is discussed at length by Gregor (1948, 1951).

Either Eq. 1.5.3 or 1.5.4 may be recognized as the well-known requirement for interphase equilibrium. We started from the flux equations here to emphasize their consistency with the laws of thermodynamics.

It is generally believed that transport processes in macroscopic membranes are sufficiently slow to justify the use of Eq. 1.5.3 or 1.5.4. For bimolecular membranes, however, the transfer rate may be limited by such interfacial reactions as desolvation and the formation of carrier complexes.

We now illustrate the utility of Eq. 1.5.3 with three examples of particular biological interest: external and internal osmotic pressure, and Donnan exclusion.

### *EX. 1.5.1. OSMOTIC PRESSURE ACROSS A SEMIPERMEABLE MEMBRANE*

An aqueous protein solution (phase 1) is completely enclosed by a membrane permeable to water but not to protein. The membrane in turn is immersed in pure water, as shown in Fig. 1.5.1. Water tends to move across the membrane into the protein solution because of its lower equilibrium water vapor pressure. Ultimately, however, this inward water transfer increases the pressure in the protein solution sufficiently to offset the lower vapor pressure. Show how this pressure can be calculated. Neglect the ionization of the protein, and the compressibility of water.

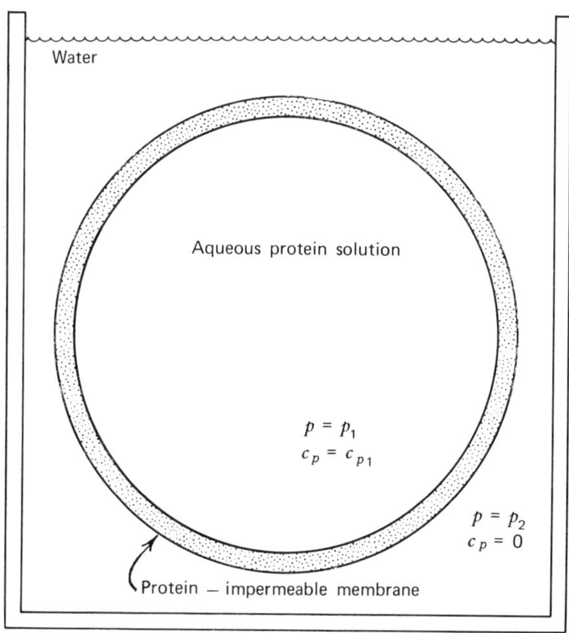


**Figure 1.5.1.** Osmotic pressure across a semipermeable membrane.

SOLUTION

Once a steady state has been reached the membrane must in equilibrium with both the internal and external liquid phases. We may then write Eq. 1.5.3 for the two solutions in the form

$$\ln\left(\frac{a_{W1}}{a_{W2}}\right)_{T,p}+\frac{\bar{V}_W}{RT}(p_1-p_2)=0 \tag{1}$$

It follows that

$$p_1-p_2=-\frac{RT}{\bar{V}_W}\ln a_{W1} \tag{2}$$

if we choose pure water at $p_2$ as the standard state. Since water vapor is very nearly ideal, we may rewrite Eq. 2 as

$$p_1-p_2=\frac{RT}{\bar{V}_W}\ln\frac{p_{W2}}{p_{W1}} \tag{3a}$$

$$\doteq\frac{RT}{\bar{V}_W}\frac{p_{W2}-p_{W1}}{p_{W1}} \tag{3b}$$

where $p_W$ = partial pressure of water vapor. The *osmotic pressure* $\pi$, defined as $p_1-p_2$, is thus seen to be approximately proportional to the fractional reduction in vapor pressure due to the presence of protein.

The proportionality constant $RT/\bar{V}_W$ is very large. Thus at 25°C

$$\frac{RT}{\bar{V}_W}=\frac{(82.05\ \text{cm}^3\text{atm/g-mole}^\circ\text{K})(298.16^\circ\text{K})}{(18.016\ \text{cm}^3/\text{g-mole})}$$

$$\doteq 1360\ \text{atm}$$

This coefficient is so large that even very modest vapor-pressure lowerings produce a measurable osmotic pressure.

For this reason osmotic pressures have been widely used in the past for estimating the molecular weights of proteins. These experiments, discussed in detail by Tanford (1959), depend upon variants of van't Hoff's limiting law for vanishingly small solute concentration:

$$\lim_{c_i\to 0}\{\pi\}=\alpha_i c_i RT \tag{4}$$

where $\pi$ is the osmotic pressure and $\alpha_i$ is a species-dependent osmotic coefficient of the order of unity.* It can be seen from Fig. 1.5.2 that Eq. 4 is of limited use for protein solutions of finite concentration.

It may also be seen from this figure that the osmotic pressures of concentrated protein solutions can be quite appreciable. Since the normal hemoglobin concentration in red cells is of the order of 33 g/100 ml, the hemoglobin osmotic pressure in the red cell is of the order of $\frac{1}{2}$ atm. The osmotic pressure due to serum proteins in the plasma bathing these cells (the "colloid osmotic pressure") is usually between 22 and 30 mm Hg, with the average normally taken as 25 mm Hg. Osmotic relations in the red cell are further complicated by ionization of the constituent proteins; they are discussed in detail in standard texts on physiology. It suffices for our present purposes to note that the cell membrane is unable to support the large equilibrium osmotic pressure suggested by the large difference between interior and external "colloid" osmotic pressures. Since the cell membrane is highly permeable to water but not to electrolytes, the cell

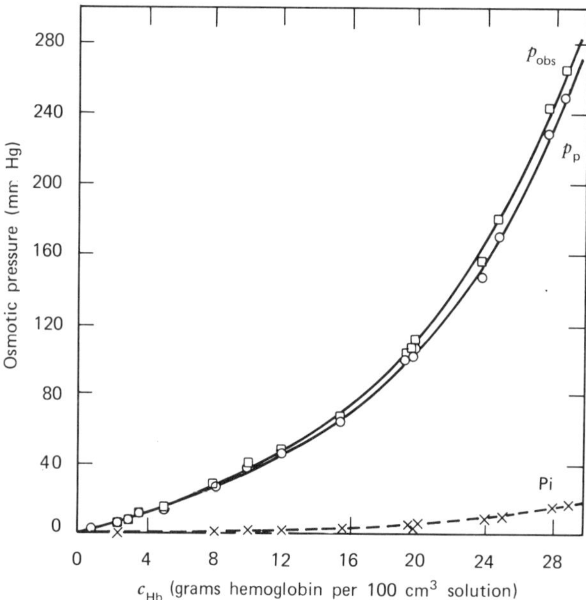


**Figure 1.5.2.** Observed osmotic pressure $p_{obs}$ of hemoglobin at 0°C, as a function of hemoglobin concentration $c_{Hb}$. Here $p_p$ is the partial pressure of the protein ions; $p_i$ is the diffusible-ion pressure difference calculated by Eq. 34 of G. S. Adair [*Proc. Roy. Soc. (Lond.)*, **A120**, 595 (1928)]. Data from G. S. Adair Proc. Roy. Soc. (Lond.) **A120**, 595 (1928).

* The origin of this simple limiting expression is discussed in most introductory texts on physical chemistry.

compensates for the unbalanced protein concentration by maintaining an offsetting imbalance in electrolyte concentration. This requires a continuing expenditure of metabolic energy, and red cells will swell and burst ("hemolyze") if any step in this metabolic process is blocked.*

## *EX. 1.5.2. OSMOTIC PRESSURES WITHIN HYDROPHILIC GELS*

Extend the previous example to explain the swelling pressures generated by the absorption of water into hydrophilic gels.

### SOLUTION

This situation differs from that of the previous example primarily in the way the expansion of phase 1, here the gel phase, is resisted. In contrast with the localized restraint produced by the external membrane, there now exists a cross-linked matrix to which the hydrophilic groups are attached. The hydrophilic groups again tend to produce water absorption, and hence swelling, by lowering the free energy of the imbibed water. Resultant stretching of the matrix elements produces an effective "internal" osmotic pressure on the absorbed water that can resist indefinite water absorption, even in the absence of an external mechanical constraint such as the surrounding membrane of the previous example. This action can be visualized by imagining the polymer chains of the matrix to act as springs tending to compress the imbibed solution. This is indicated in Fig. 1.5.3, which is a minor variation of the model originally proposed by Gregor (1948). The internal osmotic pressure is real, but not accurately measurable by usual techniques. It manifests itself whenever a mechanical restraint is applied, as when a mushroom lifts a section of sidewalk. It can be estimated, as suggested by Gregor, through vapor-pressure measurements on an uncrosslinked polymer of similar chemical nature. It should also be possible to determine it through measurement of the compressibility of the partially swollen polymer under conditions which do not hinder water movement. Normally, however, the osmotic pressure in the membrane is not known.

Here we shall consider the distribution equilibrium between a binary external solution and a gel absorbing both solute and solvent, but to differing degrees.

* Bursting is not, however, inevitable. For example, red cells in banked blood can remain intact in spite of low metabolic activity which results in $Na^+$ intrusion and $K^+$ leakage. Also, controlled swelling of red cells can result in the formation of very small "cracks" or pores through which hemoglobin can diffuse without gross rupture of the cell membrane. The product of such a process, called a red-cell ghost, is an apparently intact membrane of normal shape surrounding a colorless electrolyte solution similar to that surrounding it.

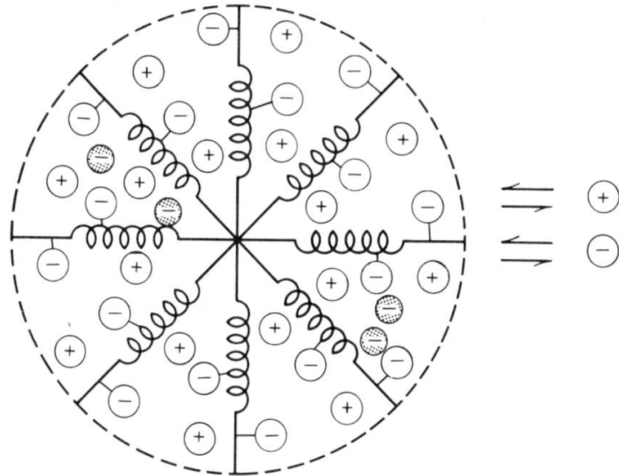

**Figure 1.5.3.** Gregor's model of a polyectrolyte gel, (shown here as a bead). The resin matrix is represented by extensible elastic elements (springs). The fixed charges, here anions, are shown as ⊖ attached to these springs. The counter-ions are shown as ⊕ free to diffuse in the gel phase. The ions of the invading electrolyte are shaded. The water molecules are not shown.

SOLUTION

We begin by noting that Eq. 1.5.3 must hold for both solute S and solvent W:

$$\ln\left(\frac{a_{W1}}{a_{W2}}\right)_{T,p} = \frac{\bar{V}_W}{RT}(p_2 - p_1)$$

$$\equiv -\frac{\bar{V}_W}{RT}\Pi \tag{1}$$

$$\ln\left(\frac{a_{S1}}{a_{S2}}\right)_{T,p} = -\frac{\bar{V}_S}{RT}\Pi \tag{2}$$

where $\Pi$ is the osmotic pressure in the membrane phase. These equations may be written in the alternative form

$$\frac{x_{S1}}{x_{S2}} = \frac{\gamma_{S2}}{\gamma_{S1}} e^{-\Pi \bar{V}_S/RT} \equiv K_D \tag{3}$$

where the $\gamma_i$ are activity coefficients.

$$\Pi = \frac{RT}{\bar{V}_W} \ln \frac{p_{W2}}{p_{W1}} \tag{4}$$

The solute and water are deliberately given nonparallel treatment here because our interest in them is different.

Equation 3 tells us that the solute distribution coefficient $K_D$ between the gel and the surrounding aqueous phase depends upon the osmotic pressure as well as the chemical nature of the two phases (appearing via the activity coefficients).*

Equation 4 relates osmotic pressure to the equilibrium partial pressures of water vapor, $p_{W2}$ and $p_{W1}$, in the external phase and gel phase, both at standard reference temperature and pressure. The internal vapor pressure $p_{W1}$ is not directly measurable, but even relatively small fractional-vapor-pressure depressions can produce large osmotic pressures, because of the large value of $RT/\tilde{V}_W$ ($\sim$1360 atm). Both direct observation of swelling pressures and estimates based on our general chemical knowledge suggest that pressures of the order of hundreds of atmospheres are common. This is discussed further in the problems.

### *EX. 1.5.3. DONNAN EXCLUSION AND ION SELECTIVITY*

A salient characteristic of some biological membranes, for example that of the red cell, and all synthetic polyelectrolytes is a very marked difference in permeability for anions relative to cations. In the case of polyelectrolytes, at least, this is due largely to the effective exclusion of the invading electrolyte and the resultant imbalance between mobile anions and cations. Show that this *Donnan exclusion* is a natural consequence of Eq. 1.5.3. Use as a basis for discussion a membrane containing sulfonic acid as fixed charges, and assume a single salt $M^+X^-$ in the external solution.

SOLUTION

We begin by looking at the qualitative aspects of this problem.

The large concentration of $M^+$ in the membrane phase results from the presence of the immobile anions attached to the matrix by covalent bonds and the need for a very close approximation to electroneutrality. As soon as a very small amount of $M^+$ diffuses out into the external solution under the influence of the concentration inequality, this solution becomes positively charged relative to the membrane, and an "electrodiffusion" flux of $M^+$ back into the membrane phase arises. A similar situation occurs for $X^-$, but in the reverse direction. Ultimately the concentration and electrodiffusion fluxes balance, so that Eq. 1.5.3 is satisfied, and for this equilibrium situation the membrane phase concentration of $M^+$ is always

* One must take solute distribution coefficients into account in the preparation of blood extenders and solutions used for perfusing organs. As discussed in the problems, it is particularly important that the *colloid osmotic pressure* be close to that of the interstitial fluid.

greater than that of $X^-$ by an amount almost exactly equal to the concentration of $R^-$. Typical membrane potentials are on the order of tens of millivolts, and these are produced by very small departures from electroneutrality: on the order of $10^{-13}$ to $10^{-14}$ $N$ (Newman, 1967).

To obtain some feeling for the order of magnitude of the exclusion of external ("invading") electrolyte, we consider as a representative case the sulfonic acid membrane used by Scattergood and Lightfoot. When immersed in a 0.1 $N$ NaCl solution at 25°C, the membrane contains

$$c_{R^-} = 1.03 \text{ g-eq/l of sulfonic acid}$$
$$c_W = 13.2 \text{ g-eq/l of water}$$
$$c_S = 0.001 \text{ g-eq/l of NaCl}$$

at equilibrium. It follows that

$$\frac{c_{Cl^-}}{c_{Na^+}} \doteq 10^{-3}$$

in the membrane and that the ratio of internal to external salt concentrations is about $10^{-2}$.

A quantitative explanation of this degree of Donnan exclusion of salt is not possible on the basis of available data, but we can estimate the relative importance of osmotic and chemical factors. To do this we first note from the results of our previous example that

$$\ln\left(\frac{a_{Se}}{a_{Sm}}\right)_{T,p} = \frac{\Pi \bar{V}_S}{RT}$$

where $a_S$ refers to

$$(a_S)_{T,p} = x_{Na^+} x_{Cl^-} \gamma_{Na^+} \gamma_{Cl^-}$$

and the $\gamma_i$ are the activity coefficients. Then

$$\frac{x_{Se}^2}{x_{Nam} x_{Clm}} = \frac{(\gamma_{Na}\gamma_{Cl})_m}{(\gamma_{Na}\gamma_{Cl})_e} e^{\Pi \bar{V}_s/RT}$$

$$\equiv K_\gamma K_\Pi \equiv K$$

Here $K$ is the equilibrium constant for the distribution, and $K_\gamma$ and $K_\Pi$ are its chemical and osmotic components. For our situation,

$$x_{Se} \doteq \frac{0.1}{55.5} = 1.8 \times 10^{-3}$$

$$x_{Nam} = \frac{1.031}{14.231} \doteq 7.25 \times 10^{-2}$$

$$x_{Clm} = \frac{10^{-3}}{14.231} \doteq 7.02 \times 10^{-5}$$

Then

$$K=\frac{(1.8\times 10^{-3})^2}{(7.25\times 10^{-2})(7.02\times 10^{-5})}=0.635$$

Thus, as is frequently true for small monovalent ions, the observed equilibrium constant is of the order of unity. The value of $K_\gamma$, however, should be substantially smaller, since the osmotic pressure is probably positive. A speculative estimate of $K_\gamma$ is given in the problems.

Data on biological membranes are much less complete, but it appears that both charged and uncharged examples are common. Thus the red-cell membrane behaves as if it has fixed positive charges with an ionization constant pK$\sim$9; it is accordingly much more permeable to anions than to cations. The membranes of nerve axons show no general preference for anions relative to cations and are probably uncharged. Both of these membranes do, however, show much more complex transport behavior than simple charged macroscopic membranes. We return to them in Section 3.

### 1.6. Useful Approximate Forms of the Diffusion Equations

The chief defect of the generalized Stefan-Maxwell equations introduced in Section 1.2 is that their use requires a great deal of information: They contain $(n/2)(n-1)$ independent diffusion coefficients and require extensive thermodynamic data to relate the chemical activities $a_i$ to the mole or mass fractions $x_i$ or $\omega_i$ and to calculate the partial molal volumes $\overline{V}_i$. For most systems of physiological interest such complete information is simply not available, and it is necessary to rely on various approximate expressions in which at least some diffusional and thermodynamic interactions are ignored. Among the most popular of these are a group of closely related *pseudobinary* relations known as the *Nernst-Planck equations*, and a set of simplified ternary relations formulated by Kedem and Katchalsky (1958) and others. These two types of flux equations are introduced briefly here.

(a) *THE NERNST-PLANCK EQUATIONS*

This term is used for a variety of flux equations similar in form to the Stefan-Maxwell relations but containing fewer diffusion coefficients. Unfortunately, various authors use slightly different forms for these equations, and no one choice can be considered standard.

We can, however, at least begin unambiguously by rewriting Eq. 1.2.21 in the form

$$\mathbf{N}_i=-c\mathfrak{D}_{im}\mathbf{d}_i+\alpha_i x_i \tag{1.6.1}$$

where

$$\mathfrak{D}_{im} = \left( \sum_{\substack{j=1 \\ \neq i}}^{n} \frac{x_j}{\mathcal{D}_{ij}} \right)^{-1}$$

is the effective binary diffusivity of $i$ through the mixture, and

$$\boldsymbol{\alpha}_i = \left( \sum_{\substack{j=1 \\ \neq i}}^{n} \frac{\mathbf{N}_j}{\mathcal{D}_{ij}} \right) \Bigg/ \left( \sum_{\substack{j=1 \\ \neq i}} \frac{x_j}{\mathcal{D}_{ij}} \right)$$

This expression contains no simplifications as yet, but it is readily simplified for many limiting situations of biological interest. It may be easily seen, for example, that it reduces to Fick's first law for a binary system.

This equation is most useful in dilute aqueous solutions where only the solvent, species W, is in large concentration. Thus, if

$$x_i \ll x_{\mathrm{W}} (i = 1, 2, \ldots, n; \neq \mathrm{W})$$

and

$$\mathbf{N}_i = O(\mathbf{N}_{\mathrm{W}})$$

then

$$\dot{\gamma}_i \approx \text{constant}; \qquad \alpha_i \doteq N_{\mathrm{W}}$$

and

$$\mathbf{N}_i \sim -c\, \mathcal{D}_{i\mathrm{W}} \left( \nabla x_i + \frac{x_i \nu_i}{RT} \mathfrak{F} \nabla \phi + \underline{\frac{\phi_i - \omega_i}{cRT} \nabla p} \right) + x_i \mathbf{N}_{\mathrm{W}} \qquad (1.6.2)$$

Note that $\mathbf{N}_{\mathrm{W}}$ can be approximated under these conditions as $c\mathbf{v}$. Equation 1.6.2 is commonly used for diffusional processes in dilute aqueous solutions, but it must always be used with care. That is, the validity of each approximation used to obtain it should be reviewed for every application.

Particular care should be used in the description of electrokinetic phenomena in charged membranes. The pseudobinary approach can be used effectively for correlating data for a single operation, such as isobaric electrodialysis. However, it has been shown by Scattergood and Lightfoot (1965) that it is not always reliable for predicting the course of one electrokinetic process in terms of pseudobinary coefficients determined by the experimental observation of another.

We now compare the utility and convenience of the pseudobinary and exact formulations by using the Nernst-Planck equations to rework some of our earlier examples. We also add one new example for membrane transport. Further applications of the pseudobinary approximations are given in the problems.

### *EX. 1.6.1. SALT DIFFUSION AND JUNCTION POTENTIALS*

Repeat Ex. 1.3.1 and 1.3.2 using the pseudobinary equations. Assume a physiological level of salt concentration.

SOLUTION

It is quite reasonable to use Eq. 1.6.2 here—both because physiological salt solutions are quite dilute, typically 0.15 to 0.2 $N$, and because water will tend to flow hydraulically to balance salt motion in the absence of forced or free convection. It is thus unlikely that solute fluxes will be dominant over the water flux. We may further neglect the pressure diffusion for almost all situations normally encountered.

We may the proceed much as in the previous example and with similar results. These are:

$$\mathbf{N}_S = -c\left[2\frac{Ð_{MW}Ð_{XW}}{Ð_{MW}+Ð_{XW}}\right]\nabla x_S + x_S\mathbf{N}_W \tag{1}$$

$$\frac{d\Phi}{dx_S} = -\frac{RT}{x_S}\left(\frac{Ð_{MW}Ð_{XW}}{Ð_{MW}+Ð_{XW}}\right) \tag{2}$$

This is a reasonable approximation to the correct answer obtained above, but it may be noted that:

1. The activity-coefficient correction has been lost.
2. It is not clear from the procedure followed that the diffusivity $Ð_{M^+X^-}$ has no effect on diffusional behavior.
3. The water flux $\mathbf{N}_W$ alone, rather than $\mathbf{N}_W+\mathbf{N}_S$, appears in the convective term.

It is thus clear that the use of the Nernst-Planck equations carries a price. For this particular situation the correct final result could have been obtained by approximating the $\alpha_i$ as $\mathbf{N}_W+\mathbf{N}_S$. There is, however, no general justification for doing this.

### *EX. 1.6.2. CONDUCTIVITY OF A 1, 1-ELECTROLYTE*

Develop an expression for the conductivity of a well-stirred solution of the salt $M^+X^-$ in water. Again assume the stirring prevents the development of any concentration gradients, but now use the pseudobinary Nernst-Planck approximations.

SOLUTION

These equations now take the very simple form

$$\mathbf{N}_{\mathrm{M}^+} \doteq -cD_{\mathrm{MW}}\left(x_{\mathrm{S}}\frac{\mathfrak{F}}{RT}\right)\nabla\phi + x_{\mathrm{S}}\mathbf{N}_{\mathrm{W}} \tag{1}$$

$$\mathbf{N}_{\mathrm{X}^-} \doteq +cD_{\mathrm{XW}}\left(x_{\mathrm{S}}\frac{\mathfrak{F}}{RT}\right)\nabla\phi + x_{\mathrm{S}}\mathbf{N}_{\mathrm{W}} \tag{2}$$

if we make the same approximations as in the last example. It then follows directly that

$$\kappa \doteq \frac{c_{\mathrm{S}}\,\mathfrak{F}^2}{RT}\left(D_{\mathrm{MW}} + D_{\mathrm{XW}}\right) \tag{3}$$

which differs from the correct result in the complete neglect of ionic interactions. This neglect leads to quite appreciable errors even at physiological salt concentrations.

### *EX. 1.6.3. DIFFUSION AND ELECTRODIFFUSION IN CHARGED MEMBRANES*

Develop an analog of Eq. 1.6.2 for the diffusion of a single salt through charged membranes in the absence of a hydraulic pressure drop. Consider the conditions of Ex. 1.5.3 as representative.

SOLUTION

One must proceed with caution in this much more complex situation, and the choice of simplifications will ultimately be dictated by experience. We therefore return to this problem in Section 3, where we take a more detailed look at membrane transport.

The usual practice is to assume that solute-membrane interactions dominate for the ions and to write

$$\mathbf{N}_i = -c\mathfrak{D}_{im}\mathbf{d}_i \qquad (i = \text{counter-ion or co-ion}) \tag{1}$$

The water flux is normally small in the absence of hydraulic flow (see below), and $x_i \ll 1$. Thus convection is normally secondary. Water flux in the absence of a pressure drop is usually ignored in pseudobinary treatments. It can, however, be approximated reasonably well by setting $\mathbf{d}_W = 0$ and neglecting interaction with the co-ion. One thus obtains

$$\frac{N_W}{N_M} \doteq \frac{x_W}{x_M + x_R(\text{Đ}_{WM}/\text{Đ}_{WR})} \tag{2a}$$

or

$$\frac{v_W}{v_M} \doteq \frac{x_M}{x_M + x_R(\text{Đ}_{WM}/\text{Đ}_{WR})} \tag{2b}$$

The system behavior is considerably more complicated in nonisobaric systems; discussion of this situation is therefore deferred to Section 3.

### (b) *THE KEDEM-KATCHALSKY RELATIONS FOR DESCRIBING MEMBRANE TRANSPORT*

In 1958 Kedem and Katchalsky proposed an approximate pair of integrated flux relations for describing solute and water transport across membranes. These have since become very popular with biologists, in part because the parameters appearing in them are directly measurable, and many of the available transport data are reported in terms of the four Kedem-Katchalsky transport coefficients, ($L_p$, $L_{pD}$, $L_{Dp}$, and $L_D$) and the reflection coefficients $\sigma$ and permeabilities $\omega$ derived from them.

These equations are strictly applicable only to thermodynamically ideal binary or pseudobinary external solutions, and they are normally written as:

$$-J_v = L_p \Delta p + L_{pD} RT \Delta c_S \tag{1.6.3}$$

$$-J_D = L_{Dp} \Delta p + L_D RT \Delta c_S \tag{1.6.4}$$

with

$$L_{Dp} \doteq L_{pD} \tag{1.6.5}$$

where $\Delta p$ and $\Delta c_i$ are the increases in pressure and concentration across the membrane, in the direction taken as positive for the fluxes. The subscript S refers to the solute. The $L$'s are the phenomenological coefficients taking the place of our $\mathcal{D}_{ij}$, and the fluxes $J_v$ and $J_D$ are defined by

$$J_v = \overline{V}_S N_S + \overline{V}_W N_W$$

$$= \left(c_S \overline{V}_S\right) v_S + \left(c_W \overline{V}_W\right) v_W \tag{1.6.6}$$

$$J_D = (c_S)_{\ln}^{-1} N_S - c_W^{-1} N_W \tag{1.6.7}$$

$$= (c_S / c_{S\ln}) v_S - v_W \tag{1.6.8}$$

Here $N_S$ and $N_W$ are the fluxes of solute and water across the membrane. Here $J_v$, the volumetric flux, is known as the *total volume flow*, and $J_D$ is known as the exchange flow. The quantities $\overline{V}_S$, $\overline{V}_W$, and $c_W$ are undefined average values in the *external* phase. The term $c_{S\ln}$ represents the logarithmic mean of terminal solute concentrations.

In addition to the four Kedem-Katchalsky coefficients, two others closely related to them have appeared frequently in the biological literature. These are the *reflection coefficient* $\sigma$ introduced by Staverman in 1952 and defined by

$$\sigma = -\frac{L_{pD}}{L_p} \tag{1.6.9}$$

and the permeability coefficient $\omega$ defined by

$$\omega = \left(L_D - \sigma^2 L_p\right) c_{S\ln} \tag{1.6.10}$$

It is common practice to characterize membrane transport behavior in terms of $L_p$, $\sigma$, and $\omega$ rather than $L_p$, $L_{pD}$, and $L_D$, because the former set of coefficients can more easily be related to the transport characteristics of greatest interest:

(i) The quantity $L_p$ directly relates the volumetric flow rate through the membrane at zero concentration difference to the hydrostatic pressure difference, and hence is a measure of hydraulic permeability.

(ii) The coefficient $\sigma$ can be considered as a measure of the membrane permselectivity, inasmuch as one can write

$$\sigma = -\left.\frac{J_D}{J_v}\right|_{\nabla c_S = 0} \tag{1.6.11}$$

If under these circumstances no solute flows across the membrane,

$$\sigma = \left(\bar{V}_W c_W\right)^{-1} \doteq 1$$

(since the volume fraction of solute is considered to be small). If, on the other hand, the flow is completely nonselective ($V_S = V_W$), then $J_D = 0$ and hence

$$\sigma = 0$$

Then $\sigma$ varies from zero for a completely nonselective membrane to (approximately) unity for an ideally selective one. It is of course also possible that the membrane is more permeable to solute than to water. For such a case, $\sigma > 1$.

(iii) $\omega$ can be considered as a measure of the membrane diffusional permeability for the solute S. Since for dilute solutions $c_W \bar{V}_W$ is approximately unity and $c_S \bar{V}_S$ is small relative to unity:

$$(J_V + J_D)c_{S\ln} = N_S\left(c_{S\ln}\bar{V}_S + 1\right) + N_W\left(\bar{V}_W - \frac{1}{c_W}\right)c_{S\ln} \doteq N_S$$

It follows on eliminating pressure from Eqs. 3 to 5 that

$$N_S \doteq (-\sigma)J_V c_{S\ln} + \mu \Delta c_S$$

so that $c\omega$ is a type of effective diffusivity for the solute through the membrane.

It remains to relate these quantities to the $\mathcal{D}_{ij}$ introduced earlier. We consider for this purpose only dilute ideal solutions with small fractional changes in concentration across the membrane. For this special case* one may write

$$\sigma = \frac{1 - K\alpha}{\phi_W(1 + K\alpha\phi_S/\phi_W)} \tag{1.6.12}$$

* It is only under these limiting conditions the Kedem-Katchalsky coefficients are consistent with the requirements of irreversible thermodynamics. This is shown rigorously in section 3. The limitations of these equations are frequently, but by no means always, of secondary importance in analyzing biological data; they can be quite serious in analyzing extracorporeal equipment.

where

$$K = x_S c_W / x_W c_S,$$

$$\alpha = (\phi_S \bar{r}_W + \phi_W)/(\phi_S + \phi_W \bar{r}_S),$$

$$\bar{r}_W = 1 + x_m \text{Đ}_{WS}/x_S \text{Đ}_{Wm},$$

$$\bar{r}_S = 1 + x_m \text{Đ}_{WS}/x_W \text{Đ}_{Sm}.$$

Here $x_i$ = *membrane*-phase mole fractions,
$c_i$ = *external*-phase molar concentrations,
$\phi_i = c_i \overline{V}_i$ = *external*-phase volume fractions.

To illustrate the nature of this relation, we consider once again the special cases of zero and ideal selectivity. For the membrane to be completely impermeable, either the solute must be totally excluded ($K=0$) or $\text{Đ}_{Sm}$ must be zero ($\alpha=0$). In either case $K\alpha=0$ and $\sigma = 1/\phi_W$ as stated above. For there to be no selectivity it is necessary that $K\alpha$ equal unity, so that

$$1 = \frac{x_S c_W}{x_W c_S}\left[\frac{c_S \overline{V}_S (1 + x_m D_{WS}/x_S D_{Wm}) + c_W \overline{V}_W}{c_S \overline{V}_S + (1 + x_m D_{WS}/x_W D_{Sm}) c_W \overline{V}_W}\right] \tag{1.6.13}$$

$$= \frac{x_S c_W}{x_W c_S}\left[\frac{1 + (x_m \overline{V}_S / D_{Wm})(c_S/x_S)}{1 + (x_m \overline{V}_W / D_{Sm})(c_W/x_W)}\right] \tag{1.6.14}$$

The simplest way in which this relation can be satisfied is if

$$K = \frac{x_S c_W}{x_W c_S} = 1$$

$$\frac{x_m \overline{V}_S}{\text{Đ}_{Wm}} = \frac{x_m \overline{V}_W}{\text{Đ}_{Sm}}$$

We arrive at this same pair of requirements later, in connection with ultrafiltration, by a quite different route. Where these conditions are met, $\sigma$ is identically zero.

It may similarly be shown that for *dilute solutions* with *small concentration and pressure differences* and a membrane thickness of $\delta$:

$$\omega = \frac{\phi_W/\delta}{\dfrac{x_W}{Ð_{SW}} + \dfrac{x_m}{Ð_{Sm}}} \tag{1.6.15}$$

The relation between these quantities and the rather widely used frictional coefficients of Spiegler can be obtained from the definitions given in connection with ultrafiltration.

## BIBLIOGRAPHY

Anderson, N. G., *Anal. Biochem.*, **23**, 72–83 (1968); see also "The Development of Zonal Centrifuges and Ancillary Systems for Tissue Fractionation," N. G. Anderson, Ed., *J. Natl. Cancer Inst. Monogr.*, **21** (1966).

Bird, R. B., C. F. Curtiss, and J. O. Hirschfelder, *Chem. Eng. Prog. Symp. Ser.*, **51**, No. 16, 69–85 (1955); see especially p. 77, Eqs. 2.15 and 2.16.

Cole, K. S., and J. W. Moore, *J. Gen. Physiol.*, **43**, 971 (1960).

Craig, L. C., *Science*, **144**, 1093 (1964).

Craig, L. C., J. D. Fisher, and T. P. King, *Biochemistry*, **4**, 311 (1965). de Groot, S. R., and P. Mazur, *Non-Equilibrium Thermodynamics*, North Holland (1962).

Evans, R. B., III, G. M. Watson, and E. A. Mason, *J. Chem. Phys.*, **16**, 1894–1902 (1962).

Fitts, D. D., *Non-Equilibrium Thermodynamics*, McGraw-Hill (1962).

Gregor, H. P., *J. Am. Chem. Soc.*, **70**, 1923 (1948); **73**, 642 (1951).

Harned, H. S., and B. B. Owen, *The Physical Chemistry of Electrolytic Solutions*, Reinhold (1950).

Helfferich, Friedrich, *Ion Exchange*, McGraw-Hill (1962).

Henderson, P., *Z. Phys. Chem.*, **59**, 118 (1907); **63**, 325 (1908); see also Hermans (1938).

Hermans, J. J., *Recent Trav. Chim.*, **57**, 1373 (1938).

Hirschfelder, J. O., C. F. Curtiss, and R. B. Bird, *Molecular Theory of Gases and Liquids*, Wiley (1954).

Katchalsky, A., and P. F. Curran, *Non-Equilibrium Thermodynamics in Biophysics*, Harv. Univ. Press (1965).

Kedem, O., and A. Katchalsky, *Biochem. Biophys. Acta*, **27**, 229 (1958).

Lavallée, Schanne, and Hébert, Eds., *Glass Micro-electrodes*, Wiley (1969).

Lightfoot, E. N., E. L. Cussler, Jr., and R. L. Rettig, *Am. Inst. Chem. Eng. J.*, **8**, 708 (1962).

Newman, J. S., *Electrochemical Systems*, Prentice-Hall (1973).

Onsager, Lars, *Phys. Rev.*, **37**, 405 (1931).

Onsager, Lars, *ibid.*, **38**, 2265 (1931).

Onsager, Lars, *Ann. N.Y. Acad. Sci.*, **46**, 21 (1945).

Robinson, R. A., and R. H. Stokes, *Electrolyte Solutions*, Second ed. (revised), Butterworths (1965).

Scattergood, E. M., and E. N. Lightfoot, "Suitability of the Nernst-Planck Equations for Describing Electrokinetic Phenomena," *Am. Inst. Chem. Eng. J.*, **110**, 175 (1965).

Scattergood, E. M., and E. N. Lightfoot, "Diffusional Interactions in an Ion-exchange Membrane," *Trans. Faraday Soc.*, **64**, No. 544, part 4. 1135–1146 (1968).

Spiegler, K.S., *Trans. Faraday Soc.*, **54**, 1409 (1958).

Stein, W. D., *The Movement of Molecules Across Cell Membranes*, Academic Press (1967).

Tanford, Charles, *Physical Chemistry of Macromolecules*, Wiley (1959).

## 2. DIFFUSION IN FREE SOLUTION

Here we build on the foundation provided by the last chapter to discuss some representative diffusional problems of biological interest. We limit ourselves to situations in which convection (bulk fluid motion) plays a secondary role and to systems free of mechanical constraints (i.e., to free solutions). Convective transport and transport through membranes are discussed in the next three sections.

The number of mass-transfer problems encountered in biology is so large that no attempt can be made to provide complete coverage. Rather we must restrict ourselves to a small number of representative examples, to illustrate most commonly encountered general behavior and to emphasize typical magnitudes of key diffusional parameters.

We begin with a brief note on the estimation of transport properties and follow this with discussions of increasing complexity: binary diffusion in nonflow systems, one-dimensional convective transport (film theory), diffusion with chemical reaction, and finally multicomponent diffusion.

### 2.1 Note on the Estimation of Diffusion Coefficients

Biological fluids range in complexity from low-density gases through dilute electrolyte solutions to concentrated protein sols and gels, and finally to heterogeneous media such as blood. Furthermore, the volumes of fluid encountered range down to cellular scales of cubic microns. In these latter cases in particular, anomalous diffusional behavior can be expected in the neighborhood of membranes and polymer chains. It is therefore difficult to summarize the diffusional behavior of biological systems, and no such attempt will be made here. Rather, we content ourselves with brief references to the literature for some particularly important systems.

The case of pulmonary gases is the simplest, and it can be described to a quite satisfactory degree of accuracy by the original Stefan-Maxwell equations. Here the $Ð_{ij}$ reduce simply to the binary diffusivities $\mathfrak{D}_{ij}$, which are composition independent to the normally used degree of approximation (see *Tr. Ph.*, Chapter 18). Values of these $\mathfrak{D}_{ij}$ are readily available both for the physiological gases $O_2$, $CO_2$, $N_2$, and $H_2O$, and for those used during anaesthesia and diagnostic procedures (*Tr. Ph.* Chapter 16). Furthermore, these mixtures are thermodynamically ideal and contain no charged species. Therefore

$$\mathbf{d}_i = \nabla x_i \tag{2.1.1}$$

and

$$\nabla x_i = \sum_{\substack{j=1 \\ j \neq i}}^{n} \frac{\mathbf{N}_j x_i - \mathbf{N}_i x_j}{c\mathfrak{D}_{ij}} \qquad (i = 1, 2, \ldots, n) \tag{2.1.2}$$

which are the classic Stefan-Maxwell equations. For multicomponent systems the nonlinearity of Eqs. 2.1.2 is awkward, and it is often desirable to use a linearized set of approximate relations. A number of essentially equivalent procedures has been developed (see, e.g., Cussler and Lightfoot, 1963; Hirschfelder, Curtiss, and Bird, 1954; Stewart and Prober, 1964; Toor, 1964). These have been found quite satisfactory for the calculation of mass transfer rates; their use is illustrated in the discussion of microelectrodes below. It should be noted that diffusional interactions in multicomponent gas mixtures tend to be substantially larger than for liquids and cannot be safely ignored.

Dilute electrolyte solutions are next simplest and are also reasonably well understood. Their most important constituents from a diffusional standpoint are dissolved gases, small nonelectrolytes and ions, and globular proteins. These solutions can usually be considered pseudo-binary, so that only the solute-water diffusivity $Ð_{i\mathrm{W}}$, is needed. For all small solutes these $Ð_{i\mathrm{W}}$ are of the order of $10^{-5}$ cm$^2$/sec.

The solubilities and diffusivities of dissolved gases, primarily $O_2$ and $CO_2$, are readily available in a wide variety of references and do not need discussion here. It must, however, be kept in mind that $CO_2$ reacts with water, hydroxyl ions, and hemoglobin, and that the hemoglobin reaction depends on the oxygen partial pressure (degree of oxygenation of hemoglobin). Both the equilibria and kinetics of these reactions are complex, but have been widely investigated. The respiratory section of the *Handbook of Physiology* is a good general reference. We return to this subject in some detail in our examples.

Electrolyte transport is an exceedingly complex subject but can be handled very simply, with usually satisfactory results for the relatively dilute electrolytes in biological solutions, by using the Nernst-Planck equations. The ion-water diffusivities $Ð_{iW}$ can be estimated from limiting ionic conductances as described in the last chapter. More extensive and authoritative discussions of electrolyte transport are given by Chapman (1969), by Newman (1973), and by Robinson and Stokes (1965); a formal treatment of electrolyte transport is provided in the examples.

The diffusivity of proteins relative to water has been extensively studied by physical chemists in past years, to help in the determination of protein structure. Much of this information is available through such standard references as Tanford (1959), but unfortunately it has largely been confined to dilute solutions.

Diffusion in particulate suspensions is complicated both by the two-phase nature of the diffusing system and by the motion of the particles relative to the suspending solvent. The case of diffusion in blood is particularly important and will be discussed at several points later in these notes; it is briefly reviewed by Spaeth in his discussion of membrane oxygenators. The usual procedure is to treat blood as a homogeneous phase, using an appropriate set of volume-averaged diffusion coefficients to describe solute transport.

Pseudohomeogeneous analyses fail, however, under two circumstances: (1) If critical system dimensions become comparable to the inter-particle spacing, and (2) If the fluid shear rates are high relative to the diffusion fluxes. In blood oxygenation equipment, for example, concentration boundary-layer thicknesses are not always greatly larger than the typical cell spacing (which is of the order of 5 $\mu$). It is therefore appropriate to take another hard look at the pseudohomogeneous diffusion models now being used to size oxygenators. In addition, it has been pointed out by Blackshear and by Keller and their associates that the rotation of red cells in a shear field can cause an appreciable convective transport of all solutes. It appears on the basis of their experiments that such *microconvective* transport can be appreciable, but that it is not likely to dominate concurrent diffusion for small solutes.*

We now conclude this subsection with a brief example involving the experimental determination of protein diffusivities.

* Microconvective transport is, however, important in the transport of formed elements (as has been shown for platelets by Turitto, Benis, and Leonard (1972) and by Grabowski, Friedman, and Leonard (1972)), and it may be significant for large molecules such as proteins.

*EX. 2.1.1 THE MEASUREMENT AND INTERPRETATION OF PROTEIN DIFFUSION COEFFICIENTS*

Compare mutual and tracer ("self-") diffusion coefficients as measured in a diaphragm cell.

SOLUTION

Mutual diffusion coefficients are measured by placing solutions of protein of two different concentrations on opposite sides of a porous diaphragm. Equilibration then takes place* by the (binary) interdiffusion of protein and water and is described locally by

$$x_P \nabla \ln a_P = \frac{\mathbf{N}_W x_P - \mathbf{N}_P x_W}{c\,Đ_{PW}} \tag{1}$$

which can be rearranged, as discussed in Chapter 1, to give:

$$\mathbf{N}_P = -c\,Đ_{PW}\left(1 + \frac{\partial \ln \gamma_P}{\partial \ln x_P}\right)\nabla x_P + x_P(\mathbf{N}_W + \mathbf{N}_P) \tag{2}$$

$$\equiv -c\mathfrak{D}_{PW}\nabla x_P + x_P(\mathbf{N}_W + \mathbf{N}_P) \tag{3}$$

where $\mathfrak{D}_{PW}$, known as the *mutual diffusion coefficient*, is just the binary diffusivity of Fick's first law. In a diaphragm cell it is normally possible to assume steady one-dimensional flow and to account for the presence of the diaphragm matrix by means of an empirical void-fraction–tortuosity correction $\epsilon/\tau$. The correction is the same for all systems and can be determined by calibration with a system of known diffusivity.

It remains to relate $N_P$ and $N_W$, and this can be done by assuming the *volumetric* flux to be zero; that is, by neglecting volume changes on mixing:

$$N_W \overline{V}_W + N_P \overline{V}_P = 0 \tag{4}$$

where $\overline{V}_i$ is the partial molal volume of species $i$. Equation 3 may then be rearranged to give

$$\mathbf{N}_p = -\frac{c(\epsilon/\tau)\mathfrak{D}_{PW}\nabla x_P}{1 - x_P\left(1 - \overline{V}_P/\overline{V}_W\right)} \tag{5}$$

* In both situations considered here, diffusional interaction with diaphragm walls is neglected. Such interaction can, however, be appreciable, especially if the pore size approaches molecular dimensions. It is always necessary to check for such interaction if reliable results are desired.

The denominator on the right is just*

$$\frac{\left(x_W\overline{V}_W + x_P\overline{V}_P\right)}{V_W} = \left(c\overline{V}_W\right)^{-1} \tag{6}$$

We may then write Eq. 5 as

$$\mathbf{N}_P = -\frac{\epsilon}{\tau}\mathfrak{D}_{PW}\left(c^2\overline{V}_W\nabla x_P\right) \tag{7}$$

which is equivalent to (see Prob. 2.□)

$$\mathbf{N}_P = -\frac{\epsilon}{\tau}\mathfrak{D}_{PW}\nabla c_P \tag{8}$$

Then for the conditions of a diaphragm cell

$$\mathfrak{D}_{PW} = Ð_{PW}\left(1 + \frac{\partial \ln \alpha_P}{\partial \ln x_P}\right) \tag{9a}$$

$$\doteq \frac{N_P\delta(\tau/\epsilon)}{\Delta c_P} \tag{9b}$$

where $\delta$ is the diaphragm thickness and $\Delta c_p$ is the protein concentration difference across the diaphragm.

The next problem to be faced is the concentration dependence of $\gamma_P$ and $Ð_{PW}$. Standard extrapolation techniques (referred to, e.g., by Keller) are, however, available for obtaining *differential coefficients*, that is, those for zero concentration difference across the diaphragm, from the *integral* coefficients obtained from Eq. 9*b*.

Tracer diffusion coefficients are measured by placing equal protein concentrations on each side of the membrane but using radioactively tagged protein on one side. Since proteins have a very high molecular weight, the properties of the tagged and normal proteins, P* and P, can be considered identical. For this situation activity coefficients are constant and

$$x_P\nabla \ln a_P = \nabla x_P = -\nabla x_{P*} \tag{10}$$

$$N_P = -N_{P*} \tag{11}$$

$$N_W = 0 \tag{12}$$

* Here advantage is taken of the facts that the *volume fraction* $\phi_i$ of any species $i$ is just $c_i\overline{V}_i$ and that the volume fractions sum to unity.

The Stefan-Maxwell equation for protein then becomes

$$\nabla x_{\mathrm{P}} = \frac{-N_{\mathrm{P}} x_{\mathrm{W}}}{c D_{\mathrm{PW}}} + \frac{N_{\mathrm{P*}} x_{\mathrm{P}} - N_{\mathrm{P}} x_{\mathrm{P*}}}{c D_{\mathrm{PP*}}} \tag{13}$$

or

$$\nabla x_{\mathrm{P}} = -N_{\mathrm{P}} \left( \frac{x_{\mathrm{W}}}{c Ð_{\mathrm{PW}}} + \frac{x_{\mathrm{P}} + x_{\mathrm{P*}}}{Ð_{\mathrm{PP*}}} \right) \tag{14}$$

Equation 11 may be integrated across the diaphragm, for pseudosteady one-dimensional operation, to obtain

$$N_{\mathrm{P}} = -c \mathfrak{D}_{\mathrm{PP*}} \frac{\epsilon}{\tau} \frac{\Delta x_{\mathrm{P}}}{\delta} \tag{15}$$

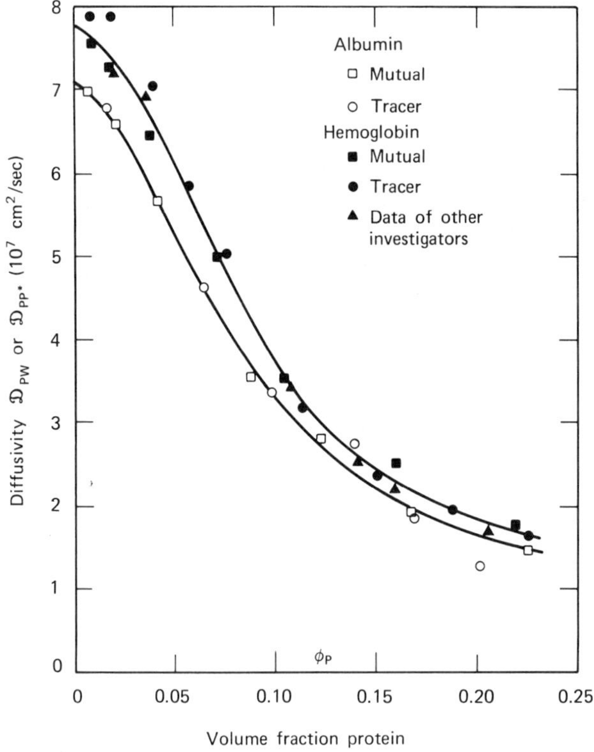


**Figure 2.1.1.** Mutual and tracer diffusion coefficients for human hemoglobin and bovine serum albumin, from Keller, Canales, and Yum (1971).

where $\delta$ is the diaphragm thickness. The observed diffusion coefficient is now the *tracer* diffusion coefficient

$$\mathfrak{D}_{PP*}=\left(\frac{x_W}{Đ_{PW}}+\frac{x_{P,tot}}{Đ_{PP*}}\right)^{-1} \qquad (16)$$

This is thus inherently quite a different quantity from the mutual diffusion coefficient defined by Eq. 9*a*:

1. It contains no thermodynamic correction.
2. However, it contains a protein-protein interaction term not appearing in the mutual coefficient.

It is simpler to measure than $\mathfrak{D}_{PW}$ because there is no concentration dependence under the measurement conditions. One thus obtains the equivalent of a differential coefficient directly.

Interestingly enough, Keller et al. (1971) find no measurable difference between $\mathfrak{D}_{PW}$ and $\mathfrak{D}_{PP*}$, for either human hemoglobin or bovine serum albumin, over a very wide range of protein concentration. This is shown by Fig. 2.1.1 which also shows that diffusivities decrease markedly with increasing protein concentration.

This situation does not occur for small solutes, where the two diffusivities differ substantially at even moderate concentration. This suggests that diffusion in concentrated protein solutions is not a simple process and provides a potentially important clue to the nature of these solutions. It appears to be an important area for further research.

## 2.2 Binary Concentration Diffusion

Since elementary treatments of binary diffusion are widely available, we limit ourselves here to two representative examples of biological interest. These build on the discussions of *Tr. Ph.*, Chapters 17 through 19, and we shall assume these to be familiar to the reader.

In our first example we review steady one-dimensional diffusion during gastric secretion of HCl and put major emphasis on the determination of species flux ratios. The water flux is regulated here by the osmotic permeability of secretory tissues to water and the rate of secretion of osmotically active substances. Here, as elsewhere in the body, water transport is passive.

In the second example we discuss diffusional behavior in glass microelectrodes. This example illustrates the fast response of small systems to concentration perturbations and lays the groundwork for later discussions of electrode junction potentials. Reference is also made to an extensive review of glass electrodes, which are widely important in medicine and biology.

### *EX. 2.2.1 INTERDIFFUSION OF* HCl *AND WATER IN STOMACH PITS*

Describe the interaction of diffusion and convection for the idealized gastric HCl secretion of Fig. 2.2.1. Assume that secretion takes place at the base of cylindrical pits, as pure HCl, and that water follows the HCl to maintain a close approach to osmotic equilibrium with the blood flowing through the surrounding tissue. Consider a special experimental situation in which the stomach itself is filled with essentially pure water, and treat the HCl secretion rate as known.

SOLUTION

The diffusion path here is clearly quite complex, but the characteristics of the system can be illustrated by assuming steady one-dimensional transfer. Then, since the system is binary, its diffusional behavior is described by

$$N_A = N_{A0}, \quad \text{a known constant} \tag{1}$$

$$N_W = N_{W0}, \quad \text{an unknown constant} \tag{2}$$

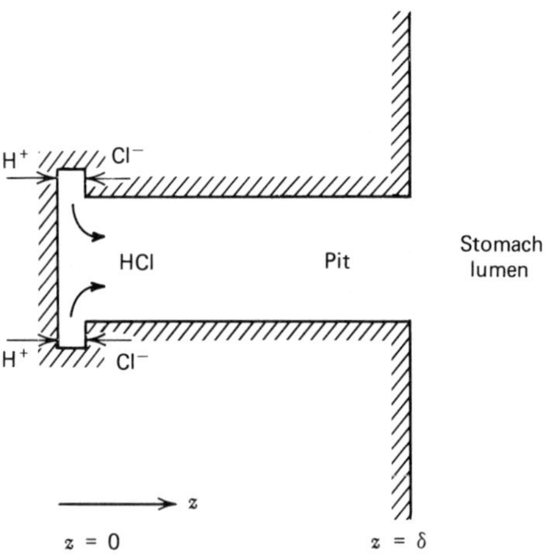


**Figure 2.2.1.** The Rehm-Schlesinger-Dennis model for diffusion during gastric HCl secretion.

$$N_{A0} = -c\mathfrak{D}_{AW}\frac{dx_A}{dz} + x_A(N_{A0} + N_{W0}) \tag{3}$$

where A refers to the HCl (acid) and W refers to water. The acid flux $N_{A0}$ is specified by the secretion rate, assumed to be known; the water flux is to be determined.

The problem statement is completed by the boundary conditions

$$\text{At } z=0, \qquad c_A = c_{A0} \tag{4}$$

$$\text{At } z=\delta, \qquad c_A = 0 \tag{5}$$

Here $c_{A0}$ is that concentration required to produce a solution isotonic with blood.

Note that we have not yet specified the flux ratio $N_{A0}/N_{W0}$; this must be determined to satisfy Eq. 2 for the acid flux and the boundary concentrations (Eqs. 4 and 5). To make this determination we may integrate Eq. 3 to obtain

$$\frac{N_{A0}}{N_{A0} - x_{A0}(N_{A0} + N_{W0})} = e^{(N_{A0}+N_{W0})\delta/c\mathfrak{D}_{AW}} \tag{6}$$

where $x_{A0}$ is the mole fraction of acid at $z=0$. Note that $x_A(\delta)=0$ for our situation. Equation 6 gives an implicit relation for the water flux which may be simplified.

For physiological conditions $N_{W0}$ will be considerably larger than $N_{A0}$, since the acid solution secreted is mostly water. Then

$$N_{A0} + N_{W0} \doteq N_{W0} = c_W v_W \doteq cv$$

and

$$N_{A0}/(N_{A0} - c_{A0}v) \doteq e^{v\delta/\mathfrak{D}_{AW}}$$

This expression was first developed by Rehm, Schesinger, and Dennis (1953), in their analysis of gastric secretion. It shows that there are both a strong convective contribution to acid transport and substantial water movement for the conditions considered here.

Water movement here (as generally in living organisms) is, however, passive and determined by osmotic gradients. It can be reversed in this problem if the osmolality of the stomach contents is sufficiently greater than that of the blood. It is also of interest that the water permeability of the secretory tissues is extremely high. We encounter similar high water permeabilities in the kidney.

Convective solute transport may also be important for other substances; see, for example, the discussion of anomalous solute drag in Prob. 4.*D*.

### *EX. 2.2.2. DIFFUSION IN THE TIP OF A MICROELECTRODE*

A glass microelectrode is initially filled with a uniform solution of a single solute (typically 3 *M* KCl). Describe diffusional behavior of the electrode on being placed in a large, initially uniform solution of the same solute at a lower concentration.*

SOLUTION

We begin by noting that microelectrodes are prepared by drawing out a small glass tube to a very fine point. This process normally produces an essentially conical shape with a very small angle of taper in the region near the tip. We may thus idealize the electrode as an infinitely long cone, if we remember that diffusional boundary layers are normally thin.

We next note that because of the small angle of taper, $\theta$, most of the diffusional resistance is within the tip as opposed to the external solution; in addition, the spherical section $r=a$ in Fig. 2.2.2 is very nearly flat across the tip. It therefore appears likely that the concentration of solute at $r=a$ is nearly uniform and equal to that in the external solution.

Our system description may then be put in the approximate form

$$\frac{\partial c_i}{\partial t} = \mathfrak{D}_{im} \frac{1}{r^2} \frac{\partial}{\partial r} r^2 \frac{\partial c_i}{\partial r} \tag{1}$$

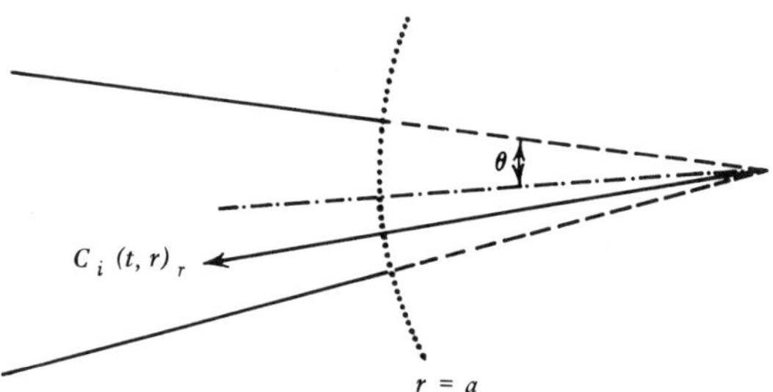


**Figure 2.2.2.** Geometry of a microelectrode tip.

* We consider the much more complex case of multicomponent solutions near the end of the chapter. The present example can, however, be used to simplify this later discussion markedly. Geometric complications are discussed in some detail by Geisler, Lightfoot, Schmidt, and Sy, and are touched on in the problems at the end of this chapter.

with

$$c_i(0,r) = c_{i\infty}, \qquad r > a \tag{2}$$

$$c_i(t,a) = c_{i0}, \qquad t > 0 \tag{3}$$

$$c_i(t,r) \to c_{i\infty}, \qquad r \to \infty,\ t \text{ finite} \tag{4}$$

where $\mathfrak{D}_{im}$ is the effective diffusivity of the solute. The concentration profile is then

$$c = \frac{1}{\eta}\left(1 - \operatorname{erf}\frac{\eta - 1}{\sqrt{4\tau}}\right). \tag{5}$$

where

$$c = (c_i - c_\infty)/(c_{i0} - c_{i\infty}), \qquad u = c\eta$$

$$\eta = r/a, \qquad \tau = t\mathfrak{D}_{im}/a^2$$

It remains to consider the orders of magnitude of the parameters to show the significance of this result. For electrodes used to probe the behavior of individual cells the following values are not unreasonable*:

$$a = 10\ \mu = 10^{-3}\ \text{cm}$$

$$\theta = 1.5°$$

$$\mathfrak{D}_{im} = 10^{-5}\ \text{cm}^2/\text{sec}$$

Note the very small angle of taper, which corresponds to a tip opening with a radius of only about $1/4\ \mu$. For these parameters an immersion time of 1 sec corresponds to

$$\tau = \frac{(1\ \text{sec})(10^{-5}\text{cm}^2)}{\text{sec}} \cdot \frac{10^6}{\text{cm}^2} = 10$$

$$\sqrt{4\tau} \doteq 6.3$$

Thus, even for relatively short times, the argument of the error function is

* Many modern electrodes are considerably smaller. Tips of the order of 1 $\mu$ are commonly used for investigating such small cells as those of heart muscle (10–15-$\mu$ diameter) or smooth muscle (5-$\mu$ diameter). (J. B. Bassingthwaighte, private communication.) These clearly will have much more rapid response times. However, double-layer and glass-porosity effects may be troublesome for such small tips.

small in the critical region near the tip. In this region we may use the relation

$$\operatorname{erf} x \sim \frac{2x}{\sqrt{\pi}} \qquad \text{as } x \to 0$$

and write

$$c = \frac{1}{\eta}\left(1 - \frac{\eta - 1}{\sqrt{\pi\tau}}\right)$$

Frequently one may assume steady operation with little loss in accuracy. We take advantage of this widely used approximation in our later discussion of multicomponent electrode potentials.

The unimportance of transient behavior can be seen even more clearly in the calculation of the rate of solute loss from the electrode:

$$-N_r\big|_{r=a} = +\mathfrak{D}_{im}\frac{\partial c}{\partial r}\bigg|_{r=a}$$

$$= \frac{c_{i0} - c_{i\infty}}{a}\mathfrak{D}_{im}\frac{\partial c}{\partial \eta}\bigg|_{\eta=1}$$

Then

$$-N_r\big|_{r=a} = \frac{\mathfrak{D}_{im}(c_{i\infty} - c_{i0})}{a}\left(1 + \frac{1}{\sqrt{\pi\tau}}\right)$$

The transient term can now be seen to disappear very rapidly.

The simple analysis given here is directly useful for electrodes immersed in large volumes of solution, for determining both salt efflux rates and concentration profiles; it has been extended by Geisler, Lightfoot, Schmidt, and Sy (1973) to boundary conditions more representative of intra-cellular conditions. It does, however, oversimplify the diffusional situation in microelectrodes, particularly for those of very small bore.

It is found that both ionic diffusion across the tip wall and selective transport along the inner wall surface can be very important, and that they can produce very complex electrokinetic effects. An example is the formation of large and as yet only incompletely understood "tip potentials." These can be as large as the order of 10 mv and can mask the behavior under study. They are strongly affected by the presence of multivalent cations which are absorbed on the glass of the electrode, and they can often be at least partially eliminated by breaking off the tip. These and many other important aspects of microelectrodes are discussed in Lavallée, Schanne, and Hébert (1969).

## 2.3. Diffusion with Chemical Reaction

In this subsection we consider several examples of diffusion with chemical reaction. We begin by discussing diffusion with instantaneous reversible reaction as an introduction to both carrier diffusion in general and the specific case of oxygenation of blood. We next consider the effect of chemical reaction on the rate of blood oxygenation in a simple but important situation. We then end the subsection with a discussion of diffusion with homogeneous zero-order reaction, which is a simple special case of very wide applicability to biological systems.

### *EX. 2.3.1 STEADY TRANSPORT OF GASES ACROSS A STAGNANT HEMOGLOBIN FILM*

Explain the difference in $O_2$ and $N_2$ transport across a stagnant film of hemoglobin solution shown in Fig. 2.3.1 (taken from the work of Keller and Friedlander, 1966). Plotted here are rates of gas transport across a hemoglobin film, under the influence of a fixed partial pressure *difference*

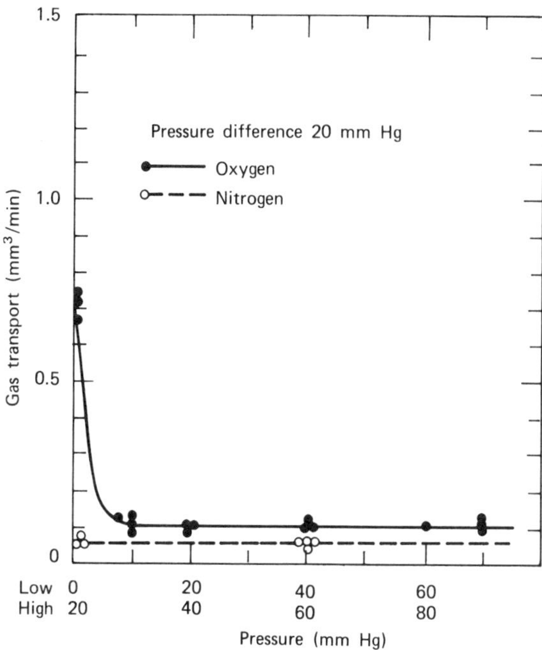


**Figure 2.3.1.** The effect of average partial pressure on transport of $O_2$ and $N_2$ through hemoglobin films (from Keller and Friedlander, 1966).

across the film, as a function of the arithmetic average of terminal partial pressures of dissolved gas.

It is easily shown that for a sparingly soluble inert gas $i$ the permeation rate is given by

$$N_i = \frac{H_i \Delta p_i \mathfrak{D}_{iB}}{\delta} = \frac{\Delta c_i \mathfrak{D}_{iB}}{\delta} \tag{1}$$

where

$N_i$ = the molar flux of species $i$,
$p_i$ = the partial pressure of $i$ in the gas at the blood-gas interface,
$c_i$ = the corresponding concentration of dissolved gas in the blood,
$H_i = c_i/p_i$, the Henry's-law constant for the gas, and
$\delta$ = film thickness

This equation predicts that pressure level will have no effect, and this prediction agrees with the data for $N_2$.

It clearly does not explain the behavior of oxygen.

SOLUTION

The additional oxygen transport results from the transport of oxygen as oxygenated hemoglobin. Oxygen combines reversibly with hemoglobin to form a series of partially oxygenated hemoglobins, each of which may diffuse in the solution. We shall simplify* this picture by assuming only one oxygenated product, oxyhemoglobin,

$$O_2 + Hb \rightleftarrows HbO_2 \tag{2}$$

with the equilibrium relation shown in Fig. 2.3.2. Then the total molar rate of $O_2$ transfer may be expressed as

$$N_{tot} = N_{O_2} + N_{HbO_2} \tag{3}$$

while a mass balance requires that

$$N_{H_2O} = 0 \tag{4}$$

$$N_{Hb} + N_{HbO_2} = 0 \tag{5}$$

Electrodiffusion and pressure diffusion may be neglected in this system, and we assume the activity coefficients to be constant.

* No appreciable error is introduced by this assumption provided that the proper dissociation relation is used. The diffusional behavior of the various hemoglobins should be very nearly the same, and activity variations will be slight. Also, as shown in *Tr. Ph.*, section 18.4, and easily verified, all the nonmoving species may be lumped with the water for diffusional purposes.

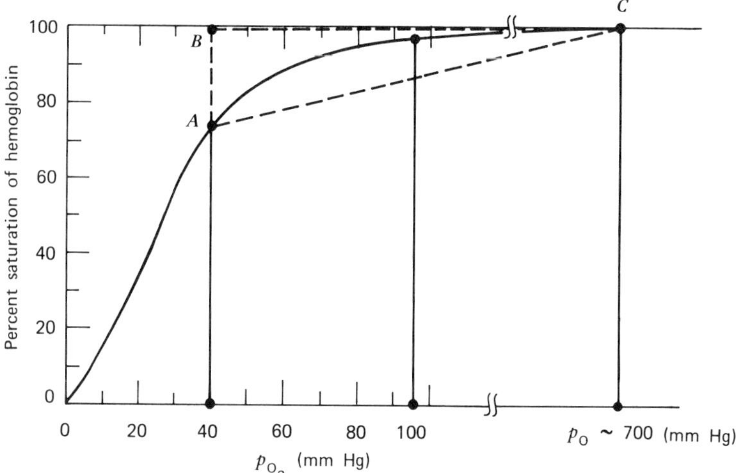


**Figure 2.3.2.** Oxygen dissociation curve for human blood ($T=38°C$, $p_{CO_2}=40$ mm Hg. The normal physiological range is between a typical venous partial pressure of 40 mm Hg and an arterial pressure of about 95 mm Hg.

We now note that the concentration of dissolved $O_2$ is extremely small, while that of $Hb+HbO_2$ is normally very large, and also that $O_2$ is very small compared with the hemoglobin molecule. Under these circumstances we can write to a first approximation.

$$Đ_{i\text{-Hb}} = Đ_{i\text{-HbO}_2} \tag{6}$$

and we can neglect the diffusional drag of $O_2$ on the water and the hemoglobins. It then follows that $c_W$ and $(c_{Hb}+c_{HbO_2})$ are very nearly constant. This justifies the asumption of constant activity coefficients and also suggests the $Đ_{ij}$ should be very nearly constant.

The Stefan-Maxwell equations may then be put in the form

$$N_{O_2} = -c\mathfrak{D}_{O_2m}\nabla x_{O_2} \tag{7}$$

with

$$\mathfrak{D}_{O_2m} = \left( \frac{x_W}{Đ_{O_2W}} + \frac{x_{Hb}+x_{HbO_2}}{Đ_{O_2Hb}} \right)^{-1}$$

and

$$N_{HbO_2} = -c\mathfrak{D}_{Hbm}\nabla x_{HbO_2} \tag{8}$$

with

$$\mathfrak{D}_{\mathrm{Hb}m} = \left( \frac{x_{\mathrm{W}}}{\text{Ð}_{\mathrm{WHb}}} + \frac{x_{\mathrm{Hb}} + x_{\mathrm{HbO}_2}}{\text{Ð}_{\mathrm{Hb\text{-}HbO}_2}} \right)^{-1}$$

The *effective binary diffusivities* $\mathfrak{D}_{im}$ should be very nearly independent of oxygen composition.

It follows that the total oxygen flux is

$$N_{\mathrm{tot}} = -c[\mathfrak{D}_{\mathrm{O}_2 m} \nabla x_{\mathrm{O}_2} + \mathfrak{D}_{\mathrm{Hb}m} \nabla x_{\mathrm{HbO}_2}] \tag{9}$$

To relate $x_{\mathrm{O}_2}$ and $x_{\mathrm{HbO}_2}$ requires kinetic data and can be quite difficult. The hemoglobin oxygenation reactions are quite rapid, however, and it is common practice to assume local equilibrium. We may then write

$$N_{\mathrm{tot}} = -c\mathfrak{D}_{\mathrm{eff}} \nabla x_{\mathrm{O}_2} \tag{10}$$

where

$$\mathfrak{D}_{\mathrm{eff}} = \mathfrak{D}_{\mathrm{O}_2 m} + \mathfrak{D}_{\mathrm{Hb}m} \frac{\partial x_{\mathrm{HbO}_2}}{\partial x_{\mathrm{O}_2}} \tag{11}$$

We find then that the transport of oxygen depends upon the shape of the hemoglobin dissociation curve and should be greatest where $\partial x_{\mathrm{HbO}_2} / \partial x_{\mathrm{O}_2}$ is largest.

The physiological importance of such *facilitated* oxygen diffusion is slight, and this is especially true in extracorporeal oxygenators ("artificial lungs"). First, physiological oxygen partial pressures are normally above 40 mm Hg, where, as can be seen from Figs. 2.3.1 and 2.3.2, the augmentation of transport is slight. Second, the hemoglobin concentration within the red cell is very large and its diffusivity correspondingly very small (see Section 2.1). Facilitated transport is thus much less effective for blood than for the relatively dilute hemoglobin solutions used by Keller and Friedlander.

From a purely practical standpoint, then, the primary value of this example is to describe what is probably the best understood case of facilitated diffusion. A similar augmentation effect is quite appreciable for $CO_2$ transport in the blood, and is primarily produced by bicarbonate ions. However, blood chemistry is exceedingly complex, and the interested reader is referred to the respiratory section of the Handbook of Physiology. A brief summary of the forms of $O_2$ and $CO_2$ existing in normal blood is provided in Table 2.3.1 by way of example and for future reference. It should be noted that oxygen and carbon dioxide each affect the equilibrium behavior of the other, through the effect of pH on

**Table 2.3.1.** Representative Oxygen and Carbon Dioxide Levels in the Blood

| | Arterial | Venous |
|---|---|---|
| Oxygen: | | |
| Tension (or partial pressure) $p_{O_2}$ mm Hg | 95 | 40 |
| Concentration, millimolar | | |
| Dissolved $O_2$[a] | 0.12 | 0.05 |
| As $HbO_2$[b] | 8.5 | 5.82 |
| Total effective | 8.6 | 5.87 |
| Carbon dioxide: | | |
| Tension (mm Hg) | 40 | 46 |
| Concentration, millimolar[c] | | |
| Dissolved $CO_2$[a] | | |
| Cells | 0.33 | 0.39 |
| Plasma | 0.71 | 0.80 |
| As $HCO_3^-$ | | |
| Cells | 4.28 | 4.41 |
| Plasma | 15.23 | 16.19 |
| As carbamino $CO_2$ (cells) | 0.97 | 1.42 |
| Total effective | 23.21 | 21.53 |

[a] The Bunsen solubility coefficients $\alpha$ for $O_2$ and $CO_2$ in human blood at 37°C are: $\alpha_{O_2}=0.023$; $\alpha_{CO_2}=0.506$ (ml STP)/(ml blood) (atm). Henry's constant for $O_2$ in normal blood at 37°C is about $7.4\times10^8$ (mm Hg) (cm$^3$)/g (mole).
[b] The molecular weight of $Hb_4$ is about 68,000, and the concentration in whole blood is about 150 g/l or about 2.2 m$M$. The molar oxygen capacity at saturation is thus about 8.8 m$M$.
[c] Based on whole blood volume.

hemoglobin dissociation and the reaction of $CO_2$ with hemoglobin. The oxygenation of blood is discussed at greater length in Section 5.2.

The assumption of local equilibrium is not justified when oxygen concentration gradients are very large, and considerable effort has been devoted to the study of nonequilibrium effects. Among the most recent papers at the time of writing are those of Bassett and Schultz (1970) and of Goddard, Schultz, and Bassett (1970). These authors, who cite much of the prior literature on oxygen transport in blood, develop a useful method for describing nonequilibrium facilitated transport by the method of matched asymptotic expansions.

The important subject of facilitated diffusion in membranes is touched on in Ex. 2.3.3.

## *EX. 2.3.2. ABSORPTION OF OXYGEN INTO STAGNANT BLOOD*

Build on the discussion of the previous example to describe the unsteady diffusion of oxygen into a deep stagnant pool of blood from pure oxygen gas, as shown in Fig. 2.3.3.

Neglect facilitated diffusion in accordance with the above discussion, and assume local equilibrium between oxygen and hemoglobin. Consider diffusion times to be sufficiently short that the absorbed oxygen does not penetrate appreciably to the bottom of the pool, but long enough that the effective penetration depth (*boundary-layer thickness*) is large relative to a red-cell diameter.

SOLUTION

For the postulated situation the blood may be considered to be a homogeneous fluid, and the diffusivity of hemoglobin may be considered zero. The effective binary diffusivity $\mathfrak{D}_{O_2m}$ will, however, differ from that in a hemoglobin solution and must be separately determined.

We may then describe oxygen transport by summing the continuity equations for oxygen and oxyhemoglobin and writing the result in the form

$$\frac{D}{Dt}(c_{O_2}+c_{HbO_2})=\mathfrak{D}_{O_2m}\nabla^2 c_{O_2} \tag{1}$$

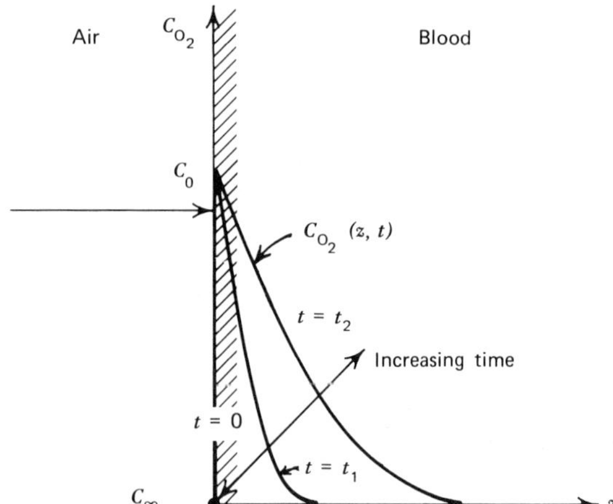


**Figure 2.3.3.** Unsteady penetration of oxygen into stagnant blood.

This is the most widely used description of oxygen transport in blood,* and we shall return to it again. Here it can be further simplified. There is no fluid motion in this system, and in the absence of end effects $c_{O_2}$ will depend only upon $z$ and $t$. The continuity equation for oxygen transfer then takes the simpler form

$$\frac{\partial}{\partial t}(c_{O_2}+c_{HbO_2})=\mathfrak{D}_{O_2m}\frac{\partial^2 c_{O_2}}{\partial z^2} \tag{2}$$

The total concentration of hemoglobin $(c_{Hb}+c_{HbO_2})$ may be considered constant; therefore $c_{HbO_2}$ depends only on $c_{O_2}$ in the absence of thermal effects and pH changes. Oxygenation is normally accompanied by significant pH changes, but these are of secondary importance and will not be considered here.† The boundary conditions are:

$$\text{At } z=0,\ c_{O_2}=c_0 \qquad \text{for } t>0 \tag{3}$$

$$\text{At } t=0,\ c_{O_2}=c_\infty \qquad \text{for } z>0 \tag{4}$$

$$\text{As } z\rightarrow\infty,\ c_{O_2}\rightarrow c_\infty \qquad \text{for } t \text{ finite} \tag{5}$$

This description is more useful than might be suspected in view of the extensive idealization, and we shall build upon it in our discussion of convective mass transfer.

It is, for example, usually possible to eliminate the restriction to pure oxygen in the gas phase, because the liquid-phase resistance normally controls. The requirement of a deep pool is not very restrictive because diffusional boundary layers in direct-contact oxygenators are typically quite thin. The requirement of no motion is quite restrictive, but as we see later, it can be removed with remarkably little change in the final expression for absorption rate.

The form of the above boundary conditions suggests a combination of variables just as for the corresponding problem without reaction (see, e.g., *Tr. Ph.*, Section 17.5). We thus postulate that

$$c_{O_2}=c_{O_2}(\zeta); \qquad \zeta=\frac{z}{\sqrt{4\mathfrak{D}_{O_2,\text{eff}}t}}$$

* See Sirs and Spaeth in Hershey (1970), as well as the Handbook of Physiology, for discussions of these points and introductions to the extensive literature in this area.

† See Sirs and Spaeth in Hershey (1970).

For convenience we also introduce the dimensionless concentration

$$\chi = \frac{c_{O_2} - c_0}{c_\infty - c_0} \tag{6}$$

Rewriting the problem statement in terms of $\chi$ and $\zeta$, we obtain

$$-2\zeta(1+m)\frac{d\chi}{d\zeta} = \frac{d^2\chi}{d\zeta^2} \tag{7}$$

with

$$m = \frac{dc_{HbO_2}}{dc_{O_2}} \tag{8}$$

and

$$\chi = 0 \qquad \text{at } \zeta = 0 \tag{9}$$

$$\chi \to 1 \qquad \text{as } \zeta \to \infty \tag{10}$$

The combination of variables is thus a success, and it remains only to integrate Eq. 7 for the above boundary conditions and the function $m(c_{O_2})$ proper for the blood and oxygenation conditions of interest. This has been done by Dindorf and Lightfoot (1973) for representative human blood, and we shall return to their analysis in the discussion of convective mass transfer.

Here, however, we look only at simpler upper and lower bounding solutions obtained by using the approximations $ABC$ and $AC$, respectively, to the dissociation curve of Fig. 2.3.2. The use of path $ABC$, for which

$$m = 0, \qquad c_\infty < c_{O_2} < c_0$$

$$m = \infty, \qquad c_{HbO_2}(c_\infty) < c_{HbO_2} < c_{HbO_2}(c_0)$$

was first suggested by Marx et al. (1960); it overestimates absorption rates, because it overestimates the oxygen-carrying capacity of the blood, but it gives very close agreement with "exact" calculations for bypass conditions. The use of path $AC$, where $m$ is constant at $M$ (equal to the change in oxyhemoglobin concentration divided by the change in dissolved oxygen concentration over the range $AC$), underestimates the absorption rate for corresponding reasons.

These two solutions can easily be applied to dissociation curves of any shape, for example to the protein binding of drugs, and we shall see that they never predict widely different mass fluxes. They are therefore frequently satisfactory for practical purposes and make more elaborate "exact" solutions unnecessary.

For path $ABC$ the calculated concentration profile is given by

$$\chi = \frac{\operatorname{erf} \zeta}{\operatorname{erf} \zeta_0} \tag{11}$$

with $\zeta_0$ given by

$$\zeta_0 e^{\zeta_0^2} \operatorname{erf} \zeta_0 = \frac{1}{\sqrt{\pi}\, M} \tag{12}$$

For path $AC$ the corresponding result is

$$\chi = \operatorname{erf}\left(\zeta\sqrt{1+M}\right) \tag{13}$$

The corresponding absorption rates are

$$N_0 \equiv N_{O_2} = -\mathfrak{D}_{O_2m} \frac{\partial c_{O_2}}{\partial z}\Big|_{z=0} \tag{14}$$

$$= (c_0 - c_\infty)\sqrt{\frac{\mathfrak{D}_{O_2m}}{\pi t}}\ \frac{1}{\operatorname{erf} \zeta_0} \quad \text{(upper bound)} \tag{15}$$

$$= (c_0 - c_\infty)\sqrt{\frac{\mathfrak{D}_{O_2m}}{\pi t}}\ \sqrt{1+M} \quad \text{(lower bound)} \tag{16}$$

The ratio of these two fluxes is then

$$\frac{N_0(\text{upper})}{N_0(\text{lower})} = \frac{1}{\sqrt{1+M}\ \operatorname{erf} \zeta_0} \tag{17}$$

It only remains to consider the possible range of this ratio.

We first note that $M$ approaches zero as the average oxygen concentration becomes large. It follows from Eq. 12 that $\zeta_0$ becomes large and therefore that $\operatorname{erf} \zeta_0$ approaches unity. The flux ratio in Eq. 17 then also approaches unity. This is the expected result, as the entire effect of chemical reaction is small under these circumstances.

We next consider the limit of very large $M$, where physical absorption becomes unimportant. Here the left side of Eq. 12 may be approximated by

$$\zeta_0 e^{\zeta_0^2} \operatorname{erf} \zeta_0 \sim \frac{2\zeta_0^2}{\sqrt{\pi}} \qquad (\zeta_0 \to 0) \tag{18}$$

Then

$$\zeta_0 \sim \frac{1}{\sqrt{2M}}, \qquad \operatorname{erf}\zeta_0 \sim \sqrt{\frac{2}{\pi M}} \tag{19a,b}$$

and

$$\frac{N_0(\text{upper})}{N_0(\text{lower})} \to \sqrt{\frac{\pi}{2}} \tag{20}$$

as $M$ becomes large. It follows that the upper and lower bounding predictions can never differ from their mean by more than about 15%. Similarly successful limiting calculations can be made for other flow situations, and it appears that the numerical integration of the diffusion equation is indicated only for complex flow situations. The increased precision obtained by considering the shape of the dissociation curve in detail is to a large degree illusory because of errors introduced in developing the idealized diffusional model represented by Eq. 1.

### *EX. 2.3.3. CARRIER TRANSPORT OF AN INSOLUBLE SOLUTE*

Consider now a situation similar to that in Ex. 2.3.1 in which the liquid solubility of the substance being transported is zero. All of the solute S must then be transported as a carrier-solute complex CS, where C refers to the carrier.

SOLUTION

The system description now takes the form:

Continuity relations:

$$N_S = N_W = 0 \tag{1}$$

$$N_{CS} = -N_C = \text{constant} \tag{2}$$

Flux equations:

$$x_{CS}\frac{\partial \ln a_{CS}}{\partial z} = -\left(\frac{x_C + x_{CS}}{c\,Đ_{C-CS}} + \frac{x_m}{c\,Đ_{CS-m}} + \frac{x_W}{c\,Đ_{CS-W}}\right)N_{CS} \tag{3}$$

$$x_C\frac{\partial \ln a_C}{\partial z} = +\left(\frac{(x_C + x_{CS})}{c\,Đ_{C-CS}} + \frac{x_m}{c\,Đ_{C-m}} + \frac{x_W}{c\,Đ_{C-W}}\right)N_{CS} \tag{4}$$

The boundary conditions are assumed to take the form

$$\text{At } z=0, \quad \frac{x_{\text{CS}}}{x_{\text{C}}}=Ky_{\text{S}0} \tag{5}$$

$$\text{At } z=\delta, \quad \frac{x_{\text{CS}}}{x_{\text{C}}}=Ky_{\text{S}\delta} \tag{6}$$

Here the subscripts S, W, C, CS, and $m$ refer to the solute, water (or other solvent), carrier, carrier-solute complex, and membrane matrix, respectively. The mole fractions $x_i$ and $y_i$ refer to the membrane and exterior solution phases, respectively, and $K$ is the equilibrium constant for the complex formation.

We now see that because of the system stoichiometry the diffusion of the moving species can be expressed in terms of effective binary diffusivities, and that there is no net convection. In the absence of more complete information, we further assume that the carrier and carrier-solute complex are chemically similar. Then their corresponding diffusivities may be considered equal, and the activity coefficients may be considered constant. It follows that

$$x_{\text{C}}+x_{\text{CS}}=x_{\text{tot}}=\text{constant} \tag{7}$$

and that

$$N_{\text{CS}}=-c\mathfrak{D}_{\text{C}m}\nabla x_{\text{CS}}$$

where

$$\mathfrak{D}_{\text{C}m}=\mathfrak{D}_{\text{CS-}m}=\left(\frac{x_{\text{C}}+x_{\text{CS}}}{Đ_{\text{C-CS}}}+\frac{x_m}{Đ_{\text{CS-}m}}+\frac{x_{\text{W}}}{Đ_{\text{CS-W}}}\right)^{-1} \tag{8}$$

We may therefore write that

$$N_{\text{CS}}=c\mathfrak{D}_{\text{C}m}\frac{x_{\text{CS}0}-x_{\text{CS}\delta}}{\delta} \tag{9}$$

It now remains to use Eqs. 5 to 7 to determine the rate of transport in terms of external solution composition. Eliminating $x_{\text{S}0}$ and $x_{\text{S}\delta}$ between these and Eq. 9, we obtain

$$\frac{x_{\text{CS}}}{x_{\text{tot}}}=\frac{Ky_{\text{S}}}{1+Ky_{\text{S}}} \tag{10}$$

and

$$N_{CS} = \frac{c\mathfrak{D}_{Cm} x_{tot}}{\delta}\left(\frac{Ky_{S0}}{1+Ky_{S0}} - \frac{Ky_{S\delta}}{1+Ky_{S\delta}}\right) \tag{11}$$

$$= \frac{c\mathfrak{D}_{Cm} x_{tot}}{\delta}\left(\frac{K(y_{S0}-y_{S\delta})}{(1+Ky_{S0})(1+Ky_{S\delta})}\right) \tag{12}$$

It follows that the rate of solute transport, $N_{CS}$, can never be larger than

$$N_{max} = \frac{c\mathfrak{D}_{Cm} x_{tot}}{\delta} \tag{13}$$

Solute transport will in addition be influenced by the presence of other materials capable of complexing with the carrier.

Not only is this type of facilitated diffusion common and important in biological systems; it probably forms the basis for all active transport. As we shall see, carriers have been tentatively identified in a few special cases, and a number of carriers have been produced capable of causing facilitated diffusion in synthetic models of biological membranes.

A wide variety of interesting and anomalous effects are observed when two or more solutes compete for the same carrier. These are described in a number of references, for example Stein (1967), and are touched on in the problems.

### *EX. 2.3.4. DIFFUSION WITH ZERO-ORDER* CHEMICAL REACTION*

Show how diffusion in a solid body with zero-order reaction can be described in terms of the nonreactive diffusion equation, and specialize the result for steady diffusion into both sides of a slab (see Fig. 2.3.4). Assume constant $c$ and $\mathfrak{D}_{im}$. Discuss the physiological significance of the development.

SOLUTION

For these assumed conditions the diffusion equation takes the simple form

$$\frac{\partial c_i}{\partial t} = \mathfrak{D}_{im}\nabla^2 c_i - R_0 \tag{1}$$

where $R_0$ is the rate of disappearance of solute $i$ per unit volume, assumed

* For an introduction to first-order reactions see Prob. III.5-L.

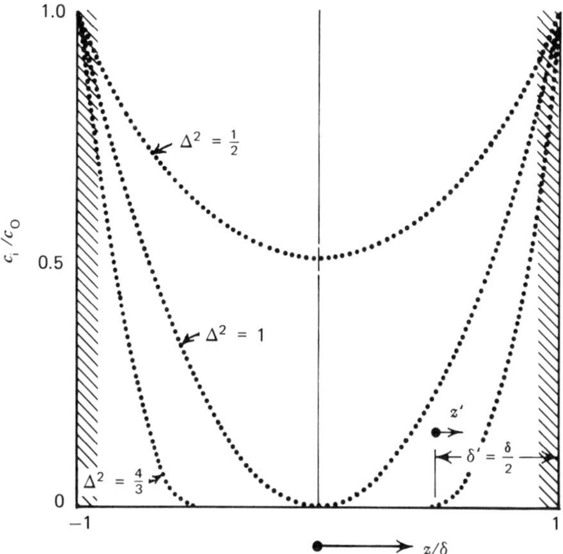


**Figure 2.3.4.** Concentration profiles for diffusion with zero-order reaction in a slab. These profiles are calculated for the boundary conditions

$$c_i(0,t)=c_0,\ \text{a constant} \qquad (-\delta<z<\delta)$$

$$c_i(-1,t)=c_i(1,t)=c_0 \qquad (t>0)$$

as functions of the parameter

$$\Delta^2=\delta^2 R_0/2c_0\mathfrak{D}_{im}$$

For $\delta^2>2c_0\mathfrak{D}_{im}/R$ there will be a solute-free core separating two reactive zones, for which

$$c_i/c_0=1+(R_0/2c_0\mathfrak{D}_{im})(z'^2-\delta'^2)$$

$$\delta'=\sqrt{2c_0\mathfrak{D}_{im}/R_0}$$

This situation will occur whenever $\Delta^2>1$. An example is shown here for $\Delta^2=\frac{4}{3}$.

constant. This is to be solved with the boundary conditions

$$c(\mathbf{r},0)=c_0(\mathbf{r}) \tag{2}$$

$$c(\mathbf{s},t)-\alpha(\mathbf{n}\cdot\nabla c)|_{\mathbf{s},t}=f_S(\mathbf{s},t) \tag{3}$$

where $\mathbf{r}$ = the position vector for any point in the body,
$\mathbf{s}$ = the position vector for prescribed points on the surface,
$\mathbf{n}$ = an outwardly directed unit normal at any point on the surface,

and $\alpha$, $c_0$, and $f$ are known functions of position and time.

For our present purposes it is now convenient to define

$$u = c_i + R_0 t \tag{4}$$

Equations 1 through 3 then take the following form:

$$\frac{\partial u}{\partial t} = \mathfrak{D}_{im} \nabla^2 u \tag{5}$$

$$u(\mathbf{r},0) = c_0(\mathbf{r})$$

$$u(\mathbf{s},t) - \alpha(\mathbf{n}\cdot\nabla u)|_{\mathbf{s},t} = f_s(\mathbf{s},t) + R_0 t \tag{6}$$

This description has thus been reduced to that for diffusion without reaction in a geometrically similar body with slightly more complex surface boundary conditions.

For the special case of $c_0 = c_s =$ constant and the slab shown in Fig. 2.3.4, these equations further simplify to

$$\frac{\partial u}{\partial t} = \mathfrak{D}_{im} \frac{\partial^2 u}{\partial z^2} \tag{7}$$

$$u(0,z) = c_0, \qquad -\delta < z < \delta \tag{8}$$

$$u(-l,t) = u(l,t) = c_0 + R_0 t, \qquad t > 0 \tag{9}$$

The solution to this problem is available from Eq. 4 or from Section 3.5 of Carslaw and Jaeger (1959). It may be written as

$$\frac{c_i}{c_0} = 1 + \frac{R_0}{2c_0\mathfrak{D}_{im}}(z^2 - \delta^2) + \frac{16R_0\delta^2}{\pi^3 c_0 \mathfrak{D}_{im}} S \tag{10}$$

with

$$S = \sum_{n=0}^{\infty} \frac{(-1)^n}{(2n+1)^3} e^{-\mathfrak{D}_{im}(2n+1)^2\pi^2 t/4\delta^2} \cos\frac{(2n+1)\pi z}{2\delta} \tag{11}$$

Since each term of the series decays with time, the concentration approaches a steady state given by

$$\frac{c_i}{c_0} \to 1 + \frac{R_0}{2c_0\mathfrak{D}_{im}}(z^2 - \delta^2) \qquad (t \to \infty) \tag{12}$$

This completes the purely mathematical aspect of our solution. It remains to consider its physical significance and biological utility. We do this only for the asymptotic limit given by Eq. 12.

We first note that the minimum concentration occurs at the center of the slab and that this cannot be negative. The above solution is therefore valid only for

$$\delta \leqslant \sqrt{\frac{2c_0 \mathfrak{D}_{im}}{R_0}} \tag{13}$$

For slabs of greater half thickness there will be at steady state a *solute-free core* separating reaction zones which are mirror images and each of half thickness $\sqrt{2c_0 \mathfrak{D}_{im}/R_0}$. This situation is indicated schematically in the accompanying figure.

Solutions of the above type are widely useful in spite of the chemical complexity of living systems. This is because metabolic activity is frequently determined primarily by the concentration of one *limiting* nutrient that is in particularly short supply at the time, and because

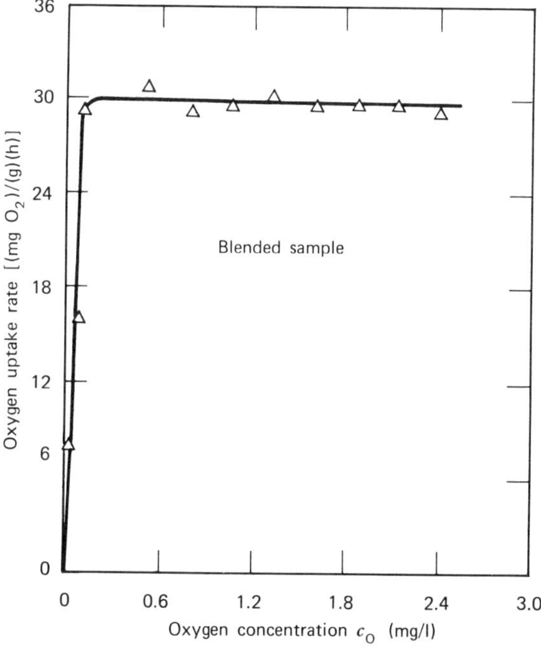


**Figure 2.3.5.** Response of partially dispersed *Zoogloea ramigera* to oxygen concentration. From Mueller (1966).

response to the concentration of this limiting species can often be satisfactorily approximated by zero-order kinetics.

Oxygen is quite commonly the limiting species because of its low solubility in aqueous solutions, and it is by far the most thoroughly investigated. Typical experimental results are shown in Fig. 2.3.5 for the oxygen response of bacterial aggregates, which are particularly convenient to use because they can be subdivided without damage. This figure, showing the response of a mechanically dispersed floc, demonstrates that zero-order response occurs down to quite low oxygen concentration (note that the oxygen solubility from air in dilute aqueous salt solutions is about 9 mg/l). It is quite probably that diffusional resistance is still appreciable at the lower oxygen concentrations even here, since complete dispersal of organisms is not feasible by the technique used to obtain these results.

Anoxic regions corresponding to the solute-free zones predicted for thick slabs in Fig. 2.3.4 are probably quite common, for example in muscle tissue during vigorous exercise. In some such cases metabolic energy can be supplied temporarily by anaerobic mechanisms, for example, the formation of lactic acid in muscle.

## 2.4. Diffusion in Multicomponent Electrolyte Solutions

In this section we develop a systematic description of the diffusional behavior of mixed electrolytes by incorporating the restraints operating in these systems with the flux equations developed in Section 1.3. We then briefly discuss the importance of diffusional interactions via representative examples. Such interactions are particularly strong for mixed salts, but they normally arise primarily from coulombic forces rather than the ion-ion diffusivities $\mathcal{D}_{ij}$.

Since all living organisms are composed primarily of quite complex mixed electrolytes separated by selectively permeable membranes, the topic under discussion is clearly an important one. However, these solutions are sufficiently dilute that *diffusional* ("frictional") interactions between ions are of secondary importance. It is therefore normally possible to use the Nernst-Planck pseudobinary relations introduced in Section 2 to calculate diffusion fluxes. We shall see, however, that diffusional interactions can easily be taken into account when sufficient information is available. It appears that they are just significant in the measurement of diffusion potential; they are often very important in membranes.

We are interested here not only in ionic concentration profiles and mass fluxes, but also in the electrostatic potentials arising from the concentration inequalities. We therefore consider as a basis for discussion an electrochemical cell containing a single region of changing concentration as shown in Fig. 1.3.2 of Chapter II. The system now contains an arbitrary number of anions and cations, but we still assume, as is normally the case in biological applications, that the small electric current necessary to measure the cell potential is negligible relative to the diffusion fluxes of the ions. The case of finite current density is considered in standard electrochemical references, for example by Newman (1967).

We are thus considering the rather general case of a cell *with transference*. Our purpose is to relate the cell potential to its diffusional behavior, and we wish to do this without neglecting any departures from thermodynamic ideality or any diffusional interactions. Cells of this type are of major biological interest but not yet well understood. The thermodynamic difficulties inherent in analyzing cells with liquid junctions are discussed in standard references (see especially Guggenheim, 1959). The diffusional behavior of cells with transference has received less attention, but recently Smyrl and Newman (1968) have shown how diffusion potentials can be calculated for thermodynamically nonideal systems if diffusional interactions between ions are neglected. That is, Smyrl and Newman's analysis is based essentially on the pseudobinary Nernst-Planck equations.

Here we relax this restriction to pseudobinary diffusion and show that all diffusional interactions can be quite simply incorporated into the description of junction potentials. Such interactions are appreciable in concentrated solutions and normally dominate diffusional behavior in the selective membranes typical of biological systems. We limit ourselves here to diffusion in free solution and defer membrane transport to a later analysis.

We treat the ion concentrations in the two electrode compartments as known functions of time, and the water velocity as a known function of position and time. The system behavior is then described completely by:

(1) A continuity equation for each ion.
(2) A diffusion (flux) equation for each ion.
(3) The requirement of electroneutrality.
(4) The requirement of no current flow.

the terminal ion concentrations $c_{il}(t)$ and $c_{i2}(t)$, *and the water velocity* $\mathbf{v}_W(\mathbf{r}, t)$, where $\mathbf{r}$ is a position vector. Specifying $\mathbf{v}_W$ is equivalent to specifying the mass or molar-average velocity, and is much more convenient for our present purposes. The water velocity profile in turn must be obtained from the equation of motion, the continuity equation, and the appropriate boundary conditions. In most cases the water velocity may be obtained to satisfactory accuracy without detailed consideration of the diffusional processes being studied.

It is now convenient to rewrite the species continuity and diffusion equations relative to the water velocity. For nonreactive systems the continuity equation may be written directly as

$$\frac{\partial c_i}{\partial t} + (\nabla \cdot c_i \mathbf{v}_W) = -(\nabla \cdot \mathbf{J}_i^0) \tag{2.4.1}$$

with

$$\mathbf{J}_i^0 = c_i(\mathbf{v}_i - \mathbf{v}_W) \tag{2.4.2}$$

We may also use water as the reference species $k$ in Eq. 1.2.21 to write

$$\mathbf{d}_i = \sum_{\substack{j=1 \\ j \neq W}}^{n} \frac{x_i}{c\mathcal{D}_{ij}} \mathbf{J}_j^0 \tag{2.4.3}$$

with the generalized driving force

$$\mathbf{d}_i = x_i \nabla_{T,p} \ln a_i + x_i \nu_i \nabla \Phi \tag{2.4.4}$$

where $\Phi$ is a dimensionless electrostatic potential defined by $\Phi = \mathfrak{F}\phi/RT$. The pressure-diffusion term has been eliminated from this equation because it is appreciable only under exceptional circumstances. The diffusion coefficients $\mathcal{D}_{iW}$ representing ion-water interaction, which are normally dominant, are incorporated into the terms $\mathcal{D}_{ii}$ via Eq. 1.2.23. Thus for the pseudobinary limit

$$\mathcal{D}_{ii} = -\mathcal{D}_{iW} \frac{x_i}{x_W} \tag{2.4.5}$$

Equations similar to 2.4.1 and 2.4.3 are widely used by electrochemists, and Eq. 2.4.5 must usually be used for lack of data.

The statements of electroneutrality,

$$\sum_{i=1}^{n} \nu_i c_i = 0 \tag{2.4.6}$$

and no current flow,

$$\sum_{i=1}^{n} \nu_i \mathbf{J}_i^0 = 0 \tag{2.4.7}$$

complete the description of the system.

The formal solution of the problem can now be completed by:

1. Developing flux expressions for the constituent salts independent of electrostatic potential and containing only salt activities, which are measurable. These flux expressions can then be used to determine the ionic concentration profiles.

2. Developing a relation for electrostatic potential in terms of the salt activities and concentration profiles, and the thermodynamic activity of the discharging ion. This ion activity in turn can be eliminated by use of the thermodynamic description of the electrode reaction, to relate measured cell potential to the salt activity and concentration profiles.

It is important to keep in mind during the subsequent analysis that the $n-1$ equations 2.4.3 are independent and that Eq. 2.4.6 reduces the number of components to $n-1$ for the $n$ individual chemical species present. Thus for a solution containing a single anion and a single cation, plus water, there are only two components, salt and water; this system is thermodynamically binary.

It will prove convenient to incorporate Eq. 2.4.7 directly into Eq. 2.4.3 by eliminating one more species, this time an ion, from the summation. Designating this species as $n$, we obtain

$$\frac{\mathbf{d}_i}{x_i} = \sum_{\substack{j=1 \\ j \neq \mathrm{W}, n}}^{n-1} \left( \frac{1}{c Ð_{ij}} - \frac{\nu_j}{\nu_n} \frac{1}{c Ð_{in}} \right) \mathbf{J}_j^0 \tag{2.4.8}$$

The choice of $n$ is again arbitrary and made for convenience as indicated below. We are now ready to calculate concentration profiles and electrostatic potential distributions using the $n-2$ independent flux relations represented by Eq. 2.4.8.

### (a) *DETERMINATION OF CONCENTRATION PROFILES*

Each of the $n-2$ components of the solution we must consider explicitly maybe considered a neutral salt of the general form $\mathrm{M}^{\nu_\mathrm{M}}_{|\nu_\mathrm{X}|}\mathrm{X}^{\nu_\mathrm{X}}_{|\nu_\mathrm{M}|}$, where $\mathrm{M}^{\nu_\mathrm{M}}$ and $\mathrm{X}^{\nu_\mathrm{X}}$ are the constituent ions. One must be an anion and the other a cation, but it is not necessary to specify formally which is which. The ionic composition of the solution, and its diffusional behavior, may always be expressed in terms of $n-1$ such salts, and once again an arbitrary choice exists.

For each salt the two ionic-flux expressions, Eqs. 2.4.8, may be combined to eliminate electrostatic potential. The result for the representative case $\mathrm{M}^{\nu_\mathrm{M}}_{|\nu_\mathrm{X}|}\mathrm{X}^{\nu_\mathrm{X}}_{|\nu_\mathrm{M}|}$ is

$$|\nu_X|\frac{\mathbf{d}_M}{x_M}+|\nu_M|\frac{\mathbf{d}_X}{x_X}=\nabla_{T,p}\ln a_M^{|\nu_X|}a_X^{|\nu_M|}$$

$$=\sum_{\substack{j=1\\ j\neq W,n}}^{n-1}\left[|\nu_X|\left(\frac{1}{c\mathcal{D}_{Xj}}-\frac{\nu_X}{\nu_n}\frac{1}{c\mathcal{D}_{Xn}}\right)\right.$$

$$\left.+|\nu_M|\left(\frac{1}{c\mathcal{D}_{Mj}}-\frac{\nu_M}{\nu_n}\frac{1}{c\mathcal{D}_{Mn}}\right)\right] \tag{2.4.9}$$

We now note that

$$\nabla_{T,p}\ln a_M^{|\nu_X|}a_X^{|\nu_M|}=\nabla_{T,p}\ln a_{MX} \tag{2.4.10}$$

where the subscript MX refers to the salt $M_{|\nu_X|}^{\nu_M}X_{|\nu_M|}^{\nu_X}$.

Further manipulations are conceptually straightforward but quite complex in detail. They are discussed in available references, for example, Stewart and Prober (1964), and in the problems for this chapter. Here we satisfy ourselves by pointing out that Eq. 2.4.9 is of the form

$$\nabla_{T,p}\ln a_{s_i}=\sum_{j=1}^{n-2}r_{ij}\mathbf{J}_{s_j} \tag{2.4.11}$$

where the component salts are $(s_1, s_2,\ldots,s_{n-2})$, and the $r_{ij}$ represent the bracketed quantities in Eq. 2.4.9, each multiplied by $x_{s_i}$ and the appropriate stoichiometric coefficient.

Equations of this form are generally inconvenient, and it is common practice to solve for the fluxes as explicit functions of the concentration gradients. The result of such a manipulation yields the expression

$$\underline{J}^0=-c\,\underline{\underline{D}}^0\nabla\,\underline{x} \tag{2.4.12}$$

where $\underline{J}^0$ and $\underline{x}$ are column vectors representing the fluxes $\mathbf{J}_i^0$ and mole fractions $x_i$ of the species represented in Eq. 2.4.9, and $\underline{\underline{D}}$ is the matrix of diffusion coefficients. The relations between the $D_{ij}^0$ of this matrix and the Stefan-Maxwell $\mathcal{D}_{ij}$ from which we started are complex and contain thermodynamic as well as diffusional parameters. In addition, the symmetry property has been lost, so that

$$D_{ij}^0\neq D_{ji}^0 \tag{2.4.13}$$

Equation 2.4.12 is, however, useful for computational purposes and is widely used.

One can also write Eq. 2.4.12 in terms of other reference velocities, for example the molar-average velocity $\mathbf{v}^{\bigstar}$:

$$J^{\bigstar} = -c\, \underline{\underline{D}}^{\bigstar}\, \nabla\, \underline{x} \tag{2.4.14}$$

For most situations of current biological interest $D_{ij}^0 \approx D_{ij}^{\bigstar}$, and the diagonal terms (for which $i=j$) dominate. Standard means are, however, available for changing reference velocities, for example, in Hirschfelder, Curtiss and Bird (1954) or in Cussler, Rettig, and Lightfoot (1963).

The $n-1$ continuity equations for a system of *n components* can then be put in the form

$$\frac{\partial \underline{c}}{\partial t} = (\nabla \cdot c\, \underline{\underline{D}}^{0}\, \nabla\, \underline{x}) - (v^0 \cdot \nabla\, \underline{c}) \tag{2.4.15}$$

with

$$\underline{c} = \begin{pmatrix} c_1 \\ c_2 \\ \cdot \\ \cdot \\ \cdot \\ c_{n-1} \end{pmatrix}$$

If one can consider the total molar concentration $c$ and the diffusion coefficients in $\underline{\underline{D}}$ to be position independent, Eq. 2.4.15 reduces to

$$\frac{\partial \underline{c}}{\partial t} = \underline{\underline{D}}^{0}\, \nabla^2\, \underline{c} - (\mathbf{v}^0 \cdot \nabla\, \underline{c}) \tag{2.4.16}$$

It has been shown by Toor (1964) and others that Eq. 2.4.16 is generally satisfactory for calculating salt fluxes. It is less successful in calculating concentration profiles, but we shall see shortly that it is not necessary, for example, to know the profiles accurately for calculating diffusion potentials. Equation 2.4.16 is almost universally used in preference to Eq. 2.4.15.

### *EX. 2.4.1. CONCENTRATION PROFILES IN THE TIP OF A MICROELECTRODE*

Use Eq. 2.4.16 and the results of Ex. 2.2.2 to obtain an approximate relation for ionic concentration profiles near the tip of a microelectrode.

Neglect the movement of the water and show that the solute concentration profiles are independent of $\underline{\underline{D}}$.

SOLUTION

The results of Ex. 2.2.2 can easily be generalized to write

$$\frac{\partial \underline{c}}{\partial t} = \underline{\underline{D}} \frac{1}{r^2} \frac{\partial}{\partial r} r^2 \frac{\partial \underline{c}}{\partial r} \tag{1}$$

and we could proceed directly on the basis of this equation. However, it was shown in Ex. 2.2.2 that transients die out rapidly. Hence operation can usually be considered as a special case of steady nonconvective diffusion, for which the general description is

$$\nabla^2 \underline{c} = 0 \tag{2}$$

We may then write, for the boundary conditions of this previous example,

$$\underline{c}(r) - \underline{c}(\infty) = [\underline{c}(a) - \underline{c}(\infty)] \frac{a}{r} \tag{3}$$

or, for any individual species:

$$\frac{c_i - c_i(\infty)}{c_i(a) - c_i(\infty)} = \frac{a}{r} \tag{4}$$

The asymptotic concentration profiles are then all predicted to be similar, so that the widely used Henderson relation (see Prob. 4.E) is obtained.

The situation is, however, more complex for unsteady diffusion or for appreciable convection. Here $\underline{\underline{D}}$ appears in the problem solution. We also see in connection with the discussion of electrode potentials below that the variation of $\underline{\underline{D}}$ with concentration can have an appreciable effect.

### (b) *DETERMINATION OF CELL POTENTIALS*

We now attempt to eliminate the molar fluxes $\mathbf{J}_j^0$ from Eqs. 2.4.3 and thus obtain a relation between electrostatic potential and the concentration profiles, which may now be considered known. To do this we seek linear combinations of the "driving forces" $\mathbf{d}_i$ such that

$$\sum_{\substack{i=1 \\ \neq W}}^{n} \frac{\alpha_i \mathbf{d}_i}{x_i} = 0 \tag{2.4.17}$$

where the $\alpha_i$ are to be solved for. It follows from Eqs. 2.4.3 and 2.4.17 that

$$\sum_{\substack{i=1\\ \neq W}}^{n} \sum_{\substack{j=1\\ \neq W}}^{n} \frac{\alpha_i}{c Ð_{ij}} \mathbf{J}_j^0 = 0 \tag{2.4.18}$$

and from Eq. 7 that

$$\sum_{\substack{i=1\\ \neq W}}^{n} \frac{\alpha_i}{Ð_{ij}} = \nu_j \qquad (j=1,2,\ldots,n,\neq W) \tag{2.4.19}$$

Since the $n-1$ Eqs. 2.4.19 are independent, we may solve for the $\alpha_i$ by Cramer's rule. Specific examples are given below.

We now note that Eq. 2.4.17 is of the form

$$\sum_{\substack{i=1\\ \neq W}}^{n} \alpha_i (\nabla \ln a_i + \nu_i \nabla \Phi) = 0 \tag{2.4.20}$$

which can be rearranged to

$$-\nabla\Phi = \frac{\sum_{\substack{i=1\\ \neq W}}^{n} \alpha_i \nu_i \nabla \ln a_i^{1/\nu_i}}{\sum_{\substack{i=1\\ \neq W}}^{n} \alpha_i \nu_i} \tag{2.4.21}$$

At this point we single out the discharging ion, for example $X^-$, for special treatment. We thus rewrite Eq. 2.4.21 in the form

$$-\nabla\Phi = \nabla \ln a_{X^-}^{1/\nu_X} + \sum_{\substack{j=1\\ \neq W\\ \neq X^-}}^{n} \frac{\alpha_j \nu_j}{\sum_{\substack{i=1\\ \neq W}}^{n} \alpha_i \nu_i} \nabla \ln \frac{a_j^{1/\nu_j}}{a_X^{1/\nu_X}} \tag{2.4.22}$$

The terms under the summation sign can be written in terms of measurable activity coefficients using only the salts chosen as solution components.

The potential between any two point can be obtained by integrating this expression. Thus for points 1′ and 2′ in the solution immediately adjacent to the electrodes one may write

$$\Phi_{1'} - \Phi_{2'} = \ln\left(\frac{a_{X2'}}{a_{X1'}}\right)^{1/\nu_X} + \sum_{\substack{j=1\\ \neq W\\ \neq X}}^{n} \int_{1'}^{2'} \frac{\alpha_j \nu_j}{\sum_{\substack{i=1\\ \neq W}}^{n} \alpha_i \nu_i} d\ln \frac{a_i^{1/\nu_i}}{a_X^{1/\nu_X}} \tag{2.4.23}$$

The ionic activity coefficients $a_X$ can now be eliminated, for a reversible electrode, by use of the half-cell potential relation

$$\Phi_i - \Phi_{i'} = \Phi^0 + \ln\left(a_{X^-}^{1/\nu_X}\right)_i \tag{2.4.24}$$

where $i$ refers to either 1 or 2, and $\Phi^0$ is the standard half-cell potential. Then the measured potential

$$\Phi_2 - \Phi_1 = \Phi_{2'} - \Phi_{1'} + \ln\left(\frac{a_{X2'}}{a_{X1'}}\right)^{1/\nu_X}$$

$$= -\sum_{\substack{j=1\\ j\neq W\\ j\neq X}}^{n} \int_{1'}^{2'} \frac{\alpha_j \nu_j}{\sum_{\substack{i=1\\ i\neq W}}^{n} \alpha_i \nu_i} d\ln \frac{a_j^{1/\nu_j}}{a_X^{1/\nu_X}} \tag{2.4.25}$$

which completes our formal development.

It remains only to consider a concrete example to illustrate the basic simplicity of these results.

### *EX. 2.4.2. THE EFFECTS OF DIFFUSIONAL INTERACTIONS, CONCENTRATION PROFILES, AND THERMODYNAMIC NON-IDEALITIES ON CELL POTENTIALS*

Discuss the calculation of junction potentials in cells of the type pictured in Fig. 1.3.1 in the light of Eq. 2.4.25. Compartments 1 and 2, now containing only two univalent cations $M^+$ and $N^+$ sharing the same anion $X^-$, are joined by a diffusion path of as yet an unspecified nature.

SOLUTION

We begin by writing Eq. 2.4.23, which for this system has the form

$$\Phi_{2'} - \Phi_{1'} = \ln\frac{a_{X2'}}{a_{X1'}} - \int_{1'}^{2'} \frac{\alpha_M}{A} d\ln\left(a_M a_X\right)$$

$$- \int_{1'}^{2'} \frac{\alpha_N}{A} d\ln\left(a_N a_X\right) \tag{1}$$

where $A = \alpha_M + \alpha_N - \alpha_X$. Since the electrodes are reversible for $X^-$, it follows that

$$\Phi_2 - \Phi_1 = -\int_{1'}^{2'} \frac{\alpha_M}{A} d\ln a_M a_X$$

$$- \int_{1'}^{2'} \frac{\alpha_N}{A} d\ln a_N a_X \tag{2}$$

which contains only measurable quantities.

Before going further we note that for zero concentration of NX, Eq. 2 reduces to

$$\Phi_2 - \Phi_1 = \int_{2'}^{1'} \left( \frac{\mathcal{D}_{MW}}{\mathcal{D}_{MW} + \mathcal{D}_{XW}} \right) d \ln a_M a_X \tag{3}$$

as in Ex. 1.3.2. For the ternary system considered here these integrals are more complex and depend upon concentration profiles.

One can proceed directly with Eq. 2, but this is in some respects inconvenient in that the potential difference given by this equation depends upon the nature of the electrode reaction as well as the diffusional process under investigation. It is therefore suggested by Smyrl and Newman (1968) that one work instead with

$$(\Phi_2 - \Phi_1)_x \equiv \Phi_2 - \Phi_1 + \ln \frac{x_{X2}}{x_{X1}}$$

The significance of $(\Phi_2 - \Phi_1)_x$ is not so simple in general; however, in the low-concentration limit where activity coefficients approach constant values, $(\Phi_2 - \Phi_1)_x$ is the potential difference across the diffusion path itself. The choice between the use of $\Phi_2 - \Phi_1$ and $(\Phi_2 - \Phi_1)_x$ is clearly arbitrary and a matter of personal preference.

To date no one appears to have made a completely unambiguous evaluation of potential for a cell of the type pictured above. However, Chapman and Lightfoot (1970) have estimated the effects of diffusional interaction for the special case shown in Fig. 1.3.1, and Smyrl and Newman (1968) have investigated the effects of concentration profiles in a wide variety of cells.

Chapman and Lightfoot considered the interdiffusion of KCl and NaCl under circumstances where $K^+$ and $Na^+$ concentrations varied linearly with each other, a situation we have shown to be closely approximated in a glass microelectrode shortly after immersion in a stirred solution. Cell potentials were then calculated for three different levels of approximation:

1. *Concentrated theory*: All multicomponent thermodynamic and diffusional effects are considered (although extrapolations of available data were necessary).

*Pseudobinary approximation*: All thermodynamic corrections are made, but ion-ion diffusional interactions are neglected. (This corresponds closely to the approximations of Smyrl and Newman.)

*Dilute theory*: Both departures from thermodynamic ideality and ion-ion diffusional interactions are ignored. (This is the common Nernst-Planck approximation.)

Representative results are shown in Table 2.4.1 for both the "concentrated", corresponding to the full Eq. 8, and pseudobinary theories.

**Table 2.4.1.** The Effects of Diffusional Interaction on Cell Potentials

| Concentration ($N$) | | Potential $V=\phi_2-\phi_1=(\Phi_2-\Phi_1)Rt/\mathfrak{F}$ mV | | |
|---|---|---|---|---|
| NaCl | KCl | Pseudo-binary | Concentrated Theory | Discrepancy |
| 0.100 | 0.001 | | | |
| 0.098 | 0.081 | 8.03 | 8.23 | +0.20 |
| 0.096 | 0.161 | 19.49 | 19.91 | 0.42 |
| 0.093 | 0.281 | 31.03 | 31.64 | 0.61 |
| 0.090 | 0.401 | 39.27 | 39.98 | 0.71 |
| 0.085 | 0.600 | 49.28 | 50.09 | 0.81 |
| 0.077 | 0.920 | 60.40 | 61.31 | 0.91 |
| 0.067 | 1.32 | 70.14 | 71.12 | 0.98 |
| 0.053 | 1.880 | 79.94 | 80.98 | 1.04 |
| 0.033 | 2.680 | 89.98 | 91.07 | 1.09 |
| 0.006 | 3.800 | 100.09 | 101.22 | 1.13 |
| 0.001 | 4.00 | 101.59 | 102.73 | 1.14 |

Thermodynamic activities were estimated using the Guggenheim equation (see Guggenheim, 1959, p. 367), and diffusion data were extrapolated from the limited data available. Both sets of estimates are open to some question, but the results of the calculations are clear: diffusional interactions have a rather small effect on cell potentials in this reasonably representative system.

In Table 2.4.2 are shown the values calculated by Smyrl and Newman for the "junction potential" $\Phi_1-\Phi_2$ for a wide variety of chemical systems and three types of concentration profiles. In all cases $X^-$ refers to chloride ion, and ion-ion diffusional interactions were neglected. That is, the results correspond to Chapman and Lightfoot's (1970) "pseudobinary" approximation. Potentials without superscripts were calculated with activity coefficients obtained from the Guggenheim equation. Those marked with asterisks were obtained using the assumption of constant activity coefficients, that is, dilute-solution theory.

The concentration profiles used to calculate the junction potentials are defined as follows:

1. *Free-Diffusion Junction.* At time zero the two solutions are brought into contact to form an initially sharp boundary in a long, vertical tube. The solutions are then allowed to diffuse into each other, and the thickness of the region of varying concentration increases as the square root of the

**Table 2.4.2.** The Effects of Concentration Profiles on Cell Potentials[a]

| | | | $(\Phi_1 - Gf_2)_x$ (mV) | | |
|---|---|---|---|---|---|
| Ion | Soln 1 | Soln 2 | Free Diffusion | Restricted Diffusion | Continuous Mixture |
| $H^+$ | 0.2 *N* | 0.1 *N* | | | −10.31 |
| $Cl^-$ | 0.2 | 0.1 | | | −11.43* |
| $K^+$ | 0.2 | 0.1 | | | 1.861 |
| $Cl^-$ | 0.2 | 0.1 | | | 0.335* |
| $K^+$ | 0 | 0.05 | −20.70 | −21.09 | −20.23 |
| $H^+$ | 0.02 | 0 | −18.50 | −18.97 | −18.02 |
| $Cl^-$ | 0.02 | 0.05 | | | |
| $K^+$ | 0 | 0.1 | −18.02 | −17.89 | −16.84 |
| $H^+$ | 0.02 | 0 | −14.05* | −14.12 | −12.90 |
| $Cl^-$ | 0.02 | 0.1 | | | |
| $K^+$ | 0 | 0.1 | −15.91 | −14.99 | −14.04 |
| $H^+$ | 0.01 | 0 | −10.85* | −10.30 | −9.09 |
| $Cl^-$ | 0.01 | 0.1 | | | |
| $K^+$ | 0 | 0.1 | | | |
| $H^+$ | 0.09917 | 0 | −27.39 | −27.48 | −27.55 |
| $NO_3^-$ | 0 | 0.05 | −26.53* | −26.62 | −26.70 |
| $Cl^-$ | 0.09917 | 0.05 | | | |
| $K^+$ | 0.1 | 0.1 | −0.157 | −0.157 | −0.157 |
| $NO_3^-$ | 0.05 | 0 | −0.423* | −0.423 | −0.423* |
| $Cl^-$ | 0.05 | 0.1 | | | |
| $Na^+$ | 0.1 | 0 | | | |
| $H^+$ | 0 | 0.05 | 28.58 | 29.64 | 28.10 |
| $ClO_4^-$ | 0 | 0.05 | 26.72* | 27.90 | 26.22 |
| $Cl^-$ | 0.1 | 0 | | | |

[a] Abstracted from Smyrl and Newman (1968).
[b] The asterisks indicate calculation for unit activity coefficients.

time. Even if the transport properties are concentration dependent and the activity coefficients are not unity, the potential of a cell containing such a junction should be independent of time.

2. *Restricted-Diffusion Junction.* The concentration profiles are allowed to reach a steady state by one-dimensional diffusion in the region between $x=0$ and $x=L$, in the absence of convection. The composition is that of one solution at $x=0$ and is that of the other solution at $x=L$. The potential of a cell containing such a junction is independent of $L$ (as well as time).

3. *Continuous-Mixture Junction*. The solute concentrations are linearly related as in Ex. 2.4.1.

The profiles calculated by method 2 differ from those by method 3 in that the authors took into account variations in activity coefficients. They did, however, once again ignore diffusional interactions between *salts* as well as the concentration dependence of the diffusivities.

It may be seen from Table 2.4.2 that the effect of concentration profile, like that of diffusional interaction, is appreciable but generally rather small. There are, however, some rather pronounced exceptions, especially in the quaternary systems. Similar exceptions can no doubt be expected for the effect of diffusional interactions.

## BIBLIOGRAPHY

### Books and Reviews

Carslaw, H. S., and J. C. Jaeger, *Conduction of Heat in Solids*, second ed., Oxford (1959).

Chapman, T. W., "Ionic Transport in Electrochemical Systems," in (*Lectures in Transport Phenomena*,) R. B. Bird, W. E. Stewart, E. N. Lightfoot, and T. W. Chapman, Am. Inst. Chem. Eng. Contin. Educ. Ser. No. 4 (1969).

Gosting, L. J., "Measurement and Interpretation of Diffusion," *Adv. in Protein Chem.* (1956).

Guggenheim, E. A., *Thermodynamics*, North Holland (1959).

Hershey, Daniel, Ed., *Blood Oxygenation*, Plenum (1970).

Hirschfelder, J. O.,, C. F. Curtiss, and R. B. Bird, *Molecular Theory of Gases and Liquids*, Wiley (1954).

Kortum, G., and J. O'M. Bockris, *Textbook of Electrochemistry*, Elsevier (1951).

Lavallée, M., O. F. Schanne, and N. C. Hebert, Eds., *Glass Micro-electrodes*, Wiley (1969).

Newman, John S., "Transport Processes in Electrolytic Solutions," *Adv. Electrochem. Electrochem. Eng.*, **5**, 87 (1967); *Electrochemical Systems*, Prentice-Hall (1973).

Robinson, R. A., and R. H. Stokes, *Electrolyte Solutions*, Butterworths, second ed. (revised) (1965).

Spaeth, E. E., "The Oxygenation of Blood in Artificial Membrane Devices,", in *Blood Oxygenation*, Daniel Hershey, Ed., Plenum (1970).

Stein, W. D., *The Movement of Molecules across Cell Membranes*, Academic Press (1967).

Tanford, Charles, *Physical Chemistry of Macromolecules*, Wiley (1959).

### Research Papers

Bassett, R. J., and J. S. Schultz, "Nonequilibrium Facilitated Diffusion of Oxygen through Membranes of Aqueous Cobaltohistidine," *Biochem. Biophys. Acta*, **211**, 194–215 (1970).

Blackshear, P. L., and Christopher Walters, "Observations of Red Blood Cells Hitting Solid Walls" in *Advances in Bio-engineering*, R. G. Buckles, Ed., Chem. Eng. Prog. Symp. Series, **67**, No. 114, 1971, pp. 60–68.

Chapman, T. W., and E. N. Lightfoot, joint meeting, Am. Inst. Chem. Eng. and Inst. Ing. Quim. Puerto Rico, San Juan, May 1970.

Cussler, E. L., Jr., and E. N. Lightfoot, *Am. Inst. Chem. Eng.* **9**, 703 (1963); 783 (1963).

Dindorf, J. A., and E. N. Lightfoot, "Mass Transfer across a Deformable Interface with Instantaneous Reversible Reaction," *Chem. Eng. Sci*., **26**, 1833–1840 (1971).

Dindorf, J. A., and E. N. Lightfoot, MS submitted.

Friedlander, S. K., and K. H. Keller, "Mass Transfer in Reacting Systems Near Equilibrium," *Chem. Eng. Sci*., **20**, 121 (1965).

Geisler, C. D., E. N. Lightfoot, F. P. Schmidt, and F. Sy, "Diffusion Effects in Liquid-filled Micropipettes: Pseudo-binary Analysis of Electrolyte Leakage," Inst. El. Eng. Trans. on Bio-med. Eng., *BME*-19, 372–375 (1972).

Goddard, J. D., J. S. Schultz, and R. J. Bassett, "On Membrane Diffusion with Near-equilibrium Reaction," *Chem. Ehg. Sci*., **25**, 665–683 (1970).

Grabowski, E. F., L. I. Friedman and E. F. Leonard, "Effects of Shear Rate in the Diffusion and Adhesion of Blood Platelets to a Foreign Surface," *Ind. Eng. Chem. Fund.* **11**, 224–232 (1972).

Henderson, P., *Z. Phys. Chem.* (Leipz.), **59**, 118 (1907); **63**, 325 (1908).

Keller, K. H., E. R. Canales, and Su Il Yum, "Tracer and Mutual Diffusion Coefficients of Proteins," *J. Phys. Chem*., **75**, 379 (1971).

Keller, K. H., and S. K. Friedlander, "Investigation of Steady-state Oxygen Transport in Hemoglobin Solution," in *Chemical Engineering in Medicine*, Chem. Eng. Prog. Symp. Ser., No. 66, **62** (1966).

Keller, K. H., *J. Gen Physiol*., **49**, 663; 681.

Marx, T. I., W. E. Snyder, A. D. St. John, and Calvin E. Moeller, "Diffusion of Oxygen into a Film of Whole Blood," *J. App. Physiol*., **15**, 1123 (1960).

Mueller, J. A., W. C. Boyle, and E. N. Lightfoot, "Oxygen Diffusion through Zoogloeal Flocs,", *Biotechnol. and Bioeng*., **X**, 331–358 (1968).

Mueller, J. A., "Oxygen Diffusion through a Pure-culture Floc of Zoogloea Ramigera," Ph.D. thesis, Univ. of Wis. (1966).

Planck, M., *Ann. Phys. Chem*., **39**, 161 (1890); **40**, 561 (1890).

Rehm, W. S., Hilda Schlesinger, and W. H. Dennis, " Effect of Osmotic Gradients on Water Transport, Hydrogen Ion and Chloride Ion Production in the Resting and Secreting Stomach," *Am. J. Physiol.,* **175**, No. 3, 473–486 (1953)).

Smyrl, W. H., and John Newman, "Potentials of Cells with Liquid Junctions", *J. Phys. Chem*., **72**, 4660–4671 (1968).

Spaeth, E. E., and S. K. Friedlander, "The Diffusion of Oxygen, Carbon Dioxide, and Inert Gas into Flowing Blood," *Biophys. J*., **7**, 827–851 (1967).

Stewart, W. E., and R. Prober, *Ind. Eng. Chem*., **3**, 224 (1964).

Toor, H. L., *Am . Inst. Chem. Eng. Journal*, **10**, 460 (1964).

Turitto, V. T., A. M Benis, and E. F. Leonard, "Platelet Diffusion in Flowing Blood," *Ind. Eng. Chem. Fund.*, **11**, 216–223 (1972).

## 3. MASS TRANSPORT ACROSS MACROSCOPIC MEMBRANES

We consider here mass transport across membranes that are thick relative to molecular dimensions, and we limit ourselves to a small number of situations peculiar to membrane transport. No effort is made at any encyclopedic coverage of this very large and fast-moving field. Rather the reader is referred to the large number of monographs and reviews appearing in the literature, and in particular to that of Lakshminaraianaiah (1965).

We begin in Section 3.1 with a brief review of the classification of membranes and a discussion of the transport and thermodynamic data needed to describe mass transfer through them. In Section 3.2 we consider nonisobaric transport of uncharged species and, as specific examples, the description of ultra-filtration and the derivation of the Kedem-Katchalsky equations referred to in Section 1. We conclude in Section 3.3 with a discussion of transport through charged membranes. This is a very important subject to which we shall return later for explanations of the behavior of biological membranes.

It may be noted that concentration diffusion in membranes, particularly facilitated transport, has already been included in the discussions of the previous chapter. In these the mechanical restraints on the membrane matrix are of little importance. Transport in ultrathin ("biomolecular") membranes is deferred to Section 4.

In all of our discussions we assume the membrane to be thin and to contain very little solute relative to that in the fluid compartments it separates. We may then neglect transients in the membrane and assume unidirectional transport. Species continuity equations are then all of the form

$$-\frac{\partial N_{iz}}{\partial z}+R_i=0 \tag{1}$$

As discussed previously, there is one independent relation of this type for each diffusing species, and is most convenient to include the membrane as one species, $m$. It is also convenient to fix the coordinate system relative to the membrane, so that $v_m=0$. To complete the system description we need the $n-1$ independent flux equations, appropriate boundary conditions, and the equations of state.

The basic flux equations are those represented by Eqs. 1.2.21 to 1.2.24, and the driving force for diffusion is that given by Eq. 1.4.9. The reference velocity used in Eq. 1.2.21 will vary with the application. Thus in our discussion of ultrafiltration it is convenient to use $\mathbf{v}_i$: the $Đ_{ij}$ are then all positive and can be related rather simply to physical models developed for ultrafiltration. In the more formal development of the Kedem-Katchalsky relations, on the other hand, it is mathematically convenient to use the membrane velocity $\mathbf{v}_m$, which is zero.

### 3.1. Classification of Membranes

There are an almost bewildering variety of membranes under investigation today, and they exhibit a very wide range of transport behavior. In any detailed discussion it is therefore necessary to classify them into more nearly homogeneous groups.

Here we follow roughly the classification scheme suggested by Lakshminaraianaiah (1965) and begin with the following rather broad groupings:

1. Bimolecular
   a. Lipid only
   b. "Doped"
2. Macroscopic
   a. Inorganic
   b. Organic

The bimolecular membranes, discussed at some length in the next chapter, have many of the properties associated with the membranes of living cells and are at present used primarily as models for cell membranes. They are all fundamentally lipid in nature, are on the order of $10^2$ Å in thickness, and, in the absence of additives, behave much like very thin sheets of liquid fats from a mass-transfer standpoint. However, the addition of small amounts of additives can give them very highly selective permeability properties and permits a surprising variety of complex dynamic mass-transfer phenomena. They are extremely tough for their thickness and may well have a future in specialized extracorporeal processing equipment. Inorganic membranes will not be discussed further in these notes, but they are of very real biological importance. The most important of these are ion-selective electrodes in general and ion-exchanging glasses in particular. The most familiar is the glass *pH* electrode, selective for the hydrogen ion. Inorganic membranes are beneficial in many applications, but the selective ionic permeability of microelectrode tips is a major source of ambiguity in present-day neurophysiological measurements. A general discussion of inorganic electrodes is provided by Lakshminaraianaiah, and the properties of glass microelectrodes are treated in detail in Lavallée et al. (1965).

We are primarily interested here in macroscopic organic membranes, and further subclassify these as follows:

1. Immobilized solvents (macroscopic pores)
2. Chemically cross-linked, or otherwise stabilized, gels
   a. Hydrophobic
      Undoped
      Doped
   b. Hydrophilic
      Few, or no, fixed charges
         Porous cellulosic
         Nonporous asymmetric cellulosic
         Organic ion exchangers

These represent a very wide variety of membranes, and we cannot hope to cover them here. The reader is therefore referred to the general references listed in the bibliography, as representative sources.

Solvents can be immobilized equally well in inorganic and organic matrices, which provide only mechanical support. Since membrane properties depend only on the solvent, discussion of these membranes belongs more properly in the last chapter. It has in fact already been noted, in particular through the bibliography, that such supported-solvent membranes have been used for carrier transport, for example of $O_2$ and $CO_2$. In addition it should be noted that there is no sharp dividing line between these and cross-linked gels. "Porous" gel membranes may thus act like immobilized solvent membranes for small solutes in that the diffusional behavior may depend primarily upon solute-solvent rather than solute-membrane interactions.

Hydrophobic membranes, particularly silicone rubbers and halogenated hydrocarbons, are widely used in blood oxygenators and hence are the subject of considerable attention.

Porous cellulosic gels have been very widely used for hemodialysis, and they have been under investigation for years as ultrafiltration membranes. Concentration diffusion through porous cellulosics has been quite systematically investigated, in particular by L. C. Craig and his colleagues (1958, 1964, 1965), for transport of biologically active materials. Asymmetric cellulosic membranes are particularly promising for reverse osmosis, but a wide variety of proprietary membranes is also available. Lonsdale (1970) and Sourirarjan (1970) have recently summarized available data on ultrafiltration and reverse osmosis. Organic ion exchangers can be prepared from various substances over a wide range of dissociation constants. They can also be prepared with both anion- and cation-exchanging groups in the same membrane, either intimately mixed or separated spatially to a controlled degree.

Fortunately the diffusional behavior of membranes can be discussed much more compactly than their thermodynamic properties, and we therefore concentrate our attention on it. Formal integration of the diffusion equations is provided in the following subsections, and we limit ourselves here to the practical problems of obtaining transport data.

The basic experimental problems can be easily summarized:

1. A large number of diffusivities, or equivalent transport parameters, is required to describe diffusional behavior.

2. Boundary-layer diffusional resistance in the solutions bathing the membranes is normally appreciable.

3. Membranes are typically deformable, unstable chemically, and very slow to respond fully to changes in environment.

In addition, the transport tends to be slow.

It should be kept in mind that almost all membrane processes are at least ternary from a diffusional standpoint, since each solute normally interacts appreciably with both the solvent and the membrane matrix. This means that, at least in principle, both solute and water fluxes must be measured in experimental determinations. We return to this topic in our examples.

Frequently, however, a complete characterization of the membrane behavior is not needed. A particularly useful example is the diffusion of nonreactive solutes through uncharged membranes, as in dialysis or hemodialysis. If, as is generally true here, convective transport is unimportant, solute transport depends primarily on terminal solute concentrations in the external phase and can be described adequately* by

$$N_i = P_i \Delta c_i \tag{3.1.1}$$

where $N_i$ = the solute flux across the membrane
$P_i$ = the solute permeability
$\Delta c_i$ = the drop in solute concentration between the two external phases.

Equation 1 is widely useful, and the permeability is easily determined experimentally; a particularly elegant and economical technique for doing this has been developed by Craig (1964). Another situation in which simple experiments give all the immediately desired information is ultrafiltration, where empirical relations between applied pressure on the one hand and flow rates and solute rejection on the other often suffice. It must, however, be kept in mind that such simple experiments do not provide a complete characterization of the membrane. One cannot, for example, predict ultrafiltration behavior from dialysis rates.

* Equation 1 is clearly of limited utility if the distribution in the membrane is nonlinear or solute-solute interactions are important.

The effect of external boundary-layer (as opposed to interfacial) resistance is normally important, at least in liquid systems, and this has been widely recognized only in recent years. In most cases it seems impractical to eliminate this resistance by effective stirring, since few membranes remain rigid under these conditions. Rather, it is desirable to measure boundary-layer effects separately and then to allow for them in obtaining membrane transport properties from the raw data. This is discussed by Smith et al (1968), by Kaufmann and Leonard (1968), and by Scattergood and Lightfoot (1968).

### *EX. 3.1.1. DIFFUSIONAL INTERACTION IN AN ION-EXCHANGE MEMBRANE*

Describe a set of experiments capable of characterizing the diffusional behavior of a cation-exchanging membrane in contact with a very dilute aqueous solution of a single electrolyte $M^+X^-$. Consider in particular the significance of the tracer diffusion coefficient widely used in the investigation of membrane transport. Consider as a specific example the investigation of Scattergood and Lightfoot (1968).

SOLUTION

There is reason to believe that the co-ion interaction with the matrix is much smaller than the corresponding counter-ion interaction. This suggests the possibility of determining the three diffusivities, $Ð_{13}$, $Ð_{14}$, and $Ð_{34}$, by working at external salt concentrations sufficiently low to prevent appreciable co-ion penetration.* Counter-ion—counter-ion interactions can, however, be important, and we consider the simplest such case, the counterdiffusion of two isotopes ("self-diffusion"). If we neglect the differences in the diffusional behavior of these two, we need only one additional diffusivity, $Ð_{11}$. Only four experiments are then needed for our purposes.

Those chosen are the determination of electroosmotic water transport, the electrical conductivity, the hydrodynamic permeability, and counter-ion self-diffusion; they have all been described qualitatively in standard references, for example Helfferich (1962).

The electroosmotic transport number of water, $t_3$, is the number of moles of water moving across the membrane on the passage of one Faraday of current through it:

$$t_3 = \frac{N_3}{\left(N_1 - \underline{N_2}\right)} = \frac{w_3 F}{i},$$

* Subscripts 1, 2, 3, and 4 refer to the mobile cation and anion, water, and the membrane matrix respectively.

or

$$t_3 = \frac{x_3\left(1/Đ_{31} + \underline{N_2/N_1Đ_{32}}\right)}{\left(x_1/Đ_{31} + \underline{x_2/Đ_{32}} + x_4/Đ_{34}\right)\left(1 - \underline{N_2/N_1}\right)} \tag{1}$$

The molar rate of water transport is defined by $w_3$, and $i$ is the current.

The conductivity $\kappa$ is the specific electrical conductivity defined by $\kappa = -I/\nabla V = \delta/RA$, or

$$\kappa = \frac{(1-\underline{N_2/N_1})(c_1F^2/RT)}{\underline{x_2/Đ_{12}} + x_3/Đ_{13} + x_4/Đ_{14} - t_3x_1/Đ_{13} - \underline{(x_1/Đ_{12} - t_3x_1/Đ_{13})N_2/N_1}} \tag{2}$$

where $I$ is the current density, $\delta$ the membrane thickness, $R$ the resistance, and $A$ the area.

The hydrodynamic permeability $D_h''$ is the volume of solution transferred across the membrane per unit area and unit pressure difference in a unit time:

$$D_h'' = -\left(N_3\bar{V}_3 + \underline{N_1\bar{V}_{12}}\right)\Delta p = -w_3/A\Delta p,$$

or

$$D_h'' = \frac{x_3c\bar{V}_3^2}{\left(x_1/Đ_{31} + \underline{x_2/Đ_{32}} + x_4/Đ_{34}\right)RT\delta}$$

$$- \underline{\frac{(x_3/Đ_{31} + x_3/Đ_{32})\bar{V}_3N_1}{\underline{(x_1/Đ_{31} + x_2/Đ_{32} + x_4/Đ_{34})\Delta p}}} \tag{3}$$

where $w_3$ is the molar rate of water transport and $\bar{V}_{12}$ is the partial molal volume of the salt.

The counter-ion self-diffusion coefficient $\bar{D}_1$ is defined by $\bar{D} = -N_1/c(dx_1/dz)$, or

$$\bar{D}_1 = \frac{1}{(x_1 + x_{1*})/\left(Đ_{11*} + \underline{x_2/Đ_{12}} + x_3/Đ_{13} + x_4/Đ_{14}\right)} \tag{4}$$

where * denotes the radioactive counter-ion.

In each case, terms involving the co-ion (dashed-underlined terms) are neglected in interpreting the experimental results. Order-of-magnitude estimates indicate that such neglect should not cause significant bias in calculation of the $Đ_{ij}$.

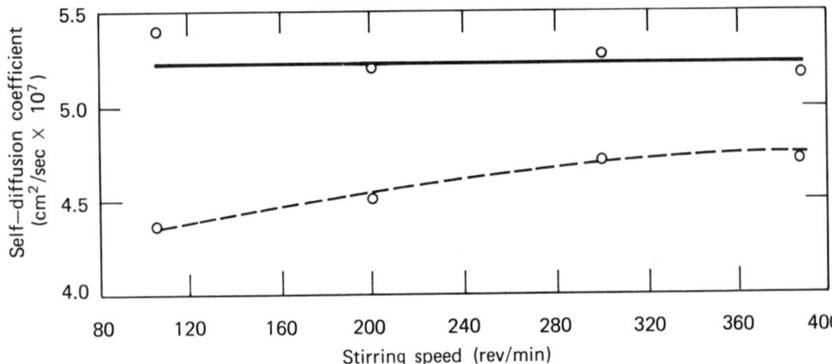


**Figure 3.1.1.** The effect of boundary-layer mass transfer on diffusion measurements. ---, experimental apparent self-diffusion coefficient $\bar{D}_{1A}$; ——, calculated, true self-diffusion coefficient $\bar{D}_1$. From Scattergood and Lightfoot (1968).

The conductivity and electroosmosis measurements are relatively* straightforward and are not discussed here. Extramembrane diffusional resistance is, however, a major problem in both permeability and diffusion measurements. This is seen most readily in the effect of cell stirrer speed on $\bar{D}_1$ shown in Fig. 3.1.1. It may also be seen from this figure that extrapolation to infinite stirrer speed is a doubtful procedure at best. Instead, the corrections shown in this figure were made from separate measurements of the boundary-layer–mass-transfer coefficients.†

The results of the above-described experiments for the dense highly charged sulfonic acid membrane used by Scattergood and Lightfoot (1968) are summarized in Table 3.1.1. along with estimated uncertainties. The corresponding Stefan-Maxwell diffusivities, obtained by simultaneous solution of Eqs. 1 to 4, are presented in Table 3.1.2. The primary entries are calculated for the species $RSO_3^-$, $Ag^+$, $Na^+$, and $H_2O$. Those in parentheses are calculated with the assumption of hydrated ions using the hydration numbers of Table 3.1.1. From these results it is seen that isotope interaction is large; that is, $Đ_{11}$ is small for both the $Ag^+$ and $Na^+$ forms of the resin. Also $Đ_{14}$ is much smaller than $Đ_{34}$, that is, the resin matrix reacts much more strongly with the counter-ion than with water. These

* It is, however, of interest to note that electrophoretic membrane movement is a problem in measuring $t_3$.

† These measurements were made via diffusion-limited electrolysis measurements on a silver plate in place of the membrane. Allowances for differences of effective diffusivity in these experiments and the membrane studies are made possible by the validity of high Schmidt number boundary-layer approximations for electrolytes:

$$Nu_{\text{loc}} = Sc^{1/3} f(\text{Re})$$

**Table 3.1.1.** Raw Data for Characterization of Membrane Transport

| Variable | 0.1 $N$ $AgNO_3$ | 0.1 $N$ NaCl |
|---|---|---|
| $c_1$ ($10^{-3}$ mole/cm$^3$) | 1.140 (±1%) | 1.031 (±1%) |
| $c_4$ ($10^{-3}$ mole/cm$^3$) | 1.135 (±1%) | 1.030 (±1%) |
| $c_3$ ($10^{-3}$ mole/cm$^3$) | 8.39 (±2%) (6.927[a]) | 13.60 (±2%) (10.685[a]) |
| $c$ ($10^{-3}$ mole/cm$^3$) | 10.665 (9.202[a]) | 15.221 (12.746[a]) |
| $x_1$ | 0.1070 (01238[a]) | 0.0676 (0.081[a]) |
| $x_4$ | 0.1065 (0.1232[a]) | 0.0676 (0.081[a]) |
| $x_3$ | 0.7865 (0.753[a]) | 0.8648 (0.0838[a]) |
| $\delta$ (cm) | 0.0197 (±2%) | 0.0196 (±2%) |
| $D_1$ ($10^{-7}$ cm$^2$/sec) | 5.53 (±2%) | 7.40 (±2%) |
| $\kappa$ [$10^{-3}$/($\Omega$) (cm)] | 3.36 (±1%) | 4.95 (±3%) |
| $D_h''$ [$10^{-9}$ cm/(sec) (psi)] | 4.7 (±6%) | 7.4 (±2%) |
| $t_3$ (mole/$F$) | 3.30 (±3%)(3.0[a]) | 7.58 (±4%)(6.08[a]) |
| $\kappa RT/F^2c_1$ ($10^{-7}$ cm$^2$/sec) | 7.8 | 12.8 |

[a] Values assumed with hydration numbers $H_{RSO_3^-} = 0.98, H_{Ag^+} = 0.3, H_{Na^+} = 1.5$.

general conclusions are not appreciably affected by the hydration numbers assigned to the counter-ions. Counter-ion interactions would, however, be weaker for less strongly cross-linked membranes.

It has been customary to attempt approximate descriptions of electrokinetic membrane processes via a variety of pseudobinary techniques (*Nernst-Planck equations*), using a smaller number of correlating parameters than required by the Onsager thermodynamic analysis. The usefulness of these in predicting the course of any one electrokinetic process from observations on another has been tested against the experimental data. They are not reliable for predictive purposes and are particularly poor for predicting the hydrodynamic permeability.

**Table 3.1.2.** Stefan-Maxwell Diffusivities

| Diffusivity ($10^{-7}$ cm$^2$/sec) | 0.1 $N$ $AgNO_3$ | 0.1 $N$ NaCl |
|---|---|---|
| $Đ_{13}$ | 29.2 (±7%) (37.2) | 13.9 (±5%) (20.7) |
| $Đ_{34}$ | 23.9 (±7%)(36.4) | 20.2 (±9%) (29.6) |
| $Đ_{14}$ | 0.94 (±4%) (1.04) | 1.28 (±10%) (1.30) |
| $Đ_{11}$ | 2.61 (±7%) (2.88) | 3.38 (±21%) (2.46) |

## 3.2. Nonisobaric Membrane Transport

In this subsection we provide a formal description of mass transport through a selectively permeable membrane under the influence of a pressure gradient. This situation is pictured schematically in Fig. 3.2.1 for a feed solution containing three solute species (indicated by ellipsoids) of various size. The permeable portions of the membrane are shown as "pores" of varying size. All species move under the combined influence of concentration and pressure gradients, and in general the smallest solutes move through the membrane the most readily. However, we shall see that selectivity is actually determined by a variety of factors, and that it is not always easy to predict. The key feature of the process is the ability of the more mobile species to move against their concentration gradient as a result of the unbalanced fluid pressure. We are thus dealing with a special case of pressure diffusion which permits separating the components of a solution without change of phase, temperature, or the chemical environment. If the excluded solutes are macromolecules, such as proteins, this separation process is known as *ultrafiltration*; if the excluded solutes are small, for example inorganic salts, it is known as *reverse osmosis*. The principle of operation is the same in both cases, and we shall not distinguish between these two processes here.

To avoid unnecessary complications we limit consideration to a binary feed solution consisting of a single solute S and a solvent W, which is

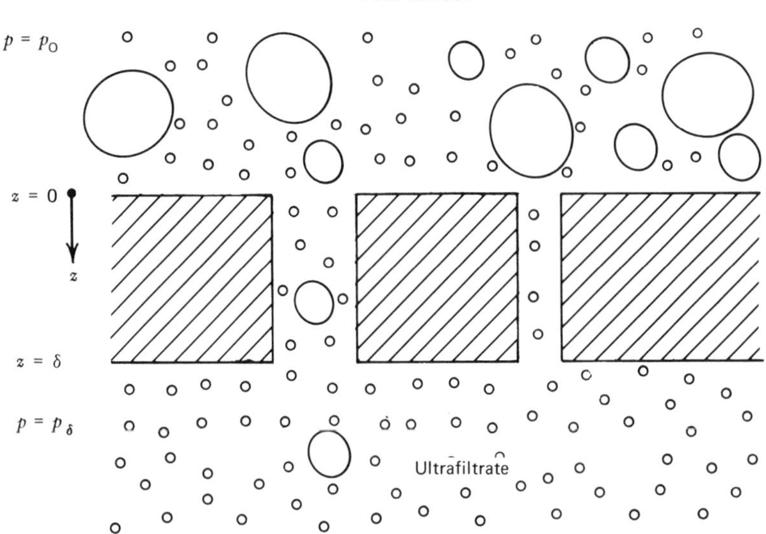


**Figure 3.2.1.** Ultrafiltration, a typical nonisobaric membrane transport process.

generally water. Formal description of this system is given by:

1. A continuity relation for each of the three species in the membrane phase. These may be expressed as

$$N_m = 0 \tag{3.2.1}$$

$$N_S = \text{constant} \tag{3.2.2}$$

$$N_W = \text{constant} \tag{3.2.3}$$

2. Flux equations for two of the three species in the membrane phase, most conveniently chosen as S and W. The most convenient forms of the flux equations vary with application, as indicated below.

3. Boundary conditions at $z=0$ and $\delta$, the "upstream" and "downstream" faces of the membrane respectively. These are only directly measurable in the external solution:

$$\text{At } z=0\,(-), \quad p=p_0; \quad y_S = y_{S0} \tag{3.2.4a;b}$$

$$\text{At } z=\delta\,(+), \quad p=p_\delta; \quad y = y_{S\delta} \tag{3.2.5a;b}$$

Here $y_S$ is the mole fraction of solute in the external solution.

If we wish to interpret the system behavior in terms of the membrane diffusional processes, we also need

4. Interfacial distribution relations. We take these from Ex. 1.5.2 in the form

$$\frac{x_S}{y_S} = K_D = \frac{\gamma_{Sy}}{\gamma_{Sx}} e^{(p-p')\bar{V}_S/RT} \tag{3.2.6}$$

$$p' - p = \frac{RT}{\bar{V}_W} \ln\left(\frac{a_{We}}{a_{Wm}}\right)_{T,p} \tag{3.2.7}$$

where $p'$ = pressure in the membrane
$p$ = pressure in the external solution
$\gamma_{Sx}$ = solute activity coefficient in membrane
$\gamma_{Sy}$ = solute activity coefficient in external solution

It follows from the discussion of Ex. 1.5.2. that $\gamma_{Sy}/\gamma_{Sx}$ is a function only of $y_S$. Furthermore since changes in osmotic pressure *resulting from changes in solution composition* are usually slight* the term $e^{-\pi\bar{V}_S/RT}$ may usually be

* Note that osmotic pressures must be of the order of hundreds of atmospheres for this exponential to be appreciably different from unity. Such *solution* osmotic pressures are rare.

considered constant. Since the activity coefficient ratio $\gamma_{Sy}/\gamma_{Sx}$ is often nearly constant as well, variations in solute distribution coefficients $K_D$ may frequently be neglected.

5. Equations of state. We need here both thermodynamic relations, for the activity coefficients as functions of composition, and some means for estimating the three independent diffusivities.

It is also necessary to define our system carefully, and we begin here.

To do this we note the membrane phase is thermodynamically binary because its equilibrium state is fixed by that of the binary external phase. It follows that the Gibbs-Duhem equation takes the form

$$x_S \nabla_{T,p} \ln a_S + x_W \nabla_{T,p} \ln a_W = 0 \tag{3.2.8}$$

and it is therefore convenient to consider as the interior phase only the imbibed water and solute. We therefore set the mole and volume fractions of membrane matrix in this internal phase equal to zero.

Later in this discussion we have to make specific assumptions about the functional dependence of the activity coefficients and transport properties, but we defer these as long as possible.

We are now ready to begin discussion of specific membrane processes: ultrafiltration and reverse osmosis.

For the situation, pictured in Fig. 3.2.1, the "downstream" solution, or product, is produced entirely by the ultrafiltration process. It follows that the solute mole fraction in the product is

$$y_{S\delta} = \frac{N_S}{N_S + N_W} \tag{3.2.9}$$

where $N_S$ and $N_W$ are the solute and water fluxes through the membrane. The terminal external pressures and the upstream solute concentration are considered known. The system behavior is therefore completely specified.

We now choose to specify $k = i$ in Eq. 1.2.21. The Stefan-Maxwell equations can then be written as

$$x_S \nabla_{T,p} \ln a_S + \frac{\phi_S}{cRT} \nabla p = R_{SW}(x_S N_W - x_W N_S) - r_S N_S \tag{3.2.10}$$

$$x_W \nabla_{T,p} \ln a_W + \frac{\phi_W}{cRT} \nabla p = R_{SW}(x_W N_S - x_S N_W) - r_W N_W \tag{3.2.11}$$

Here

$$R_{SW} = R_{SW} = 1/c\,Đ_{SW},$$

$$r_S = x_m/c\,Đ_{Sm},$$

$$r_W = x_m/c\,Đ_{Wm}.$$

These resistance factors prove more convenient than the diffusivities $Ð_{ij}$, especially in the case of $r_S$ and $r_W$, which avoid introducing the membrane molecular weight.

Before going on to a discussion of real membranes we consider the limiting case of zero penetration, for which completely unambiguous results may be obtained.

## (a) *LIMITING BEHAVIOR OF AN IDEAL MEMBRANE*

If the membrane is completely impermeable to salt, then the internal phase, as above defined, is completely water. Then

$$x_W = c\bar{V}_W = 1$$

and

$$x_S = N_S = 0$$

Any ultrafiltrate will now be pure water, and Eq. 3.2.10 is no longer needed; Eq. 3.2.11 reduces to

$$\frac{d}{dz}\left(\ln(a_W)_{T,p} + \frac{\bar{V}_W}{RT}p\right) = -r_W N_W \qquad (3.2.12a)$$

or

$$\ln(a_W)_{T,p} + \frac{\bar{V}_W}{RT}p\bigg|_0^\delta = -N_W\int_0^\delta r_W\,dz \qquad (3.2.12b)$$

The left side of Eq. 3.2.12*b* can be evaluated in the external phase, since this expression must be continuous across the interface. We may then write

$$N_W = \left(\frac{1}{\delta\bar{r}_W}\right)\left[\ln\left(\frac{a_{W0}}{a_{W\delta}}\right)_{T,p} + (p_0 - p_\delta)\frac{\bar{V}_W}{RT}\right] \qquad (3.2.13)$$

where*

$$\bar{r}_W = \frac{1}{\delta}\int_0^\delta r_W\,dz \qquad (3.2.14)$$

* The hydraulic resistance varies with membrane density and increases with crushing pressure on the membrane. It should be possible to relate it to membrane structure, but not much seems yet to have been done along these lines.

It remains to fix $a_{W\delta}$. Normally we are interested only in producing liquid water at $z=\delta$, and in this case it will be pure water. Then since $c \doteq 1/\overline{V}_W$,

$$N_W \doteq \frac{1}{cRT\delta\bar{r}_W}(p_0 - p_\delta - \pi) \tag{3.2.15}$$

where $\pi = -cRT\ln a_{W0}$ is the osmotic pressure of the feed solution. We then find that a liquid ultrafiltrate can be produced only when the hydrostatic pressure drop exceeds the osmotic pressure of the feed. This is true only for a completely selective membrane. For a real membrane the situation is considerably more complicated, as we show shortly. First, however, we consider the behavior of a nonideal membrane qualitatively.

### (b) *BEHAVIOR OF NONIDEAL MEMBRANES*

The behavior of a nonideal membrane is indicated in Fig. 3.2.2. If the membrane is at all effective, the ultrafiltrate will contain less solute than the feed solution. The bases for this selectivity are:

1. Unfavorable distribution of solute into the membrane.
2. Stronger diffusional interaction of solute than of solvent with the membrane matrix.

In general the solute concentration decreases, as shown, with increasing distance into the membrane, and the solute flux is determined by a

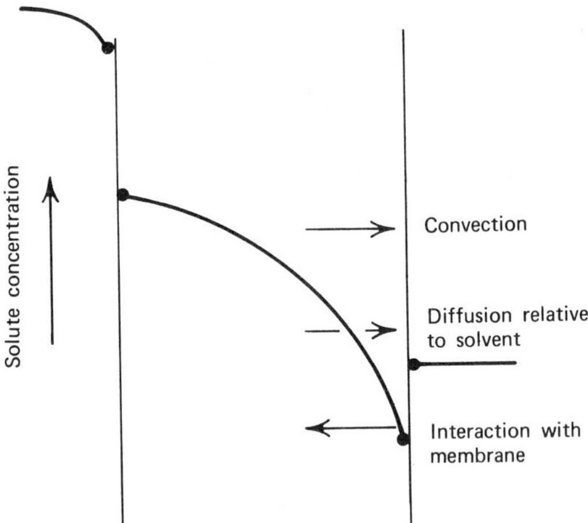


**Figure 3.2.2.** Mechanism of solute rejection.

balance between convection, diffusion relative to the solvent, and diffusional interaction between solute and membrane matrix. The solvent flux is normally determined primarily by diffusional interaction between it and the membrane. The membrane effectiveness can then be increased by:

1. Decreasing the equilibrium solute concentration in the membrane. In this way "convective" transfer—transport by bulk motion—can be reduced.
2. Increasing $r_S$ relative to $r_W$, that is, increasing the relative diffusional interaction of solute with the membrane matrix.
3. Decreasing $Ð_{SW}$ to decrease the diffusion of solute relative to solvent.

These qualitative statements are necessarily oversimplifications, and it is important to extend them with a quantitative description of the actual diffusion process.

This is necessarily more complicated than with ideally selective membranes, because we can no longer work with the sum of the chemical and the hydrostatic potential $[\ln(a_i)_{T,p} + p\bar{V}_i/RT]$. Rather, it is now necessary to separate concentration and pressure to determine the concentration profiles and pressure distribution in the membrane. These results can then be used, together with the interfacial boundary conditions, to relate conditions in the feed solution and the ultrafiltrate.

We begin by simply adding Eqs. 3.2.10 and 3.2.11, and making use of Eq. 3.2.8 to obtain

$$\nabla p = -(r_S N_S + r_W N_W)cRT \tag{3.2.16}$$

Note once again that we consider our system to be the matrix-free membrane phase. Equation 3.2.16 describes the "hydraulic" permeability of the membrane.

We next eliminate the pressure between Eqs. 3.2.10 and 3.2.11 to obtain a mass-transfer relation reminiscent of Fick's first law:

$$\mathbf{N}_S = -c\mathfrak{D}_{Sm}\nabla x_S + x_S\mathbf{N} \tag{3.2.17}$$

where

$$c\mathfrak{D}_{Sm} = \frac{1}{R_{SW} + r_S}\left(\frac{\partial \ln a_S}{\partial \ln x_S}\right)_{T,p} \tag{3.2.18}$$

$$\mathbf{N} = \frac{R_{SW}(\mathbf{N}_W + \mathbf{N}_S) + c\bar{V}_S(r_W\mathbf{N}_W + r_S\mathbf{N}_S)}{R_{SW} + r_S} \tag{3.2.19}$$

The terms $(\partial \ln a_S/\partial \ln x_S)_{T,p}$ and $c\bar{V}_S$ are thermodynamic in nature and can in principle be determined independently of transport processes in the membrane. Both these and the transport properties should normally be nearly concentration independent for the usual situation of $c_S \ll c_W$. In fact, for most applications of biological interest,

$$\frac{\partial \ln a_S}{\partial \ln x_S} \doteq 1; \qquad c\bar{V}_S \doteq \bar{V}_S/\bar{V}_W$$

We may then expect to obtain a first-order description of ultrafiltration by integrating Eqs. 3.2.16 and 3.2.17 across the membrane using average values of these transport and equilibrium parameters.*

The concentration profile inside the membrane is then given, implicitly, by

$$\frac{zN}{c\mathfrak{D}_{Sm}} = \ln\left(\frac{N_S - x_S N}{N_S - x_{S0} N}\right)$$

and the *intramembrane* pressure drop is related to the mass fluxes by

$$p'_0 - p'_\delta = (r_S N_S + r_W N_W)cRT\delta \tag{3.2.20}$$

It remains to relate the internal compositions and pressures to their external counterparts via the relations

$$x_{S0} = y_{S0} K_D; \quad x_{S\delta} = \left(\frac{N_S}{N_S + N_W}\right) K_D \tag{3.2.21}$$

and

$$p'_0 - p'_\delta = p_0 - p_\delta - \frac{RT}{\bar{V}_W}\left[\ln\left(\frac{a_{W\delta}}{a_{W0}}\right) - \ln\left(\frac{a_{W\delta}}{a_{W0}}\right)'\right] \tag{3.2.22}$$

Here primed quantities are for the membrane, and unprimed quantities for the external solution.

The membrane permeation rate, product composition, and intramembrane concentration profile are then given, respectively, by

$$r_S N_S + r_W N_W = \frac{p_0 - p_\delta}{cRT\delta} - \frac{(\pi_e - \pi_m)\bar{V}_W}{RT\delta} \tag{3.2.23}$$

* In practice, however, this tends to underestimate the effectiveness of ultrafiltration, because it neglects compression of the membrane, which becomes appreciable at commonly encountered pressures. [See for example Harriott and Michelsen (1968)].

$$r_S N_S + r_W N_W = \frac{\partial \ln a_S}{\partial \ln x_S}\left[\frac{Y_{S\delta} - K_D y_{S\delta}(N/N_W)}{Y_{S\delta} - K_D y_{S0}(N/N_W)}\right] \tag{3.2.24}$$

$$\frac{z}{\delta} = \ln \frac{Y_{S\delta} - K_D Y_S(N/N_W)}{Y_{S\delta} - K_D Y_{S0}(N/N_W)} \tag{3.2.25}$$

with

$$\pi_e = \left(RT/\bar{V}_W\right)\ln(a_{W\delta}/a_{W0}),$$

$$\pi_m = \left(RT/\bar{V}_W\right)\ln(a_{W\delta}/a_{W0})',$$

$$Y_{S\delta} = y_{S\delta}/(1 - y_{S\delta}) = N_S/N_W,$$

and

$$\frac{N}{N_W} = \frac{R_{SW}(1 + Y_{S\delta}) + c\bar{V}_S(r_W + r_S Y_{S\delta})}{R_{SW} + r_S}.$$

This completes our formal description. It should be noted, however, that one quantity appearing in these equations, the internal osmotic correction $\pi_m$, is not directly measurable. Fortunately, it is usually much smaller than the external correction $\pi_e$, because the membrane phase normally is more nearly pure water than the external solutions in these circumstances. In fact, as we shall see, both osmotic corrections are normally small. Note also that $\bar{V}_W \doteq 1/c$.

It is now of interest to consider representative limiting cases:

1. $K_D \to 0$. This is the situation already described by Eq. 3.2.13; Equation 3.2.14 shows formally how one can allow for membrane compression.

2. $p_0 - p_\delta \to \infty$. Here the osmotic corrections, which remain finite, are overshadowed by the pressure drop, and

$$cRT\delta(r_S N_S + r_W N_W) = p_0 - p_\delta$$

At the same time the argument of the logarithm must become very large, so that

$$Y_{S\delta} \to K_D y_{S0}\left[\frac{R_{SW}(1 + Y_{S\delta}) + c\bar{V}_S(r_W + r_S Y_{S\delta})}{R_{SW} + r_S}\right]$$

In the absence of membrane compression this represents the minimum possible value of $Y_{S\delta}$ and hence the maximum possible solute rejection.

It is clear from this limiting expression that $Y_{S\delta}$ will generally be decreased by a decrease in the distribution coefficient $K_D$. Thus for very small solute concentrations

$$\frac{y_{S\delta}}{y_{S0}} \sim K_D\left(\frac{R_{SW}+c\overline{V}_S r_W}{R_{SW}+r_S}\right)$$

This is not unexpected.

It may also be seen that purification can be obtained with no solute rejection, that is, $K_D=1$. Thus for this case, in the dilute solution limit

$$\frac{y_{S\delta}}{y_{S0}} = \frac{R_{SW}+c\overline{V}_S r_W}{R_{SW}+r_S}$$

There will then be separation if

$$\frac{r_S}{r_W} > c\overline{V}_S = \frac{\overline{V}_S}{\overline{V}_W}$$

3. $p_0 \rightarrow p_\delta$: This situation can perhaps be explained best by looking at behavior when

$$y_{S\delta} \rightarrow y_{S0}$$

Now the argument of the logarithm approaches unity, and the right side of the equation approaches zero. At the same time, both $\pi_e$ and $\pi_m$ approach zero, since there is no concentration difference across the membrane. It follows that $p_0 \rightarrow p_\delta$.

Conversely, if $p_0 \rightarrow p_\delta$ the degree of purification will also approach zero. This is because the diffusion of solute relative to solvent is now sufficiently rapid to prevent concentration differences from developing.

We are now ready to consider some specific examples.

### *EX. 3.2.1. SELECTIVITY AND CONCENTRATION PROFILES FOR FRICTIONALLY INDUCED PURIFICATION*

Describe the ultrafiltration behavior of a system for which

$$K_D = 1.0, \quad y_{S0} = 0.01, \quad \frac{\overline{V}_S}{\overline{V}_W} = 4$$

$$\frac{r_S}{r_W} = 25\frac{\overline{V}_S}{\overline{V}_W}, \quad \frac{R_{SW}}{r_W} = 25$$

These conditions correspond roughly to the experimental measurements obtained by Ginzburg and Katchalsky (1963), and later by Kaufmann and Leonard (1968), for aqueous glucose solutions through cellophane.

SOLUTION

For $K_D = 1.0$ the membrane osmotic pressure is zero. Then $\pi_e - \pi_m = 0$ and

$$\frac{(p_0 - p_\delta)/cRT}{\partial \ln a_S / \partial \ln x_S}\left(4 + \frac{25}{100 y_{S\delta} + 1}\right) = \ln\left(\frac{0.768 y_{S\delta} - 3.2 y_{S\delta}^2}{0.968 y_{S\delta} - 0.00234}\right) \tag{1}$$

$$r_S N_S + r_W N_W = \frac{p_0 - p_\delta}{cRT\delta} \tag{2}$$

The behavior predicted by these equations can be seen to agree with the qualitative remarks made earlier in that the degree of purification rises from zero at no flow toward a high-flow asymptotic value. In the high-flow limit the ultrafiltrate sugar concentration is given by

$$y_{S\delta} \rightarrow \frac{0.00234}{0.968} \doteq 0.00242$$

which is slightly less than one-fourth of $y_{S0}$.

### EX. 3.2.2. SELECTIVITY AND CONCENTRATION PROFILES FOR THERMODYNAMICALLY INDUCED SOLUTE REJECTION

We now take as a specific but hypothetical example a membrane for which

$$\beta = 0.1, \quad y_{S0} = 0.01, \quad \frac{r_W}{r_S} = \frac{\bar{V}_W}{\bar{V}_S}$$

The concentration profile may now be expressed as

$$\frac{z}{\delta} = \ln\left(\frac{y_{S\delta} - 0.1 y_S}{y_{S\delta} - 0.1 y_{S0}}\right) \Big/ \ln\left(\frac{0.9 y_{S\delta}}{y_{S\delta} - 0.1 y_{S0}}\right) \tag{1}$$

The relation between the pressure drop and the terminal concentrations is

$$\frac{p_0 - p_\delta}{cRT} = \ln\left(\frac{0.9 y_{S\delta}}{y_{S\delta} - 0.1 y_{S0}}\right) \Big/ \left[\frac{\bar{V}_S}{\bar{V}_W} + \frac{R_{SW}}{r_W} + \ln\left(\frac{a_{W\delta}}{a_{W0}}\right)_e - \ln\left(\frac{a_{W\delta}}{a_{W0}}\right)_m\right] \tag{2}$$

This relation is plotted in Fig. 3.2.3. It is clear from Eq. 2 that the limiting ratio of product solute concentration to feed concentration is given by

$$y_{S\delta}/y_{S0} \rightarrow 0.1$$

as $P_0 - P_\delta \rightarrow \infty$.
As discussed above, this is the expected result. For finite pressure drops, however, it remains to consider the osmotic corrections. This requires thermodynamic data, and for convenience we assume here that the activity coefficient of water is concentration independent in both phases. It follows that

$$\ln\left(\frac{a_{W\delta}}{a_{W0}}\right)_e - \ln\left(\frac{a_{W\delta}}{a_{W0}}\right)_m \doteq \ln\left(\frac{y_{W\delta}}{y_{W0}}\frac{x_{W0}}{x_{W\delta}}\right)$$

$$\doteq \ln\left(\frac{1-y_{S\delta}}{1-y_{S0}}\frac{1-0.1y_{S0}}{1-0.1y_{S\delta}}\right) \qquad (3)$$

The membrane phase contributes little to this osmotic correction because it is much more nearly pure water than the corresponding external solution. Since $y_S << 1$ over the whole possible concentration range, we may rewrite the osmotic correction as

$$\ln\left(\frac{a_{W\delta}}{a_{W0}}\right)_e - \ln\left(\frac{a_{W\delta}}{a_{W0}}\right)_m \doteq \ln\left[1+0.9(y_{S0}-y_{S\delta})\right]$$

$$\doteq 0.9(y_{S0}-y_{S\delta}) \qquad (4)$$

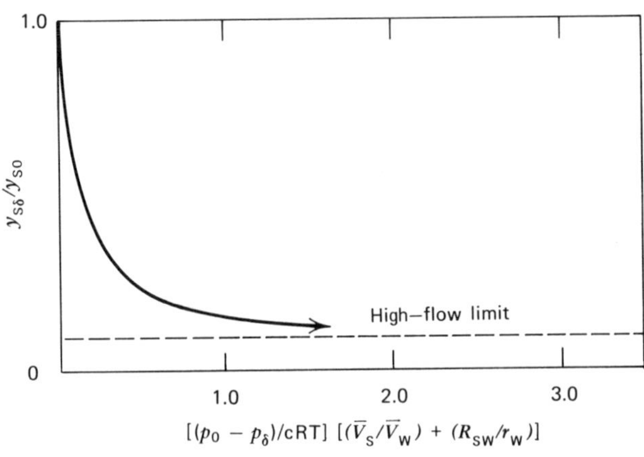


**Figure 3.2.3.** Effect of pressure drop on solute rejection for the conditions of Ex. 3.2.2.

It can be seen that this will normally be a small correction. To approach the high-flow purification potential of the membrane will require a dimensionless pressure drop of the order of unity. The osmotic correction, however, can never exceed 0.008. It should be kept in mind, however, that concentration boundary layers in the "upstream" solution can result in rather large $y_{S0}$ even if the solute concentration in the bulk of the feed solution is small. This problem of "concentration polarization" is discussed in Chapter IV, Section 2. Such osmotic corrections can then be large. In any event the purification–pressure-drop relation is qualitatively much like that of the previous example of frictionally induced purification.

### 3.3. The Integrated Flux Equations of Kedem and Katchalsky

In this section we develop formal integrations of the membrane transport equations in terms of space-averaged values of the $Đ_{ij}$. This is a quite different approach than that of the last section, where we started with specific assumptions about the concentration dependence of the thermodynamic and transport properties. The absence of such assumptions makes it impossible to calculate concentration profiles. However, we are able by this approach to develop rather generally applicable formal relations between mass fluxes and directly measurable properties: compositions and pressures in the solutions bathing the membrane.

The two approaches are thus complementary, and they become identical in the limit of small fractional changes in composition and pressure across the membrane.

Our specific purpose is to justify the Kedem-Katchalsky equations introduced earlier, and to show their relation to the Stefan-Maxwell diffusion coefficients on which all our analyses have been based. To do this we start with "Kirkwood formulation" in which all species velocities are referred to the membrane, which is considered stationary. Equation 1.2.21 may then be written in the form

$$\nabla_{T,p} \ln a_i + \frac{\bar{V}_i}{RT}\,\Delta p = \sum_{j=\mathrm{S,W}} \frac{\mathbf{N}_j}{cĐ_{ij}} \qquad (i=\mathrm{S,W}) \tag{3.3.1}$$

Since the $N_j$ are constant across the membrane, this expression can be integrated formally to

$$\ln\left(\frac{a_{i\delta}}{a_{i0}}\right)_{T,p} + \frac{\bar{V}_{im}}{RT}(p_\delta - p_0) = \sum_{j=1}^{2} \left(\frac{\delta}{cĐ_{ij}}\right)_m N_j \tag{3.3.2}$$

where

$$\left(\frac{\delta}{c\mathcal{D}_{ij}}\right)_m = \int_0^{\delta} \frac{dz}{c\mathcal{D}_{ij}} \equiv r_{ij} \tag{3.3.3}$$

and $\bar{V}_{im}$ is the corresponding mean partial molal volume. Since the dimensionless potential $\mu^*_i = (\ln(a_i)_{T,p} + \bar{V}_{im}p/RT)$ is continuous across the membrane-solution interface, the left side of Eq. 3.3.2 may be taken to represent external activities and pressures—which are directly measurable. This is the primary advantage of the Kedem-Katchalsky approach. However, the space-average transport properties of Eq. 3.3.3 depend upon the concentration and pressure distributions within the membrane, and thus depend upon the nature of the transport processes taking place as well as the terminal conditions. They are ambiguous except in the limit of small fractional changes in pressure and composition; this is the fundamental weakness of the approach used.

The chemical driving forces may be rewritten in terms of mole fractions and the corresponding activity coefficients. Thus

$$\ln \frac{a_{S0}}{a_{S\delta}} = \frac{a_{S0} - a_{S\delta}}{a_{S\ln}} \equiv \frac{x_{S0} - x_{S\delta}}{x_{S\ln}} \kappa_{\gamma S} \tag{3.3.4}$$

where the subscript ln refers to the logarithmic mean of terminal values. The correction $\kappa_{\gamma S}$ for variation in external activity coefficients, defined by Eq. 3.3.4, reduces to unity for small fractional changes in external solute concentration, or for ideal external solutions. We may similarly write for water

$$\ln \frac{a_{W0}}{a_{W\delta}} = -\int_{a_{S0}}^{a_{S\delta}} \frac{x_S}{x_W} \, d\ln a_S = \frac{x_{S0} - x_{S\delta}}{x_{W\ln}} \kappa_{\gamma W} \tag{3.3.5}$$

Again $\kappa_{\gamma W}$ tends to unity as the fractional concentration change or departures from ideality become small. We may now put these results into Eq. 3.3.2 to obtain

$$r_{SS}N_S + r_{SW}N_W = (x_{S\delta} - x_{S0}) \frac{\kappa_{\gamma S}}{x_{S\ln}} + (p_\delta - p_0) \frac{\bar{V}_{Sm}}{RT} \tag{3.3.6}$$

$$r_{SW}N_S + r_{WW}N_W = -(x_{S\delta} - x_{S0}) \frac{\kappa_{\gamma W}}{x_{W\ln}} + (p_\delta - p_0) \frac{\bar{V}_{Wm}}{RT} \tag{3.3.7}$$

The solute and solvent fluxes can then be expressed as linear combinations of the solute concentration and pressure differences between the two external solutions. Since $J_v$ and $J_D$ (Eqs. 1.6.6 to 1.6.8) are simply weighted sums of $N_S$ and $N_W$, they must also be linearly related to the concentration and pressure differences. The general form of Eqs. 1.6.3 and 1.6.4 is thus obtained. The detailed relations between the Kedem-Katchalsky coefficients and the $Ð_{ij}$ are complex and are left to the problems. We merely note here that the reciprocal relation, Eq. 1.6.5, holds only for ideal systems. In biological situations, however, departures from thermodynamic ideality are often small compared with uncertainties of measurement.

The inability of the foregoing formulation to describe solute-solute interactions, electrodiffusion, and facilitated diffusion is generally more serious in practice than the ambiguities mentioned above. It should also be kept in mind that the $Ð_{SS}$ which play an important role in this formulation are typically strongly concentration dependent. From Eq. 1.2.23,

$$\frac{1}{Ð_{SS}} = -\frac{1}{x_S}\frac{x_m}{Ð_{Sm}} + \frac{x_W}{Ð_{SW}}$$

It is generally observed that $x_S/Ð_{SS}$ is constant to a first approximation. Because of this the coefficient $L_D$, for example, is in general strongly concentration dependent and awkward to use for computational purposes. As indicated in Section 1.6(*b*), it is normally more convenient to work with the set $L_p$, $\omega$, and $\sigma$.

## 3.4. Transport across Charged Membranes

Here we consider briefly the effects of fixed charges, that is, charged sites attached to the membrane matrix. Transport in systems of this type is of widespread biological importance and has been much studied. It is inherently multicomponent in nature and typically involves many diffusing species. Most analyses of transport in charged membranes have, however, been limited by lack of thermodynamic and transport data to the pseudobinary Nernst-Planck approach introduced earlier in connection with diffusion in free solution. Many of these analyses are quite old and are reviewed in standard texts, for example Helfferich (1962). However, all such pseudobinary treatments are fundamentally incomplete and must be viewed with suspicion. In particular, as pointed out by Scattergood and Lightfoot (1968), it is dangerous to predict transport behavior for boundary conditions different from those for which available pseudobinary diffusion coefficients have been obtained. The electrochemistry of porous membranes has been reviewed more recently by Sollner (1969).

Diffusion in charged membranes differs from diffusion in free solution in these essential respects:

1. *One charged species, the membrane m, is immobilized, so that*

$$N_m = 0 \tag{3.4.1}$$

2. *The concentration of the fixed ions is determined by the characteristics of the membrane and is usually considered to be constant.*

3. *The generalized driving force* $\mathbf{d}_i$ *is given by Eq. 1.4.9 for the mobile species.*

4. *Membrane transport is effectively steady and one-dimensional.*

In other respects the two situations are essentially similar. It is, for example, normally possible to assume electrical neutrality for the membrane phase as a whole, to a very high order of accuracy. (This is *not*, however, always true for the bimolecular membranes discussed in the next chapter.)

We may provide a compact but still widely useful description of membrane transport by the following modifications of Eqs. 2.4.1 to 8. First we use the stationary matrix rather than water as the reference species, so that the fluxes $\mathbf{N}_i$ replace the $\mathbf{J}_i^\circ$ of Section 2.4, and we specialize the description to steady one-dimensional transport. Equations 2.4.1 to 2.4.4 may then be replaced by

$$\frac{\partial N_i}{\partial z} + R_i = 0 \qquad (i = 1, 2, \ldots, n, \neq m) \tag{3.4.2}$$

$$\frac{\partial}{\partial z}\left(\ln(a_i)_{T,p} + \frac{\nu_i \mathfrak{F}}{RT}\phi + \frac{\overline{V}_i}{RT}p\right) = \sum_{\substack{j=1 \\ \neq m}}^{n} \frac{N_i}{c\mathcal{D}_{ij}}$$

along with Eq. 3.4.1. Equations 2.4.6 and 2.4.7 take the modified forms

$$\sum_{\substack{i=1 \\ i \neq m}}^{n} \nu_i c_i = -\nu_m c_m, \qquad \sum_{i=1}^{n} \nu_i N_i = 0$$

where $c_m$ is normally considered to be constant.

This is essentially the formulation used to obtain the Kedem-Katchalsky relations in the last section, and it is not normally convenient for computation: once again we must consider the concentration dependence of the diffusivities, and most particularly the $\mathcal{D}_{ii}$.

The most commonly considered situation is that for which only diffusional interaction with the membrane is considered. Then

$$Ð_{im} \ll Ð_{ij}, \qquad j \neq i, m$$

so that

$$Ð_{ii}^{-1} = \frac{x_m}{x_i} \frac{1}{Ð_{im}}$$

and

$$N_i = -cÐ_{im} x_i \frac{\partial}{\partial z} \left[ \ln(a_i)_{T,p} + \frac{\nu_i \mathfrak{F}}{RT}\phi + \frac{\bar{V}_i}{RT} p \right] \tag{3.4.6}$$

This is another example of the Nernst-Planck approximation introduced in Section 1.6. We shall see that it is not always adequate for the interpretation of experimental observations, but it does offer the advantage of containing only a single diffusional parameter, the effective binary diffusivity $Ð_{im}$. For most situations considered to date, $Ð_{im}$ appears to be concentration independent to a first approximation. We consider immediately below two examples where Eq. 3.4.6 provides useful insight, and we use the results of these examples in the next chapter.

It does appear that solute-solute interactions are important in biological as well as artificial membranes, but it is not clear how to take these interactions into account. We therefore content ourselves with three relatively simple examples.

### *EX. 3.4.1. THE USSING RELATION AND UNIDIRECTIONAL ION FLUXES*

To avoid active transport during the measurement of ion permeabilities through cell membranes it is common practice to poison the membrane—that is, to inactivate the enzymes responsible for active transport. To determine if this procedure has been successful it is common practice to resort to tracer experiments which, with the aid of a mass balance for the ion in question, permit a determination of the fluxes of ions moving across the membrane in each direction.

Develop a relation between these *unidirectional* fluxes and the ion-concentration and electrostatic-potential differences across the membrane.

SOLUTION

Since it is normally impossible to obtain all of the diffusivities required by Eq. 3.4.3, we begin with the Nernst-Planck approximation, Eq. 3.4.6. We also follow the suggestion of Ussing and write this equation in the form

$$N_i = -c\mathfrak{D}_{im}x_i \frac{d\ln a_i^{(\text{tot})}}{dz} \tag{1}$$

where

$$a_i^{(\text{tot})} = (a_i)_{T,p} e^{\nu_i(\mathfrak{F}/RT)\phi} \tag{2}$$

It follows that

$$N_i = -\frac{c\mathfrak{D}_{im}}{\gamma_i} e^{-\nu_i(\mathfrak{F}/RT)\phi} \left( \frac{da_i^{(\text{tot})}}{dz} \right) \tag{3}$$

If we now consider interdiffusion of an "untagged" ion $M^+$ and a radioactive isotope $M^{*+}$ we may write

$$\nu_{M^+} = \nu_{M^{*+}}$$

and

$$\mathfrak{D}_{Mm} \doteq \mathfrak{D}_{M^*m} \tag{4}$$

It follows that

$$\frac{N_M}{N_{M^*}} \doteq \frac{da_M^{(\text{tot})}}{da_{M^*}^{(\text{tot})}}$$

$$= \frac{a_{M2}^{(\text{tot})} - a_{M1}^{(\text{tot})}}{a_{M^*2}^{(\text{tot})} - a_{M^*2}^{(\text{tot})}} \tag{5}$$

where the subscripts 1 and 2 refer to the two solutions surrounding the membrane. We may now consider $M^+$ ions initially on opposite sides of the membrane as distinct "isotopes" and set

$$a_{M1}^{(\text{tot})} = a_{M^*2}^{(\text{tot})} = 0 \tag{6}$$

For this situation $\mathfrak{D}_{Mm} = \mathfrak{D}_{M^*m}$ exactly and

$$-\frac{N_{M1\to2}}{N_{M2\to1}} = \frac{(\gamma_M x_M)_2}{(\gamma_M x_M)_1} e^{\nu_M(\phi_2-\phi_1)(\mathfrak{F}/RT)} \tag{7}$$

where $N_{M1\to2}$ = flux of ions originally in compartment 1 toward compartment 2 and similarly for $N_{M2\to1}$. These *unidirectional fluxes* can be determined by putting an actual tracer in one compartment. In this case $\mathfrak{D}_{Mm}$ and $\mathfrak{D}_{M^*m}$ differ slightly because of the different atomic weights of the two isotopes.

As we later see Eq. 9 does provide a convenient test for passive transport; it has been very widely used. It is important to note in this connection that it does not require either electroneutrality for the membrane phase as a whole, or even immobilization of the fixed charges, and it does not require a knowledge of the diffusion coefficients. It is therefore very widely applicable. Note that the individual ion activity coefficients appearing in this equation can be eliminated by considering the electrodes required to complete the circuit, just as for the diffusion in free solution.

### *EX. 3.4.2. BI-IONIC POTENTIALS*

Two binary salt solutions sharing a single anion but containing different cations are placed on opposite sides of a membrane. Assume for convenience that: (1) the membrane is completely impermeable to anions, (2) both cations are univalent,* (3) water movement and current flow are negligible, and (4) the concentration $c_{R^-}$ of fixed charges is uniform across the membrane.

We also limit discussion to the Nernst-Planck approximation. Our system description then takes the form

$$\frac{d\ln a_M}{dZ} + \frac{\mathfrak{F}}{RT}\frac{d\phi}{dZ} = -\frac{N_M}{c_M \mathfrak{D}_{Mm}} \tag{1}$$

$$\frac{d\ln a_N}{dZ} + \frac{\mathfrak{F}}{RT}\frac{d\phi}{dZ} = -\frac{N_N}{c_N \mathfrak{D}_{Nm}} \tag{2}$$

with

$$N_M + N_N = 0 \tag{3}$$

$$c_M + c_N = c_R \tag{4}$$

and where $z$ is the direction of transport of M.

We may first eliminate electrostatic potential $\phi$ between these equations to obtain

$$N_M = \frac{\mathfrak{D}_{Mm}\mathfrak{D}_{Nm}}{c_M\mathfrak{D}_{Mm} + c_N\mathfrak{D}_{Nm}} c_M c_N \left( \frac{d\ln a_M}{dZ} - \frac{d\ln a_N}{dZ} \right) \tag{5}$$

or

$$N_M = \frac{\mathfrak{D}_{Mm}\mathfrak{D}_{Nm} c_M (c_R - c_M)}{c_M(\mathfrak{D}_{Mm} - \mathfrak{D}_{Nm}) + c_R \mathfrak{D}_{Nm}} \frac{d\ln(a_M/a_N)}{dc_M} \frac{dc_M}{dZ} \tag{6}$$

* Helfferich discusses a more general case in which the first two restrictions are removed and the cations are of different charge; the simpler situation discussed here will, however, prove sufficient for our purposes.

Here the activities and concentrations refer to the membrane phase. All of the terms on the right side of Eq. 6 depend only on $c_M$ and can in principle be evaluated from thermodynamic measurements, just as for diffusion in free solution. If we may assume that the activity coefficients and diffusivities are constant, the integration may be carried out explicitly to give

$$N_M = \frac{c_R \mathfrak{D}_{Mm} \mathfrak{D}_{Nm}}{\delta \mathfrak{D}_{ln}} \tag{7}$$

where

$$\mathfrak{D}_{ln} = \frac{\mathfrak{D}_{Mm} - \mathfrak{D}_{Nm}}{\ln(\mathfrak{D}_{Mm}/\mathfrak{D}_{Nm})} \tag{8}$$

is the logarithmic mean of the two effective binary diffusivities for the ions, and $\delta$ is the membrane thickness. Note that the ion flux does not depend on the ion concentrations in the external solution; this is because the terminal membrane concentrations are independent of solution concentration—in the limit of *an ideally permselective membrane*, that is, one containing no co-ion. The effect of counter-ion interactions on exchange rates has been investigated by Gottlieb (1968).

We can now eliminate the ion fluxes and relate the potential between the two solutions to their compositions and the membrane properties. If the membrane-phase activity-coefficient ratio is constant, then

$$\phi_M - \phi_N = -\frac{RT}{\mathfrak{F}} \ln\left[\frac{\mathfrak{D}_{Mm}}{\mathfrak{D}_{Nm}}\left(\frac{a_M}{a_N}\right)_{ext}\left(\frac{\gamma_N}{\gamma_M}\right)_m\right] \tag{9}$$

Here $\phi_M - \phi_N$ = the potential in the MX solution relative to that of the NX

$a_M, a_N$ = activities of $M^+$ and $N^+$ in their respective solutions, and

$\gamma_M, \gamma_N$ = membrane-phase activity coefficients *referred to the prevailing pressure in the membrane* (and thus containing an osmotic-pressure correction).

Bi-ionic potentials are fundamentally complex and cannot be simply described for nonideal membranes. However, here the qualitative explanation is simple: The more mobile ion tends to move faster and thus builds up a charge on the far side of the membrane, to retard it and speed up the less mobile ion. The potential developed depends on the mobility ratio of the two ions, the concentration ratio of the two solutions, and the equilibrium constant $[(\gamma_M/\gamma_N)_{ext}(\gamma_N/\gamma_M)_m]$ for the exchange reaction.

*EX. 3.4.3. SALT DIFFUSION AND DIFFUSION POTENTIALS IN CHARGED MEMBRANES*

Repeat Ex. 1.3.1 and 1.3.2 for a system in which the diffusion path is a nonideal charged membrane of uniform charge density $c_{R^-}$. Do not neglect diffusional interactions, but assume the fixed charge density is considerably greater than that of the co-ion, here $X^-$, and that the solutions bathing the membrane are at essentially the same hydrostatic pressure.

SOLUTION

Because of the small difference in osmotic corrections and the absence of a net hydrostatic head it is reasonable to neglect the pressure gradients of Eq. 3.4.3. We may then write

$$\frac{d\ln a_M}{dz}+\frac{\mathfrak{F}}{RT}\frac{d\phi}{dz}=N_S(r_{MM}+r_{MX})+N_W r_{MW} \tag{1}$$

$$\frac{d\ln a_x}{dz}-\frac{\mathfrak{F}}{RT}\frac{d\phi}{dz}=N_S(r_{XX}+r_{MX})+N_W r_{XW} \tag{2}$$

$$\frac{d\ln a_w}{dz}+0=N_W r_{WW}+N_S(r_{MW}+r_{XW}) \tag{3}$$

where $r_{ij}=1/c\mathcal{D}_{ij}$. The requirement of no current flow has been used to replace $N_M$ and $N_X$ by the salt flux $N_S$.

We may next eliminate the electrostatic potential between Eqs. 1 and 2 to obtain

$$\frac{d\ln a_S}{dz}=N_S(r_{MM}+2r_{MX}+r_{XX})+N_W(r_{MW}+r_{XW}) \tag{4}$$

This may be rearranged to

$$N_S=-c\mathcal{D}_{S,\text{eff}}\frac{\partial\ln a_S}{\partial\ln x_S}\frac{dx_S}{dz}+\kappa x_S(N_S+N_W) \tag{5}$$

where

$$c\mathcal{D}_{S,\text{eff}}=[-x_S(r_{MM}+2r_{MX}+r_{XX}-r_{MW}-r_{XW})]^{-1}$$

$$\kappa=-\frac{r_{MW}+r_{XW}}{x_S(r_{MM}+2r_{MX}+r_{XX}-r_{MW}-r_{XW})}$$

Note that Eq. 5 becomes equivalent to Eq. 3.4.6 when solute membrane interactions become dominant.

It remains, however, to determine the ratio $N_S/N_W$ to calculate mass fluxes from Eq. 5. To do this we take advantage of the relation

$$x_S \frac{\partial \ln a_S}{\partial z} + x_W \frac{\partial \ln a_w}{\partial z} = 0 \tag{6}$$

and eliminate the concentration gradients between Eqs. 3 and 4 to obtain

$$\frac{N_W}{N_S} = -\frac{x_S(r_{MM}+2r_{MX}+r_W)-x_W(r_{MW}+r_{XW})}{x_W r_{WW} - x_S(r_{MW}+r_{XW})} \tag{7}$$

Equations 5 and 7 complete the description of mass transfer, except for boundary conditions and equations of state.

The diffusion potential can now be calculated from the concentration profiles, which may be considered known, by the use of Eq. 2.4.17. It is only necessary to replace $\mathbf{J}_i^0$ by $N_i$, and the reference species from water to the membrane matrix, to use the results of Section 2.4 here. We may thus write (after considerable rearrangement):

$$\frac{\mathfrak{F}}{RT}(\phi_2-\phi_1) = \ln\frac{a_{X2}}{a_{X1}} + \int_1^2 \left[\left(\frac{\alpha_M}{\alpha_X-\alpha_M}\right) + \frac{x_S}{x_W}\left(\frac{\alpha_W}{\alpha_X-\alpha_M}\right)\right] d\ln a_S \tag{8}$$

where the $\alpha_i$ are as defined by Eq. 2.4.19. The term in $a_{X2}/a_{X1}$ can be eliminated by addition of the electrode potentials as in Eq. 2.4.25. Although complicated by inclusion of the stoichiometric coefficient $\alpha_W$ for water, Eq. 8 shows that the diffusion potential depends only on terminal concentrations—as it should for a thermodynamically binary system. Furthermore for an uncharged membrane ($x_M = x_R$) the term $\alpha_W$ is identically zero, and Eq. 8 takes the same form as for a binary free solution (see Ex. 1.3.2 or Eq. 4, Ex. 2.4.2).

Equation 8 also simplifies for a highly charged membrane. Thus if $c_X \ll c_M$ and diffusional reactions with the membrane matrix dominate diffusional behavior, then

$$\frac{\mathfrak{F}}{RT}(\phi_2-\phi_1) = \ln\frac{a_{X2}}{a_{X1}} - \int_1^2 t_M\, d\ln a_S \tag{9}$$

where

$$t_M = \frac{c_M Đ_{MR}}{c_M Đ_{MR} + c_X Đ_{XR}} \tag{10}$$

and the subscript R refers to the matrix with its associated fixed charges $R^-$. This limiting equation differs from its free-solution counterpart only in the weighting of the diffusivities by the (local) concentration of the mobile ions. It appears likely that the remaining diffusional interactions are also important, but there are very few (if any) systems for which sufficient data are available to determine their effects quantitatively.

Equation 9 could have been obtained directly from Eq. 3.4.6 by the elimination of pressure gradients, and such pseudobinary approaches are normally used for the interpretation of membrane potentials. The approach used in this example does, however, offer the advantage of showing each simplification explicitly.

## BIBLIOGRAPHY

### Books and Reviews

Altgelt, K. H., and J. C. Moore, "Gel Permeation Chromatography," in *Polymer Fractionation*, M. J. R. Cantow, Ed., Academic Press (1967). (Mass transport in gels.)

American Physiological Society, *Symposium: Biological and Artificial Membranes*, Federation Proceedings **27**, 6, 1249-1309 (1968).

Clarke, H. T., Ed., *Ion Transport Across Membranes*, Academic Press (1954).

Crank, John, and G. S. Park, *Diffusion in Polymers*, Acad. Press (1968).

Helfferich, Friedrich, *Ion Exchange*, McGraw-Hill (1962).

Katchalsky, A., and P. F. Curran, *Non-equilibrium Thermodynamics in Biophysics*, Harv. Univ. Press (1965).

Katz, Bernard, *Nerve, Muscle, and Synapse*, McGraw-Hill (1966).

Keller, K. H., and E. F. Leonard, *Application of Chemical Engineering to Problems in Biology and Medicine, Part I: Engineering Analysis of the Functions of Blood*, Am. Inst. of Chem. Eng. Twelfth Adv. Semin., Am. Inst. Chem. Engrs. (1969).

Lakshminaraianaiah, N., *Transport Phenomena in Artificial Membranes, Chem. Rev.* **65**, 491 (1965).

Lavallée, M., O. F. Schanne, and N. C. Hebert, Eds., *Glass Micro-electrodes*, Wiley (1969).

Lonsdale, H. K., *Separation and Purification by Reverse Osmosis*, in *Prog. in Sep. and Purif.*, **3**, 191 (1970).

Merten, Ulrich, *Desalination by Reverse Osmosis*, Mass. Inst. of Tech. Press (1966).

Sollner, Karl, "The Electrochemistry of Porous Membranes, with Particular Reference to Ion Exchange Membranes and Their Use in Model Studies of Biophysical Interest," *J. Macromol. Sci.—Chem.*, **A-3**, 1–84 (1969).

Sourirarjan, S., *Reverse Osmosis*, Acad. Press (1970).

Stein, W. D., *The Movement of Molecules Across Cell Membranes*, Academic Press (1967).

## Periodical and Secondary References

Brian, P. L. T., *Ind. Eng. Chem. Fund.*, **4**, 439 (1965).

Craig, L. C., *Science*, **144**, 1093 (1964).

Craig, L. C., J. D. Fisher, and T. P. King, *Biochem.*, **4**, 311 (1965).

Craig, L. C., W. Konigsberg, A. Stracher, and T. P. King, *The Proteins*, Methuen (1958).

Eisenman, George, "Ion Permeation of Cell Membranes and Its Models," Am. Physiol. Soc. Symp., *op. cit. sup.*, p. 1249.

Eisenman, George, S. M. Ciani, and Gabor Szabo, *op. cit.*, 1289.

Evans, R. B., III, G. M. Watson, and E. A. Mason, *J. Chem. Phys.*, **16**, 1894–1902 (1962).

Ginzburg, B. Z., and A. Katchalsky, *J. Gen. Physiol.*, **47**, 403–418 (1963).

Gottlieb, M. H., "The Rates of Interexchange of Ions across Fixed-charge Membranes in Bi-ionic Systems," *Biophys. J.* **8**, 1413–1425 (1968).

Harriott, P., and D. L. Michelsen, OSW Res. Dev. Report, Grant 14-01-0001-750 (Dec. 1967).

Hemmingsen, E., and P. F. Scholander, *Science*, **132**, 1379 (1960).

Henderson, W. E., and C. Sliepcevich, *Chem. Eng. Prog. Symp. Ser.*, No. 24, **55**, p. 145.

Kaufmann, T. G., and E. F. Leonard, *Am. Inst. Chem. Eng.*, **14**, 110–117 (1968).

Kedem, O., and A. Katchalsky, *Biochem. et Biophys. Acta*, **27**, 229 (1958).

Keller, K. H., and S. K. Friedlander, *J. Gen. Physiol.*, **49**, 663 (1966).

Kirkwood, J. G., in Clarke, *op. cit. sup.*

Marangozis, J., and A. I. Johnson, *Can. J. Chem. Eng.*, **40**, 231 (1962).

Merten, Ulrich, *Desalination by Reverse Osmosis*, Mass. Inst. of Tech. Press (1966).

Newman, John S., *Electrochemical Systems*, Prectice-Hall (1973); "Transport Processes in Electrolyte Solutions," *Adv. Electrochem. Eng.*, **5**, 87 (1967).

Raridon, R. J., L. Dresner, and K. A. Kraus, *Desalination*, **1**, 210 (1966).

Scattergood, E. M., and E. N. Lightfoot, "Diffusional Interaction in an Ion-exchange Membrane," *Trans. Faraday Soc.*, **64**, No. 544, Part 4, 1135–1146 (April, 1968).

Smith, K. A., C. K. Colton, E. W. Merrill, and L. B. Evans, "Conective Transport in a Batch Dialyzer," *The Artificial Kidney*, R. L. Dedrick, K. B. Bischoff, and E. F. Leonard, Eds., *Chem. Eng. Prog. Symp. Series*, **64**, No. 84 (1968), pp. 45–58.

Spiegler, K. S., *Trans. Faraday Soc.*, **54**, 1409 (1958).

Ussing, H. H., "The Distinction by Means of Tracers between Active Transport and Diffusion," *Acta Physiol. Scand.*, **19**, 43–56.

## 4. MASS TRANSPORT ACROSS BIOLOGICAL MEMBRANES AND BIMOLECULAR LIPID MODELS

We now turn our attention to the very important and extremely challenging subject of mass transport through biological membranes. We summarize some of the more important and better understood material in this fast-changing area, but make no attempt at complete coverage. A much more comprehensive treatment is provided in the three-volume work *Membranes and Ion Transport* edited by E. E. Bittar (1971).

We begin with a brief review of the probable structure of natural membranes followed by a description of their characteristic transport behavior. We then proceed to a more detailed discussion of mass transfer in the chemically well-defined *black lipid* bilayers widely used as models for natural membranes.

## 4.1. The General Properties and Probable Structure of Natural Membranes

The structure and organization of living cells has already been touched on in the introductory section of these notes and illustrated by a drawing of a pancreatic cell taken from Mahler and Cordes. It can be seen from this drawing, Fig. 2.1.2 of Chap I, that selectively permeable membranes play a key role in the control of the complex integrated chemical processes taking place in living cells. We are particularly concerned here with the outer boundary, the cell (or plasma) membrane. This appears to be a highly organized structure consisting of polar lipids and proteins. It is not homogeneous, but is a mosaic of different functional units, differing slightly in structure, highly selective and specialized in individual cases, and controlling selective permeability, the active transport of nutrients and ions, and other functions of less relevance to the present discussion. Plasma membranes can, for example, serve as sites for such integrated enzymatic processes as glycolysis in yeasts. The general characteristics of cell membranes are reviewed very effectively by Stein (1967), and the reader interested in a general introduction is referred to his monograph for a more extended discussion.

Our knowledge of the cell membrane can be summarized as follows:

1. Nearly all cell membranes so far examined in the electron microscope show the "unit membrane" structure of two dark osmiophilic layers separated by a light layer.

2. Cell membranes isolated and analyzed biochemically consist for the most part of cholesterol, phospholipid, and protein. Typical examples of these lipids are shown in Fig. 4.1.1.

3. Model studies show that the osmiophilic regions are the charged portions of the phospholipid molecules and proteins.

4. A particular cell membrane, that of the nerve myelin sheath, has been shown by x-ray crystallography to be composed of a bimolecular layer of lipid molecules with the hydrocarbon chains forming an orderly array,

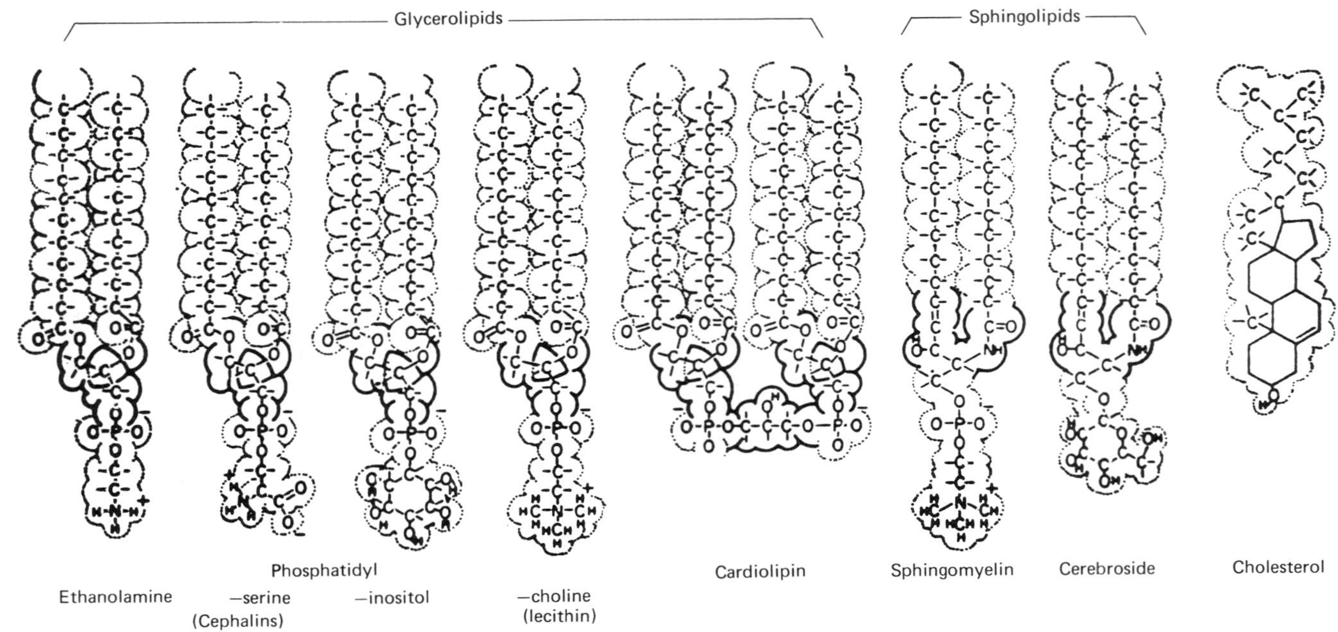


**Figure 4.1.1.** Chemical formulas of structurally important lipids, showing also the approximate spatial outlines of the molecular configuration. From J. B. Finean, *Chemical Ultrastructure in Living Tissues*, Thomas, Springfield, Ill. (1961).

directed perpendicularly to the plane containing the charged phosphatidyl groups and the proteins.

5. Many components of cell membranes spontaneously form arrays of lipid leaflets which can be shown by electron microscopy to be bimolecular.

On the basis of the above evidence it had been generally believed that cell membranes consisted of bimolecular lipid leaflets surrounded by external protein layers. This model has, however, been continually challenged, and there now appears to be strong evidence that it is at least not always a reasonable description of membrane structure. A *globular* structure was proposed as an alternate by Fernandez-Moran in 1962, on the basis of high-magnification electron-microscopic studies, and this proposal has since been expanded substantially by Green and his associates (1963, 1965, 1966, 1967, 1969). The two models are compared schematically in Fig. 4.1.2.

It is not appropriate to present detailed arguments here about the relative merits of these two models. Suffice it to say that there is evidence favoring both and suggesting that neither alone is satisfactory. Furthermore, it appears that advance in understanding enzyme structure and action will rapidly provide much more meaningful bases for the examination of membrane models. For example, it is pointed out by Triggle that the importance of nonpolar interactions on tertiary enzyme structure, catalysis, and ligand binding suggests substantial interpenetration of lipid by protein. Rapid progress can be expected in this area, which is clearly an

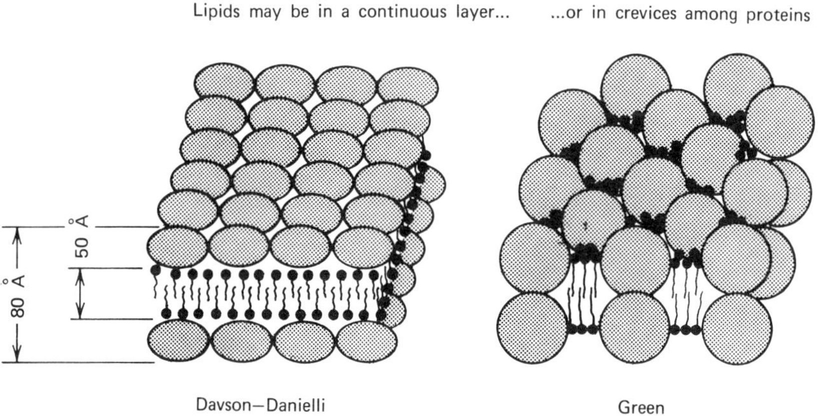


**Figure 4.1.2.** Comparison of the leaflet and globular models of cell membranes. The large spheroids in this figure are globular proteins, and the tadpole-like figures are phospholipids. From *Chem. Eng. News*, May 25, 1970, p. 42.

exciting one. Much of the transport behavior of cell membranes can, fortunately, be explained simply on the basis of their extreme thinness and their general chemical composition. We concentrate here on these simpler aspects.

## 4.2 The General Mass-Transport Characteristics of Natural Membranes

This discussion is separated from that of transport through macroscopic membranes because it differs from that more familiar process in several ways. The most important differences in diffusional behavior arise from the extremely small membrane thicknesses involved, and this is true irrespective of the specific model used for describing membrane structure. Another major difference, however, is in the much more complex nature of the transport processes and in particular their heavy reliance on facilitated diffusion. Finally there is the phenomenon of active transport, for which there is as yet no counterpart in nonliving systems. Active transport is a special case of carrier transport, or facilitated diffusion, in which solutes are moved against electrochemical gradients directly by means of energy-releasing chemical reactions, normally the partial hydrolysis of adenosine triphosphate (ATP) or similar compounds. The identification of carriers for facilitated diffusion and active transport has just begun and makes an exciting story, which is deferred to the next section. It is of interest here, however, to note that the first carrier identifications were reported in 1967 by Andreoli et al., by Lev and Buzhinsky, and by Mueller and Rudin. Nonetheless, the key characteristics of the carrier molecules had already been properly deduced on the basis of indirect evidence, surely a remarkable achievement.

From a purely practical standpoint the measurement of cellular mass transport also differs substantially from measurements in larger systems. In addition to constraints provided by the small size and labile nature of typical systems of interest, there is the near impossibility of measuring intramembrane compositions. Hence it is the custom to report transport data in terms of a permeability $P_S$ defined by

$$N_S = P_S \Delta c_S \tag{4.2.1}$$

where $N_S$ is flux of solute S across the membrane, and $\Delta c_S$ is the corresponding drop in solute concentration between the two solutions. Equation 4.2.1 neglects both diffusional interactions with other solutes and convection. This is reasonable for concentration diffusion of most nonreactive solutes, because they are present in extremely small concentration. Hence for them diffusion is effectively pseudobinary, and convection is important only if the net mass flux across the membrane is very large. *The neglect is not reasonable for water*, and we shall return to this special case shortly, along with those of forced diffusion and carrier transport.

For a macroscopic membrane and a linearly distributing solute one could write

$$P_S \doteq \frac{\mathfrak{D}_{Sm} K_{DS}}{\delta} \tag{4.2.2}$$

where $\mathfrak{D}_{Sm}$ is the effective binary diffusivity of solute S in the membrane, $K_{DS}$ is the equilibrium ratio of solute concentration in the membrane to that in the external solution, and $\delta$ is the membrane thickness. In bimolecular membranes, however, the situation is somewhat different; here the "diffusion" path is only a very few molecular diameters even for the smallest solutes, and the activation step looms large in the overall transport process. This is particularly important in the carrier transport of electrolytes, where the ion must first lose its hydration shell and then enter into a carrier complex on moving into the membrane. The importance of interfacial processes must be kept in mind in the analysis of transport through molecular membranes, and we return to this point in connection with the action of *ionophores* (ion carriers). As a practical matter, however, $P_S$ is reasonably independent of concentration level and transport rate for many solutes. Indeed we find that simple continuum models such as Eq. 4.2.2 provide useful insight into a variety of membrane transport processes.

### (a) *MEMBRANE PERMEABILITY AND SOLUBILITY CHARACTERISTICS OF DIFFUSING SOLUTES*

Thin as they are, biological membranes behave very much like a layer of liquid oily material. Thus they are highly permeable to fat-soluble materials, and in general the addition of each methyl group (—$CH_2$—) to members of a homologous series approximately doubles the permeability. Conversely, the permeability is decreased by the addition of moieties such as hydroxyl groups (—OH) which can form hydrogen bonds with the water in the solutions surrounding the membranes. This characteristic behavior is illustrated in Fig. 4.2.1, taken from Stein (1967), which shows that there is a general positive correlation between membrane permeability and solute oil solubility.* There are, however, some substantial departures from this trend.

Water is a very important special case and shows qualitatively different behavior from the other solutes under discussion. It can be seen from Fig.

* There is a corresponding decrease in permeability with increasing hydrogen-bonding capability of the solute. See Stein for a quantitative discussion.

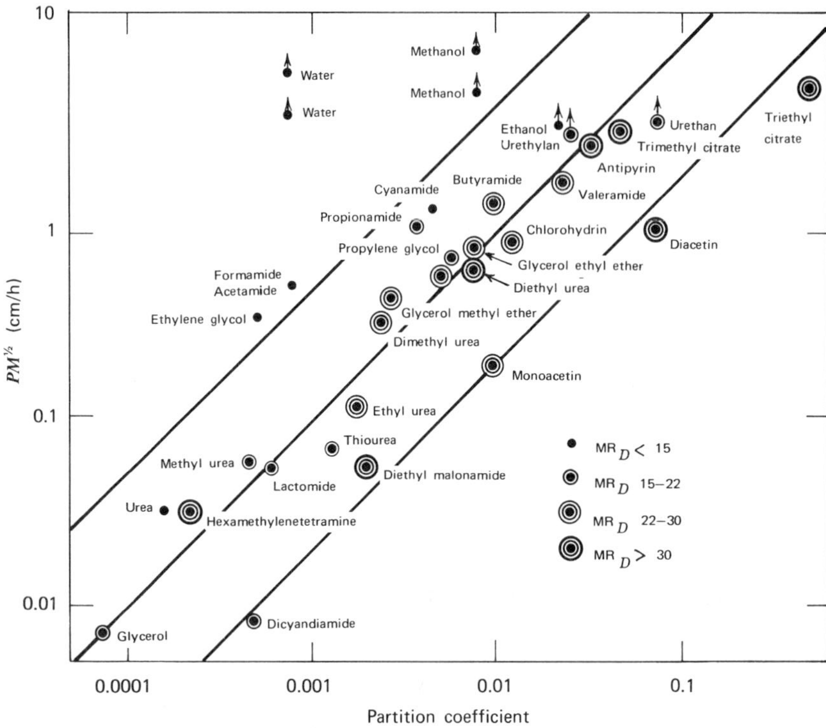


**Figure 4.2.1** The permeability of cells of *C. ceratophylla* to organic nonelectrolytes of different oil solubility and different molecular size. Ordinate: $PM^{\frac{1}{2}}$ ($P$ in cm $h^{-1}$); abscissa: olive-oil–water partition coefficients. $MR_d$ is the molar refraction of the molecules depicted, a parameter proportional to the molecular volume. Taken from R. Collander, *Physiol. Plant.*, **2**, 300–311 (1949).

4.2.1 that its permeability is anomalously high. Some insight into this anomalous behavior can be obtained by comparing the hydraulic and diffusional permeabilities of plasma membranes. To do this we begin by treating the membrane as a homogeneous phase and writing

$$N_W = -c\mathcal{D}_{Wm}\left[\left(1+\frac{\partial \ln \gamma_W}{\partial \ln x_W}\right)\frac{dx_W}{dZ} + x_W\frac{\bar{V}_W}{RT}\frac{dp}{dZ}\right] + x_W N_W \quad (4.2.3)$$

If we now remember that the external phases are nearly pure water and that water is only very slightly oil soluble, we can expect the total molar concentration $c$ and the activity coefficient $\gamma_W$ to be essentially constant throughout the membrane. We also expect the fractional change in $x_W$ to be small. We can then write

$$N_W = -Ɖ_{Wm} K_{DS} \frac{dc_W}{dz}$$

$$-Ɖ_{Wm} K_{DS} \frac{\bar{V}_W}{RT} c_W \frac{dp}{dZ} + x_W N_W \quad (4.2.4)$$

where $c_W$ is the external water concentration in equilibrium with local membrane composition. This expression can be integrated directly to give

$$N_W = -P_W^{(l)}\left(\Delta c_W + \frac{1}{RT}\Delta p\right) + (x_W)_{av} N_W \quad (4.2.5)$$

where $P_W$ is the permeability defined by Eq. 4.2.1, and

$$P_W^{(l)}/RT \equiv L_P^{(l)} \quad (4.2.6)$$

is the *hydraulic permeability*. The superscript ($l$) indicates a restriction to the lipid phase.

If the membrane were a simple lipid sheet, then $x_W \ll 1$, and the term $(x_W)_{av} N_W$ could be dropped from Eq. 4.2.5.

Observed values of $P_W$, $RTL_P$, and their ratio are shown in Table 4.2.1, which is abstracted from Stein. For a few of the biological membranes the ratio ($P_W/RT\ L_P$) is indeed of the order of unity, and the homogeneous lipid model appears not unreasonable. One would also expect a ratio higher than unity for hydrophilic membranes, where $x_W$ is large. However, the very high ratios observed for many of the biological membranes are very much in disagreement with the model.* It follows that the membrane cannot act as a homogeneous lipid phase for water transport in these cases.

These observations have been explained in the past by assuming the membrane to contain aqueous holes or "pores" through which water can move hydraulically, in parallel with the diffusional flow described by Eq. 4.2.5. This hydraulic flow in turn can be described by

$$(N_W)_{pore} \doteq -K\Delta p \quad (4.2.7)$$

since the fluid filling the "pores" will be primarily water. We may then

* It may further be noted from the results for toad bladder that water transport is affected by flow conditions in the microcirculation. However, biological water transport appears always to be *passive*, that is, produced entirely by concentration and pressure gradients and by diffusional interaction with other species.

**Table 4.2.1.** Values of the Hydraulic (Osmotic) Permeability Coefficient $RTL_p$ and the Diffusional Coefficient $P_W$, Together with the Derived Values for the Equivalent Pore Radius $r$, for the Penetration of Water across the Membranes of a Number of Cells and Tissues[a]

| Cell or tissue | $RTL_P$ ($10^{-4}$ cm sec$^{-1}$) | $P_W$ ($10^{-4}$ cm sec$^{-1}$) | $RTL_P/P_W$ | Derived value of $r$ (Å) |
|---|---|---|---|---|
| Single cells: | | | | |
| Amoeba | | | | |
| Frog, ovarian egg | 0.37 | 0.23 | 1.61 | 2.1 |
| Frog, body cavity egg | 89.1 | 1.28 | 70 | 30 |
| Xenopus, body cavity egg | 1.30 | 0.75 | 1.74 | 2.8 |
| Zebra fish, ovarian egg | 1.59 | 0.90 | 1.77 | 2.8 |
| Zebra fish, shed egg | 29.3 | 0.68 | 43 | 23 |
| Human (adult) erythrocyte | 0.45 | 0.36 | 1.25 | 1.3 |
| Squid, axon (axolemma) | 116 | 41 | 2.9 | 4.1 |
| | | | 7.8 | 8.5 |
| Tissues: | | | | |
| Frog, gastric mucosa | | | 20 | 15 |
| Toad, bladder: | | | | |
| No vasopressin | 4.1 | 0.95 | 4.3 | 8.5 |
| With vasopressin | 188 | 1.6 | 118 | 40 |
| Synthetic membranes: | | | | |
| Bimolecular lipid membrane | 8.3–14.4 | 2.3 | 3.6–6.3 | 5.0–7.5 |
| Dialysis tubing | 380 | 10.9 | 35 | 23 |
| Wet gel | 1200 | 19.2 | 62.5 | 31 |

[a] Literature references are provided by Stein (1967).

write for the two-phase parallel system of lipid sheets and aqueous pores that

$$(N_W)_{av} = -(1-\epsilon)P_W^{(l)}\Delta c_W - [(1-\epsilon)L_P^{(l)} + \epsilon K]\Delta p \tag{4.2.8}$$

where $\epsilon$ is the fraction of membrane cross section occupied by pores and

$(N_W)_{av}$ is the observable (surface-average) water flux across the membrane. We may now define

$$P_W = (1-\epsilon)P_W^{(l)} \tag{4.2.9}$$

$$L_P = (1-\epsilon)L_P^{(l)} + \epsilon K \tag{4.2.10}$$

where $P_W$ and $L_P$ are the observed diffusional and hydraulic permeabilities. For this model $L_P$ is always greater than $P_W/RT$, as is required by the data.

It remains to consider the probable structure of the pores. This cannot be done rigorously, but one can gain some insight by assuming them to be circular in cross section, and that the flow obeys Poiseuille's law. Then it is easily shown that

$$(N_W)_{\text{pore}} = -\frac{R^2 \Delta P}{8\mu\delta \bar{V}_W}, \qquad K = \frac{R^2}{8\mu\delta \bar{V}_W} \tag{4.2.11}$$

It is this relation, or modifications of it for "wall effects," coupled with estimates of $\epsilon$, that was used to estimate the "pore radii" of Table 4.2.1. Since there is no good reason for Poiseuille's law or any simple modification of it to hold at these small dimensions, these sizes are primarily useful to provide qualitative insight into water transport.

The above model is certainly not convincing in detail, and it has been subjected to much criticism. It fails to allow for diffusional resistance outside the membrane, which lowers $P_W$ relative to $L_P$ and which is more important for larger cells; and it does not take into account secondary water transport mechanisms, for example pinocytosis and electroosmosis. External diffusional resistance is now thought to be particularly important. However, the model is probably correct in suggesting that many membranes are heterogeneous from a diffusional standpoint and that the characteristic dimensions of the hydrophilic regions are quite small.

Water transport is discussed in some depth by Dick in Bittar (1971), Volume 3.

### (b) *TRANSPORT OF SUGARS AND ELECTROLYTES*

It may be noted that many physiologically important substances, in particular sugars and electrolytes, do not appear in the above tables and figures. This is because the transport of these substances cannot be explained by Eq. 4.2.1, that is, in terms of simple diffusional transport. Cell membranes show very marked selectivities for individuals sugars and ions and frequently cause them to be transported in the direction of increasing thermodynamic activity. In addition, the transport of sugars and ions is often "coupled" in the sense that the ratio of sugar to ion transport is a constant.

Much of this otherwise puzzling behavior can be explained in terms of facilitated diffusion, previously discussed. Thus it appears that sugar transport is always accomplished by a carrier mechanism and that this is frequently, but not always, true of ions as well. The coupling of sugar and ion transport, *cotransport*, can be explained by postulating that the carrier must be simultaneously complexed to a sugar molecule and an ion. Transport against an activity gradient can be explained either by coupling with the transport of a second solute moving toward a region of lower thermodynamic potential (*secondary active transport*; see Stein, 1967) or by coupling with an energy-consuming chemical reaction (*primary active transport*).

We consider here the movement of ions across cell membranes by diffusional processes in the basence of energy-releasing chemical reactions, that is, by *passive transport*. Stein gives a sound and thorough review of passive ion transport, which we shall follow rather closely here.* First, however, we review briefly the diffusional models used to calculate ion permeabilities and to differentiate between active and passive resistance.

Most permeability data appear to have been analyzed on the basis of the pseudobinary Nernst-Planck equations in the form

$$N_i = -c_i \mathfrak{D}_{im}\left(\frac{d\ln a_i}{dZ} + \frac{\nu_i \mathfrak{F}}{RT}\frac{d\phi}{dZ}\right) \tag{4.2.13}$$

The ion permeabilities have in turn been estimated by assuming the observed membrane potential to be the sum of membrane potentials for the individual ions weighted by their relative fluxes across the membranes.

To avoid active transport during the measurement of ion permeabilities it is common practice to poison the membrane—that is, to inactivate the enzymes responsible for active transport. To determine whether this procedure has been successful it is common practice to resort to tracer experiments and the Ussing relation developed in the last chapter.

Representative permeability data are shown in the Tables 4.2.2 and 4.2.3, taken from Stein (1967). It can be seen that there are surprisingly large differences within a series of apparently similar ions and that the permeabilities of smaller (hydrated) ions are generally larger. It may also be seen from Table 4.2.3 that cation permeabilities across sheep erythro-

* Much more detailed and up-to-date information is, however, provided by specialized monographs listed in the bibliography.

**Table 4.2.2.** Relative Entrance of Ions into Frog Muscle and into Kidney Slices[a,b]

| Series | Relative permeability | |
|---|---|---|
| | Frog muscle | Kidney slices |
| Cation | | |
| KCl | 100 | 100 |
| RbCl | 38 | 25 |
| CsCl | 0 | 5 |
| NaCl | 0 | 3 |
| LiCl | 0 | 16 |
| $CaCl_2$ | 0 | — |
| $MgCl_2$ | 0 | — |
| Anion | | |
| KCl | 100 | — |
| KBr | 63 | — |
| $KNO_3$ | 17 | — |
| K phosphate | 4 | — |
| $KOOCCH_3$ | 3 | — |
| $KHCO_3$ | 1 | — |
| $K_2SO_4$ | 0 | — |

[a] Value for KCl set equal to 100.
[b] Data of E. J. Conway, *Symp. Soc. Exp. Biol.*, **8**, 297 (1954) and Whittembury et al., *Nature*, **187**, 699 (1960), as collected by Stein (1967).

cytes are very low.* By way of comparison, water permeabilities across (human) erythrocyte membranes are about $5 \times 10^{-3}$ cm/sec—or $10^7$ times higher. It may also be noted that the unidirectional flux ratios do not generally compare well with the prediction of Ussing. The exception is $K^+$ transport for the "HK" sheep where the experimental and computed flux ratios differ by only about 3%. Since the ion transport is believed to be passive in all cases (because the enzyme systems responsible for active transport were poisoned), the discrepancies are believed due to strong diffusional interaction between ions attempting to move in opposite directions. The actual situation is probably much more complicated than that in synthetic ion exchangers, however, and it appears that even passive cation permeabilities are affected by cell metabolic processes.

* See problems. The assumptions used in the determination of ion permeabilities are discussed by Harris.

**Table 4.2.3.** Passive Cation Movements across Sheep Erythrocyte Membranes (Strophanthidin-Poisoned)[a]

| Parameter | Unit | Value for High potassium (HK) strain | Value for Low potassium (LK) strain |
|---|---|---|---|
| | Sodium Ions | | |
| External conc. | m$M$ | 165 | 165 |
| Internal conc. | m$M$ | 37 | 137 |
| Passive efflux | mmole/(liter red cells)(h) | 2.65 | 5.0 |
| Passive influx | mmole/(liter red cells)(h) | 2.8 | 3.9 |
| Flux ratio (experimental) | — | 0.95 | 1.28 |
| Flux ratio (computed) | — | 0.15 | 0.56 |
| Mean permeability coefficient | cm/sec | $4.4\times10^{-10}$ | $3.7\times10^{-10}$ |
| | Potassium Ions | | |
| External conc. | m$M$ | 5.0 | 5.0 |
| Internal conc. | m$M$ | 121 | 17.4 |
| Passive efflux | mmole/(liter red cells)(h) | 0.67 | 1.53 |
| Passive influx | mmole/(liter red cessl)(h) | 0.04 | 0.12 |
| Flux ratio (experimental) | — | 16.8 | 12.7 |
| Flux ratio (computed) | — | 16.3 | 2.4 |
| Mean permeability coefficient | cm/sec | $0.84\times10^{-10}$ | $5.7\times10^{-10}$ |

[a] Data of D. C. Tosteson and J. F. Hoffman, *J. Gen. Physiol.*, **44**, 169 (1964); flux ratios calculated by Stein (1967) from the Ussing relation (Ex. 3.4.1, Eq. 7).

Red-cell membranes are of particular interest in that anions, particularly chloride, exhibit much higher permeabilities than cations. This selectivity is pH dependant and appears to be in part due to a Donnan exclusion resulting from the presence of fixed positive charges in those regions of the membrane permeable to ions. The pH dependence is shown, for example, by increase in sulfate permeability with increasing acidity (decreasing pH). It can be explained by assuming that the fixed charges are produced by the ionization of weak bases:

$$R+H^+\rightleftharpoons RH^+$$

Passow was able to explain observations of chloride-sulfate exchange by postulating these bases to have an effective concentration of 3 $N$ and a dissociation constant (pK) of 9. This concentration is typical of that in synthetic ion exchangers, and a pK of 9 is consistent with the high probability that the bases are amino groups. The observed behavior is thus in reasonable agreement with the postulate of Donnan exclusion.

By way of contrast, squid-axon membranes exhibit relative $K^+$, $Cl^-$, and $Na^+$ permeabilities of about 1:0.4:0.04. This small difference between the $K^+$ and $Cl^-$ permeabilities can be explained by assuming an absence of fixed charges in the axon membrane, but no simple diffusional explanation can be given for the large difference between $K^+$ and $Na^+$ permeabilities. Squid-axon membranes appear to be representative of both nerve and muscle membranes, which in general behave quite differently from red-cell membranes.

Red-cell membranes also differ strongly from nerve membranes in the temperature coefficient of permeability, as shown in Table 4.2.4, taken from Stein. Activation energies for diffusion in water and synthetic ion exchangers are also presented in this table for comparison purposes. It may also be noted that activation energies for uncatalyzed homogeneous reactions are typically of the order of 20 to 25 kcal.

We find then that activation energies for passive cation transport in red-cell membranes are high for free diffusion but well below expected values for chemical reactions. They are, however, in good agreement with activation energies for synthetic phospholipid membranes. One can explain them in terms of the ion-exchange data shown by postulating the diffusion paths to correspond to a very highly cross-linked exchanger (i.e., to have very small "pores"), but this is simply speculation at present.

Squid-axon membranes, on the other hand, show very low activation energies and, in one of the cases shown, no detectable activation energy at all. This once more points up the difference between these two types of membranes and is consistent with the assumption of negligible fixed charge in the axon membrane. There is, however, no simple diffusional explanation for activation energies lower than those in water, and we must accept the fact that we do not understand these observations.

We must, in fact, conclude that passive ion diffusion in membranes is not a well-understood process and that a great deal of work remains to be done in this area. This is at present an exceedingly active research field, however, and one should be watching constantly for new developments.

## 4.3. Synthetic Bimolecular Lipid Membranes

Stable bimolecular lipid membranes can be made routinely, for example, by collecting a chloroform-methanol solution of an appropriate lipid mixture on a small loop or frame and either evaporating or dissolving

**Table 4.2.4.** Temperature Coefficient of the Passive Cation Permeability of Various Cell Membranes[a]

| Cell studied | Flux | $Q_{10}$[b] | Activation energy $E_a$ (kcal/mole) |
|---|---|---|---|
| Human erythrocyte | Linear component of $K^+$ influx | 2.3 | 15 |
| Human erythrocyte | $K^+$ efflux | 2.5 | 15.8 |
| Human erythrocyte | $K^+$ efflux | 2.0 | 12.4 |
| Human erythrocyte | $Na^+$ influx | 3.0 | 20.2 |
| Human erythrocyte | $Na^+$influx | 2.3 | 15 |
| Ascites tumor cells (mouse) | $Na^+$ influx | 2.5 to 4.0 | 17 to 25 |
| Ascites tumor cells (mouse) | $Na^+$ influx | 1.2 | 3.3 |
| Ascites tumor cells (mouse) | $K^+$ efflux | 4.1 | 26 |
| Nerve cell (squid axon) | $Na^+$ influx | 1.2 to 1.6 | 3.3 to 9.7 |
| Nerve cell (squid axon) | $K^+$ efflux | 1.1 | 1.6 |
| Nerve cell (squid axon) | Passive component of $K^+$ influx | 1.0 | 0 |
| Synthetic membrane phospholipid micelles | $K^+$ efflux | 2.3 | 15 |
| Free diffusion in water | $K^+$ | 1.26 | 4.2 |
| Free diffusion in water | $Na^+$ | 1.3 | 4.7 |
| Diffusion in ion-exchange resins | $K^+$ | 1.4 | 6.5 |
| Diffusion in ion-exchange resins | $Na^+$ | 1.3 | 5.22 |
| Diffusion in ion-exchange resins (highly cross-linked) | $Na^+$ | 1.6 | 8.62 |

[a] Taken from a variety of sources by Stein (1967).
[b] Here $W_{10}$ is the ratio by which the permeability changes when the temperature is increased by 10°C.

away the solvent. As the film thins, it begins to exhibit color patterns due to optical interference, and then begins to show "black" areas where the film is too thin to exhibit such patterns. The membrane in the "black" region is two molecules thick. There is then an abrupt transition to the colored regions which are about 100 times thicker. The bilayers are invisible in transmitted light but have a faint gray sheen in reflected light at large angles of incidence. They are tough, resilient, and self-sealing on puncture. They can be formed easily with a little experience, and it is believed that they can be made indefinitely stable. Yet electron micrographs show that they are only 60–90 Å thick.

### (a) *THE GENERAL NATURE OF BIMOLECULAR LIPID MEMBRANES*

When prepared from pure lipids, such membranes exhibit a resistance of about $10^8$ Ω $cm^2$ and a capacitance of about 1 $\mu F/cm^2$. This is about what one would expect from a thin oil film, and the structure is almost certainly a bimolecular leaflet much like the central portion of the Davson-Danielli model of the plasma membrane.

It is of particular interest that the membrane properties can be transformed qualitatively by incorporation of other materials in small amounts, as by adsorption from the surrounding solution. For example:

1. The adsorption of as yet unidentified water-soluble molecules extracted from bacterially fermented egg white, retina, and white matter from the brain lowers their resistance and renders them electrically excitable: the membrane resistance shifts reversibly and regeneratively between two stable values in response to suprathreshold voltage stimuli (Mueller, Rudin, Tien, and Wescott, 1964).
2. Membranes containing chlorophyll and carotene can exhibit photovoltaic effects.
3. The incorporation of macrocyclic polypeptides and ethers makes them selectively permeable to individual ions.

It is thus clear that additives of various sorts to the basic lipid bilayer are necessary to produce the most interesting mass-transfer characteristics of natural membranes.

We concentrate our attention here on the carrier transport of ions, which is of particular current interest, and we shall base much of our discussion on the American Physiological Society symposium on biological and artificial membranes (see Bibliography). The properties of "black" lipid membranes have since been reviewed by many others, including Goldup, Ohki, and Danielli (1970); Wolman (1970); and Triggle (1970). Ion transport in membranes is reviewed in several sources, the most exhaustive of which appears to be Bittar (1971).

The fact that synthetic lipid membranes can be induced to exhibit much of the highly selective ion transport characteristic of living membranes, by the addition of chemically identified carriers (ionophores), is one of the exciting scientific developments of recent years. Possible mechanisms for such selective ion transport are shown in Fig. 4.3.1 (from Eisenman, 1968).

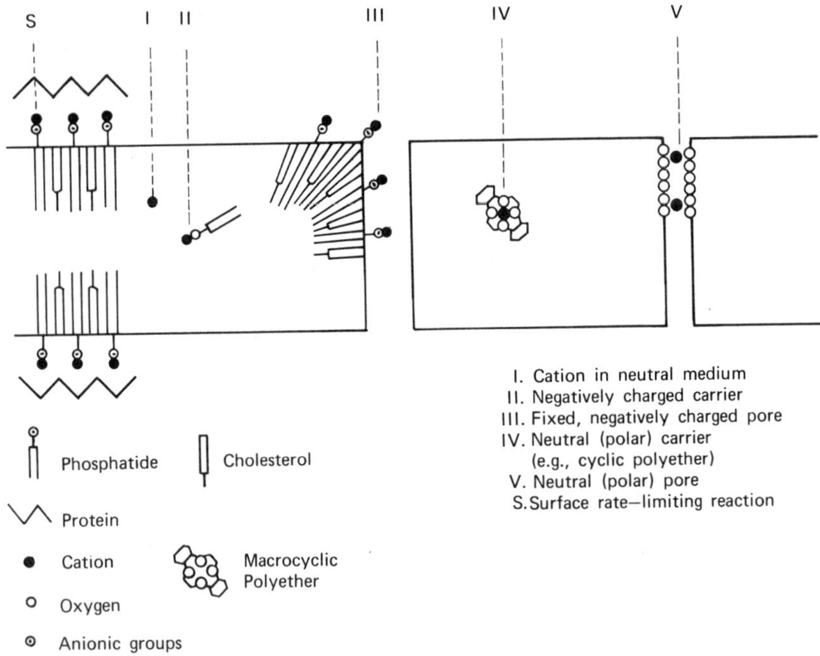


**Figure 4.3.1.** Possible ion-transport mechanisms. From Eisenman (1968).

These are:

1. Free diffusion of ions through the lipid layer.
2. Ion diffusion through charged "pores".
3. Ion diffusion through neutral "pores".
4. Carrier diffusion in a complex with an electrically neutral carrier.
5. Carrier diffusion in an uncharged complex.

In addition to these transport mechanisms, Eisenman points out the possibility that transport rates may be controlled by surface reactions (indicated by S in his figure). There is good reason to believe that all these transport mechanisms, and also control by surface reactions, occur.

(1) Ions do have finite solubilities in hydrocarbons and appear to be responsible for the leakage currents in insulating oils. However, the high electrical resistance of artificial bimolecular lipid membranes in the absence of "carriers" suggests that such permeation is small.

(2, 3) The existence of pores has long been postulated to explain transport of water and ions, but there is little, if any, direct evidence to back up this postulate. The classic arguments for the pore model are reviewed by

Stein (1967) and have been touched on above.

(4, 5) It has recently been shown that many neutral and charged macrocyclic compounds are extremely effective carriers for inorganic ions. These materials, discussed further below, can increase ion permeabilities several orders of magnitude when they themselves are present in only nanomolar concentrations in the surrounding aqueous solutions. The first reports of such carrier transport were by Moore and Pressman at the University of Pennsylvania in 1964 and concerned macrocyclic peptide antibiotics including valinomycin and the gramocidins. In 1967 C. J. Pedersen of the du Pont Company showed that structurally much simpler macrocyclic polyethers have similar properties. Pressman (1968) has since suggested that these materials, which he calls *ionophores*, be categorized in three classes depending on the charge of the carrier molecule and the nature of the ring structure. His method of classification is illustrated in Fig. 4.3.2:

1. Neutral carrier with a completely covalently bonded ring.
2. Charged carrier with a ring closed by hydrogen bonds.
3. Charged carrier with a completely covalently bonded ring.

A fourth class, neutral carriers with hydrogen-bond ring closure, is clearly also possible, but so far not reported.

Many laboratories, including that of Lardy at Wisconsin, are now actively engaged in research on these ionophores, and it has been found that they can produce some of the most complex and characteristic mass-transfer processes of living cells. Thus Meuller and Rudin and their associates in Philadelphia have shown (1964, 1967) that artificial lipid

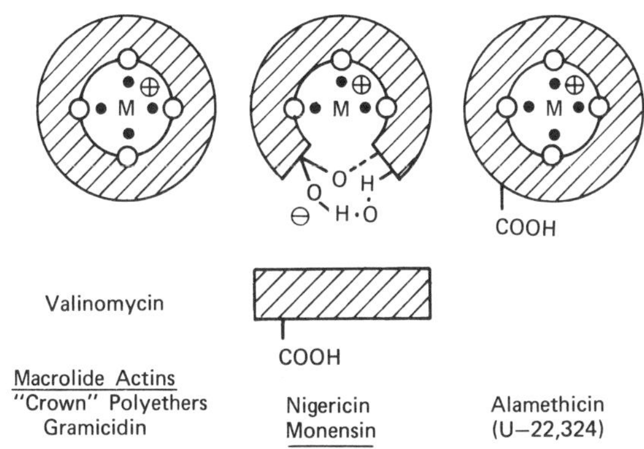


**Figure 4.3.2.** Classification of carriers: three subclasses of ionophores. The structures of the ionphore complexes underlined have been establishhed by x-ray crystallographic analysis. From Eisenman (1968).

membranes doped with well-characterized ionophores exhibit many of the anomalous current-voltage characteristics of nerve membranes. The Pressman (1968) and Lardy (1968) groups have shown that *some of these ionophores can induce ion transport against electrochemical gradients by an energy-requiring process*. A number of workers have shown that the effect of ionophores on cations is very specific—most, for example, being more effective carriers for potassium than for sodium ions.

### (b) *FACILITATED DIFFUSION OF IONS ACROSS CELL MEMBRANES: THE ACTION OF IONOPHORES*

The effects of several macrocyclic compounds on the resistance of lipid membranes bathed in KCl solutions are shown in Figs. 4.3.3 and 4.3.4 from Tosteson (1968). The membranes are lipid bilayers reconstituted from

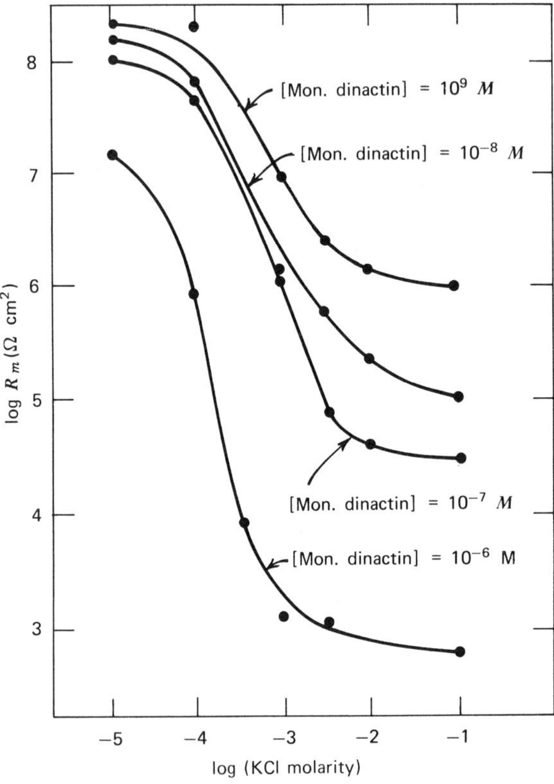


**Figure 4.3.3.** Effects of ionophore (monactin dinactin) and salt (KCl) concentrations on the resistance of thin lipid membranes. From Tosteson (1968).

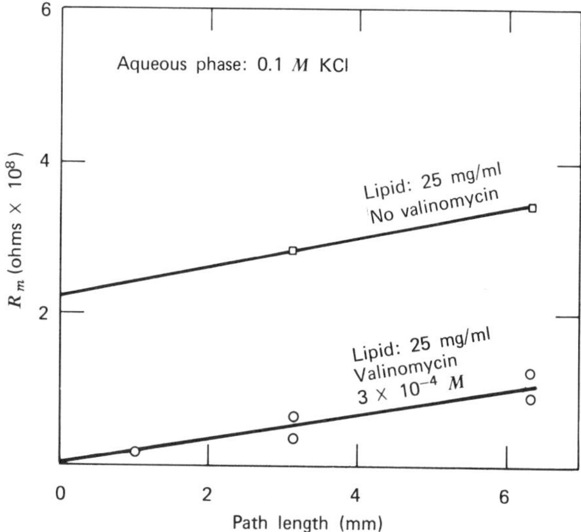


**Figure 4.3.4.** Effect of thickness on the resistance of lipid membranes. From Tosteson (1968).

extracts of HK-sheep red-cell membranes.* All but one of the curves shown in Fig. 4.3.3 are sigmoid in shape, indicating that the macrocyclic carriers become essentially saturated with potassium ions at the highest KCl concentrations used.

It appears that the primary effect of these carriers is to reduce the interfacial resistance. In Fig. 4.3.4 are shown the resistances of thick lipid layers with and without valinomycin, which can be seen to have a very large effect on the intercept (reflecting the limiting resistance of a bilayer), but none on the slope.

We next note that the ionophores under consideration typically show marked preferences for specific cations and a very high degree of specificity for cations relative to anions. This was shown, for example, by Mueller and Rudin in 1967 using valinomycin, enniatin, dinactin, and gramicidin in bilayer lipid membranes.

When identical monovalent salts of different concentrations were placed on opposite sides of such membranes, resting potentials of 54–58 mV per decade of concentration ratio were observed, with the concentrated solution negative. This is very nearly the Nernst potential for a membrane permeable only to cations, and suggests that the anion/cation permeability

* It has also been shown by Tosteson (1968) that these ionophores exhibit similar behavior in sheep red-cell membranes.

ratio is very small. In addition, if such membranes are used to separate solutions of different alkali-metal chlorides at the same concentration, very large bi-ionic potentials are observed. These are shown relative to $Li^+$ (the

**Table 4.3.1.** The Effects of Macrocyclic Antibiotics on the Permselectivity of Thin Lipid Membranes (from Mueller and Rudin, 1967)[a]

| | | Li | Na | Cs | K | Rb |
|---|---|---|---|---|---|---|
| Bi-ionic potential $E_{AB}$ (mV) | Val | 0 | 8 | 135 | 151 | 172 |
| | Enn | | 0 | 21 | 90 | |
| | Din | 0 | 5 | 35 | 85 | 110 |
| | Gram | 0 | 15 | 55 | 60 | 65 |
| Selectivity coefficient $K_{AB}=P_A/P_B$ | Val | 1 | 1.4 | 210 | 395 | 920 |
| | Enn | | 1 | 2.3 | 37 | |
| | Din | 1 | 1.2 | 4.1 | 30 | 82 |
| | Gram | 1 | 1.8 | 8.8 | 11 | 13 |
| Single-ion conductivity ratios $\kappa_A/\kappa_B$(35°C) | Val | 1 | 1.2 | 50 | >200 | >300 |
| | Gram | 1 | | | 20 | |
| Bi-ionic conductivity ratio $\kappa_{A+B}/\kappa_A$ | Val | 2.0 | 1.8 | 0.63 | 0.5 | 0.44 |
| | Gram | | | | 0.5 | |
| Single ion activation energies (kcal/mole) | Val | | 61.10 | | 30 | |
| | Din | | 35 | | 20 | |
| | Gram | | | | 28 | |

[a]Val = 3 unit valinomycin, $10^{-6}$ g/cm$^3$; Din = Dinactin, $10^{-6}$ g/cm$^3$; Enn = Enniatin B, $10^{-5}$ g/cm$^3$; Gram = Gramicidin a, $10^{-6}$ g/cm$^3$. Observed bi-ionic potentials ($E_{AB}$) at 0.05 $M$. The selectivity coefficients ($K_{AB}$) are derived from $E_{AB}$ by $K_{AB}=P_A/P_B=[A_0]/[B_i]\exp E_{AB}F/RT$, where $P_A$, $P_B$ are the permiability constants for the ions A and B; $[A_0]$, $[B_i]$ are the ion activities in the outside and inside compartments, and in all cases in the Table $B_i$ is $Li^+$; $F/RT$ has the usual meaning, The observed single conductances, $\kappa_A$, for ion A inside and outside at 0.05 $M$ are given as the ratio with respect to $\kappa_{Li^+}$.

least permeable) in Table 4.3.1. Here the $P_i$ are the estimated permeabilities for the ions $i$. Note* that valinomycin-doped membranes show selectivities (permeability ratios) of nearly 400 for $K^+$ relative to $Na^+$. Qualitatively similar results are given by Pressman (1968), who shows, however, a wide variation in ion preferences among known carriers.

Tosteson (1968) shows that, at least for his experimental conditions, this selective permeability is due not so much to selectivity in complex formation as to variation in the mobility of the ion-carrier complex. It is now clear from x-ray analysis of the $K^+$-nonactin complex that the "backbone" of the carrier molecule is bent on itself much like the seam on a baseball. A naked potassium ion at the center of the ball interacts with all four carbonyl and all four ether oxygens in the carrier. The methyl groups are all directed outward, and the complex has quasispherical symmetry. Tosteson suggests that variations in the electrical field about such complexes with the size of the central ion are responsible for the variations in ion permeability. Quite different behavior is observed for the polyethers, which appear to be rigid molecules.

Tosteson also showed, as indicated in Fig. 4.3.5, that the Ussing relation held to a reasonable degree over a very wide range of flux ratios for

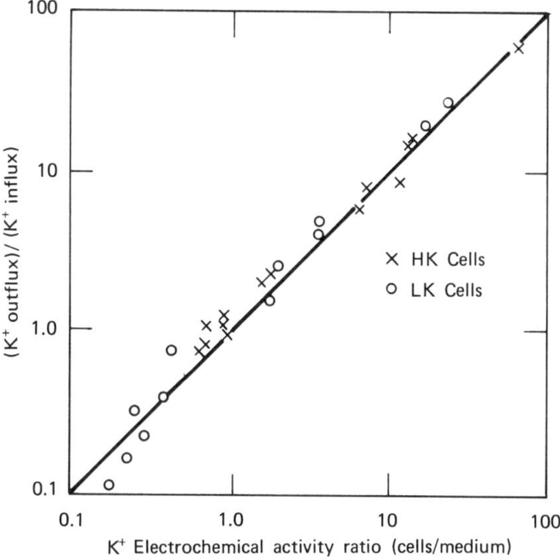


**Figure 4.3.5.** Monactin-dinactin-induced $K^+$ fluxes in sheep red cells. Flux ratio and electrochemical activity ratio. From Tostesin (1968).

*It is also of interest to note that the activation energies shown here are very high.

monactin-dinactin-induced $K^+$ transport across both HK and LK sheep red-cell membranes. This is a very strong indication of passive transport resulting from gradients in the electrochemical potential.

However, the macrocyclic antibiotics and polyethers have also been shown to affect active ion transport, that is, transport against electrochemical gradients at the expense of metabolic energy:

1. A number of ionophores, including depsipeptides, macrotetralides, polypeptides, and polyethers induce active alkali-metal-ion transport into mitochondria.
2. Monocarboxylic polyethers, including nigericin, dianemycin, and the monensins, cause loss of $K^+$ from mitochondria, especially in the presence of the neutral antibiotics described above.
3. At least one polyether, dibenzo-18-crown-6, prevents other ionophores from inducing $K^+$ uptake, but does not affect retention.

Pressman (1968) has suggested a specific model for these effects, shown in Fig. 4.3.6, which builds on earlier models of active transport reviewed by Stein (1967). He visualizes the mitochondrial ion pump as existing in series with a portion of the membrane which acts as a barrier to ion penetration.

In this model valinomycin (V), which can induce active transport, serves to transport the ion $M^{\oplus}$ across the barrier as the valinomycin complex $M^{\oplus}\cdot V^*$. In this process $M^{\oplus}$ gives up its water of hydration, and the valinomycin undergoes a conformational change, indicated here by the

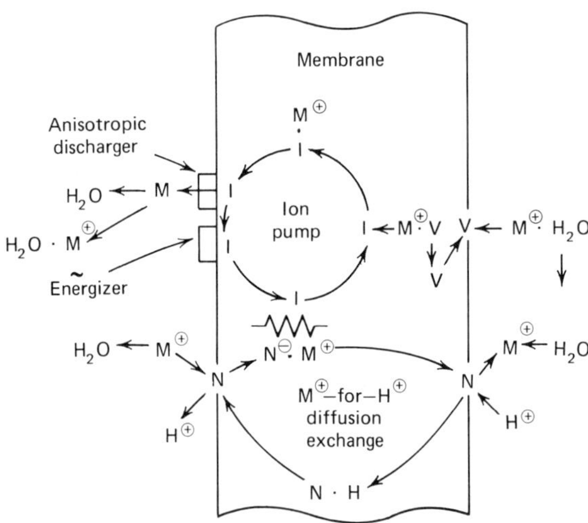


**Figure 4.3.6.** Hypothetical scheme for an ionophore-containing mitochondrial electrogenic cation pump. From Pressman (1968).

asterisk. Once inside the barrier, V* releases $M^{\oplus}$ to the active form I of the pump ionophore. The $M^{\oplus}\cdot I^*$ complex so formed moves across the membrane where $M^{\oplus}$ is discharged, and the ionophore is released in the form $I_0$, which has little affinity for $M^{\oplus}$. To complete the cycle $I_0$ is returned to the active configuration I by an energizer using either an oxidizable substrate or ATP.

Nigericin can form the analogous complex $N^{\ominus}M^{\oplus}$, but cannot release $M^{\oplus}$ to I, since the latter cannot supply the proton required by the nigericin. Nigericin can therefore only contribute to passive transport in this system.

## BIBLIOGRAPHY

### Books and Reviews

Altgelt, K. H., and J. C. Moore, "Gel Permeation Chromatography," in *Polymer Fractionation*, M. J. R. Cantow, Ed., Academic Press (1967).

American Physiological Society, "Symposium: Biological and Artificial Membranes," *Fed. Proc.*, **27**, 6, 1249–1309 (1968).

Barnes, C. D., and Christopher Kircher, *Readings in Neurophysiology*, Wiley (1968).

Bittar, E. E., Ed., *Membranes and Ion Transport*, Wiley-Interscience (1971) (three volumes).

Buckles, R. G., Ph.D. thesis, Mass. Inst. of Tech. (1966).

Clarke, H. T., Ed., *Ion Transport Across Membranes*, Academic Press (1954).

deGroot, S. R., and P. Mazur, *Non-equilibrium Thermodynamics*, North Holland (1962).

Dick, D. A. T. D., "Water Movement in Cells," in Bittar (1971), Vol. 3.

Goldup, A., S. Ohki, and J. F. Danielli, "Black Lipid Films," *Recent Progress in Surface Science*, J. F. Danielli, M. D. Rosenberg, D. A. Cadenhead, and A. C. Riddiford, Eds., Vol. 3, Academic Press (1970).

Green, D. E., and Goldberger, R. F., *Molecular Insights into the Living Process*, Academic Press (1967), Chap. 10.

Helfferich, Friedrich, *Ion Exchange*, McGraw-Hill (1962).

Hendler, R. W., *Protein Biosynthesis and Membrane Biochemistry*, Wiley (1968).

Hirschfelder, J. O., C. F. Curtiss, and R. B. Bird, *Molecular Theory of Gases and Liquids*, Wiley (1954).

Katchalsky, A., and P. F. Curran, *Non-equilibrium Thermodynamics in Biophysics*, Harvard Univ. Press (1965).

Katz, Bernard, *Nerve, Muscle, and Synapse*, McGraw-Hill (1966).

Keller, K. H., and E. F. Leonard, *Application of Chemical Engineering to Problems in Biology and Medicine, Part I: Engineering Analysis of the Functions of Blood*, Am. Inst. Chem. Eng. Twelfth Adv. Sem. (1969).

Merten, Ulrich, *Desalination by Reverse Osmosis*, Mass. Inst. Tech. Press (1966).

Stein, W. D., *The Movement of Molecules Across Cell Membranes*, Academic Press (1967).

Triggle, D. J., "Some Aspects of the Role of Lipids in Lipid-protein Interactions and Cell Membrane Structure and Function," in *Recent Progress in Surface Science*, Vol. 3, J. F.

Danielli, M. D. Rosenberg, D. A. Cadenhead, and A. C. Riddiford, Eds., Academic Press (1970), page 273.

Wolman, Moshe, "Structure of Biological Membranes: The Laminar vs. the Globoid Concept," *Recent Progess in Surface Science,* Vol. 3, J. F. Danielli, M. D. Rosenberg, D. A. Cadenhead, and A. C. Riddiford, Eds., Academic Press (1970), page 262.

## Periodical References

Andreoli, T. E., M. Tieffenberg, and D. C. Tosteson, *J. Gen. Physiol.*, **50**, 2527 (1967).

Brian, P. L. T., *Ind. Eng. Chem. Fund.*, **4**,439 (1965).

Craig, L. C., *Science*, **144**, 1093 (1964).

Craig, L. C., J. D. Fisher and T. P. King, *Biochem.*, **4**,311 (1965).

Craig, L. C., W. Konigsberg, A. Stracher, and T. P. King, *The Proteins*, Methuen (1958).

Eisenman, George, "Ion Permeation of Cell Membranes and Its Models," in American Physiological Society (1968), p. 1249.

Eisenman, George, S. M. Ciani, and Gabor Szabo, in American Physiological Society (1968), p. 1289.

Evans, R. B., III, G. M. Watson, and E. A. Mason, *J. Chem. Phys.*, **16**, 1894–1902 (1962).

Fernandez-Moran, H., *Res. Publ.* A.R.N.M.D., **40**,235 (1962).

Fernandez-Moran, H., in *Macromolecular Complexes*, M. V. Edds, Jr., Ed., Ronald Press (1961).

Green, D. E., *Isr. J. Med. Sci*, **1**,1187 (1965).

Green, D. E., and S. Fleisher, *Biochim. Biophys. Acta*, **70**,554 (1963).

Green, D. E., and J. F. Perdue, *Proc. Natl. Acad. Sci. U. S.*, **55**,1295 (1966).

Green, D. E., and A. Tzagoloff, *J. Lipid Res.*, **7**,587 (1966).

Green, D. E., D. W. Allman, E. Bachmann, H. Baum, K. Kopaczyck, E. F. Korman, S. H. Lipton, D. H. MacLennan, D. C. McConnell, J. F. Perdue, J. S. Rieske, and A. Tzagoloff, *Arch. Biochem. Biophys.*, **119**,312 (1967).

Green, D. E., and D. H. MacLennan, *Bioscience*, **19**,213 (1969).

Ginzburg, B. Z., and A. Katchalsky, *J. Gen. Physiol.*, **47**,403–418 (1963).

Hemmingsen, E., and P. F. Scholander, *Science*, **132**,1379 (1960).

Henderson, W. E., and C. Sliepcevich, *Chem. Eng. Prog. Symp. Ser.*, No. 24,**55**,145.

Kaufman, T. G., and E. F. Leonard, *Am. Inst. Chem. Eng.*, **14**,110–117 (1968).

Kedem, O., and A. Katchalsky, *Biochem. Biophys. Acta*, **27**,229 (1958).

Keller, K. H., and S. K. Friedlander, *J. Gen. Physiol.*, **49**,663 (1966).

Kirkwood, J. G., in Clarke (1954).

Lardy, Henry, in *American Physiological Society*, (1968), p. 1278.

Lev, A. A., and Buzhinsky, E. P., *Cytology* (USSR), **9**,102 (1967).

Mackay, D., and P. Meares, *Trans. Faraday Soc.*, **55**,1221 (1959).

Moore, C., and B. C. Pressman, *Biochem. Biophys. Res. Commun.*, **15**,562 (1964).

Mueller, P., and D. O. Rudin, *Biochem. Biophys. Res. Commun.*, **26**, 398 (1967).

Mueller, P., D. O. Rudin, H. T. Tien, and W. C. Wescott, in *Recent Progress in Surface Science*, J. F. Danielli, Panhurst, and A. C. Riddiford, Eds., Academic Press (1964).

Onsager, Lars, *Phys. Rev.*, **37**, 405 (1931).

Onsager, Lars, *ibid.*, **38**, 2265 (1931).

Onsager, Lars, *Ann. N. Y. Acad. Sci.*, **46**, 21 (1945).

Pedersen, C. J., in *American Physiological Society* (1968), p. 1305.

Pressman, B. C., in *American Physiological Society* (1968), p. 1283.

Spiegler, K. S., *Trans. Faraday Soc.*, **54**, 1409 (1958).

Tosteson, D. C., in *American Physiological Society* (1968), p. 1269.

Ussing, H. H., *Acta Physiol. Scand.*, **19**, 43–56.

## 5. CONVECTIVE MASS TRANSFER

Here we discuss mass-transfer situations for which convective solute transport, represented by the term $(\mathbf{v} \cdot \nabla c_i)$ in the species continuity equations, is of major importance. Our purpose here is to model relatively well-defined transport processes of particular interest in the body and extracorporeal circuits. The analysis of more complex systems is deferred to the next chapter and is based in part on the discussions here.

The high Schmidt numbers characteristic of biological systems and the predominance of laminar forced-convection flows frequently permit the use of very simple asymptotic solutions to mass-transport problems. Even where this is not possible, satisfactory numerical solutions can almost always be obtained for models justified by available data. In almost all situations the solution techniques are relatively simple extensions of well-known procedures,* and here it is only necessary to show how the magnitudes of the governing parameters and the ultimate goals of the calculation can be used to advantage in developing a solution strategy.

It is particularly important to recognize the impact of biological considerations, some of them quite subtle, on what are normally considered purely engineering aspects of equipment design. These are important in modeling the mass-transfer process itself, in determining the magnitudes of governing parameters, and in choosing the equipment configuration.

To emphasize the importance of physiology and biochemistry, we organize this section around specific examples. We begin in Section 5.1 and 5.2 with dialysis and ultrafiltration as representative nonreactive processes. We then turn to the more complex cases of blood oxygenation and oxygen supply to tissues, where mass transfer is accompanied by a complex set of chemical reactions.

* The effectiveness of asymptotic approximations for the estimation of mass-transfer rates is, for example, reviewed by Lightfoot (1969).

## 5.1. Dialysis

In both the kidney and its partial extracorporeal substitute, the hemodialyzer, unwanted solutes and water are removed from blood by a combination of dialysis and ultrafiltration. These operations can be described much more simply in the dialyzer than in the kidney, where a number of other transport operations occur simultaneously with dialysis, and we confine ourselves here primarily to this simpler case.

A representative hemodialysis system is shown in Fig. 5.1.1, taken from Babb, Grimsrud, Bell, and Layno (in Hershey, 1967). This system and its operation are too complex for detailed discussion, either here or in Section 6, and the interested reader must therefore refer to the literature. An excellent qualitative summary of artificial kidney technology as of 1968 has been published by Leonard and Dedrick. Current developments are covered by the periodic reviews of the National Institute of Arthritis and Metabolic Diseases (NIAMD).

The heart of this system, though in some respects one of the simpler components, is the dialyzer itself. In it, circulating blood is put in contact with a buffered electrolyte across a cellulosic membrane permitting the passage of water and "small" solutes by diffusion. Normally water is ultrafiltered from the blood by maintaining it at a greater hydrodynamic pressure than the dialyzate stream. This ultrafiltration, which is discussed in the next section, is too slow to have an appreciable direct effect on solute mass transfer, either within the membrane or in the adjacent mass-transfer boundary layers. We are therefore dealing with what is in some respects a conventional case of convective mass transport between fluids separated by a semipermeable membrane, at net mass transfer rates too small to affect velocity profiles.

As in all such operations, the major technical problems are to provide a large effective mass-transfer surface with adequate mechanical support and uniform flow distribution, at acceptable pressure drops and costs. All of these factors, including cost in the case of chronic treatment, are important in hemodialysis.

There are in addition, however, a large number of clinical and physiological factors, discussed for example by Leonard and Dedrick. Particularly important to our present discussion are the need to minimize extracorporeal blood volume and blood losses, the undesirability of pumping blood to high pressures or subjecting it to high local shear rates, and the tendency for blood in extracorporeal circuits to clog small passages and form clots* in stagnant regions.

* This is a serious problem even though heparin is universally used to minimize coagulation problems.

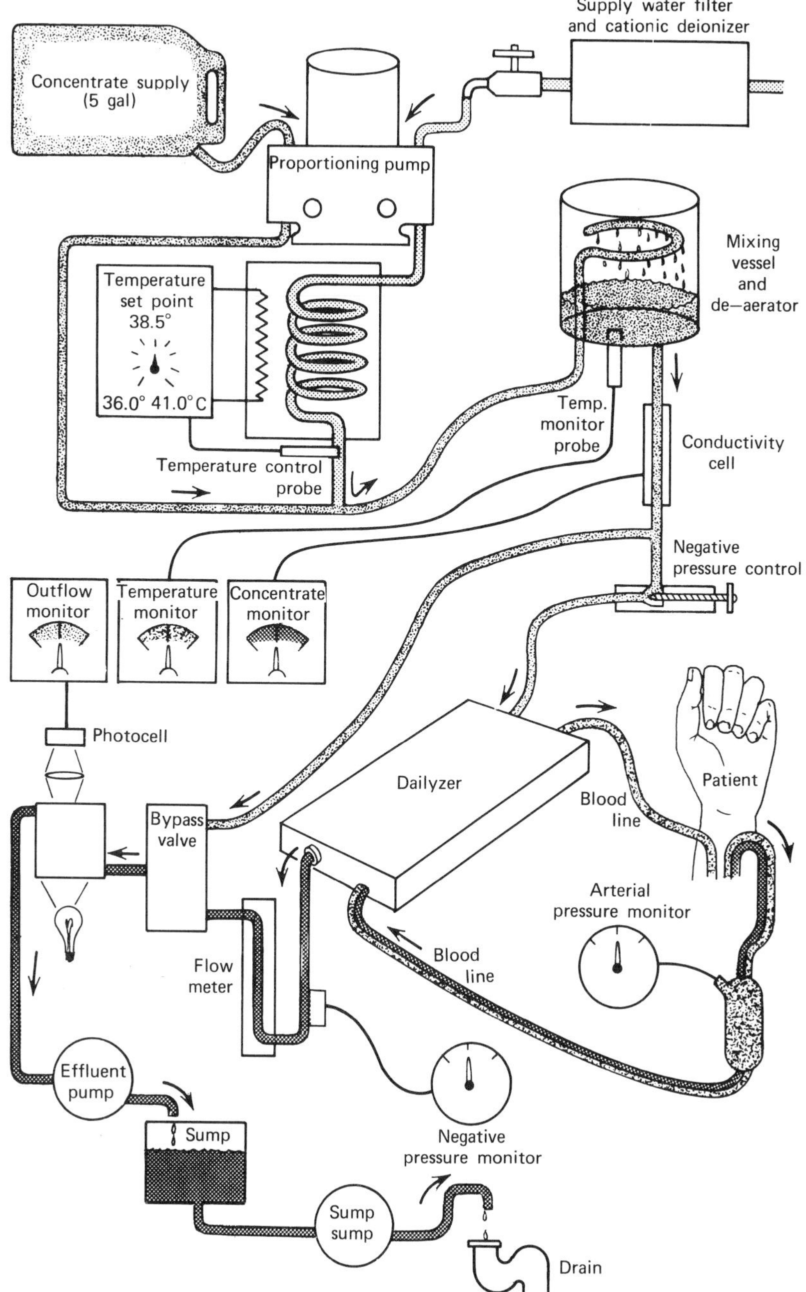


**Figure 5.1.1.** Babb-Grimsrud dialysis system. From A. L. Babb, Lars Grimsrud, R. L. Bell, and S. B. Layno, in Hershey (1967).

Because of the volume and shear-rate limitations, all designs considered seriously to date use laminar blood flow in passages of small hydraulic radius. At the time of writing, most systems confine the blood between parallel sheets, either flat or in flattened tubes rolled in a spiral with a screen spacer. There has, however, been increasing interest in putting the blood through a bundle of small-diameter tubes. As we see shortly, each of these configurations has its advantages as well as its limitations.

Before these can be discussed in a meaningful way, however, one must provide at least a semiquantitative description of the convective mass-transfer process taking place. We do this here for the widely used countercurrent flat-sheet design of Fig. 5.1.2. For simplicity we restrict ourselves here to the simplified situation of Fig. 5.1.3, for which the dialyzate-side mass-transfer resistance may be considered position independent. Such behavior is closely approximated in many dialyzers through frequent disruption of dialyzate flow. We next note from the results of Chapter I, Section 3 that the velocity entrance length is very short for low-Reynolds-number flow in narrow passages and, from the low solute concentrations in plasma, that we are dealing with a dilute pseudobinary solution.† We may then write for a counterflow situation:

*Blood*:

$$v_x \frac{\partial c_i}{\partial x} = \mathfrak{D}_{im} \frac{\partial^2 c_i}{\partial y^2} \tag{5.1.1}$$

$$\text{At } x=0, \quad c_i = c_{i0} \tag{5.1.2}$$

$$\text{At } y=0, \quad \frac{\partial c_i}{\partial y} = 0 \tag{5.1.3}$$

$$\text{At } y=B, \quad -\mathfrak{D}_{im} \frac{\partial c_i}{\partial y} = K_c (c_i - \bar{c}_i) \tag{5.1.4}$$

*Dialyzate*:

$$Q_D \frac{\partial \bar{c}_i}{\partial x} = 2WK_c (c_i - \bar{c}_i) \tag{5.1.5}$$

$$\text{At } x=L, \quad \bar{c}_i = \bar{c}_{i0} \tag{5.1.6}$$

† The important case of protein-bound solutes can be handled by the methods of Section 5.3.

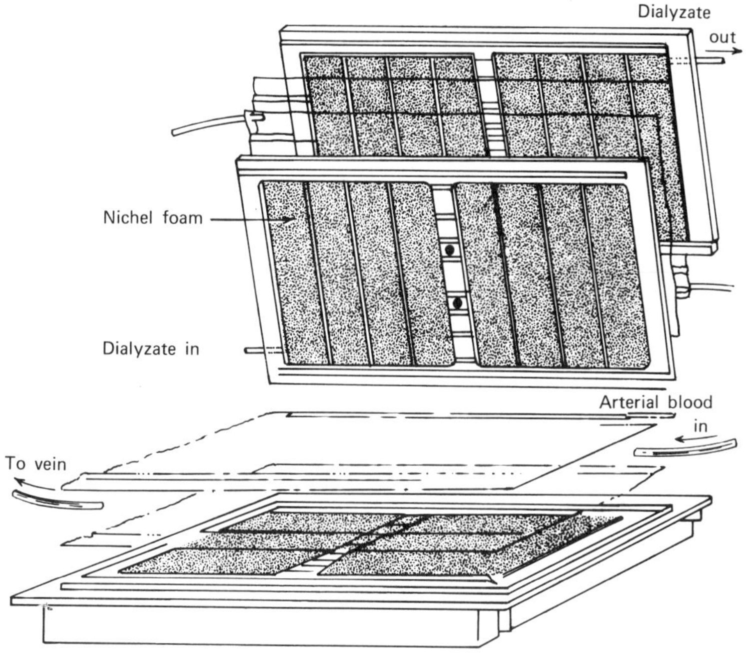


**Figure 5.1.2.** Sketch of the arrangement of the foam nickel plates in the Babb-Grimsrud dialyzer. From A. L. Babb, Lars Grimsrud, R. L. Bell, and S. B. Layno, in Hershey (1967).

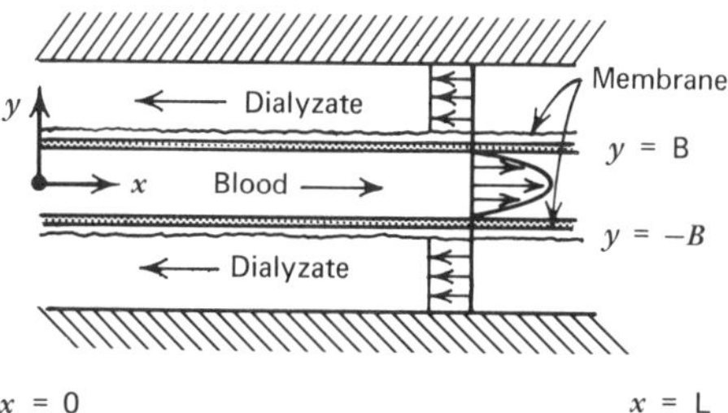


**Figure 5.1.3.** Flow model of a simple dialyzer.

Here $c_i = c_i(x,y)$ = local solute concentration in the blood
$\bar{c}_i = \bar{c}_i(x)$ = solute concentration in the dialyzate
$v_x$ = velocity of blood (in the $x$ direction)
$= v_{max}[1-(y/B)^2]$ where the blood can be assumed Newtonian
$Q_d$ = volumetric flow rate of the dialyzate stream, taken as positive in the $x$ direction
$B$ = half the distance between the sheets enclosing the blood
$W$ = the width of the blood path, in the $z$ direction
$K_c$ = the overall mass-transfer coefficient for the membrane and dialyzate boundary layer

Note that for counterflow, $Q_D$ is negative.

If these equations are dimensionally analyzed, they contain four independent parameters, for example:

$v_{max}L/\mathfrak{D}_{im} \equiv \text{Pé}$ = length Péclet number,
$K_c B/\mathfrak{D}_{im} \equiv \text{Sh}_m$ = membrane dialyzate Sherwood number,
$(v_{max}WL/Q_D)(L/B) \equiv R$ = modified flow ratio,
$c_{i0}/\bar{c}_{i0}$.

The large number of these dimensionless groups, plus the mass-transfer boundary condition (Eq. 5.1.4), delayed the solution of these equations.

Limiting solutions for small $R$ were obtained in 1966 by Lars Grimsrud and A. L. Babb. For this situation the dialyzate composition remains constant, and Eqs. 5.1.5 and 5.1.6 are unnecessary. This solution was useful for typical dialysis conditions at that time and is valid for cross flow as well as either co- or countercurrent operation. Such conditions are, however, wasteful of dialyzate. Tien (see Hershey, 1967) provided an approximate boundary-layer solution to Eqs. 5.4.1 through 5.4.6 in 1967. It does not seem to have been widely used, perhaps because it covers a rather restricted range of conditions. Ramirez, Lewis, and Mickley solved the full set of equations numerically, which appears to be the most practical approach, in 1970, and have since extended their calculations to consider variation of the dialyzate resistance with position.

Ramirez et al. found good agreement with prediction using a Cobe Kiil hemodialyzer and cuprophane membranes. This is, however, by no means representative, and measured performance is often well below predicted values. This appears to result from inability to hold design clearance* between sheets—both locally, because of inadequate support, and over the dialyzer width. As a result, a disproportionate fraction of the blood stream flows through the thicker regions. This type of behavior is illustrated in

* It should be noted in this respect that design clearances are typically near 0.2 mm, while sheet dimensions are tens of centimeters.

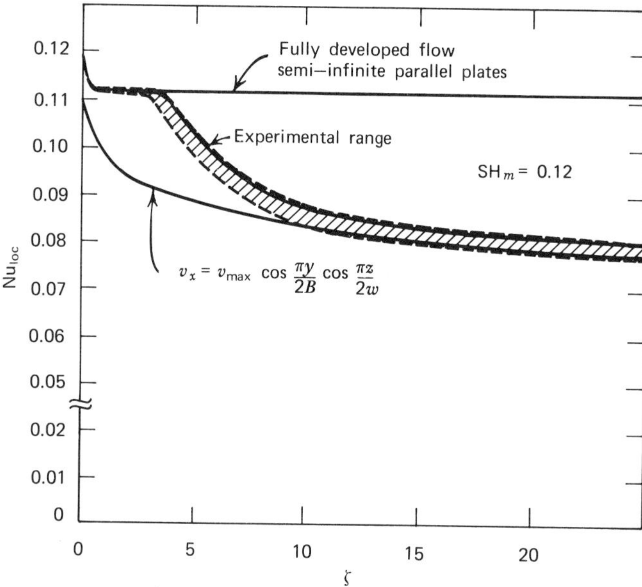


**Figure 5.1.4.** A comparison of predicted (solid lines) and experimental (shaded area) mass-transfer behavior in a test dialyzer (from Grimsrud and Babb, 1966). Here $\zeta$ is a dimensionless distance into the dialyzer, defined by $\zeta = x\mathfrak{D}_{im}/v_{max}B^2$, and $Nu_{loc}$ is the local Nusselt number for the "blood" stream based on the half thickness $B$. Here a saline solution is used in place of blood.

Fig. 5.1.4 for a test dialyzer used by Grimsrud and Babb. For small (dimensionless) lengths $\zeta$, the observed performance is in close agreement with their approximate solution of Eqs. 5.1.4; this indicates a very nearly uniform flow distribution. For larger $\zeta$, local mass-transfer performances drop toward the prediction for a parabolic velocity distribution across the flow (in the $z$ direction), a convenient approximation for severe maldistribution of flow. The flat-sheet geometry is intrinsically a difficult one to maintain in practice, especially in the presence of the pressure gradients needed for ultrafiltration. Tubular geometries are better in this respect, and the use of blood-filled capillary tubes is particularly attractive: here the transmembrane pressure drop tends to preserve the desired geometry. It does, however, appear that clogging is a problem for a significant fraction of patients if the tubes are small enough to provide efficient mass transfer, and that regions of high shear at the tube entrances may be damaging.†

† It may be noted that the tubular geometry is used in the kidney—and that the diameters of kidney tubules are so small as essentially to eliminate lateral mass-transfer resistance. It has not proven possible even to approach this behavior in extracorporeal circuits, however.

In general the major mass-transfer problems in dialyzer design appear to be mechanical: to maintain close, uniform spacing and adequate manifolding. Cross flow is nearly as effective as counterflow from a theoretical standpoint and facilitates mechanical design considerably. This and relative ease of maintaining uniform clearances are the chief advantages of the coil dialyzers. These, however, typically require rather high inlet blood pressures.

There is not space in this limited review to discuss the merits of individual designs, and the reader is referred to the NIAMD bibliographies for up-to-date sources of such information. It is, however, important to note that the modeling of idealized systems is not a serious problem. Emphasis should be on the realization of biologically acceptable designs in a practical apparatus.

## 5.2. Ultrafiltration

The removal of large solutes from water by filtration across semipermeable membranes occurs generally at the arterial ends of vascular beds, and most particularly in the glomeruli of the kidneys. This process, known as ultrafiltration, is also an important part of the treatment of uremia by hemodialysis, and it is increasingly used for the concentration of proteinaceous solutions. It is thus important from a practical standpoint. It is also of interest from a fundamental point of view in that it represents a situation intermediate between filtration of macroscopic particles from true solutions and the fractionation of true solutions by reverse osmosis. It has all the essential characteristics of both processes, so that they can be treated as limiting special cases of ultrafiltration.

The essential nature of ultrafiltration has already been indicated in Fig. 3.2.1. To describe this process one must consider both (1) the finite fluid velocity through the semipermeable membrane forming part of the system boundary and (2) the resulting *concentration polarization* required to produce back diffusion of the rejected solute. Diffusional processes within a semipermeable membrane under nonisobaric conditions have already been discussed in Section 3.2, and we shall concentrate on the boundary-layer aspects here (see Fig. 5.2.1).

In principle, the many papers on boundary-layer mass transfer with finite wall velocities (see Kozinski, 1970, and bibliography therein) can be applied to ultrafiltration. In practice, however, the ranges of permeation rates and physical properties are so different that these papers are of limited value. The large effects of protein concentration on both viscosity and diffusivity, and the very large Schmidt numbers characteristic of protein solutions, are particularly important.

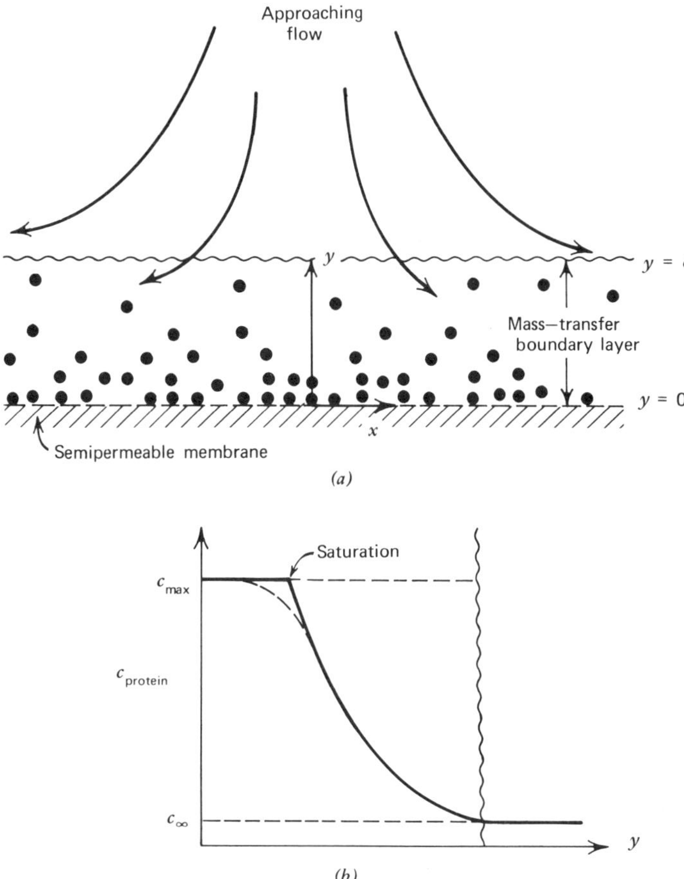


**Figure 5.2.1.** Boundary-layer aspects of ultrafiltration. Uniform boundary-layer thickness as shown here is observed near a stagnation point or about a spinning disc. (*a*) Semipictorial representation: a spinning disc or stagnation flow. (*b*) Schematic protein concentration profile. Drawings from Lightfoot, Safford, and Stone (1971).

We believe that the analyses and experiments that are most useful at present are those dealing with monodispersed well-characterized proteins in simple flow situations. We therefore base our discussion on the boundary-layer analyses of Kozinski and Lightfoot (1971, 1973), but note that the work of Brown (1971) et al. is similar and complementary.

Ultrafiltration in vascular beds has yet to be thoroughly studied and therefore are not discussed here. Velocity profiles have been calculated for tube flows of Newtonian fluids of constant viscosity by Kozinski, Schmidt, and Lightfoot (1970), but no physically realistic analysis for diffusional phenomena in capillary beds appears to be available. Older work is reviewed briefly by Aroesty and Gross (1970).

We further limit discussion to proteins that are essentially completely rejected by the membrane; an extension to the more general case of partial rejection can be obtained by combining the analyses summarized below with the membrane aspects of ultrafiltration discussed previously. This would be useful in many applications, but has yet to be done.

Our problem now reduces to determining the relation between wall permeation rate and applied pressure drop, and if possible obtaining a geometry-insensitive correlation. More specifically, we would like to find out why the filtration rate approaches an asymptotic limit as the pressure drop increases. This behavior, illustrated for bovine serum albumin and a rotating-disc system in Fig. 5.2.2, is the chief limitation on productivity in

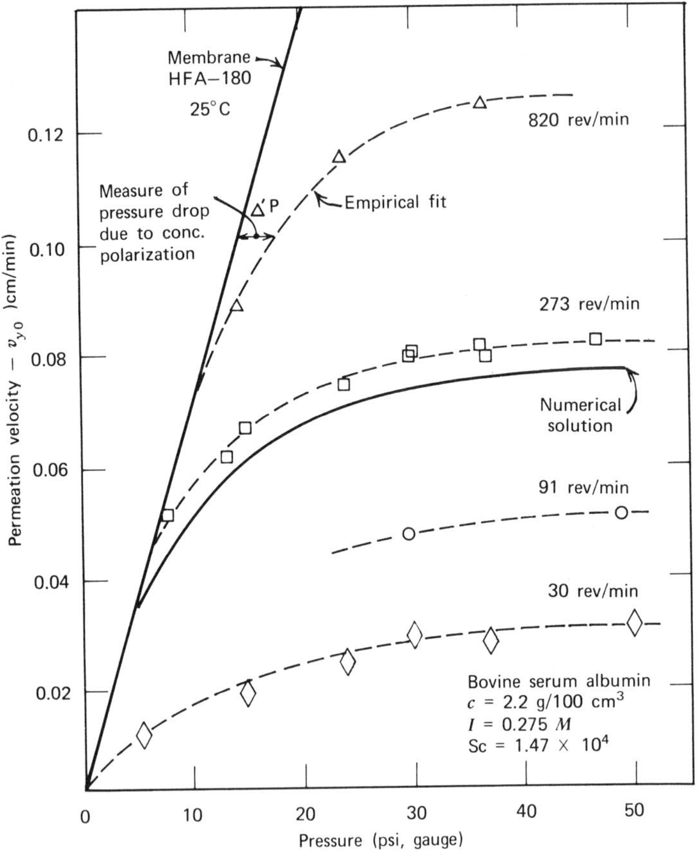


**Figure 5.2.2.** Ultrafiltration across a membrane on a spinning disc (from Kozinski, 1971).

commercial protein processing equipment. Concentration polarization is probably much less severe during hemodialysis and *in vivo*, but its importance has yet to be investigated quantitatively for these applications.

It is shown by Kozinski and Lightfoot (1973) that this behavior can be described simply in terms of convection and diffusion. Furthermore, since for protein solutions the Schmidt number $\mathrm{Sc}>10^4$, the boundary-layer approximation can normally be used. Then for steady flow in a generalized two-dimensional system* such as that suggested by Acrivos (1962) and illustrated by a simple example in Fig. 5.2.1, we may write

Continuity:
$$\frac{\partial v_x}{\partial x}+\frac{\partial v_y}{\partial y}=0 \tag{5.2.1}$$

Motion:
$$\frac{\partial}{\partial y}\mu_{\text{eff}}\frac{\partial v_x}{\partial y}=0 \tag{5.2.2}$$

Diffusion:
$$v_x\frac{\partial c_i}{\partial x}+v_y\frac{\partial c_i}{\partial y}=-\frac{\partial}{\partial y}\mathbf{J}_{iy}^{\star} \tag{5.2.3}$$

with the boundary conditions

$$\text{At } y=0, \quad v_x=0 \tag{5.2.4}$$

$$v_y=\kappa[p_0-p_\delta-\pi] \tag{5.2.5}$$

$$J_{py}^{\star}+c_p v_y^{\star}=0 \tag{5.2.6}$$

where the subscript $p$ refers to protein.

$$\text{As } y\to\infty, \quad c_p\to c_{p\infty} \tag{5.2.7}$$

$$\tau_{yx}\to\tau_\infty \tag{5.2.8}$$

Here $\kappa$ is the membrane permeability, $\pi$ is the osmotic pressure of the protein solution at the membrane-solution interface, $p_0$ is the corresponding hydrodynamic pressure, $p_\delta$ is the pressure on the far side of the membrane, $c_{p\infty}$ is the protein concentration in the feed solution, and $\tau_\infty$ is

* Here $x$ is the distance measured along the surface in the direction of the immediately adjacent streamlines, and $y$ is distance from the surface measured along local surface normals. See Fig. 5.3.7. The generalization to arbitrary three-dimensional flows is straightforward and is discussed by Stewart (1963). The definition of boundary-layer thickness is identical to that for the constant-property case (see Stewart or the review by Lightfoot, 1969).

the shear stress outside the concentration boundary layer (also presumed known*).

It now only remains to develop relations for the diffusion fluxes $J^{\bigstar}_{iy}$ in terms of the appropriate driving forces for mass transfer and to find suitable equations of state. We begin by writing flux expressions for such boundary-layer situations as that shown in Fig. 5.2.1.

Although even monodisperse protein solutions normally are multicomponent in nature, because of the need for buffers and a finite ionic strength,† small solutes tend to move with the same velocity as the water. Then from a diffusional standpoint the system can be considered a pseudobinary mixture of solvent W (mainly water) and protein P, with the pseudobinary diffusivity $Ɖ_{Wp}$ defined as

$$\frac{1}{Ɖ_{Wp}} = \left( \sum_{\substack{i=1 \\ \neq p}}^{n} \frac{x_i}{Ɖ_{ip}} \right) \Big/ \sum_{\substack{i=1 \\ \neq p}}^{n} x_i \tag{5.2.9a}$$

and

$$d_{Wy} = -d_{py} = \frac{x_W x_p}{Ɖ_{Wp}} (v_{py} - v_{Wy}) \tag{5.2.10}$$

where the $d_{iy}$ are the $y$ components of the generalized driving forces $\mathbf{d}_i$, and

$$x_W = \sum_{\substack{i=1 \\ \neq p}}^{n} x_i$$

If the protein solution is always fluid throughout the boundary layer, then $\partial p / \partial y = 0$ in this region and‡

$$J^{\bigstar}_{iy} = -\mathfrak{D} x d p_W \frac{\partial c_i}{\partial y} \qquad \text{(unsaturated solution)} \tag{5.2.11}$$

where $\mathfrak{D}_{pW} = Ɖ_{pW}(1 + \partial \ln \gamma_p / \partial \ln x_p)$. If, however, a gel or precipitate forms adjacent to the membrane, a force is transmitted to the protein from

* The matching with the constant-property velocity profile outside the concentration boundary layer is discussed by Kozinski. The determination of $\tau_\infty$ is discussed briefly in the example at the end of this section.

† Proteins are typically unstable at very low solution ionic strength.

‡ It is shown by Kozinsky and Lightfoot (1971) that the effects of density variations are of second order.

the membrane, and a hydrodynamic pressure gradient will develop within the concentration boundary layer. The protein now is subjected to an effective body force

$$\mathbf{g}_p^{(s)} = \frac{1}{\rho_p} \nabla p$$

as required by Eq. 1.4.3. Diffusion is thus most conveniently described by

$$J^{\bigstar}_{\mathrm{W}y} = -Ɖ_{\mathrm{W}p} \left[ \left( 1 + \frac{\partial \ln \gamma_p}{\partial \ln x_p} \right) c \frac{\partial x_{\mathrm{W}}}{\partial y} + \frac{\bar{V}_{\mathrm{W}}}{RT} \frac{\partial p}{\partial y} \right] \tag{5.2.12}$$

This expression covers a surprising range of behavior:

1. It reduces to Eq. 5.2.11 for the relatively dilute solutions characteristic of reverse osmosis: this is the situation pictured for the unsaturated region in Fig. 5.2.1*b*.
2. It yields

$$J^{\bigstar}_{\mathrm{W}y} = -\left( \frac{Ɖ_{\mathrm{W}p} \bar{V}_{\mathrm{W}}}{RT} \right) \frac{\partial p}{\partial y} \tag{5.2.13}$$

for mechanical filtration, with $Ɖ_{\mathrm{W}p} \bar{V}_{\mathrm{W}}/RT$ representing the permeability of the "saturated" region—the filter cake. In this limiting case the concentration drops very rapidly in the "unsaturated" region, and the concentration profile approaches a step function.
3. It must be used in its entirety for ultrafiltration at high levels of concentration polarization. This is a complex situation, and there may be a gradual transition from "free" solution to a cake of precipitate, as indicated by the dashed line of Fig. 5.2.1*b*.

To date, only the limiting cases of pure concentration and pure pressure diffusion have been treated in the literature. For the most part we assume pure concentration diffusion in our analysis below, but it is important to note that it is difficult to distinguish between these two driving forces in practice. For example, the water flux through the membrane depends only upon the sum of the chemical and hydrodynamic potentials of water at the membrane surface.

Finally, we need equations of state relating $\mu_{\mathrm{eff}}$, $\mathfrak{D}_{p\mathrm{W}}$, and the osmotic pressure $\pi$ to protein the concentration. Of these, $\mathfrak{D}_{p\mathrm{W}}$ is the most difficult to measure, but it also must be kept in mind that many protein solutions exhibit non-Newtonian behavior.

Once these parameters have been obtained—or estimated—one can integrate Eqs. 5.2.1 to 5.2.3 for any given flow situation outside the concentration boundary layer. This external flow, which is geometry dependent, determines both the shear-stress distribution $\tau_\infty(x)$ and the pressure distribution $p(x,\infty)$.

It is, of course, possible to work directly with the partial differential equations shown, and this has been done for slit flows by Brown et al. (1971). However, one can, under favorable circumstances, both simplify the calculation procedure and increase the utility of the results by generalizing a variable-property similarity transformation developed by Acrivos (1962) to high mass-transfer rates. To do this one seeks the conditions under which

$$\chi = \chi(\eta) \tag{5.2.14}$$

where

$$\chi = \frac{c_p - c_{p0}}{c_{p\infty} - c_{p0}} \tag{5.2.15}$$

$$\eta = \frac{y}{\delta(x)} \tag{5.2.16}$$

and $\delta(x)$ is the local mass-transfer boundary-layer thicknes. The protein concentration $c_{p0}$ at the membrane surface is treated as known and constant during this development,* and the definition of $\delta$ is completed below.

If one puts Eq. 5.2.14 into the system description, one obtains†

$$\left[v_{y0}\delta - H(\eta)\delta\frac{\partial}{\partial x}(\dot{\gamma}_\infty \delta^2)\right]\frac{d\chi}{d\eta} = \mathfrak{D}_{p\mathrm{w}\infty}\frac{d}{d\eta}D\frac{d\chi}{d\eta} \tag{5.2.17}$$

with

$$\chi(0) = 0; \qquad \chi(\infty) = 1$$

and where‡

$$H(\eta) = \int_0^\eta \int_0^{\eta'} \frac{\mu_\infty}{\mu}\, d\eta''\, d\eta'.$$

$$\gamma_\infty = \tau_\infty/\mu_\infty; \quad D = \mathfrak{D}_{p\mathrm{w}}/\mathfrak{D}_{p\mathrm{w}\infty}.$$

* As we shall see, it is reasonable to make this assumption under a limited but very useful range of conditions.

† It is implied by Eq. 5.2.14 that the transport properties also are functions only of $\eta$.

‡ Note that $H = \eta^2/2$ for systems of constant viscosity.

We now complete our definition of the boundary-layer thickness by setting

$$\delta \frac{\partial}{\partial x} \dot{\gamma}_{\infty} \delta^2 = 6\mathfrak{D}_{pW\infty}; \quad \delta(0) = 0 \qquad (5.2.18a;b)$$

Equation 5.2.18 may then be integrated to give

$$\delta = \frac{\left(9\mathfrak{D}_{pW\infty} \int_0^x \sqrt{\dot{\gamma}_{\infty}}\, dx\right)^{1/3}}{\sqrt{\dot{\gamma}_{\infty}}} \qquad (5.2.19)$$

and Eq. 5.2.17 may be put in the form

$$\left[\left(\frac{v_{y0}\delta}{\mathfrak{D}_{pW\infty}}\right) - 6H(\eta)\right] \frac{d\chi}{d\eta} = \frac{d}{d\eta} D \frac{d\chi}{d\eta} \qquad (5.2.20)$$

Our similarity transformation is thus successful if

$$\frac{v_{y0}\delta}{\mathfrak{D}_{pW\infty}} = c_W = \text{constant} \qquad (5.2.21)$$

We return to this point shortly, and show that Eq. 5.2.21 is consistent with our boundary conditions for many situations of interest.

We now note that all effects of physical-property variation are concentrated in Eq. 5.2.20 (via $H$ and $D$) while all effects of geometry* are concentrated in Eq. 5.2.19. The significance of this statement becomes clearer if we write Eq. 5.2.6 as

$$\frac{v_{y0}\delta}{\mathfrak{D}_{pW\infty}} = -\frac{c_{p0} - c_{p\infty}}{c_{p0}} \left.\frac{d\chi}{d\eta}\right|_{\eta=0} \qquad (5.2.22)$$

Thus $c_W$ should indeed be constant if $c_{p0}$ is independent of $x$. We have now generalized the treatment of *Tr. Ph.*, Section 21.6 to arbitrary geometries† and variable $\mu$, $\mathfrak{D}_{pW}$, and osmotic pressure. We may complete the formal

* The product $v_{y0}\delta$ is shown below to be geometry independent. See also the review by Lightfoot (1969).

† Equations 5.2.14 to 5.2.22 are applicable to three-dimensional geometries on suitable generalization of Eq. 5.2.19, as described by Stewart (1963) or Lightfoot (1969).

development by defining a local mass-transfer coefficient

$$k_{c\,\mathrm{loc}} = \frac{J^{\star}_{p0}}{c_{p0} - c_{p\infty}} \tag{5.2.23}$$

and the corrresponding Nusselt number

$$\mathrm{Nu}_{\mathrm{loc}} = k_{c\,\mathrm{loc}} D / \mathfrak{D}_{p\mathrm{W}} \tag{5.2.24}$$

*where D is now any convenient reference length.* We may now relate this Nusselt number to the concentration profile, for negligible pressure diffusion, by using Eqs. 5.2.23 and 5.2.11:

$$\mathrm{Nu}_{\mathrm{loc}} = -\left.\frac{d\chi}{d\eta}\right|_{\eta=0} \frac{D}{\delta} \tag{5.2.25}$$

and to the ultrafiltration rate, from Eq. 5.2.6, as

$$\mathrm{Nu}_{\mathrm{loc}} = -\frac{v_{y0} D}{\mathfrak{D}_{p\mathrm{W}}} \frac{c_{p0}}{c_{p0} - c_{p\infty}} \tag{5.2.26}$$

Equation 5.2.25 permits calculation of $\mathrm{Nu}_{\mathrm{loc}}$ from $\chi$, and Eq. 5.2.26 permits determination of ultrafiltration rate from $\mathrm{Nu}_{\mathrm{loc}}$.

It follows from Eqs. 5.2.20 and 5.2.22 and from the assumed boundary conditions that

$$\left.\frac{d\chi}{d\eta}\right|_{\eta=0} = f\left(\frac{c_{p0}}{c_{p\infty}}, \text{ equations of state}\right) \tag{5.2.27}$$

and it is readily shown* that

$$\frac{D}{\delta} = \mathrm{Sc}^{1/3} f\,(\mathrm{Re}, \text{geometry}) \tag{5.2.28}$$

where $\mathrm{Re} = DV/\nu$, and $V$ is any convenient characteristic velocity. The determination of $\chi(\eta)$ is discussed in two papers by Kozinski and Lightfoot (1971, 1973) and in Kozinski's thesis (1970). It is sufficient here to state that $\chi(\eta)$ can be determined in a straightforward way and that

$$\mathrm{Nu}_{\mathrm{loc}} = A\,\mathrm{Sc}^{1/2}\mathrm{Re}^{1/3} f\,(\mathrm{Re}, \text{geometry}) \tag{5.2.29}$$

* See, for example, Lightfoot.

with

$$A = A\left(\frac{c_{p\infty}}{c_{p0}}, \text{equations of state}\right)$$

Furthermore, the function $f$ is obtainable from the constant-property low-mass-transfer limiting results of boundary-layer analysis (see, e.g., Lightfoot, 1970).

It is now, however, time to remember the constraint provided by Eq. 5.2.5, which expresses an additional relation between the ultrafiltration rate $v_{y0}$ and the boundary protein concentration $c_{p0}$ (through its effect on the osmotic pressure $\pi$). The nature of this constraint is seen most clearly for a protein giving negligible osmotic pressure and for negligible pressure diffusion. For this situation the local ultrafiltration rate is proportional to $p(x,\infty)$, which in general shows a different $x$ dependence from the mass-transfer boundary-layer thickness $\delta$. Then Eq. 5.2.6 generally requires a position-dependent boundary protein concentration $c_{p0}$; this invalidates the similarity transformation, and hence our above expressions, as a generally applicable solution. There are, however, two important special circumstances for which $c_{p0}$ may be considered constant:

1. Systems for which both $\delta$ and $p(x,\infty)$ are constant. These include the flow about spinning discs, and axisymmetric or two-dimensional stagnation flows. These are of limited practical importance, but they are very useful for examining the structure and behavior of protein boundary layers.
2. Systems of arbitrary geometry and flow operated at very large $p_0 - p_\delta$. This situation is very complex from a diffusional standpoint, but very important from a practical one.

We begin here with a discussion of the simpler and less ambiguous of these cases.

The spinning disc is a particularly attractive system from an experimental standpoint and has recently been used by Kozinski (1971) to measure the ultrafiltration behavior of bovine serum albumin. Typical experimental results of this work are shown in Fig. 5.2.2 along with a numerical solution of Eqs. 5.1.20 and 5.1.22 for the concentration dependence of $\mu$, $\mathfrak{D}_{pw}$, and $\pi$ observed for this protein.*

* Details are provided by Kozinski (1971), and in abbreviated form by Kozinski and Lightfoot (1971). In this three-dimensional system the definition of $\delta$ is somewhat more complex than that given by Eq. 5.2.19; see Stewart (1963) or Lightfoot (1969)—or Eq. 7 of Ex. 5.2.1.

In each case the ultrafiltration rate at first increases linearly with applied pressure, and the flow is limited primarily by membrane resistance. As the pressure increases, however, the apparent permeability of the system falls, and the ultrafiltration rate approaches an asymptotic limit, which is higher for higher stirrer speeds. This behavior is closely approximated by the above boundary-layer model, as indicated by the typical rather good agreement between observation and prediction shown in Fig. 5.2.2 for a rotational speed of 273 rev/min. It is found experimentally that a dynamic equilibrium is quickly established and that there is negligible hysteresis in the absence of protein denaturation.

In its essentials, then, the reason for this behavior is clear: the increase of boundary concentration with pressure drop results from the need for a larger concentration driving force if diffusion is to balance the increasing convective protein flow. The higher boundary concentration in turn retards water flow, first by increasing the boundary osmotic pressure and ultimately through the formation of a solid, or semisolid, cake of appreciable hydraulic resistance.

There appear to be at least three possible reasons for an asymptotic limit to the permeation rate:

1. The osmotic coefficient of the protein ($\pi / c_P RT$) can rise more rapidly with concentration than the effective average diffusivity in the boundary layer falls. Increases in hydrostatic pressure drop are then ultimately nearly compensated by a corresponding increase in osmotic pressure. This appears to be the situation in the albumin experiments of Kozinski (1970).
2. The solubility of the protein can be reached near the membrane surface, with the result that a more or less porous solid cake forms on the membrane. Any further increase in driving pressure merely thickens the cake without increasing the ultrafiltration rate. This is the model used in Kozinski and Lightfoot's first paper (1971), which appears to give a useful semiquantitative description of the behavior for a wide variety of proteins.
3. A semisolid gel can form which produces an appreciable need for pressure diffusion and a consequent drop in the effective driving force.

The behavior is actually much the same in all cases, and it is therefore not always necessary to distinguish between them.

It is important to note that in all three, the boundary protein concentration approaches something very close to an asymptotic limit independent of both geometry and flow conditions. Our similarity transformation should therefore be successful at or near the asymptotic flow limit, and it should be possible to predict this limit in one system from measurements on another. That is,

$$\frac{v_{m1}}{v_{m2}} = \frac{\delta_2}{\delta_1} \tag{5.2.30}$$

where $v_{mi}$ is the maximum permeation rate for system $i$, and $\delta_i$ is the corresponding boundary-layer thickness as defined by Eq. 5.2.19 (or its three-dimensional equivalent).

At the time of writing no carefully controlled test of Eq. 5.2.29 had been made. However, Kozinski was able to show reasonable consistency between his results for bovine serum albumin on a spinning disc and fragmentary data for ultrafiltration in a commercial cell. More rigorous tests of Eq. 5.2.29 are highly desirable.

*EX. 5.2.1. PREDICTION OF ULTRAFILTRATION CAPACITY IN AN UNTESTED SYSTEM*

From a practical standpoint the most important characteristic of an ultrafiltration system is the maximum filtration rate of which it is capable:

$$v_{\max} = \lim_{(p_0 - p_\delta)\to\infty} v_{y0}$$

Show how this maximum permeation rate can be estimated for slit flow from measurements on the same protein with a spinning disc.

SOLUTION

We begin by assuming the protein concentration at the membrane-solution interface to approach the same limiting value over the entire filtration surface for all geometries. We next recognize that the boundary-layer assumptions then hold for the two systems under consideration, so that both are described by Eqs. 5.2.20 and 5.2.22.

Then since $d\chi/d\eta|_{\eta=0}$ is the same for both systems, we may write Eq. 5.2.29 as

$$\frac{(v_{\max})_s}{(v_{\max})_d} = \frac{\mathrm{Nu}_s^\bullet}{\mathrm{Nu}_d^\bullet} \tag{1}$$

Here the subscripts $s$ and $d$ refer, respectively, to slit and disc, and the superscript $\bullet$ refers to conditions, including the *boundary-layer shear stress*, at *maximum filtration rate*.

The Nusselt number for the slit, $\mathrm{Nu}_s^\bullet$, is particularly easy to describe, and we therefore concentrate our attention on it. Referring to Fig. 5.1.3, we may write for the region $-B<y<B$:

$$\tau_{yx}|_B = -\tau_{yx}|_{-B} = -B\frac{\partial \mathcal{P}}{\partial x} \tag{2}$$

and therefore that

$$\dot{\gamma}_\infty = -\frac{B}{\mu}\frac{\partial \mathcal{P}}{\partial x} \tag{3}$$

Here the generalized boundary-layer coordinates $x$ and $y$ are simply the corresponding rectangular coordinates $x$ and $y$ of the figure. Equation 3 contains only the very reasonable assumption that the mass-transfer boundary layer is thin relative to the channel width, and it is therefore quite reliable. However, it is inconvenient, because local pressure gradients are not easily measured directly. It is therefore desirable to relate $\dot{\gamma}$ to some more accessible quantity before using it in Eq. 19 for calculation of $\delta$. In principle this should be done by matching the boundary-layer velocity with a solution of the equation of motion for the bulk of the fluid; such a matching procedure is discussed by Kozinski and Lightfoot in their first paper (1971).

In practice, however, it is probably satisfactory, and certainly much simpler, to neglect the tangential fluid velocity $v_x$ within the concentration boundary layer relative to that in the bulk of the fluid. This is a natural extension of the simplifications suggested in Section I.3.3 and leads here to*

$$\dot{\gamma}_\infty = 3\langle v\rangle / B \tag{4}$$

where $\langle v\rangle$ is the flow-average velocity in the slit.

Equation 5.2.19 then takes the simple form

$$\delta = \left(\frac{3\mathfrak{D}_{pW\infty} Bx}{\langle v\rangle}\right)^{1/3} \tag{5}$$

It therefore follows from Eqs. 5.2.25 and 5.2.26 that

$$-(v_{y0})_s = \left.\frac{d\chi}{d\eta}\right|_0 \frac{c_{p0} - c_{p\infty}}{c_{p0}} \mathfrak{D}_{pW\infty}\left(\frac{\langle v\rangle}{3\mathfrak{D}_{pW\infty} Bx}\right)^{1/3} \tag{6}$$

Note that on the right only the term $(\langle v\rangle / 3\mathfrak{D}_{fW} Bx)^{1/3}$ is geometry dependent.

* See, for example, Problem 2.E, p. 62, *Tr. Ph.* This is the limiting shear rate for constant viscosity and negligible interfacial velocity. It also assumes fully developed flow throughout the duct, as is reasonable if $L/B \gg 1$ and the Reynolds number is small (see Section 3.5 for a discussion of this situation.

The corresponding expression for the spinning disc cannot be obtained quite so easily, since this flow field is fully three-dimensional. However, it may be shown that here*

$$\mathrm{Nu}_d = -\left.\frac{d\chi}{d\eta}\right|_{\eta=0} \frac{1}{6^{1/3}} \mathrm{Re}^{1/2}\mathrm{Sc}^{1/3} \tag{7}$$

where $\mathrm{Re} = D^2\omega/\nu$
with $D$ = disc diameter
$\omega$ = angular velocity of disc in radians per unit time.

If this result is put into Eq. 5.2.26 and the result is compared with Eq. 6 of this example, we find

$$\frac{(v_{y0})_s}{(v_{y0})_d} = 2^{1/3}\left(\frac{\langle v\rangle^2 \nu_\infty}{B^2 x^2 \omega^3}\right)^{1/6} \tag{8}$$

This is essentially the expression used by Kozinski (1970) to predict ultrafiltration rates for bovine serum albumin in Amicon filtration cells.

## 5.3. Gas Exchange in Blood

In addition to their obvious physiological importance, studies of oxygen and carbon dioxide transfer in blood are needed both to improve diagnostic procedures and for more reliable design of extracorporeal oxygenation equipment. These transfer processes also pose a very interesting problem in strategy to the medically oriented engineer because of their extreme complexity. Those interested in the prediction of transfer rates must accept the present large uncertainties in our knowledge of kinetic, diffusional, and flow behavior—and take these into account in their calculation procedures. At the same time it is necessary continually to reassess these procedures and to determine the most critical areas for further refinement.

Here we begin by reviewing briefly kinetic, thermodynamic, and diffusional behavior at the molecular and cellular levels to provide a basis for assessing the continuum models used at present for predicting mass-transfer rates. We then introduce the most widely used model and go on to describe what we believe to be the most useful prediction procedures. We

* See, for example, Levich (1962), Section 11. This result can also be obtained directly from the velocity profile using Stewart's formula, which is included in Lightfoot's review (1969). Note again that $d\chi/d\eta|_0$ is geometry independent.

put particular emphasis on boundary-layer simplifications and approximations yielding upper and lower bonds for oxygenation rates, and we do this for two reasons:

1. These techniques give quite acceptable accuracy, if one takes into account the doubtful nature of the diffusional models in current use.
2. Their relative simplicity provides insight into the nature of the oxygenation processes and facilitates the rational design of "artificial lungs."

We also, however, review the results of numerical calculations, which are particularly useful for the important case of duct flows. For a more complete discussion see Spaeth's review (1970).

### (a) *MOLECULAR ASPECTS OF BLOOD OXYGENATION: THERMODYNAMICS AND KINETICS*

Most important here are the reactions of oxygen with hemoglobin and myoglobin, and of carbon dioxide with water, hemoglobin, and plasma proteins. The metabolic oxidation reactions are also of fundamental importance, but these have been treated more casually by those interested in mass-transfer rates.

The equilibria in the $O_2$-hemoglobin-$CO_2$-water system have been very extensively investigated and are much better understood than the kinetic aspects. Such thermodynamic information is readily available in standard references, for example the *Handbook of Physiology*, and does not warrant discussion here. It is, however, worth pointing out that the $O_2$ and $CO_2$ equilibria are coupled both by the change of hemoglobin pK with degree of oxygenation and by the decrease of oxygen affinity with decreasing pH. The second of these, the *Bohr effect*, is particularly important. However, in most oxygenation calculations such coupling is ignored. When this is done, oxygenation equilibria are adequately expressed over the physiological range by the empirical relation

$$S = \frac{\left(p_{O_2}/p_c\right)^n}{1+\left(p_{O_2}/p_c\right)^n} \tag{5.3.1}$$

where $S$ is the fractional saturation of hemoglobin, $p_{O_2}$ is the partial pressure of oxygen, and $p_c$ and $n$ are functions of temperature, pH, and the chemical nature of the hemoglobin or myoglobin involved (see Eq. 5.3.5 and Ex. 5.4.1).

The kinetics of the gas-exchange reactions are also relatively well understood in buffered hemoglobin solutions; the reader is referred to the symposium volume "Blood Oxygenation" edited by Hershey (1970), and in particular to the contribution of Sirs, for supplementation of older references. It is particularly important to note that the reaction of $CO_2$ with water to form carbonic acid is very slow. This reaction must be catalyzed by carbonic anhydrase to proceed at a physiologically significant pace, and this enzyme occurs naturally only inside the red blood cells. Sirs has suggested that this segregation of carbonic anhydrase, together with the permselectivity of red-cell membranes and the coupling of the gas-exchange reactions referred to above, can produce some very interesting transient behavior during oxygenation. This appears to be beneficial physiologically and also provides a simple unambiguous example of oscillatory behavior in reacting systems.

The consumption of oxygen by metabolic reactions is normally taken to be zero order.

Diffusional transport of $O_2$ and $CO_2$ in both hemoglobin solutions and blood has been particularly carefully investigated by Spaeth and Friedlander (1967) and earlier by Keller and Friedlander (1966). Transport of $O_2$ by diffusion of oxygenated hemoglobin can be important at low oxygen partial pressures and in dilute hemoglobin solutions. It does not seem to be important at physiological oxygen concentrations. It also appears from these and other results that in most situations of current interest the rates of Hb-$O_2$ dissociation reactions do not have a major effect on oxygen-transport rates. This is not true at very high oxygen fluxes, however—for example, in very thin films—and nonregation of carbonic anhydrase, together with the permselectivity of red-cell memequilibrium transport has been investigated in some detail, notably by Goddard, Schulz, and Bassett (1970) and by Bassett and Schulz (1970). See also the discussion of capillary mass transfer in Section 5.4.

### (b) *CELLULAR ASPECTS OF BLOOD OXYGENATION AND THE DEVELOPMENT OF A CONTINUUM MODEL*

This is an area of very extensive research activity, and we must limit ourselves severely. Much wider coverage can be obtained from such sources as Hershey (1967, 1970) and the forthcoming proceedings of the 1973 International Symposium on Tissue Oxygenation. The contributions in Sirs (in Hershey, 1970) and of Bruley et al. (1971), which have already been cited, deal primarily with the kinetic and diffusional aspects.

The kinetics of the exchange reactions have been extensively investigated in both blood and red-cell suspensions, and much of the recent work is discussed and referenced in the symposium proceedings (Hershey, 1973). The interpretation of such experiments must, of course, be somewhat ambiguous because of the complex geometry of the red cell and the complexity of flow conditions. It is, however, clear that reaction rates are lower for intact cells than for hemoglobin solution and that intracellular diffusional resistance most probably does not account for all of the decrease. It is therefore widely believed that the red-cell membrane provides a high resistance to oxygen transfer, improbable as this may seem in view of its very small thickness and lipid nature. Unfortunately, neither boundary-layer resistances outside the red cells nor the effects of aggregation or segragation of cells appears to have been examined very carefully in connection with these kinetic experiments.

The motion of red cells relative to plasma can also produce a wide variety of convection patterns with characteristic dimensions on the order of a red-cell diameter. Two of these are indicated schematically in Figs. 5.3.1 and 5.3.2.

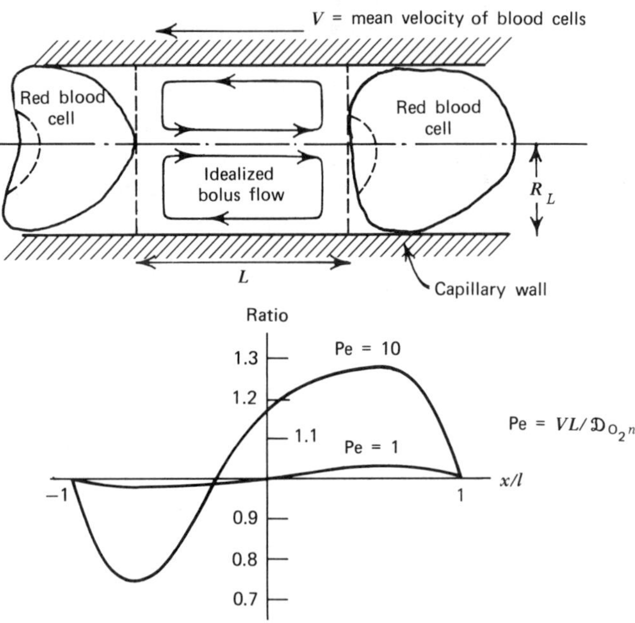


**Figure 5.3.1.** The idealized capillary bolus flow of Aroesty and Gross, and their predictions of the effect of convection on oxygen transport to surrounding tissue. The flow situation is shown from the viewpoint of an observer moving with the red cells. The graph shows the effect of convection on the local mass-transfer rate for $R_L = L$; the ordinate is the ratio of the rate for convection and diffusion to that for diffusion only.

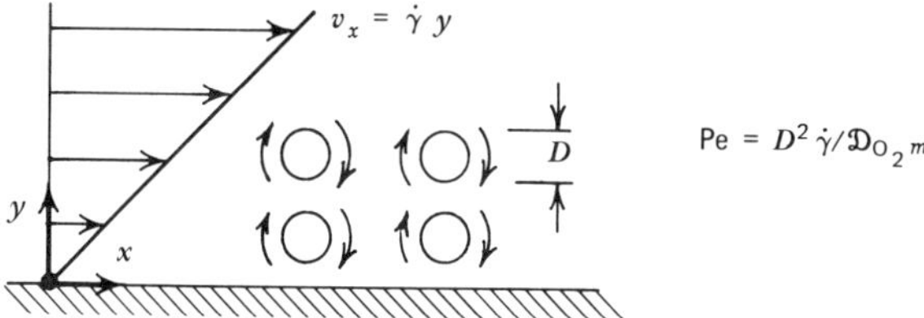


**Figure 5.3.2.** Convection patterns produced by red cells in a shear field. The red-cell spacing is much exaggerated for illustrative purposes.

The capillary bolus flow of Fig. 5.3.1 has been investigated with particular care, and typical results of a recent analysis by Aroesty and Gross (1970) are shown. The net effect of such convection is to increase mass-transfer rates above those for diffusion alone. The relative importance of convection is, however, strongly dependent on the capillary Péclet number, which is probably always below 10 in situations of biological interest. It follows that oxygen transfer in capillaries can be adequately described by ignoring convection and assuming plug flow.

Convection in larger ducts is even more complex, and both translational and rotational motions of red cells have been noted.* The rotational motion shown in Fig. 5.3.2, first pointed out by Blackshear (1965), is representative. Its importance depends upon the rotational Péclet number shown and thus is greatest at high shear rates. Experiments by Keller and his associates show that this, and other microconvective motions which may occur simultaneously, can have an appreciable effect on mass transfer at very high shear rates. These effects are, however, small for conditions presently envisaged for extracorporeal oxygenators, and they are usually ignored.†

The tendency of red cells to migrate away from confining walls must also be considered whenever the cell-free layer has a thickness $\delta_c$ approaching the diffusional boundary-layer thickness $\delta_{AB}$. We see below that $\delta_{AB} \gg \delta_c$ for many typical situations today, but that this cannot be expected universally.

In summary, the particulate nature of blood appears at present to be of secondary importance. It is therefore customary to describe oxygen transfer in terms of the following pseudobinary continuum model:

$$\frac{Dc_{\text{HbO}_2}}{Dt} = R_{\text{HbO}_2} \tag{5.3.2a}$$

* See for example, Spaeth (in Hershey, 1970).

† They are, however, important for larger solutes and hydrodynamic bodies such as platelets. See Turitto et al. (1972) and Grabowski et al. (1972).

$$\frac{Dc_{O_2}}{Dt} = \mathfrak{D}_{O_2m}\nabla^2 c_{O_2} + R_{O_2} \tag{5.3.2b}$$

$$\left(\frac{\partial c_{HbO_2}}{\partial c_{O_2}}\right)_{T,\mathrm{pH}} = m(c_{O_2}) \tag{5.3.3}$$

where $H$ is the hematocrit. Equation 5.3.2a states that hemoglobin cannot diffuse; it is based on the findings of Keller and Friedlander (1966) cited earlier in connection with facilitated diffusion, but now in addition ignores the convective contribution of cell roation. Equation 5.3.2b assumes oxygen diffusion to follow the binary form of Fick's law with a position-independent diffusivity, and Eq. 5.3.3 assumes local equilibrium between dissolved oxygen and hemoglobin. These equations may now be combined to give

$$(1+m)\frac{Dc_{O_2}}{Dt} = \mathfrak{D}_{O_2m}\nabla^2 c_{O_2} \tag{5.3.4}$$

which is a generalization of Eq. 1, Ex. 2.3.2 to a much wider range of conditions.

Rather extensive experimentation, reviewed briefly by Spaeth, indicates that $\mathfrak{D}_{O_2m}$ is reasonably independent of both oxygen partial pressure and concentration gradients, but strongly dependent on hematocrit. At 37°C it is not unreasonable to assume the values in Table 5.3.1. A brief look at Spaeth's tabulation is, however, sufficient warning that these figures are not highly precise; deviations of the order of 30%, and perhaps much more, can be expected. This should be kept in mind in developing calculation procedures.

If we keep the above uncertainty in mind and restrict ourselves to the physiologically interesting range of oxygen pressures

$$20 \text{ mm Hg} < p_{O_2} < 750 \text{ mm Hg}$$

**Table 5.3.1.** Effective Diffusivity of Oxygen

| Solution | $\mathfrak{D}_{O_2m}$ $(10^5)\,\mathrm{cm^2/sec}$ |
|---|---|
| Water | 3.0 |
| Plasma | 2.0 |
| Normal blood | 1.4 |
| Packed red cells | 0.8 |

we can use Hill's approximation for Eq. 5.3.3:

$$m \doteq n\frac{c_{\text{sat}}}{c_H}\frac{\left(c_{\text{O}_2}/c_{\text{H}}\right)^{n-1}}{\left[1+\left(c_{\text{O}_2}/c_{\text{H}}\right)^n\right]^2} \tag{5.3.5}$$

where $n$ and the reference oxygen concentrations are functions of temperature, pH, and the type of blood. For normal human blood,

$$n = 2.66$$

$$c_{\text{H}} = 3.73 \times 10^{-8}\ (\text{moles O}_2)/\text{cm}^3$$

$$c_{\text{sat}} = 20.5H \times 10^{-6}\ \text{moles}/\text{cm}^3$$

where $H$ is the hematocrit expressed as the volume fraction of red cells.

We now turn to a description of oxygenation, and we organize the remaining discussion so as to emphasize the salient characteristics of this process. We begin by determining the sensitivity of oxygenation rates to the shape of the hemoglobin dissociation curve and then look at the effect of geometry. Our dicussion is essentially a condensation of the work of Dindorf, Lightfoot, and Solen (1971). The reader is also referred to work by Buckles et al. (1968), Spaeth (1970), Villaroel et al. (1971), Weissman and Mockros (1967), and White (1969), cited in the bibliography.

Referring to Fig. 2.3.2, we note that the true saturation curve lies between the linear approximation $AC$ and the step function $ABC$. Furthermore, use of $A$ and $ABC$ instead of the true dissociation curve will provide lower and upper bounds, respectively, about the estimate of the blood oxygenation rate obtained from this true curve. The reasons are the same as those given in Ex. 2.3.2, and we can see that very frequently the upper and lower bounding solutions differ little from each other. This latter point is important from at least the following standpoints:

1. It justifies neglect of the Bohr effect in our calculations.
2. It simplifies recalculation for different types of blood.
3. It permits simple extension of our calculation procedures to other cases of reversible protein binding, as may for example occur in hemodialysis.

We therefore want to compare upper and lower bounding approximations for a wide variety of systems. We begin with an overall view of the linear, or lower-bounding, approximation, as here we can obtain a general solution by analogy in terms of the corresponding solutions for nonreactive systems.

### (c) *THE LINEAR (LOWER-BOUNDING) APPROXIMATION*

Consider a nonreactive pseudobinary system with characteristic length $L$, fluid velocity $V$, and solute concentrations $c_{i0}$ and $c_{i\infty}$. Let the dimensionless concentration profile be given by

$$\Pi_i = \frac{c_i - c_{i0}}{c_{i\infty} - c_{i0}} = \Pi_0\left[\left(\frac{LV\rho}{\mu}\right), \left(\frac{\mu}{\rho \mathfrak{D}_{im}}\right), \mathbf{r}, t\right] \tag{5.3.6a}$$

where $\mathbf{r}$ is the position vector and $t$ is the time.

Then from Eq. 5.3.4 the corresponding oxygen profile is

$$\Pi_{O_2} = \frac{c_{O_2} - c_0}{c_\infty - c_0} \doteq \Pi_0\left[\left(\frac{LV\rho}{\mu}\right), \left(\frac{\mu(1+M)}{\rho \mathfrak{D}_{O_2m}}\right), \mathbf{r}, t\right] \tag{5.3.6b}$$

for the same geometry and boundary conditions. Here $M$ is a suitable average value of $m$, most simply taken as the slope of the line $AC$, Fig. 2.3.2.

The local rate of oxygen transfer across the boundary into the blood is

$$(N_{O_2})_0 = -\mathfrak{D}_{O_2m} \left.\frac{\partial c_{O_2}}{\partial y}\right|_{y=0} \tag{5.3.7}$$

where $y$ is distance into the blood along a normal to the boundary. If we define a local mass-transfer coefficient in terms of the characteristic driving force $c_\infty - c_0$, so that

$$(N_{O_2})_0 = k_{c\,\text{loc}}(c_0 - c_\infty) \tag{5.3.8}$$

then the dimensionless mass-transfer coefficient or Nusselt number is

$$\text{Nu}_{\text{loc}} = \left(\frac{k_{c\,\text{loc}} D}{\mathfrak{D}_{O_2m}}\right) = \left.\frac{d\Pi_0\left(LV\rho/\mu, \mu(1+M)/\rho\mathfrak{D}_{O_2m}\right)}{dy^*}\right|_{y^*=0} \tag{5.3.9}$$

with $y^* = y/L$ and $M$ is the average value of $m$ as defined in Ex. 2.3.2. Equation 5.3.9 then permits the calculation of oxygenation rates wherever the functional dependence of $\text{Nu}_{\text{loc}}$ on Schmidt number is known for the corresponding nonreactive situation.

The fundamental simplicity of Eq. 5.3.9, and its utility, are shown in the discussions of boundary-layer mass transfer immediately below. We return to the linear approximation for description of pulmonary oxygenation in the next section.

(d) *DIRECT-CONTACT OXYGENATORS: MASS TRANSFER ACROSS FREE INTERFACES*

At the time of writing it is nearly universal medical practice to oxygenate blood by direct contact with an oxygen-rich gas. Examples include various types of bubble oxygenators, which are probably the most common, and spinning-disc oxygenators, of which one example is shown in Fig. 6.3.1.

These are nearly all characterized by the high Schmidt numbers, low interfacial stresses, and short contact times required by the penetration theory, and they can be described in terms of the generalized penetration theory developed in papers by Angelo and Lightfoot and by Stewart, Angelo, and Lightfoot (see Lightfoot, 1969). We therefore begin by considering a surface of arbitrary shape.

For convenience, one may define a non-orthogonal coordinate system that moves with the interface. Here, $y$ measures the distance to the interface, and the coordinates $u$ and $w$ form a non-orthogonal grid imbedded in the surface. Fig. 5.3.3 shows the behavior of such a grid for coalescing spheres.

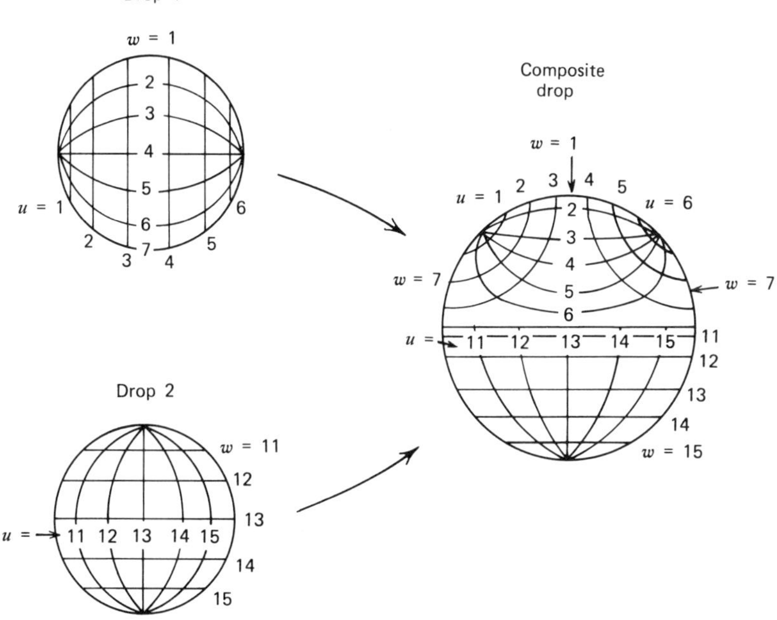


**Figure 5.3.3.** The behavior of imbedded coordinates: the coalescence of drops on bubbles. Differential surface elements $dudw$ are defined in terms of imbedded coordinates $(u, w)$ which move, and deform, with the surface velocity. From Stewart, Angelo, and Lightfoot.

We now note that mass-transfer descriptions in this non-orthogonal coordinate system can be greatly simplified by making the following boundary-layer assumptions:

1. Diffusive fluxes are normal to the nearest interfacial element.
2. The tangential velocity in the boundary layer is essentially that of the nearest interfacial element.
3. The thickness of the boundary layer is small relative to the local radii of surface curvature.

These assumptions become exact in the limit of large Schmidt numbers and short contact times, and they are closely approximated in direct-contact oxygenators.* It should be noted that the first two statements imply that any differential surface element may be considered independently, since mass transfer between adjacent surface elements is neglected.

One may then write Eq. 5.3.4 for the area $dS \equiv s\,du\,dw$ of any differential surface element $du\,dw$ as:

$$(1+m)\left[\frac{\partial c_{O_2}}{\partial t} - y\left(\frac{\partial \ln s}{\partial t}\right)_{u,w}\frac{\partial c_{O_2}}{\partial y}\right] = \mathfrak{D}_{O_2 m}\frac{\partial^2 c_{O_2}}{\partial y^2} \tag{5.3.10}$$

The term $(\partial \ln s/\partial t)_{u,w}$ is the fractional rate of expansion of the element surface and is a function only of time for a given element. The significance of $s$ is discussed by Lightfoot and by Stewart, Angelo, and Lightfoot. We merely note here that it is proportional to element area.

We now define a dimensionless dissolved-oxygen concentration $\Pi$ as in Eq. 5.3.6:

$$\Pi = \frac{c_{O_2} - c_0}{c_\infty - c_0}$$

where $c_0$ is the concentration in equilibrium with the oxygenating gas, and $c_\infty$ that in the blood far from the interface. Since the liquid-phase mass-transfer resistance is generally controlling in oxygenators, and since gas and bulk blood compositions change relatively slowly with position, we may write to a first approximation

$$c_{O_2} = c_0, \qquad y = 0 \tag{5.3.11}$$

$$c_{O_2} \to c_\infty, \qquad y \to \infty \tag{5.3.12}$$

* It is, however, possible that these boundary layers may overlap—for example, in a thin film between two gas bubbles.

throughout the effective lifetime of a given surface element. As shown by Dindorf et al. (1971), it is then possible to make a similarity transformation and to write Eqs. 5.3.10 to 5.3.12 as

$$-2\eta(1+m)\frac{d\Pi}{d\eta}=\frac{d^2\Pi}{d\eta^2} \tag{5.3.13}$$

with

$$\eta=\frac{y}{\delta(t,u,w)} \tag{5.3.14}$$

$$\delta(t,u,w)=\left[\frac{4\mathfrak{D}_{O_2m}}{s^2}\int_0^t s^2dt\right]^{1/2} \tag{5.3.15}$$

and the boundary conditions

$$\Pi(0)=0, \qquad \Pi(\infty)=1 \tag{5.3.16}$$

Equations 5.3.13 to 5.3.16 represent a complete problem statement and generalize Ex. 2.3.2 to a distorting surface.

It is important to note that fluid-mechanic parameters occur only in the definition of the boundary-layer thickness $\delta$. Similarly, the shape of the equilibrium curve is described solely by the factor $1+m$ in Eq. 5.3.13. Thus, the effects of surface distortion and chemical reaction have been separated, and since $\delta$ is the same for the reacting and nonreacting cases, Eq. 5.3.13 reduces to its nonreactive counterpart as $m$ approaches zero.

One may now define a time-averaged Nusselt number *based on a reference area,* $dS_0$:

$$[\overline{\mathrm{Nu}}^{\,0}]_{\mathrm{loc}}=-\Pi'(0)\sqrt{\frac{D^2}{\mathfrak{D}_{O_2m}t}}\sqrt{\frac{1}{t}\int_0^t\left(\frac{s}{s_0}\right)^2dt} \tag{5.3.17}$$

where

$$-\Pi'(0)=\frac{1}{\int_0^\infty e^{-\int_0^p 2(1+\mathrm{m})qdq}\,dp} \tag{5.3.18}$$

is just $d\Pi/d\eta|_{\eta=0}$. Once again, all of the effects of chemical reaction are concentrated in $\Pi'(0)$, and all of the fluid-mechanical behavior in the integral of Eq. 5.3.17.*

* It is interesting to note that all of the effects of surface distortion can be expressed simply as the temporal root mean square area of the surface element.

The diffusional aspect of the problem is then given by Eq. 5.3.18, and it is convenient to summarize here the limiting behavior:

| Approximation | $-\Pi'(0)$ | Ha |
|---|---|---|
| Linear (lower bound) | $\sqrt{4(1+M/\Pi)}$ | $\sqrt{1+M}$ |
| Step function (upper bound) | $\sqrt{4/\Pi}\ /\mathrm{erf}\,\zeta_0$ | $1/\mathrm{erf}\,\zeta_0$ |

Once again, $M$ is the average value of $m$, namely, $\Delta c_{HLO_2}/\Delta c_{O_2}$. Here Ha is the Hatt number, defined as the calculated rate of oxygen absorption divided by that in the absence of chemical reaction, and $\mathrm{erf}\,\zeta_0$ is defined by

$$\zeta_0 e^{\zeta_0^2}\mathrm{erf}\,\zeta_0 = \frac{1}{M\sqrt{\Pi}} \tag{5.3.19}$$

These results are equivalent to those of Ex. 2.3.2, but the $\Pi'(0)$ are higher by a factor of 2, as they represent time-average rather than instantaneous values. It remains only to compare these limiting results with the exact solution of Eq. 5.3.18, which must be obtained by numerical means.

Such a comparison is provided in Fig. 5.3.4 for a venous oxygen tension $p_\infty$ of 30 mm Hg and a wide range of gas compositions (represented by $p_0$). It may be seen that for typical surgical conditions ($p_0 \sim 700$ mm Hg) the upper bounding solution is very close to the exact result; this is emphasized in Fig. 5.3.5 where the corresponding concentration profiles are shown.

For normal human blood with $H = 0.45$ the exact solution to Eq. 5.3.18 has been shown by Dindorf et al. to be expressible as

$$-\Pi'(0) = \left[44.6 + \sqrt{3.81\times 10^3 - (83.5\ \text{mm Hg}^{-1})p_\infty}\ \ \text{mm Hg}^{-a}\right]p_0^a \tag{5.3.20}$$

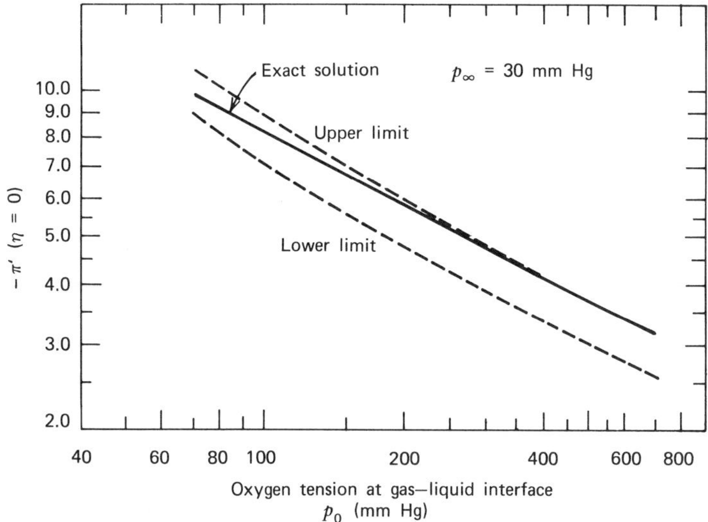


**Figure 5.3.4.** Comparison of dimensionless oxygen fluxes as calculated by various means. In each case $\Pi'(0)$ is given by Eq. 5.3.18. From Dindorf, Lightfoot, and Solen (1971).

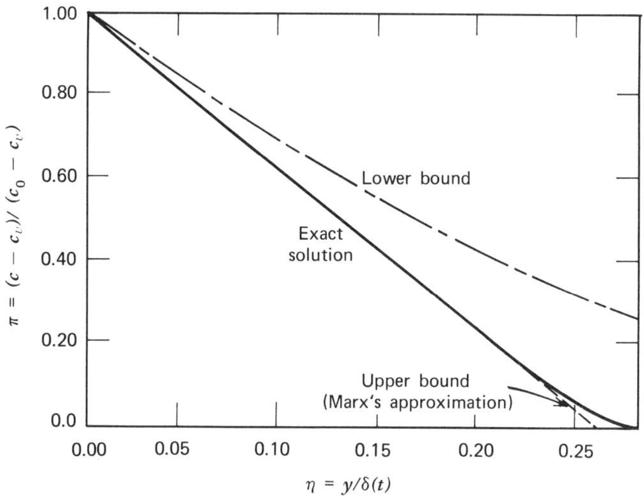


**Figure 5.3.5.** Calculated dissolved oxygen profiles for typical heart-lung bypass conditions ($p_0$=700 mm Hg; $p_\infty$=20 mm Hg).

where $a = -0.5[\sin(1.02 + p_\infty/67.1 \text{ mm Hg})]$, over the range

$$20 < p_\infty < 40 \text{ mm Hg}$$

$$70 < p_0 < 760 \text{ mm Hg}.$$

This expression is accurate to within 2% over the indicated range.

We now consider a specific numerical example to illustrate the utility of this development.

### *EX. 5.3.1. FLOW OF BLOOD OVER A DOUBLE CONE*

Estimate the steady-state oxygenation capacity of the system shown in Fig. 5.3.6 in which blood is poured down over a packing particle with the shape of a double cone. Assume for the sake of simplicity that the blood flow is Newtonian, laminar, and uniform laterally.

SOLUTION

The rate of mass transfer can be obtained by means of Eq. 5.3.17 by (1) calculating the mass of $O_2$ absorbed by a single surface element in passing over the cone, and (2) multiplying by the number of surface elements

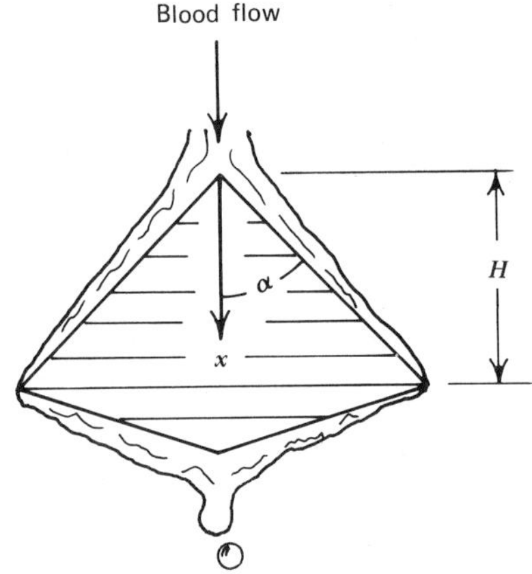


**Figure 5.3.6.** Blood flow over a double cone.

passing any reference point on the cone in a unit time. It is convenient to define the reference area $dS_0$ of an element so that it occupies a region $dx_0$ when $x=H$, and to use $x=H$ as a reference position, where $dS=dS_0$ for any element.

At steady state the total rate of oxygen absorption $\dot{m}_{O_2}$ is equal to

$$\dot{m}_{O_2}=\frac{dS_0}{dt}\frac{dm_{O_2}}{dS_0}$$

where $dS_0/dt$ = rate at which surface elements $dS_0$ pass any point, for example $x=H$

$dm_{O_2}/dS_0$ = mass of $O_2$ absorbed by each element during its passage over the cone.

From Eq. 5.3.17 we may write

$$\frac{dm_{O_2}}{dS_0}=\left(\overline{\mathrm{Nu}_0}\,\frac{\mathfrak{D}_{O_2m}}{D}\right)(c_0-c_\infty)T \tag{2a}$$

as

$$\frac{dm_{O_2}}{dS_0}=-\Pi'(0)(c_0-c_\infty)\sqrt{\mathfrak{D}_{O_2m}T}\sqrt{\frac{1}{T}\int_0^T\left(\frac{s}{s_0}\right)^2 dt} \tag{2b}$$

where $T$ is the time required for the elements to traverse the cone.

To evaluate Eqs. 1 and 2 we need the surface velocity of the fluid. If the film of blood is thin we may write for the surface velocity:

$$v_s=Ax^{-2/3} \tag{3}$$

where

$$A=(9/32\pi^2)^{1/3}\left(\frac{g\cos\alpha}{\nu}\right)^{1/3}\left(\frac{Q}{\tan\alpha}\right)^{2/3} \tag{4}$$

and $Q$ is the volumetric flow rate of blood. Then

$$T=\frac{(3/5A)H^{5/3}}{\cos\alpha} \tag{5}$$

and

$$\frac{dS_0}{dt}=(2\pi H\tan\alpha)v_s|_{x=H}$$

$$=2\pi AH^{1/3}\tan\alpha \tag{6}$$

It now remains only to evaluate the integral in Eq. 2b.

We begin by noting that

$$\frac{s}{s_0} = \frac{dS(x)}{dS(H)} = \frac{x}{H}\frac{dx}{dx_0} \tag{7}$$

where $dx$ is the region occupied by an element at an arbitrary position and $dx_0$ is the value of $dx$ at $x = H$. Since an element must take the same time to pass each point on the surface,

$$\frac{dx}{dx_0} = \frac{v_s(x)}{v_s(H)} = \left(\frac{H}{x}\right)^{2/3} \tag{8}$$

We then find from Eqs. 3, 7, and 8 that

$$\frac{s}{s_0} = \left(\frac{x}{H}\right)^{1/3} = \left(\frac{t}{T}\right)^{1/5} \tag{9}$$

It follows that Eq. 2b takes the form

$$\frac{dm_{O_2}}{dS_0} = -\Pi'(0)(c_0 - c_\infty)\sqrt{\mathfrak{D}_{O_2m}T}\sqrt{\int_0^1 \theta^{2/5}d\theta} \tag{10}$$

$$= -\Pi'(0)(c_0 - c_\infty)\sqrt{\mathfrak{D}_{O_2m}T}\sqrt{5/7} \tag{11}$$

The extent of mass transfer on the lower cone will be much less and can be neglected. The lower surfaces of packing particles are typically rather inactive, as the fluid surface elements are shrinking here.

### (e) *MEMBRANE OXYGENATORS: MASS TRANSFER ACROSS FIXED INTERFACES*

Although direct-contact oxygenators are widely used at present, membrane oxygenators are believed to do less damage to blood. They are therefore quite promising, and it is desirable to develop convenient methods for describing their behavior quantitatively. For many possible designs boundary-layer approximations are useful, and we shall review these briefly. However, for the important case of flow in long ducts, concentration boundary layers can become quite thick, and other techniques have to be used. We shall also touch on these.

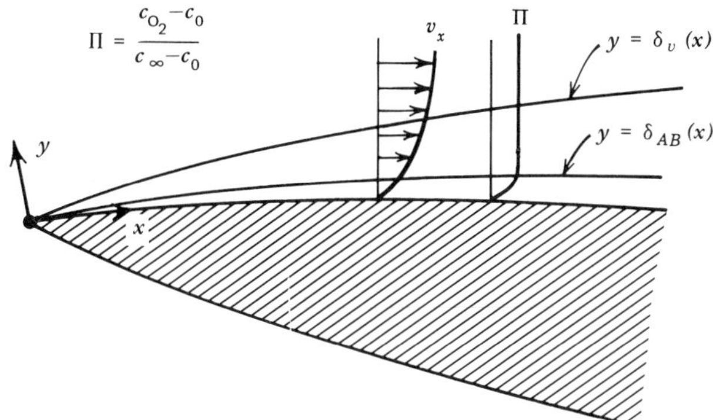


**Figure 5.3.7** Velocity and concentration boundary layers about a two-dimensional body, for systems of large Schmidt number. Here the bulk of the concentration change takes place in a concentration boundary layer $[y<\delta_{AB}(x)]$ much thinner than the velocity boundary layer $[y<\delta_v(x)]$. The shear stress within the concentration boundary layer is

$$\tau_{yx} \doteq -\mu \frac{\partial v_x}{\partial y} \doteq -\tau_0(x) = \mu \dot{\gamma}_0$$

Here $x$ is the distance measured along the surface in the primary direction of flow, and $y$ is measured normal to the surface into the fluid. For a more complete description and extension to three dimensions see Stewart (1963) or Lightfoot (1969).

We begin by considering boundary layers about an arbitrary two-dimensional body as shown in Fig. 5.3.7. The behavior here is analogous to that discussed in ultrafiltration and can be described by

$$-3(1+m)\eta^2 \frac{d\Pi}{d\eta} = \frac{d^2\Pi}{d\eta^2} \tag{5.3.20}$$

with

$$\eta = \frac{y}{\delta_{AB}}$$

$$\delta_{AB} = \frac{\left(9\mathfrak{D}_{O_2m} \int_0^x \sqrt{\dot{\gamma}_0}\, dx\right)^{1/3}}{\sqrt{\dot{\gamma}_0}} \tag{5.3.21, 5.3.22}$$

where $\Pi$ and $\dot{\gamma}_0$ are as defined in the figure.

The local Nusselt number is now given by

$$\mathrm{Nu}_{\mathrm{loc}} = -\Pi'(0) \frac{D}{\delta_{AB}} \tag{5.3.23}$$

as before. This equation can be rearranged to give:

$$\mathrm{Nu_{loc}} = -\Pi'(0)(\mathrm{Re\,Sc})^{1/3} \frac{\dot{\gamma}_0^{1/2}}{\left[9(V/D^2)\int_0^x \sqrt{\dot{\gamma}_0}\, dx\right]^{1/3}} \tag{5.3.24}$$

Once again the effects of chemical reaction are concentrated entirely in $\Pi'(0)$, while the dependence on flow conditions and geometry is given by the quantity in braces.

It is possible to obtain upper and lower bounding solutions here also, and these are given by

| Approximation | $-\Pi'(0)$ | Ha |
|---|---|---|
| Linear (lower bound) | $(1+M)^{1/3}/\Gamma(\frac{4}{3})$ | $(1+M)^{1/3}$ |
| Step function (upper bound) | $[\Gamma(\frac{4}{3})\Gamma(\frac{1}{3},\eta_v^3)]^{-1}$ | $[\Gamma(\frac{1}{3},\eta_v^3)]^{-1}$ |

For the upper bound $\eta_v$ is found from

$$M\Gamma(\tfrac{1}{3})\eta_v^3 e^{\eta_v^3}\Gamma(\tfrac{1}{3},\eta_v^3) = 1$$

where $\Gamma(x,y)$ is the incomplete gamma function.

As before, the bounding solutions become identical in the limit of small $M$, and in the limit of large $M$ one finds

$$\lim_{M\to\infty}\left\{\frac{[\Pi'(0)]_{\text{upper}}}{[\Pi'(0)]_{\text{lower}}}\right\} = \frac{\sqrt[3]{3}}{\Gamma(\frac{4}{3})} = 1.62 \tag{5.3.25}$$

Thus at most bounding solutions differ by about 30% from their mean.

Since the general trends for the exact and bounding solutions are similar to those shown earlier for mass transfer across mobile interfaces, no extensive comparison is given here. However, the following sample values that correspond to $p_0 = 760$ mm Hg and $p_v = 37$ mm Hg are indicative:

$$[-\Pi'(0)]_{\text{upper}} = 2.156$$

$$[-\Pi'(0)]_{\text{exact}} = 2.147$$

$$[-\Pi'(0)]_{\text{lower}} = 1.754$$

Once again the upper bounding solution closely approximates the exact solution, while the lower bound is quite conservative. We return to discuss these results in the example at the end of the section, but first we briefly consider developed boundary layers in ducts.

Here discussion is limited to steady* flow in cylindrical ducts of circular cross section with uniform oxygen concentration at the blood surface. The appropriate diffusion equation for this situation is

$$2(1+m)\left[1-(r^*)^2\right]\frac{\partial \Pi}{\partial z^*}=\frac{1}{r^*}\frac{\partial}{\partial r^*}\left(r^*\frac{\partial \Pi}{\partial r^*}\right) \tag{5.3.26}$$

where $\Pi$ is as previously defined and

$$r^*=\frac{r}{R}, \qquad z^*=\frac{\mathfrak{D}_{O_2m}z}{R^2\langle v\rangle}=\frac{4z}{D\text{Pé}}$$

Here $\langle v\rangle$ is the flow-average velocity, and $\text{Pé}=D\langle v\rangle/\mathfrak{D}_{O_2m}$.

An "exact" solution to these equations can be obtained numerically using a finite-difference technique. This was done by Weissman and Mockros (1967) using an approximation to the equilibrium curve for cattle blood. Colton and Drake (1971) and Buckles et al. (1968) solved similar problems, but used a constant wall flux as a boundary condition. The results of such an "exact" solution for constant wall concentration are shown in Fig. 5.3.8 along with some very simple bounding solutions developed by Dindorf, Lightfoot, and Solen (1971). The lower bounding solution is easily adapted to even quite complex boundary conditions and can be seen to be reasonably accurate. The upper bound here is not so useful.

We now turn to a final example of the utility of boundary-layer theory in oxygenation calculations.

## *EX. 5.3.2. FLOW OF BLOOD TRANSVERSE TO A CYLINDER*

Estimate the effectiveness of the situation shown in Fig. 5.3.9, assuming the dissolved-oxygen concentration to be constant over the blood-solid interface.

SOLUTION

Even though there is a separated flow over the rear of the cylinder, especially at high Re, there is reason to expect that the mass transfer will

* Pulsatile flows have been analyzed by Villaroel and associates (1971), and appear to enhance mass transfer. The effects of secondary flows in coiled tubes have been described by Weissman and Mockros (1967).

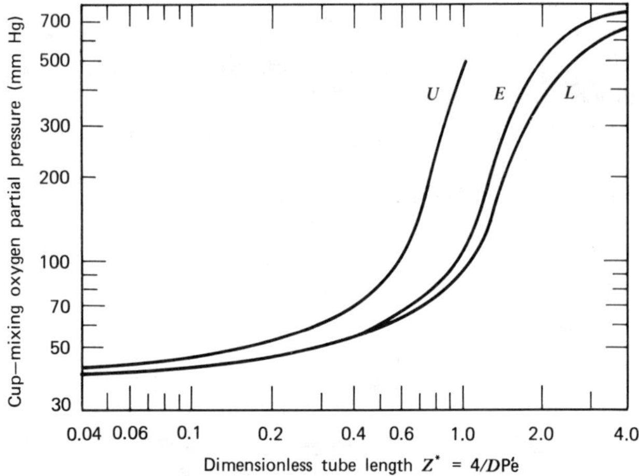


**Figure 5.3.8.** Average oxygen tension as a function of dimensionless tube length for flow in a long cylindrical tube. The upper bound ($U$), lower bound ($L$), and exact solution ($E$) are shown for bypass conditions, where $p_0 = 760$ mm Hg and $p_v = 37$ mm Hg. From Dindorf, Lightfoot, and Solen (1971).

be dominated by typical boundary-layer behavior. We thus expect to find that

$$\mathrm{Nu}_m = -\Pi'(0)\mathrm{Pe}^{1/3} f(\text{geometry, Re}) \tag{1}$$

However, the last factor is not available from theoretical studies, so we rely on experimental data. To do this we begin by noting that the well-known

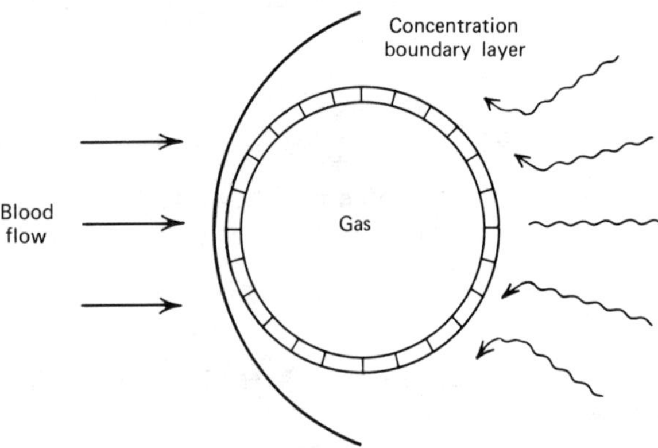


**Figure 5.3.9** Blood flow across a cylinder.

heat-transfer correlation (see, e.g., *Tr. Ph.*, Fig. 13.3.3-1) can be closely approximated by

$$\mathrm{Nu}_m \doteq 0.63\mathrm{Re}^{1/2}\mathrm{Pr}^{1/3} \qquad (\mathrm{Pr}>\tfrac{1}{2},\ 10^2<\mathrm{Re}<10^5) \tag{2}$$

whereas for a nonreactive system the thermal analog of Eq. 2 is

$$\mathrm{Nu}_m = \frac{1}{\Gamma(\frac{4}{3})}\mathrm{Re}^{1/2}\mathrm{Pr}^{1/3}\frac{f}{\mathrm{Re}^{1/6}} \tag{3}$$

It follows that

$$\frac{f}{\mathrm{Re}^{1/6}} = 0.56 \tag{4}$$

We may then write Eq. 2 in the form

$$\mathrm{Nu}_m = 0.56\mathrm{Re}^{1/2}\mathrm{Sc}^{1/3}[-\Pi'(0)]$$

Values of $\Pi'(0)$, dependent only upon $p_v$, $p_0$, and the chemical composition of blood, can be obtained from the discussion above. Thus for the typical bypass conditions of $p_0 = 760$ mm Hg and $p_v = 37$ mm Hg, one finds that

$$-\Pi'(0) = 2.147$$

The calculation of the absorption rate is now straightforward.

## 5.4. Oxygenation of Tissue

Oxygen transfer through metabolizing tissue is frequently the rate-limiting step in physiological processes, and the need to maintain a continuous adequate oxygen supply has had a profound effect on the development of multicellular organisms. This is particularly true in the brain, where permanent damage can result from even a modest reduction of normal oxygen supply over a short time (see for example Knisely, Reneau, and Bruley, 1970). The intrinsic importance of this topic is therefore sufficient reason for its inclusion here. It also, however, provides an introduction to a rather wide variety of mass-transfer studies based on a similar mathematical framework.

The critical nature of oxygen transfer is indicated in Fig. 5.4.1, which shows that the death of nervous tissue can result from occlusion of a single capillary, here in the form of a loop. Because the effective oxygenation radius is so small, actively metabolizing tissue requires a very extensive

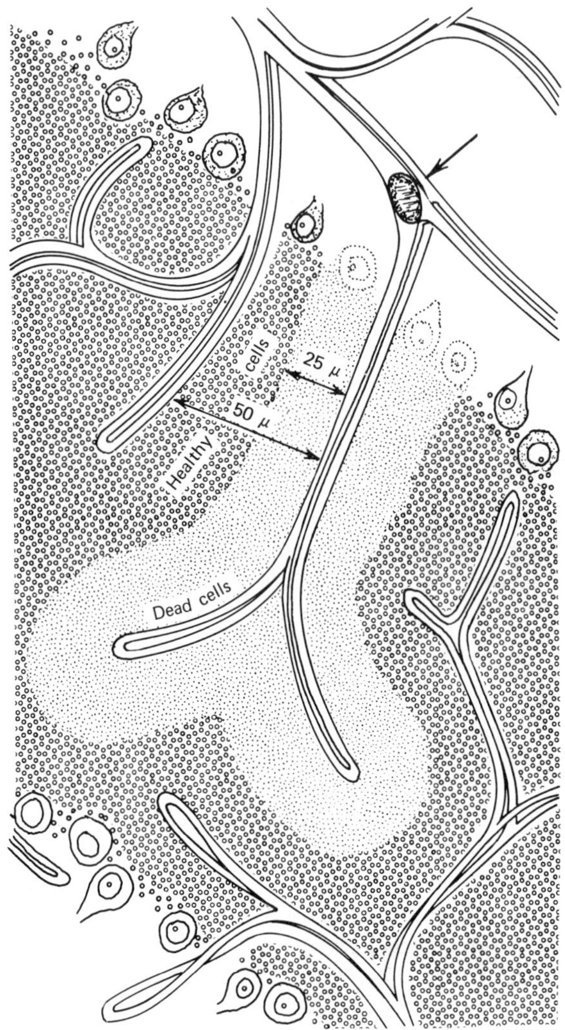


**Figure 5.4.1.** The effects of occlusion of an artery in the cerebellum of the opossum by a Lycopodium spore (arrow); figure from E. Scharrer, "Blood Vessels of Nervous Tissue," *Quart. Rev. Biol.* **19**, 308 (1944). Note the sharply delimited border between the healthy nerve cells and the dead. Reproduced by permission from Knisely, Reneau, and Bruley (1970).

capillary network. As indicated in Chapter I, Section 3.6, the anatomy of such networks has been investigated since about the time of Leeuwenhoek. Most present analyses of tissue oxygenation are, however, based primarily on the much more recent work of August Krogh (1930). Krogh's careful experiments showed that for many vascular beds the capillaries were very nearly parallel and that the spacing between them was remarkably uniform. Krogh's *tissue-cylinder* model, as later slightly modified by Thews (1967), is shown in Fig. 5.4.2, and the uniformity of spacing is indicated by the examples of Table 5.4.1.

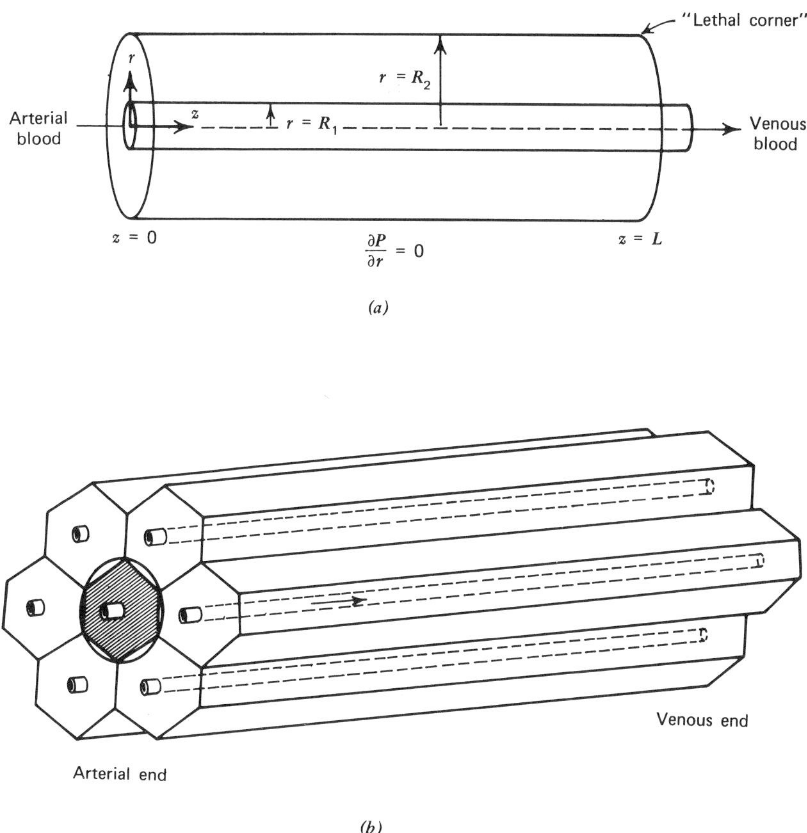


**Figure 5.4.2.** (*a*) Krogh tissue cylinder. (*b*) Krogh tissue cylinder arrangement as modified by Thews (1967). The arrangement of cylinders circumscribing hexagonal prisms is one which includes all of the tissue.

**Table 5.4.1.** Results of Counting of Capillaries in Striated Muscle (From Krogh, 1930).

| Animal | Muscle | Capillaries/mm$^2$ | Standard deviation |
|---|---|---|---|
| Horse | *M*. gastrocnemius | 3350 | $\pm 31$ |
| Dog | *M*. semimembranosus | 4630 | $\pm 53$ |

Almost all* serious quantitative analyses of tissue oxygenation have been based on this cylindrical model and on pseudohomogeneous forms of the diffusion equation for capillary and tissue. Building on the results of the previous section, we may then write

$$(1+m)\left(\frac{\partial \bar{p}_{O_2}}{\partial t} + v_z \frac{\partial \bar{p}_{O_2}}{\partial z}\right) = \frac{1}{r}\left(\frac{\partial}{\partial r}\mathfrak{D}^{(r)}_{O_2B}\left(r\frac{\partial \bar{p}_{O_2}}{\partial r}\right)\right) + \frac{\partial}{\partial z}\mathfrak{D}^{(z)}_{O_2}\frac{\partial \bar{p}_{O_2}}{\partial z} \qquad (0<r<R_1) \tag{5.4.1}$$

$$\frac{\partial p_{O_2}}{\partial t} = \frac{1}{r}\left(\frac{\partial}{\partial r}\mathfrak{D}^{(r)}_{O_2T}\left(r\frac{\partial p_{O_2}}{\partial r}\right)\right) + \frac{\partial}{\partial z}\left(\mathfrak{D}^{(z)}_{O_2T}\frac{\partial p_{O_2}}{\partial z}\right) + R_{O_2}K \qquad (R_1<r<R_2) \tag{5.4.2}$$

with the matching conditions

At $\qquad r=R_1 \qquad p_{O_2}=\bar{p}_{O_2}$

$$\mathfrak{D}^{(r)}_{O_2T}\frac{\partial p_{O_2}}{\partial r} = \mathfrak{D}^{(r)}_{O_2B}\frac{\partial \bar{p}_{O_2}}{\partial r}$$

where $\bar{p}_{O_2}, p_{O_2}$ = local oxygen tension in blood and tissue respectively

$\mathfrak{D}_{O_2B}, \mathfrak{D}_{O_2T}$ = effective binary oxygen diffusivity in blood and tissue, respectively (which may take on different values in the radial and axial directions)

* See, however, the work of Longo, Power, and Forster (1969) for a very different point of view. These researchers suggest, that the major blood-phase resistance in placental exchange is kinetic.

$-R_{O_2}$ = volumetric oxygen consumption rate in tissue (moles per unit volume and time), usually assumed constant

$K = p_{O_2}/c_{O_2}$ = Henry's constant $\doteq 7.4\times 10^8$ (mm Hg) cm$^3$/g-mole at 37°C

$m = dc_{HbO_2}/dc_{O_2}$, as in the previous section.

The most widely used boundary conditions are:

$$\text{At } z=0, \qquad \bar{p}_{O_2} = p_0(t), \qquad 0<r<R_1 \tag{5.4.5}$$

$$\text{At } t=0, \qquad \bar{p}_{O_2}, p_{O_2} = p_i, \qquad \left\{ \begin{array}{c} 0<z<L \\ 0<r<R_2 \end{array} \right\} \tag{5.4.6}$$

$$\text{At } r=0, \qquad \frac{\partial \bar{p}_{O_2}}{\partial r} = 0, \qquad 0<z<L \tag{5.4.7}$$

$$\text{At } r=R_2, \qquad \frac{\partial p_{O_2}}{\partial r} = 0, \qquad 0<z<L \tag{5.4.8}*$$

$$\text{At } z=0, L, \qquad \frac{\partial p_{O_2}}{\partial z} = \frac{\partial \bar{p}_{O_2}}{\partial z} = 0, \qquad 0<r<R_2 \tag{5.4.9, 10}$$

It may be seen immediately that this system is very complex, and that substantial simplifications are in order when one considers the degree of idealization represented by the physical model.

First, however, it should be pointed out that equations very similar to the above have been widely used to describe solute transport in much larger and less well-defined body regions (e.g., Goresky, Ziegler and Bach (1970); Perl and Chinard (1968); Levitt (1971)). Here each of the terms in the description may be important in specific situations. It should also be noted that the situation modeled is very similar to those encountered in the analysis of chromoatographic columns and fixed-bed reactors, on which there are large literatures; little use appears yet to have been made of these prior developments.

Returning to the specific case of tissue oxygenation, we note the following parameters are representative* (see Reneau, Bruley, and Knisely, 1969):

*In the presence of an anoxic region this equation must be replaced by

$$\text{At } r=\gamma R, \qquad \bar{p}_{O_2} = \frac{\partial \bar{p}_{O_2}}{\partial r} = 0 \tag{5.4.8 a,b}$$

where $\gamma R$ is the outer limit of the actively metabolizing region.

* For grey matter in the brain. The $\mathfrak{D}_{O_2m}$ are from Section 5.3. Reneau et al. used $\mathfrak{D}_{O_2B} = 1.12\times 10^{-5}$ cm$^2$/sec and $\mathfrak{D}_{O_2T} = 1.7\times 10^{-5}$ cm$^2$/sec.

$$R_1 = 3\mu, \qquad R_2 = 30\ \mu, \qquad L = 180\ \mu$$

$$\langle v \rangle = 400\ \mu/\text{sec}$$

$$\mathfrak{D}_{O_2T} = \mathfrak{D}_{O_2B} = 1.4 \times 10^{-5}\ \text{cm}^2/\text{sec}$$

$$R_{O_2} = -3.72 \times 10^{-8}\ \text{g-mole}/(\text{cm}^3)(\text{sec})$$

For these dimensions axial diffusion should be of only secondary importance in the tissue and negligible in the capillary. This expectation is borne out in calculations by Reneau et al. (1967), who found that axial diffusion should normally be important only near the arterial end of the tissue cylinder, where conditions are of little physiological interest. It is also clear that the characteristic response times $t_c$ should be small. For example, if we use the linear approximation to the hemoglobin dissociation curve (see

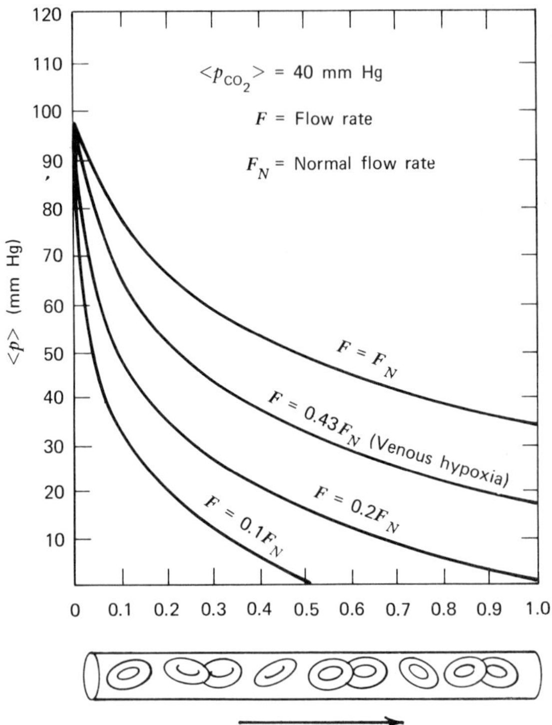


**Figure 5.4.3.** Axial oxygen partial-pressure profiles in the capillary as a result of reduced flow rates. All other conditions are normal. From Reneau, Bruley, and Knisely (1969).

Section 5.3) and the thermodynamic data of Table 2.3.1, we find:*

$$\text{Capillary:}\quad t_c \equiv (1+M)\frac{R_1^2}{\mathfrak{D}_{O_2B}}$$

$$\doteq 40\frac{9\times10^{-8}\ \text{cm}^2}{1.4\times10^{-5}\ \text{cm}^2/\text{sec}} \doteq 0.25\ \text{sec}$$

$$\text{Tissue:}\quad t_t \doteq \frac{R_2^2}{\mathfrak{D}_{O_2B}} \doteq \tfrac{2}{3}\ \text{sec}$$

The response times are therefore small even compared to holdup time, and the tissue response time is small compared to the 3–20 minutes believed necessary for anoxia-induced death of nervous tissue (see Knisely, Reneau, and Bruley, 1970). They are also short compared to the response time of the brain to changes in arterial oxygen tension. This time is typically of the

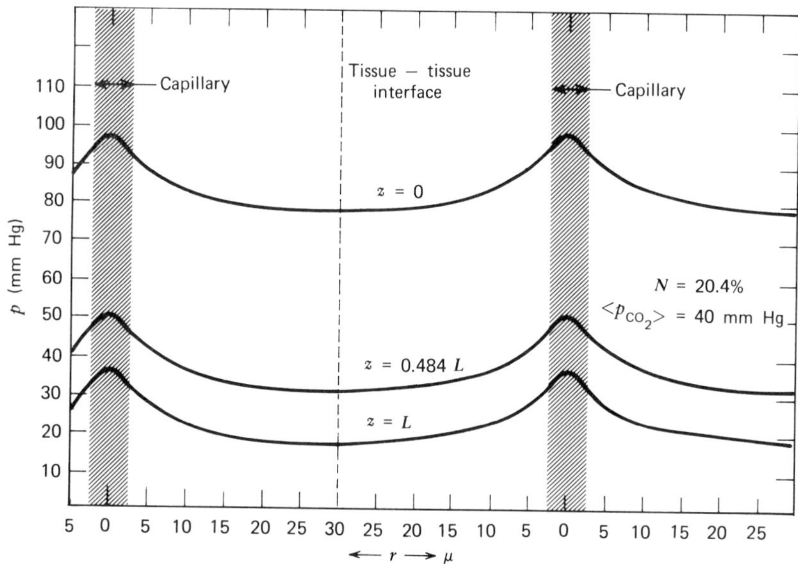


**Figure 5.4.4.** Calculated radial oxygen partial-pressure profiles for a healthy human under normal conditions. The radius of the tissue cylinder is 30 $\mu$. From Reneau, Bruley, and Knisely (1969).

* These are somewhat arbitrary, but generally transients are largely eliminated in a dimensionless time $t\mathfrak{D}_{\text{eff}}/R^2 \sim 1$. See, for example, *Tr. Ph.*, Chapter 11.

order of 1 min and appears to be determined by autoregulatory mechanisms.

For most purposes, then, it is reasonable to neglect both transients and axial diffusion in the description of tissue oxygenation.† Representative results obtained by Reneau et al. (1967) on this basis are shown in Figs. 5.4.3 to 5.4.5. Figures 5.4.3 and 5.4.4 show that rather low oxygen tensions can be expected at the venous end of the tissue cylinder even under conditions of normal good health. Figure 5.4.5 shows that even modest decreases in capillary blood flow can cause local anoxia and resultant death of tissue. The utility of this procedure is shown in Fig. 5.4.6, where the steady-state solution of Eqs. 5.4.1 and 5.4.2, as obtained by Reneau et al., is compared with previously published data.

These steady-state results can be obtained very readily, by taking advantage of the specific characteristics of the system, and most particularly the position independence of $R_{O_2}$

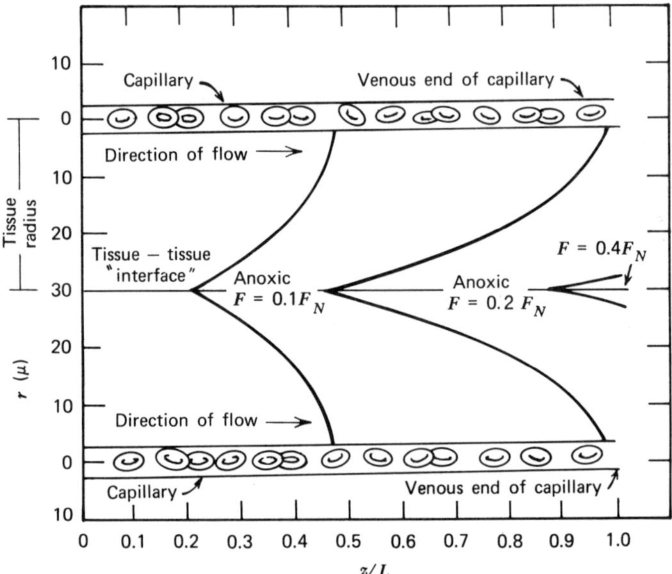


**Figure 5.4.5.** Calculated anoxic areas of tissue between parallel capillaries having concurrent flow as a function of reduced capilary flow rates. All other conditions are normal. Note that these vessels need not be plugged to give local anoxic areas; anoxic areas (volumes) develop even when the blood is moving. Significantly, with sufficiently reduced flow rate, even though the blood is moving, the venous end of the capillary wall (as well as the elements between capillaries) begins to be anoxic. From Reneau, Bruley, and Knisely (1969).

† Flow pulsations are of about 1-sec period and must be considered in any unsteady diffusional analysis.

First we may integrate Eq. 5.4.1, with either Eq. 5.4.8 or 8a and 8b as boundary conditions to give

$$p_2 - p_1 = \frac{R_2^2 R_{O_2} K}{4\mathfrak{D}_{O_2 T}} \gamma^2 (\ln \gamma^2 - 1) + 1 \tag{5.4.11}$$

$$\left.\frac{\partial p}{\partial r}\right|_{r=R_2} = -\left(\frac{R_2^2 R_{O_2} K}{2\mathfrak{D}_{O_2 T}}\right)\gamma^2 - 1 \tag{5.4.12}$$

with

$$\gamma = \frac{R_2}{R_1} \qquad \text{(no anoxic core)} \tag{5.4.13}$$

or

$$\gamma^2(\ln \gamma^2 - 1) + 1 = \frac{p_1 \mathfrak{D}_{O_2 T}}{R_2^2 R_{O_2} K} \qquad \text{(anoxic region present)} \tag{5.4.14}$$

where $p_1$ is the oxygen tension at $r = R_1$ and $p_2$ that at $r = R_2$.

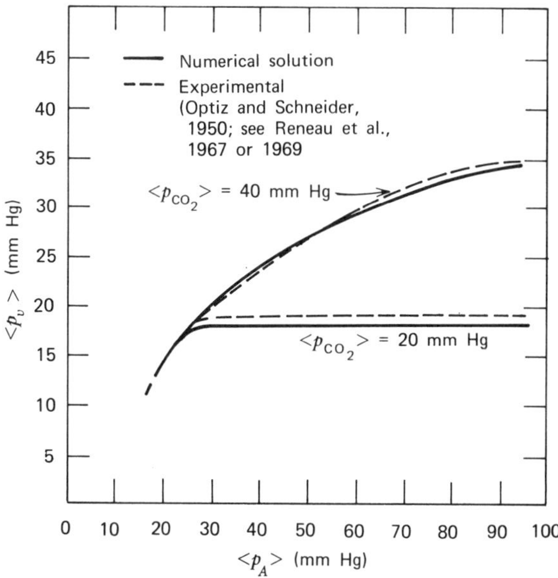


**Figure 5.4.6.** Comparison plot of arterial oxygen partial pressure of the capillary vs. corresponding venous partial pressure for $\langle p_{CO_2} \rangle = 40$ mm Hg and $\langle p_{CO_2} \rangle = 20$ mm Hg. From Reneau, Bruley, and Knisely (1969).

To obtain the radial distribution of oxygen tension in the capillary we assume for the moment that $m$ is constant and note that to a first approximation oxygen flux from the capillary surface is position independent. The solution to this problem is known (Carslaw and Jaeger, 1959, Section 7.8). Furthermore beyond an entrance length

$$z_e \doteq 0.15 R_2^2 v (1 + m_{avg}) / \mathfrak{D}_{O_2T} \doteq 15\ \mu \tag{5.4.15}$$

the pressure drop across the capillary is within 10% of the limiting value:

$$\bar{p}_{O_2}(r,t,z) - \bar{p}_{O_2}(R_1,t,z) = -\frac{K\left[(N_{O_2})_r|_{r=R_1}\right](r^2 - R_1^2)}{2R_1 \mathfrak{D}_{O_2B}} \tag{5.4.16a}$$

$$= \frac{K R_{O_2}(R_2{}^2 - R_1)^2 (r^2 - R_1{}^2)}{4R_1{}^2 \mathfrak{D}_{O_2B}} \tag{5.4.16b}$$

or

$$(\bar{p}_{O_2})_b - (\bar{p}_{O_2})_{R_1} = \frac{K R_{O_2}(R_2{}^2 - R_1{}^2)}{8 \mathfrak{D}_{O_2B}} \tag{5.4.17}$$

where once again the subscript $b$ refers to bulk or cup-mixing value and the subscript $R_1$ refers to conditions at the capillary wall. Note that $m$ has now disappeared. Since the entrance length is short compared with the length of the capillary we may drop the restriction to constant wall flux. In addition, since the change in $m$ across the capillary is small we can also relax the restriction to constant $m$ for all practical purposes.

The axial distribution of oxygen is most easily calculated from the material balance

$$\frac{d}{dz}(c_{O_2} + c_{HbO_2})_b = -2(N_{O_2})_r|_{r=R_1} / R_1 v \tag{5.4.18}$$

with

$$(N_{O_2})_r|_{r=R_1} = -\mathfrak{D}_{O_2T} \left.\frac{\partial p}{\partial r}\right|_{r=R_1} \tag{5.4.19}$$

For the normal situation of no anoxic region

$$(N_{O_2})_r|_{r=R_1} = -R_{O_2}(R_2^2 - R_1)/2R_1 \tag{5.4.20}$$

and total bulk oxygen concentration drops linearly with distance.

Calculation of the axial oxygen profile is, in any event, surprisingly unimportant. In the absence of anoxic regions one obtains the same predicted venous and lethal-corner oxygen tensions for the above model as for complete axial mixing of capillary blood.

This result follows directly from the postulate of zero-order kinetics for oxygen consumption in the tissue cylinder. It is not, however, unusual to find that a wide variety of dissimilar models are almost equally effective for correlation of data—or that it is difficult to distinguish between postulated models solely on the basis of system response to changes of input conditions.

The steady-state description is now complete; its utility is illustrated in the example below.

The transient behavior of this system has been investigated both experimentally and mathematically by Bruley, Reneau, Bicher, and Knisely (1971), among others. Experimental response times have been found to be on the order of a minute, which is somewhat longer than predicted above. Much of this delay appears, however, to result from autoregulatory mechanisms. It is also found by these authors that transient descriptions based on the complete Eqs. 5.4.1 and 5.4.2 are quite expensive in computer time; as a result, the authors have made extensive use of very crude lumped-parameter approximations. The simplifications suggested in our discussion provide an intermediate approach, providing tractable equations which should give reliable results well within the uncertainties of the model. Equation 5.4.16a is particularly attractive; since capillary response times are so much shorter than those for tissue, it should be suitable for any physiologically interesting transient calculations.

### *EX. 5.4.1: OXYGEN TENSION AT STEADY STATE*

Arterial blood as described in Table 2.3.1 flows into a tissue cylinder with the characteristics given in the above discussion. Estimate steady-state oxygen tensions in the venous blood and at the "lethal corner."

SOLUTION:

The first step is clearly to calculate the total available oxygen concentration at the venous end of the tissue cylinder from Eq. 5.4.18 and Table 2.3.1:

$$
\begin{aligned}
c_{\text{tot}}(L) &= 8.6\times 10^{-6}\text{g-mole/cm}^3 \\
&\quad - \frac{180\,\mu}{400\,\mu/\text{sec}}\left[3.72\times 10^{-8}\text{g-mole/(cm}^3)(\text{sec})\right](99) \\
&= (8.6-1.66)\times 10^{-6}\text{g-mole/(cm}^3)(\text{sec}) \\
&= 6.94\times 10^{-6}\text{g-mole/(cm}^3)(\text{sec})
\end{aligned}
$$

This quantity in turn can be related to dissolved oxygen concentration by the relation

$$c_{\text{tot}} = c_{O_2} + c_{\text{HbO}_2} = c_{O_2} + c_{\text{sat}}\left(\frac{x^n}{1+x^n}\right) \tag{1}$$

where

$c_{\text{sat}} = 8.82\,\text{m}M$ is the total oxygen capacity of hemoglobin at saturation

$$x = c_{O_2}/c_H$$

with

$$c_H \doteq 3.73\times 10^{-2}\,\text{m}M$$

$$n = 2.66$$

The values of $c_{\text{sat}}$, $c_H$, and $n$ are characteristics of the blood and are strongly dependent on the hematocrit, pH, $T$, and $p_{CO_2}$. Those used here are representative of normal physiological conditions (see Section 5.3). Equation 1 is consistent with both Eqs. 5.3.1 and 5.3.5, which are two different forms of Hill's approximation to the hemoglobin-oxygen dissociation curve (see the *Handbook of Physiology*). This simple expression is quite reliable for oxygen tensions above about 20 mm Hg.

It is now convenient to rewrite Eq. 1 in the form

$$x^n = \frac{\beta - \alpha x}{1 - \beta + \alpha x} \tag{2}$$

where

$$\alpha = c_H/c_{\text{sat}} = 4.23\times 10^{-3}$$

$$\beta = c_{\text{tot}}/c_{\text{sat}} = 0.782.$$

Since $\alpha x \ll \beta$, Eq. 2 may be easily solved by an iterative technique in which the $m$th iteration is given by

$$x_m = \left(\frac{\beta - \alpha x_{m-1}}{1 - \beta + \alpha x_{m-1}}\right)^{1/n} \tag{3}$$

and with $x_0 = \beta/(1-\beta)$. It is thus quickly found that

$$x \doteq 1.6, \qquad c_{O_2}(z=L) \doteq 6\times 10^{-8}\,(\text{mole O}_2)/\text{cm}^3$$

It follows from Henry's law that

$$p_{O_2 b}(z=L) \doteq 44\ \text{mm Hg}$$

which is slightly above the "normal" value of 40 mm Hg.

To complete our solutions it is only necessary to calculate the drops in oxygen tension across the venous ends of the capillary and tissue cylinder. From Eq. 5.4.17 we find*

$$(\bar{p}_{O_2})_b - (\bar{p}_{O_2})_{R_1} = \frac{(7.4\times 10^8)(3.72\times 10^{-8})[(900-9)\times 10^{-8}]}{8(1.4\times 10^{-5})} \text{ mm Hg}$$

$$= 2.2 \text{ mm Hg}$$

which is in close agreement with Fig. 5.4.4. From Eq. 5.4.11 we find

$$p_{O_2}(R_1) - p_{O_2}(R_2) = \frac{(7.4)(3.72)(9\times 10^{-8})}{4(1.4\times 10^{-5})}(100 \ln 100 - 100 + 1)$$

$$= 15.9 \text{ mm Hg}$$

Then the oxygen tension at the lethal corner is calculated to be

$$p_{LC} = 44 - 2.2 - 15.9 \doteq 26 \text{ mm Hg}$$

or not much more than half the tension in the venous blood. It may now be seen that while venous partial pressures (shown in Fig. 5.4.6) depend only on the metabolic rate, partial pressures in the lethal corner depend heavily on diffusional behavior.

For a quite different approach to this problem see Longo, Power, and Foster (1969).

## BIBLIOGRAPHY

### Books and Reviews

American Institute of Chemical Engineers, Chem. Eng. Prog. Symp. Ser.:
No. 66: *Chemical Engineering in Medicine* (1966)
No. 84: *The Artificial Kidney* (1968)
No. 99: *Mass Transfer in Biological Systems* (1970)
No. 114: *Advances in Bio-engineering* (1971)

* Note that the change in pressure from centerline to wall is twice this amount.

Colton, C. K., *A Review of the Development and Performance of Hemodialyzers*, Artificial Kidney–Chronic Uremia Program, NIAMD, Natl. Inst. Health, Bethesda, Md. (1967).

Dindorf, J. A., E. N. Lightfoot, and K. A. Solen, "Prediction of Blood Oxygenation Rates," in American Institute of Chemical Engineers No. 114, pp. 75–87 (1971).

Hershey, Daniel, Ed., *Blood Oxygenation*, Plenum, N.Y. (1970); *Chemical Engineering in Biology and Medicine*, Plenum, (1967).

Knisely, M. H., D. D. Reneau, Jr., and D. F. Bruley, "The Development and Use of Equations for Prediciting the Limits on the Rates of Oxygen Supply to the Cells of Living Tissues and Organs," *Angiology*, **20** (supplement), No. 11, 1–56 (1970).

Leonard, E. F., and R. L. Dedrick, *The Artificial Kidney. Problems and Approaches for the Engineer*, in American Institute of Chemical Engineers, No. 84, (1968).

Lightfoot, E. N., "Estimation of Heat and Mass Transfer Rates," American Institute of Chemical Engineers, Continuing Education Series, No. 4, *Lectures in Transport Phenomena* (1969).

NIAMD, Proceedings of the Annual Contractors Conferences of the Artificial Kidney Program (Yearly).

Lightfoot, E. N., R. E. Safford, and D. R. Stone, "Biological Applications of Mass Transport," *CRC Critical Reviews of Bio-engineering*, **1**, 1 (1971).

Norman, J. C., et al., Eds., *Organ Perfusion and Preservation*, Appleton-Century-Crofts (1968).

Spaeth, E. E., *Critical Reviews in Bio-engineering*, in press.

## Periodical and Secondary References

Acrivos, A., "On the Solution of the Convection Equation in Boundary-layer Flows," *Chem. Eng. Sci.*, **17**, 457–465 (1962).

Aroesty, J., and J. F. Gross, "Convection and Diffusion in the Micro-circulation," *Microvasc. Res.*, **2**, 247 (1970).

Bassett, R. J., and J. S. Schultz, "Non-equilibrium Facilitated Diffusion of Oxygen through Membranes of Aqueous Cobaltohistidine," *Biochim. Biophys. Acta*, **211**, 194–215 (1970).

Blackshear, P. L., Jr., "On Transport of Heat, Mass, and Momentum Transfer in Blood Due to Particulate Motion." *Bioeng. Memo No. One*, Department of Mech. Eng., Univ. of Minnesota, Minneapolis, July 1915.

Brown, C. E., M. P. Tulin, and Peter van Dyke, "On the Gelling of High Molecular Weight Impermeable Solutes during Ultrafiltration," in American Institute of Chemical Engineers, No. 114, pp. 174–180 (1971).

Bruley, D. F., D. D. Reneau, H. I. Bicher, and M. H. Knisely, "Modelling Cerebral Tissue Oxygenation with Auto-regulation," *Workshop on Oxygen Transport in Tissue*, Dortmund, West Germany (July 1971).

Buckles, R. G., E. W. Merrill, and E. R. Gilliland, *Am. Inst. Chem. Eng. J.*, **14**, 703–707 (1968).

Carslaw, H. S., and J. C. Jaeger, *Conduction of Heat in Solids*, 2nd ed., Oxford (1959), Section 7.8.

Colton, C. K., and R. F. Drake, *Chem. Eng. Prog. Symp. Series*, No. 114, 67 (1971).

Gill, W. N., C. Tien, and D. W. Zeh, "Analysis of Continuous Reverse Osmosis Systems for Desalination," *Int. J. Heat Mass Transfer*, **9**, 907 (1966).

Gill, W. N., "Concentration Polarization Effects in a Reverse Osmosis System," *IEC Fund.*, **4**, 433 (1965).

Goddard, J. D., J. S. Schulz, and R. J. Bassett, "On Membrane Diffusion with Near-equilibrium Reaction," *Chem. Eng. Sci.*, **25**, 665–683 (1970).

Goresky, C. A., W. H. Ziegler, and G. G. Bach, "Capillary Exchange Modeling," *Circ. Reg.*, **27**, 739 (1970).

Grabowski, E. F., L. I. Friedman, and E. F. Leonard, "Effects of Shear Rate on the Diffusion and Adhesion of Blood Platelets to a Foreign Surface," *Ind. Eng. Chem. Fund.* **11**, 224 (1972).

Grimsrud, Lars, and A. L. Babb, "Velocity and Concentration Profiles from Laminar Flow of a Newtonian Fluid in a Dialyzer," in American Institute of Chemical Engineers, No. 66, p. 19 (1966).

Keller, K. H., "Effect of Fluid Shear on Mass Transport in Flowing Blood," *Fed. Proc. Suppl.* (in press).

Keller, K. H., and S. K. Friedlander, "The Steady-state Transport of Osxygen through Hemoglobin Solutions," in American Institutute of Chemical Enggineers, No. 66, p. 19 (1966).

Kozinski, A. A., "Ultra-filtration of Protein Solutions," Ph.D. Thesis, Univ. of Wisc. (1970).

Kozinski, A. A., and E. N. Lightfoot, "Ultra-filtration of Proteins in Stagnation Flow ," *Am. Inst. Chem. Eng. J.*, **17**, 81 (1971).

Kozinski, A. A., and E. N. Lightfoot, "Ultra-filtration of Proteins," *Am. Inst. Chem. Engrs. J.*, **18**, 1030–1040 (1972).

Kozinski, A. A., F. P. Schmidt, and E. N. Lightfoot, "Velocity Profiles in Porous-walled Ducts," IEC Fund., **9**, 502 (1970).

Krogh, August, *The Anatomy and Physiology of Capillaries*, Yale (1922; second ed., 1930).

Levich, V G., *Physico-chemical Hydrodynamics*, Prentice-Hall (1962).

Levitt, D. G., "Theoretical Model of Capillary Exchange, etc.," *Am. J. Physiol.*, **220**, 250 (1971).

Longo, L. D., G. G. Power, and R. E. Forster, II, "Placental Diffusing Capacity for Carbon Monoxide at Varying Partial Pressures of Oxygen," *J. Appl. Physiol.*, **26**, No. 3, 360 (1969).

Lonsdale, H.K., "Separation and Purification by Reverse Osmosis,"*Progr. Sep. Purif.*, 3, 119 (1970).

Perl, W., and F. P. Chinard, "A Convection-diffusion Model of Indicator Transport through an Organ," *Circ. Res.*, **22**, 273 (1968).

Ramirez, W. F., M. C. Mickley, and D. W. Lewis, "Characteristics of Parallel-plate Dialyzers, *Trans. Am. Soc. Int. Organs*, **26**, 292 (1970); "Mathematical Modelling of a Kiil Dialyzer," in *Am. Inst. Chem. Engrs.*, No. 114, pp. 116–127 (1971).

Reneau, D. D., D. F. Bruley, and M. H. Knisely, "Mathematical Simulation of Oxygen Activity in the Brain," in gen. ref. 5 above (1967); *Am. Inst. Chem. Engrs. J.*, **15**, 916–925 (1969).

Sirs, J. A., "The Interaction of Carbon Dioxide with Rate of Exchange of Oxygen by Red Blood Cells," pp. 116–136 in Hershey (1970).

Spaeth, E. E., "The Oxygenation of Blood in Artificial Membrane Devices," pp. 276–305 in Hershey (1970).

Spaeth, E. E., and S. K. Friedlander, "The Diffusion of Oxygen, Carbon Dioxide, and Inert Gases in Flowing Blood," *Biophys. J.*, **7**, 827 (1967).

Stewart, W. E., "Forced Convection in Three-dimensional Flows," *Am. Inst. Chem. Engrs. J.*, **9**, 528 (1963).

Stewart, W. E., J. B. Angelo, and E. N. Lightfoot, "Forced Convection in Three-dimensional Flows. II. Asymptotic Solutions for Mobile Interfaces," *Am. Inst. Chem. Engrs. J.*, **5**, 771–768 (1970).

Thews, G., "Gaseous Diffusion in the Lungs and Tissues," in *Physical Bases of Circulatory Transport*, E. B. Reeve and A. Guyton, Eds., Saunders (1967).

Turitto, V. T., A. M. Benis, and E. F. Leonard, "Platelet Diffusion in Flowing Blood," *Ind. Eng. Chem. Fund.*, **11**, 216 (1972).

Ulanowicz, R. E. , and G. C. Frazier, Jr., "The Transport of Oxygen and Carbon Dioxide in Hemoglobin Systems," *Math. Biosci.*, **7**, 111–129 (1970).

Villaroel, Fernando, C. E. Lanham, K. B. Bischoff, T. M. Regan, and J. M. Calkins, "Gas Transfer to Gas Flowing in Semipermeable Tubes under Steady and Pulsatile Flow Conditions," in American Institute of Chemical Engineers, No. 114, pp. 96–104 (1971).

Weissman, M. H., and L. F. Mockros, *Proc. Am. Soc. Civ. Engrs*., **93**, 225 (Dec. 1967); *Med. Biol. Eng.* **7**, 169 (1969).

White, D. A., *Chem. Eng. Sci.,* **24**, 369–376 (1969).

## 6. MASS TRANSFER AND THE MACROSCOPIC BALANCES

Almost all interesting mass-transfer problems are far too complex for detailed solution, and one must normally begin with a judicious simplification of the geometry and gross flow distribution. The macroscopic mass balance provides a convenient framework at this level of modeling, but it must normally incorporate a large amount of physical information; this in turn must be obtained by preliminary analysis and experimentation. The description of the system may then be completed by detailed analysis of critical regions in the model—for example, by the application of boundary-layer theory. In favorable situations the prediction of the system behavior requires only the specification of such primary parameters as the system size and the compositions and flow rates of blood and other input streams; normally, however, the description contains one or more parameters that must be evaluated empirically.

The primary goals in modeling are to increase the reliability of prediction to an acceptable level and to decrease the number of empirical parameters to the smallest possible. Success requires an appreciation of the orders of magnitude of the governing parameters, familiarity with the means available for detailed analysis, and, most difficult, good judgement in selecting a basic model.

Modeling is clearly a heuristic process, and one cannot lay out detailed procedures in advance. It is, however, possible to review characteristic orders of magnitude of key parameters and to discuss previous modeling efforts. One can thus provide at least general guidelines for future efforts, and that is the purpose of this section. We begin in Section 6.1 with a general discussion of dimensional analysis and scaling, and then proceed in Section 6.2 to a review of lumped-parameter modeling in mammalian systems. In Section 6.3 we show the need for distributed-parameter modeling, and, finally, in Section 6.4 we give a very brief introduction to modeling at the molecular level.

It must, however, be kept in mind that one cannot set the requirements for a model unless one has specific goals in mind: modeling, however important, is never more than a component of some design effort. The role of all aspects of transport phenomena in design is discussed in the next chapter, the conclusion to this text.

## 6.1 Scaling and Dimensional Analysis in Biological Systems

In this section* we compare the gross physiological behavior of different animals and plants to show the effects of body size. Such a comparison is clearly of immediate practical utility in the interpretation of animal experiments, and we shall see that it also provides useful insight into mass transfer in human bodies. This section thus provides an introduction to the lumped-parameter modeling of Section 6.2. We are interested in changes in metabolic rate, anatomy, heart rate, and body chemistry. We shall find both impressive correlations of these factors with body size and some strong deviations; both will prove instructive.

Perhaps the most surprising similarity is in shape, as evidenced by specific surface. It is shown in Fig. 6.1.1 that the outer specific surface of vertebrates is about twice that of a sphere with the same volume, over a mass ratio of $10^7$. The larger points on this graph represent beech trees and show that this surprising regularity in shape extends even to plants. The success of this correlation cannot be easily explained in terms of such surface-dependent processes as heat loss or mass diffusion.

The volume fraction of the body occupied by individual organs is essentially independent of body size; this is shown in Fig. 6.1.2 for the lungs of vertebrates, and it can be seen here that even such marine mammals as the porpoise and the whale do not show significant deviations. Similar relations hold for the dimensions of other organs. Individual species differences do occur, however, and many of these are not unexpected. Thus the hearts of racehorses and greyhounds are larger than the usual 0.5 to 0.6% of body mass.

* Much of this discussion is taken from an entertaining and instructive paper by Schmidt-Nielsen cited in the bibliography. The interested reader is referred to this and to the cited works of Kleiber and of Stahl.

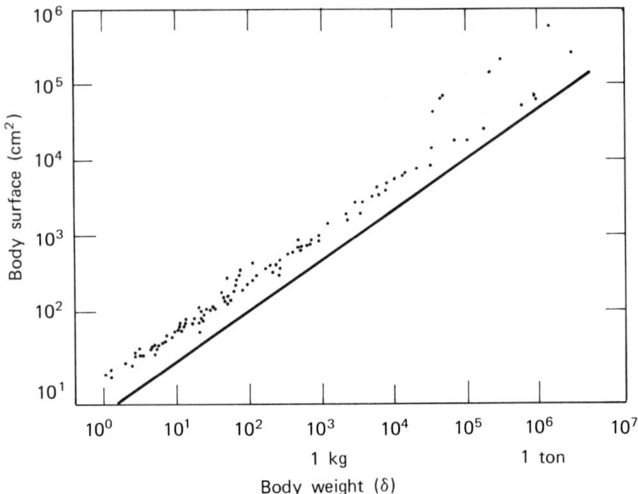


**Figure 6.1.1.** Body surface of vertebrates in relation to body weight. The solid line represents the surface of a sphere of density 1.0. The larger points in the upper right-hand corner represent beech trees. From Schmidt-Nielsen (1970).

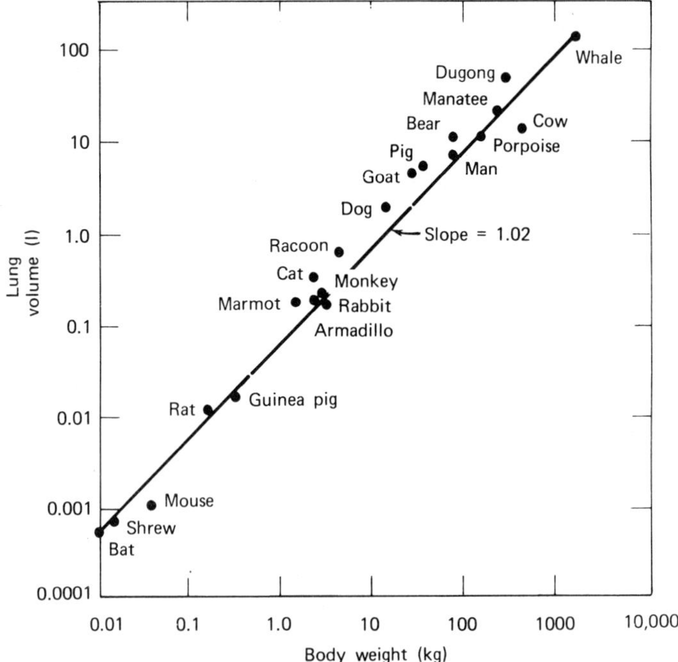


**Figure 6.1.2.** Lung volume of mammals is scaled in simple proportion to body size, as indicated by a slope of the regression line of nearly 1.0. From S. M. Tenney and J. E. Remmers, *Nature*, **197**, 54 (1963).

Such anatomic regularity cannot extend to all levels, however, since living organisms operate under a number of scale-sensitive restraints. Among these is the need of terrestial species for mechanical support to resist gravitational forces, which results in proportionately heavier skeletons for larger animals. This effect is surprisingly small, however, and skeletal mass is found to increase with only the 1.13 power of body weight; this must be considered quite a design achievement for the larger species. Somewhat more subtle, and not completely understood, are the effects of hydrostatic pressure in animals such as the giraffe.

More important to us, however, is the increase in specific body metabolic rate with decrease in body size shown in Fig. 6.1.3. This is a truly remarkable correlation, extending over 18 orders of magnitude in mass, and showing that total metabolic rate tends to vary with the $\frac{3}{4}$ power of body weight. There is no known explanation for this behavior, which is observed even for very small organisms where heat and mass transfer resistance are negligible, and which has a major impact on anatomic detail and blood chemistry. The efficiency of energy conversion in mammals is, however, essentially size independent.

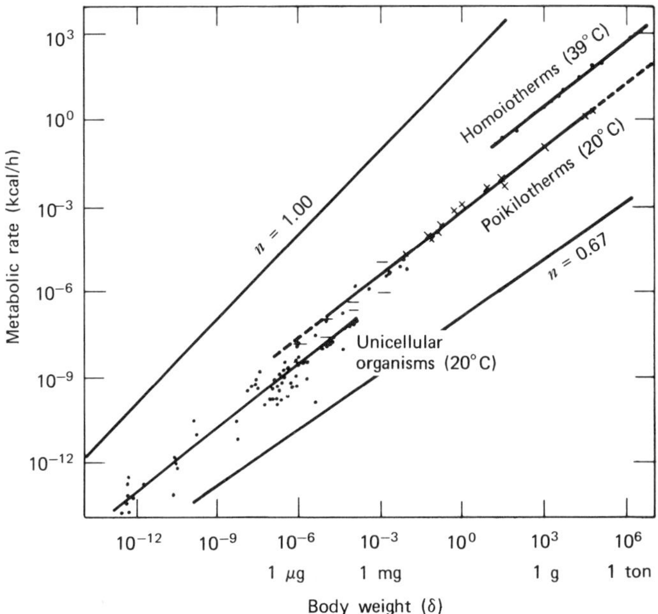


**Figure 6.1.3.** Metabolic rates of various organisms in relation to body weight. Note that each mark on the coordinates denotes a 1000-fold difference in magnitude. From Hemmingsen (1960).

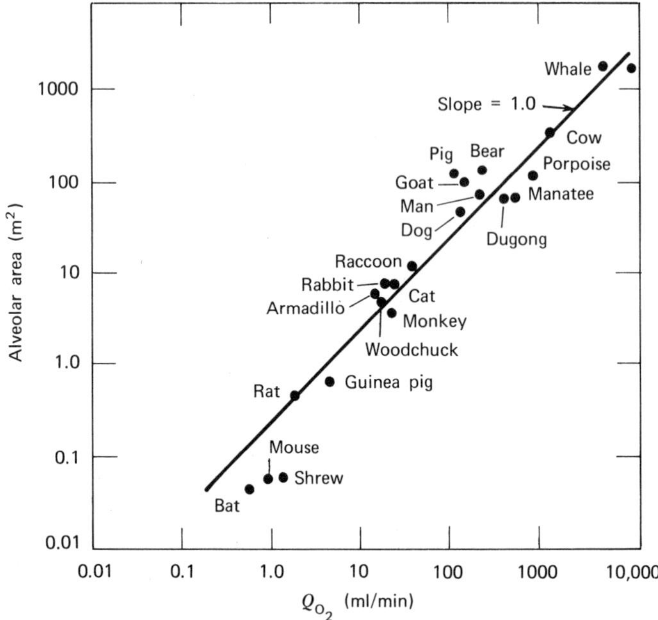


**Figure 6.1.4.** The diffusion area of the mammalian lung is scaled in simple proportion to the rate of oxygen consumption. From Tenney and Remmers (1963).

Fig. 6.1.4 shows that the alveolar, or diffusion, area of the lung varies in proportion to oxygen consumption rate rather than exterior body surface area. This suggests that the mass-transfer effectiveness of alveolar surface is size independent. Similarly the capillary density is greater in small animals, but according to Schmidt-Nielsen (1970) this increase is insufficient to meet the greater metabolic demand. There is accordingly a systematic change in blood chemistry with size.

This effect of size on mammalian blood takes two forms, as shown in Figs. 6.1.5 and 6.1.6: (1) the hemoglobin of smaller animals releases oxygen more readily at a given pH, and (2) the release of $O_2$ on pH decrease or $pCO_2$ increase (the Bohr effect) is greater for smaller animals. The general trend is clear from these figures, but Fig. 6.1.6 also shows two interesting deviations. First, it may be seen that the Bohr effect is unexpectedly large for horses, which is not unreasonable in an animal accustomed to the exertion of running for long periods. In contrast, the chihuahua, a relatively new breed of dog developed for fashion rather than function, exhibits an abnormally small Bohr effect. This would appear to be the result of rapid selective breeding and a disadvantage to the dog.

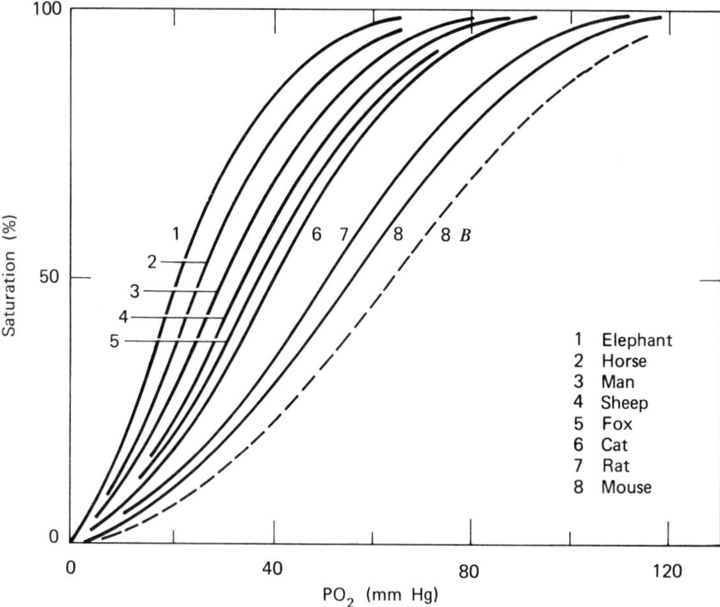


**Figure 6.1.5.** Dissociation curves of oxyhemoglobin from a number of mammals show that the blood of smaller mammals will unload oxygen under a higher oxygen pressure than that of larger mammals. This helps in the delivery of sufficient oxygen to the tissues to maintain the high metabolic rates of small animals. The dashed curve ($8B$) indicates the effect of acid (Bohr effect) on mouse blood (curve 8).

The trends shown in these figures, interesting in themselves, can be used both for quantitative predictions and to provide qualitative insight into mass transfer and metabolism. This is considered in the first example below.

*EX. 6.1.1. THE DIMENSIONAL ANALYSIS OF DRUG DISTRIBUTION*

Shown in Figs. 6.1.7 and 8 are plasma concentrations of the anticancer drug methotrexate (MTX) as a function of time since injection into the system, taken from Dedrick, Bischoff, and Zaharko (1970). Figure 6.1.7 shows data for various dosage levels and a variety of species without any attempt at scaling. The dispersion of data is almost bewildering. Figure 6.1.8 shows the same data adjusted for dosage level, body weight, and circulation time. The correlation is now excellent by biological standards,

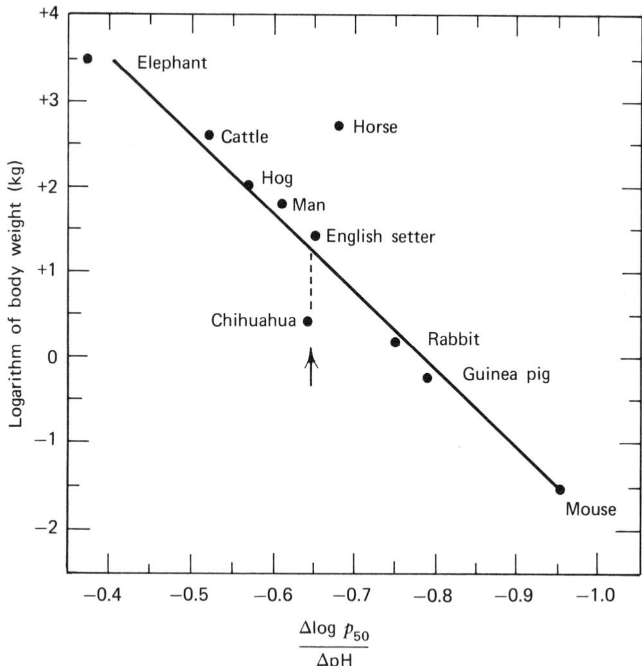


**Figure 6.1.6.** The effect of acid on the unloading of oxygen from the blood (Bohr effect) is greater in small than in large animals. From Riggs (1960).

and very similar behavior is observed, in this modified coordinate system, for the five widely varying species tested.

Discuss the significance of this result in the light of the above text.

SOLUTION

It must be admitted at the outset that the Dedrick-Bischoff-Zaharko correlation shown here is essentially empirical and that it owes its success to a number of simplifying factors:

1. MTX distributes itself linearly among the various aqueous and fatty regions of the body and is not highly fat soluble.
2. It is eliminated principally by kidney clearance, which is *flow limited.*
3. The time scale of MTX elimination is large relative to either typical residence times in vascular beds or blood circulation times (about 1 sec and 1 min, respectively, in adult human beings).

It is instructive to consider each of these factors in turn.

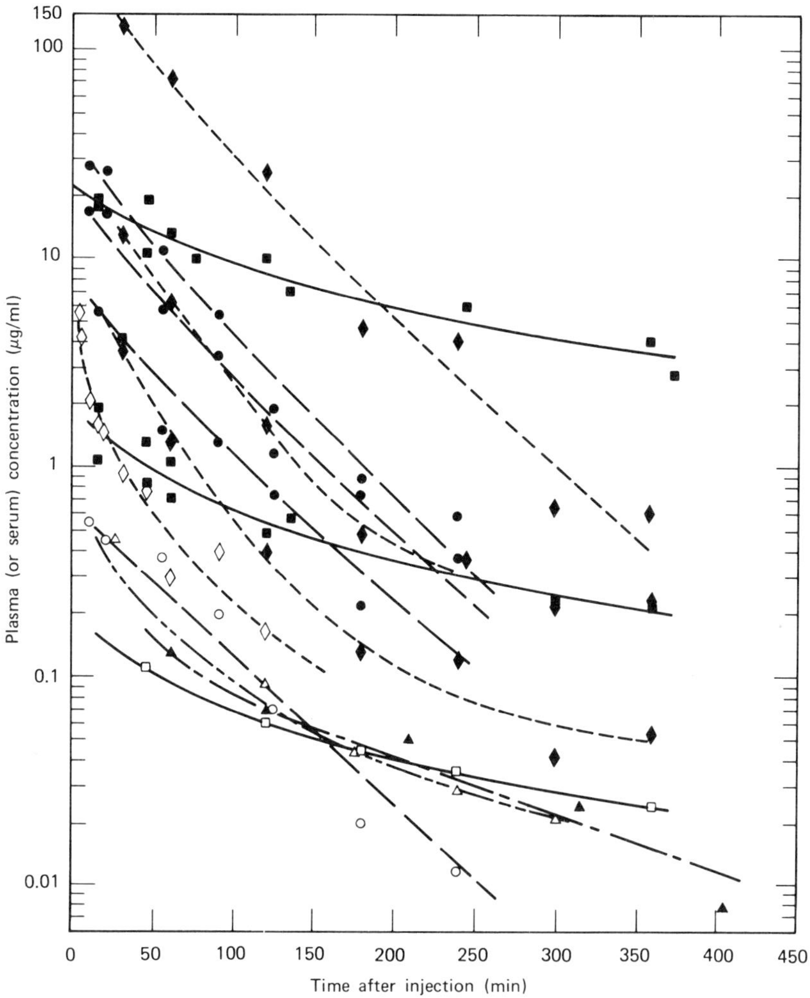


**Figure 6.1.7.** Plasma (or serum) concentration of methotrexate after intravenous or intraparietal injection. The species are indicated by: mouse - - - - ; rat — — — ; monkey - —— - ; dog —- - — ; man ———— . From Dedrick, Bischoff, and Zaharko (1970).

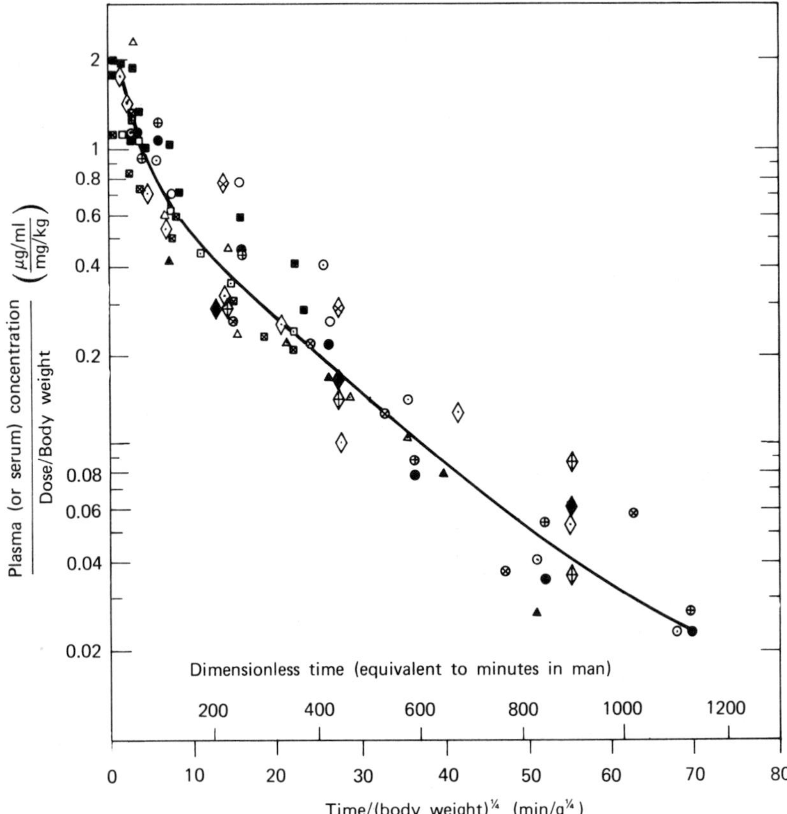


**Figure 6.1.8.** Graphical correlation plasma (or serum) concentration of methotrexate in mouse, rat, monkey, dog, and man. From Dedrick, Bischoff, and Zaharko (1970).

The linearity of MTX distribution clearly suggests scaling the ordinate by dividing the plasma concentration by the dose per unit mass of the test animal or patient,* to normalize concentrations within any species. Interspecies correlation, which could be complicated by differences in the composition of body lipids, is aided here by the low lipid solubility of the drug.

Mass-transfer resistance within vascular beds is very unlikely to be important for the time scales of interest here. As explained in the text, vascular beds have developed the ability to strip arterial blood of oxygen (or, in the lungs, to return it) in one pass; their mass-transfer effectiveness

* The ordinate could easily be made dimensionless, with negligible change in results, by using mass units for the plasma concentration.

is accordingly very high for the much slower mass-transfer processes considered here.* In fact, since our time scales are long even relative to the blood circulation time, one might expect the plasma concentration to be very nearly position independent.

One may then expect elimination to be controlled by the flow rate through the kidneys, since kidney clearances† are usually to a first approximation proportional to renal flow rate and independent of solute concentration. It remains to estimate the effect of size on renal flow rates.

We begin by assuming the fraction of cardiac output flowing through the kidneys to be size independent and remembering that metabolic rate varies as about the $\frac{3}{4}$ power of body mass. Since both the fraction of oxygen removed by vascular beds and the respiratory quotient are essentially size independent, it follows that the volumetric output of blood from the heart varies as

$$Q_B \propto M_{\text{tot}}^{0.75} \tag{1}$$

where $M_{\text{tot}}$ is total body mass. If, as indicated above, the blood volume $V_B$ is proportional to $M_{\text{tot}}$, then the mean circulation time is

$$\bar{t} \equiv \frac{V_B}{Q_B} \propto M_{\text{tot}}^{1/4} \tag{2}$$

We should then expect our scaled plasma concentration to depend only on $\tau \equiv t/\bar{t}$, where $t$ is the time, and also that

$$\tau \propto M_{\text{tot}}^{-1/4} \tag{3}$$

This is just what is observed in Fig. 6.1.8. For well-equilibrated plasma and a time-independent clearance we would also expect an exponential decrease of MTX concentration with time. This also is in reasonable agreement with the data.

We find then that the Dedrick-Bischoff-Zaharko correlation is entirely consistent with Schmidt-Nielsen's short treatise on comparative physiology summarized above. Predictions are not, however, always so successful. Dedrick et al, rank the following factors in decreasing order of predictability:

* In adult humans both oxygenation and deoxygenation times are on the order of 1 sec (see Section 5.4 and Ex. 6.3.2).

† See Ex. 6.2.2 for a definition of clearance, and Ex. 6.3.3 for a discussion of kidney behavior.

1. Blood flow and metabolic rates.
2. Thermodynamic activity coefficients (e.g., in describing protein binding and lipid solubility).
3. Kinetics (e.g., the presence and effectiveness of individual enzymes and enzyme systems).

We have been dealing here with a situation dominated by factors in category 1; we now turn our attention to a situation in category 3, which is all too timely at the moment of writing.

*EX. 6.1.2. HOW MUCH* LSD *FOR AN ELEPHANT?*

Schmidt-Nielsen (1970) cites an unfortunate experiment in which 297 mg of the drug LSD was injected into an elephant to simulate the condition of musth. This dose was calculated on the basis of the 0.1 mg/kg needed to enrage a cat. To quote Schmidt-Nielsen (who was not responsible for this experiment): "the elephant immediately started trumpeting and running around, then he stopped and swayed; 5 minutes after the injection he collapsed, went into convulsions, defecated, and died."

Comment on the scaling aspects of this experiment.

SOLUTION

This is clearly a complex situation of very limited predictability. However, one can say that equal mass of drug per unit body mass is on its face a doubtful basis for scale-up.

It is clear from the previous example that the drug retention time will be much greater for a 3000-kg elephant if elimination is flow limited. In fact, the time scale will be increased by a factor of

$$\left(\frac{2970}{2.6}\right)^{1/4} = 5.8$$

and the exposure to the drug will be correspondingly more severe. This calculation alone suggests that the dose was too high.

However, the extremely rapid death of the elephant suggests that some more subtle species difference was operating. This is confirmed by the fact that the dose necessary to send a 70-kg human on a "trip" is only 0.2 mg, or less than that needed for a 2.6-kg cat. Unfortunately for the test elephant, cats have a high species-specific LSD tolerance not shared by either elephants or humans.

This example is thus primarily useful as a warning against oversimplified interpretations of animal experiments, which should be carried out cautiously.

One important aspect of such a test program, the prediction of transient drug concentrations ("pharmacokinetics"), is, however, amenable to engineering analysis; it is discussed in the next section.

## 6.2 Lumped-Parameter Modeling and the Macroscopic Balances

Even a casual study of mammalian anatomy and physiology shows that mass-transfer resistance is normally distributed in a complex way throughout organs and even smaller units. Hence the detailed description of biological mass-transfer processes tends to be rather formidable mathematically. Such detailed treatments are frequently desirable, and much of these notes is devoted to them. Fortunately, however, they are not always necessary—or justified by the amount of information available.

It is common experience that the salient features of even rather complex processes can be described in terms of highly simplified *lumped-parameter* models. In these, flow patterns and geometry are idealized, and the mass-transfer resistance is assumed to be concentrated in very thin regions, such as phase boundaries. A classic case from the chemical engineering literature is the analysis of packed-column mass-transfer processes (see, e.g., *Tr. Ph.*, Fig. 22.5-3). Such simplification may make the *a priori* prediction of system behavior impossible, and this can be a severe drawback. However, in many biological applications, lack of anatomical and transport-property data makes *a priori* prediction impossible in any event. Moreover, as in Ex. 6.1.1, we are frequently interested in time scales sufficiently long that the mass-transfer resistance within vascular beds is negligible. Under such flow-limited conditions surprisingly powerful predictions can be made; these are discussed in the examples.

In all cases anatomic simplification is the first step in system description, since we cannot deal directly with the classical medical representation shown in Fig. 6.2.1*a*. Rather, one must split the body into a limited number of regions connected by major blood vessels, as shown in Fig. 6.2.1*b*. Each "compartment" in turn is subdivided into its major diffusionally active components. One common subdivision is indicated in Fig. 6.2.1*c*, but others may be more convenient in individual cases.

It remains to characterize the mass-transfer behavior of the compartments and their subdivisions, and this can be done in several ways. The simplest conceptually is to use a purely stochastic approach, that is, to develop empirical relations between the output concentrations from a given compartment and the input to it. Stochastic techniques are discussed in Ex. 6.2.1.

The simplest physical model is to consider the contents of each subcompartment, for example, the capillary blood of Fig. 6.2.1*c*, to be well mixed. The macroscopic mass balance for such a system takes the form, for any diffusing species $i$:

$$\frac{dm_{i,\text{tot}}}{dt} = -\Delta w_i + w_i^{(m)} + r_{i,\text{tot}} \tag{6.2.1}$$

where $m_{i,\text{tot}}$ = total mass of species $i$ in the subsystem
$-\Delta w_i$ = net rate of input of $i$ by flow through entering and leaving streams
$w_i^{(m)}$ = net rate of input across mass-transfer surfaces
$r_{i,\text{tot}}$ = net rate of formation of $i$ by chemical reaction within the subsystem.

Models of this type have proven very useful and are discussed in Ex. 6.2.2 and 6.2.3.

Another sometimes useful simplification is to assume plug flow in the blood and to neglect diffusional fluxes parallel to blood flow in the tissue. This approach has been used for example by Goresky et al. (1970) and by Levitt (1971); it leads to equations very similar to those describing chromatographic operations and can be extended to quite complex parallel networks. Here one must use Eq. 6.2.1 for differential length segments of the compartment shown in Fig. 3.2.1*c*, rather than the system as a whole. This procedure is similar in its essentials to that described in *Tr. Ph.*, Ex. 22.6-2, and is not further considered here. A much more complex problem of this type is the description of nerve impulse transmission. The reader interested in this critically important subject is referred to Cole (1968).

Situations where details of the velocity profile are important are discussed in Section 6.3. Further references to the rapidly growing field of mass-transfer modeling are given by Lightfoot, Safford, and Stone (1971).

### *EX. 6.2.1. STOCHASTIC ANALYSIS OF BLOOD-FLOW DISTRIBUTION*

One of the most effective means of characterizing flow and diffusional behavior of the circulatory system is by analysis of system response to the injection of short pulses of tracer solutes (e.g., dyes or radioactive substances). Among the first to realize the potentialities of this technique were Nicholes, Warner, and Wood (1964), some of whose results are shown in

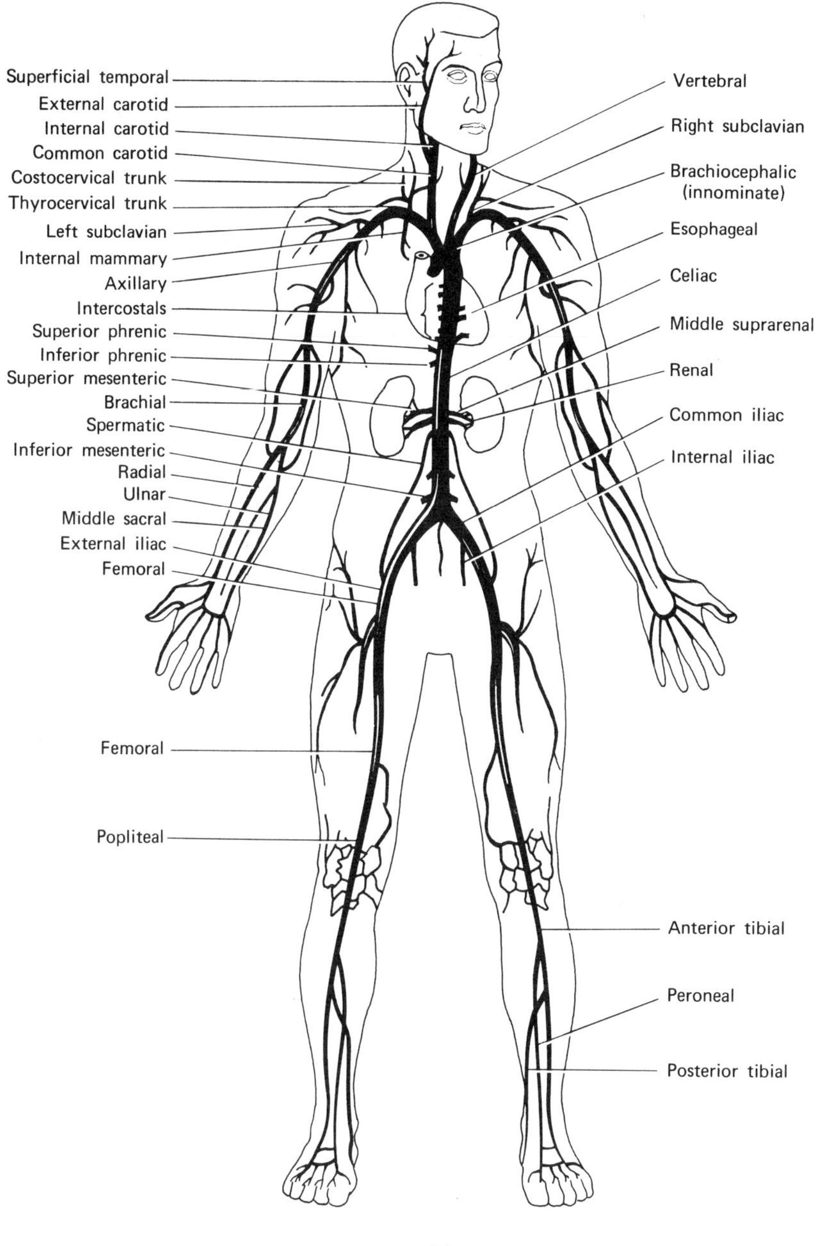


*(a)*

**Figure 6.2.1.** (*a*) Schematic drawing of the arterial system. Taken from Chaffee, E. E. and E. M. Greisheimer, *Basic Physiology and Anatomy*, J. B. Lippincott Co., Fig. 232 (1964).

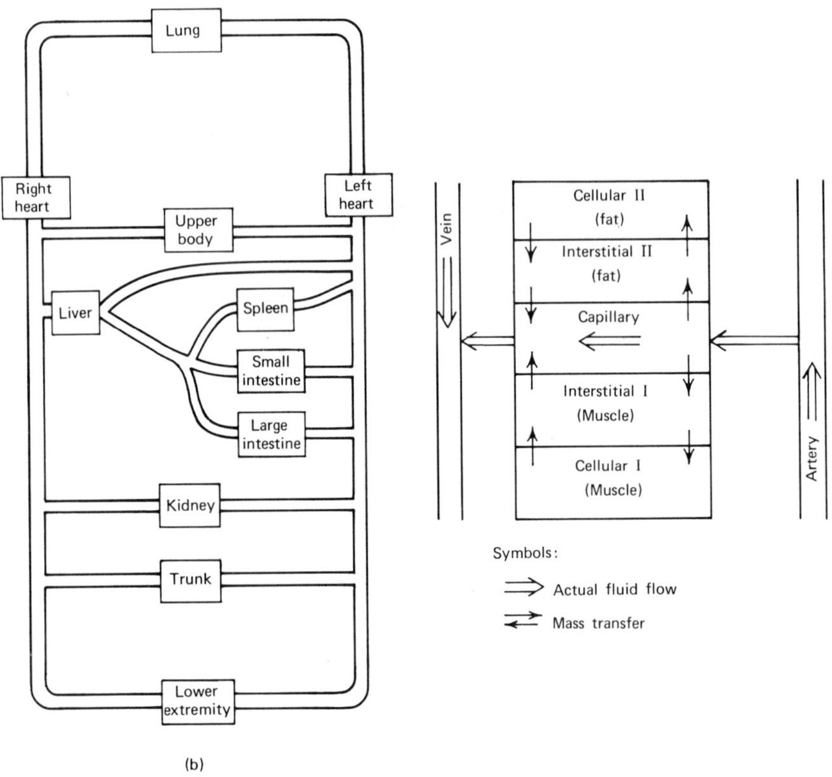


**Figure 6.2.1** (*continued*). (*b*) Multi-compartment anatomic approximation. From Bischoff (1967). (*c*) Compartmental details. From Bischoff (1962).

Figs. 6.2.2 to 6.2.4. Discuss the utility of pulse-response techniques in the light of these figures.

SOLUTION

*Analysis in the absence of recirculation.* We begin with the relatively simple case of Fig. 6.2.2, which shows the dispersion of a dye tracer as it traverses the pulmonary circulation of a dog. Here the input (*a*) is the dye concentration as a function of time, measured at the pulmonary artery, and the output (*b*) is measured at the aorta. The primary peak in each of the four experimental curves represents the first traverse of dye past the measurement point; the smaller secondary peaks, or humps, result from the *closed-loop* nature of the flow (recirculation). Analysis of the system response is much facilitated by eliminating recirculation effects. Fortunately, these effects are frequently small until well after the primary concentration peak has been reached, and it has been found empirically that the concentration typically falls off very nearly exponentially with time in the latter stages of open-loop operation. The usual procedure, then, is to plot the logarithm of the exit concentration against time and look for a linear region in the descending part of the curve. If such behavior is found, the equivalent open-loop concentration distribution is approximated by extrapolating this linear behavior to infinite time.

Once recirculation effects have been eliminated in this way, the input and output curves can easily be used to calculate the blood flow rate through the system and the effective system volume.

We first note that the total mass passing any point in the system over a sufficiently long time interval must equal the mass $M_0$ of dye injected:

$$M_0 = \int_0^\infty Qc_d\,dt \tag{1}$$

where $Q$ is instantaneous volumetric flow rate, $c_d$ is the dye concentration, and $t$ is time elapsed after dye injection. If the flow is steady (and it is usually so *considered*—see Bassingthwaighte, 1970), one may write

$$Q = \frac{\int_0^\infty c_d\,dt}{M_0} \tag{2}$$

Equation 2 provides a convenient and widely used means of estimating flow rates in intact systems.

For steady flows of nonreactive solutes, the fluid volume $V$ between inlet and outlet can be calculated from the equation*

$$\frac{V}{Q} \equiv \bar{t} = \frac{Q}{M_0}\int_0^\infty t[(c_d)_o - (c_d)_i]\,dt \tag{3}$$

* This expression is valid for systems of arbitrary geometry and rigid impermeable walls. Its remarkable generality seems first to have been recognized by Spalding (1958); its range of validity is discussed in some detail by Lightfoot, Safford, and Stone (1971).

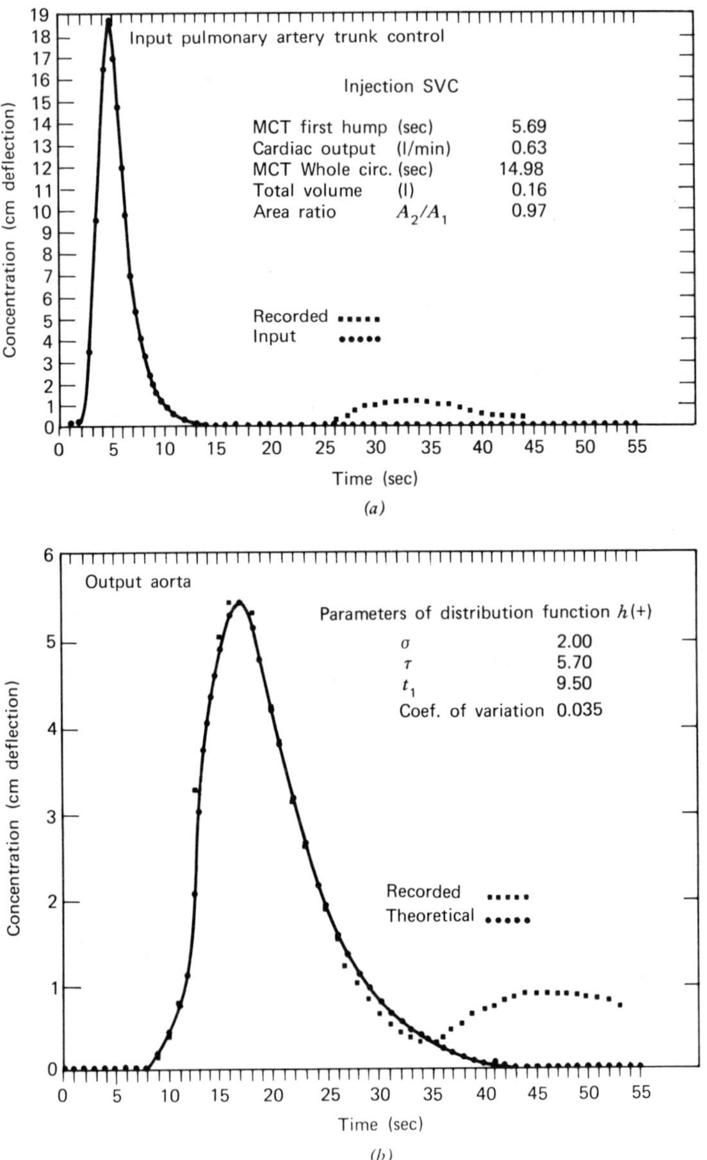


**Figure 6.2.2.** Comparisons of curves for circulation of indicator from *SVC* (superior vena cava) as recorded from aortic root of dog (*b*) with theoretical aortic curves obtained by convolving pulmonary-artery curve (*a*). *MCT* stands for mean circulation time. From Nicholes, Warner, and Wood (1964).

Here the subscripts $o$ and $i$ represent conditions at the outlet and inlet respectively. This expression can be simply modified for solutes which distribute linearly into stationary accessible tissue.

If we define the $n$th moment of concentration with respect to time by

$$\mu_n = \int_0^{\infty} t^n c_d \, dt \tag{4}$$

we see that the integrals in Eqs. 1 and 3 represent the zeroth and first moments. Higher moments may also be calculated, but even a small dispersion of data at large $t$ makes moments higher than the third difficult to determine.

It remains to characterize the response of the system to solute input, and this is usually most conveniently done in terms of the response $h(t)$ to a unit input pulse at time zero, defined by

$h(t)$ = the fraction of tracer particles with a transit time $t$ in the system, between outlet and inlet.

If a mass $M_0$ of tracer is injected impulsively at time zero at the inlet, the outlet concentration is*

$$(c_d)_0 = \frac{M_0}{V} h(t) \tag{5}$$

For such time-varying input concentrations as are shown in Fig. 6.2.2, the corresponding result is†

$$(c_d)_0 = \int_0^t [c_d(t-\lambda)]_i h(\lambda) \, d\lambda \tag{6}$$

It remains to obtain a convenient mathematical characterization of $h$.

The theoretical output curves in Fig. 6.2.2 were calculated from the input curves by convolution with the *lagged* normal probability distribu-

* Provided the system responds linearly to solute concentration, as is normally the case with inert tracers.

† Usually written as $(c_d)_i * h$ and known as the *convolution* of $(c_d)_i$ with $h$ (see Bassingthwaighte, 1970; Bischoff, 1967; or Lightfoot, Safford, and Stone, 1971).

tion function

$$h(t) = \frac{1}{\sigma\sqrt{2\pi}} e^{-\frac{1}{2}(t-t_1)^2/\sigma^2} - Th'(t) \tag{7}$$

where $\sigma$ = standard deviation of the normal part of the distribution
$t_1$ = mean time of this normal part, and
$T$ = time constant of the lag term
$h'(t)$ = derivative of $h$ with respect to time.

Since the mass of tracer, $M_0$, is required to normalize the system response, Eq. 7 contains four parameters. The theoretical curves of the lower figure were obtained by selecting optimal values of $\sigma$, $T$, and $t_1$ via a digital computer program; it can be seen here that the fit is excellent.* It was also found by comparing the areas under the upper and lower curves that essentially the same amount of dye passed the two recording sites on the "first circulation." The success of this conceptually simple approach is due to the comparative symmetry of the pulmonary circulation.

In the systemic circulation the presence of many major parallel flow paths, some of them quite short, makes the concept of a single "circulation time" artificial. This can be seen rather simply by comparing the first and second humps of the curves in Fig. 6.2.2. In (*a*), for example, the area under the second hump, which represents the first recirculation, is much smaller than that under the first hump, which represents the total dye input. The ratio is in fact only 0.38 (and 0.78 following ATP injection, which stimulates flow through the extremities). This behavior results from some pathways so long that dye traversing them does not contribute significantly to the first recirculation hump.

*Analysis of more complex systems.* Nicholes, Warner, and Wood (1964) were able to obtain a satisfactory description of the entire circulation by an extension of their above-mentioned curve-fitting procedure to the type of system diagrammed in Figure 6.2.3. This system is described formally by

$$c_4 = \sum_{i=1}^{n} F_i(c_3 * h_i) \tag{8}$$

$$c_3 = (c_4 * h_{43}) \tag{9}$$

* These parameters may also be determined from the moments of Eq. 4. The zeroth moment provides normalization, and the first gives the mean transit time $\bar{t}$. The second gives a measure of the dispersion and hence $\sigma$, while the third describes skewness. It is hard to obtain reliable values of more than four parameters for one "compartment." Other three- and four-parameter expressions are cited by Bassingthwaighte (1970).

where $F_i$ = fraction of cardiac output through path $i$
$h_i$ = the effective unit pulse response for path $i$, and
$h_{43}$ = the effective unit response of the series system comprised by the left heart, lungs, and right heart.

The distribution functions $h_i$ can be determined from tracer measurements in the aorta and major veins draining the organ systems; the $F_i$ can then be selected to give the best possible fit between Eq. 8 and actual observations. Nicholes et al. were able to obtain a satisfactory fit using four paths as indicated in Figure 6.2.4.

The above-described work appears to be the first dynamic description of the blood flow distribution in the entire circulation and is an impressive accomplishment. In fact, it anticipates some of the much later pharmacokinetic modeling discussed in the next section. The approach used by these authors is, however, limited to systems with a small number of identifiable pathways. It will not, for example, work to describe the

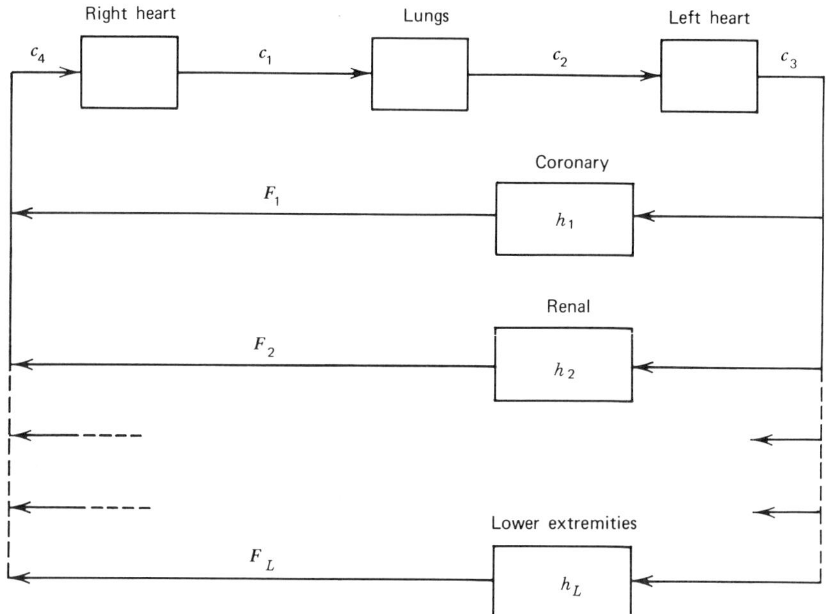


**Figure 6.2.3.** Schematic representation of the mathematical model of mixing in the circulation, and representation of numerical approximation of Eq. 5 for computer solution. Here $\Delta D$ is set equal to period of the heart cycle. From Nicholes et al. (1964).

$$c_4(t) = \sum_{I=1}^{L} F(I) \sum_{J=1}^{J(t)} \sum_{K=1}^{K(t)} c_3(J) H(I, J+K-1) \Delta D$$

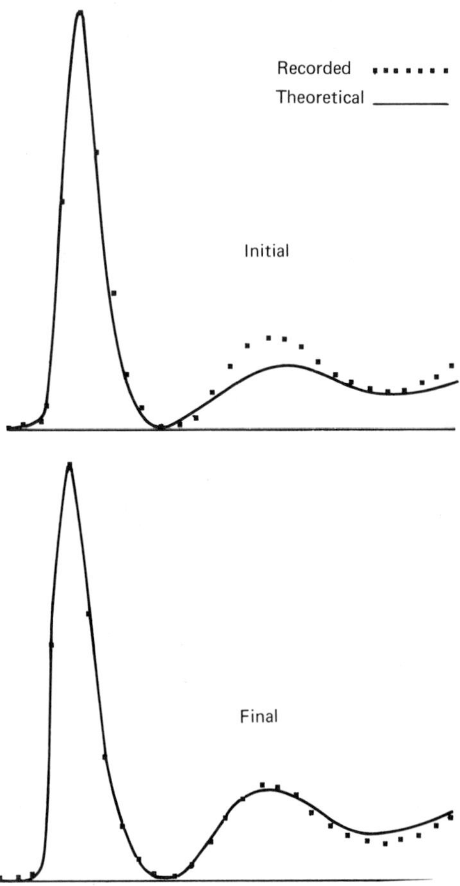


**Figure 6.2.4.** Comparison of computer solutions (from Nicholes et al., 1964) for indicator concentration in aorta ($c_3$) using true values (points) and arbitrary values (solid line, upper panel) for the fraction of cardiac output traversing each organ system. In the lower panel is a comparison of the predicted $c_3$ after the computer program has optimized the fit by adjusting flow fractions automatically to those shown in the column labelled "Final".

Fraction of Cardiac Output

| | Recorded | Theoretical | |
|---|---|---|---|
| Pathway | True value | Initial (estimate) | Final (prediction) |
| Coronary | 0.060 | 0.390 | 0.064 |
| Renal | 0.200 | 0.050 | 0.199 |
| Hepatic and S. V. C | 0.440 | 0.300 | 0.426 |
| Slow pathways | 0.300 | 0.560 | 0.320 |

distribution of flow within individual organs. (Consider, for example, the kidney with $10^6$ nephrons.) For such systems it is desirable to look for common features in the dispersive systems encountered, and it appears that some exist:

1. It is found that for a given flow system the relative degree of dispersion is flow independent. More precisely, $h(t/\bar{t})$ is independent of $Q$.

2. Each pathway through an organ has a transport function with relative dispersion and skewness similar to those for an artery.

Both these statements are useful first approximations, but neither is wholly reliable.

The normalized pulse response of the arterial system of the human leg is shown in Figure 6.2.5 for a variety of flow rates through the external iliac artery. Careful examination of this figure shows no detectable effect of flow rate (and even a cursory examination shows the poor reproducibility

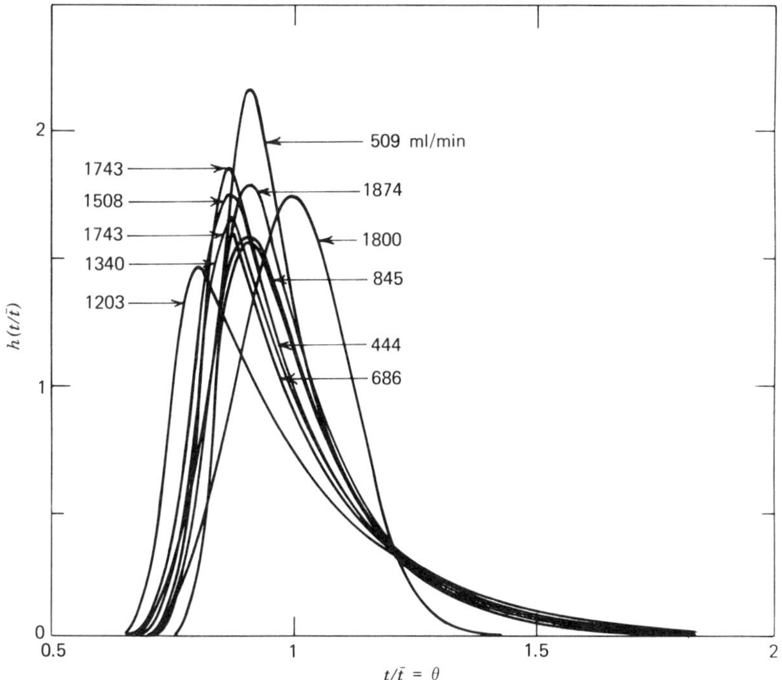


**Figure 6.2.5.** Normalized pulse response of the arterial system of the human leg at different steady flow rates through the external iliac artery. From Bassingthwaighte (1970).

of the measurements). This situation is thus seen to follow the first "rule of thumb" above. On the other hand, for the conditions of Figure 6.2.2, ATP infusion markedly affects the flow distribution in the circulation. Similar effects result from exercise and a very wide variety of other stimuli which also affect flow rate. This statement must therefore be used with caution.

Greenleaf et al. (1968) have developed a means of analyzing parallel-pathway systems based on the assumption that

$$h_i(t) = \frac{\bar{t}_j}{\bar{t}_i} \cdot h_j\left(\frac{\bar{t}_j}{\bar{t}_i} \cdot t\right)$$

Here the $h_j$ and $h_i$ are the pulse responses of paths $i$ and $j$, and $\bar{t}_j$ and $\bar{t}_i$ are the corresponding mean residence times. This is an active research area and a very promising one. Other attractive possibilities for stochastic techniques are discussed by Aris (1965) and Bassingthwaighte (1970).

### *EX. 6.2.2. TWO-COMPARTMENT LUMPED-PARAMETER MODELS*

Compare Bell, Curtiss, and Babb's diffusion-limited model for hemodialysis (shown in Fig. 6.2.6) with Bischoff and Dedrick's flow-limited model for drug clearance (Fig. 6.2.7).

SOLUTION

*The Hemodialysis Model of Bell et al.* One of the major contributions of the Babb-Grimsrud group (Babb et al., 1967) to hemodialysis was to show that the removal of small solutes could be described by a simple two-

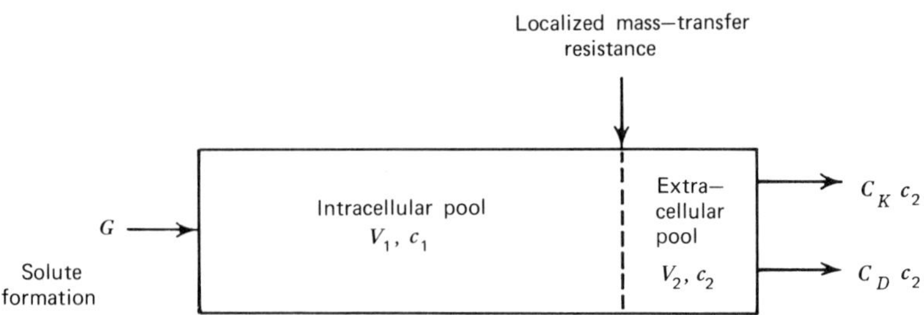


**Figure 6.2.6.** The dialysis model of Bell, Curtiss, and Babb; the clearance $C_K$ or $C_D$ is defined as rate of flow through kidney or dialyzer, respectively, times the fractional rate of solute removal.

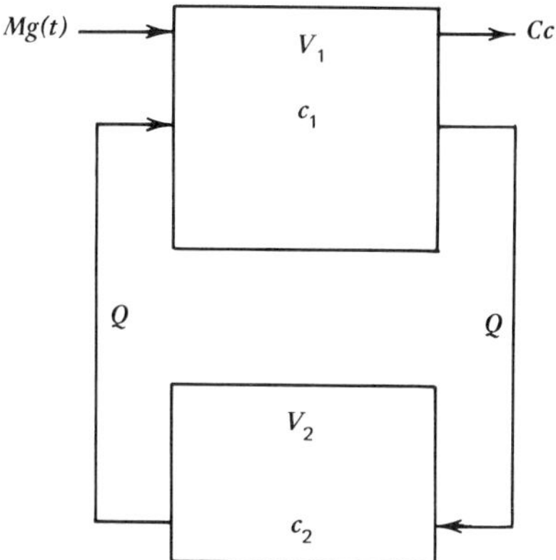


**Figure 6.2.7.** The two-compartment drug-elimination model of Bischoff and Dedrick. Drug input is to well-perfused visceral tissue (compartment 1) at a schedule $g(t)$, and removal is by kidney clearance $C$ from this region. The drug is simultaneously transported to the more poorly perfused tissue (compartment 2) by blood flow at volumetric rate $Q$.

compartment model of the body (see Fig. 6.2.6):

$$V_1 \frac{dc_1}{dt} = G - K(c_1 - c_2) \tag{1}$$

$$V_2 \frac{dc_2}{dt} = K(c_1 - c_2) - C_t c_2 \tag{2}$$

where $V_1$ is the volume of the combined interstitial and intra-cellular space; $V_2$ is the volume of the blood; $c_1$ is the average concentration of a given solute species in the combined interstitial and intracellular space; $c_2$ is the average solute concentration in the blood; $G$ is the total rate of "production" of the solute species by ingestion plus body metabolic processes; $K$ is the rate constant for transfer from compartment 1 to 2; $C_t = C_K + C_D$ during dialysis, but $C_t = C_K$ in the absence of dialysis; $C_K$ is the clearance of the kidney; and $C_D$ is the clearance of the dialyzer. (Clearance is defined in Fig. 6.2.6.) Equations 1 and 2 may be readily solved for any pair of initial conditions for any period over which $C_t$ is consant. The result may be expressed formally as

$$\underline{c} = \left( \underline{c}_0 + \underline{\underline{A}}^{-1} \underline{B} \right) e^{\underline{\underline{A}}t} - \underline{\underline{A}}^{-1} B \tag{3}$$

where

$$\underline{c} = \begin{pmatrix} c_1 \\ c_2 \end{pmatrix} = \underline{c}(t - t_0) \tag{4}$$

$t$ = time since start of operation

$\underset{\sim}{c} = \underset{\sim}{c}(t_0)$, the initial concentration vector ($t_0 = 0$ for hemodialysis; $t_0 = t_D$ for the period following dialysis)

$$\underline{\underline{A}} = \begin{pmatrix} -K/V_1 & K/V_1 \\ (K - C_t)/V_2 & -K/V_2 \end{pmatrix}$$

$$\underline{B} = \begin{pmatrix} G/V_1 \\ 0 \end{pmatrix}$$

At the beginning of dialysis it is assumed that $c_1 = c_2$. The course of dialysis can then be determined from Eq. 3 for any known solute concentration in the blood. The calculated conditions at the end of dialysis can then be used as initial conditions for the postdialysis period.

It was found that values of $K$, $V_1$, $V_2$, $G$, and $C_t$ obtained by fitting Eq. 3 to one experimental concentration-time relation* permitted accurate prediction of the course of future dialyses. This is shown by the results summarized in Fig. 6.2.8. Here the lines represent model predictions based on experimental results for the first dialysis (up to 1 day in the figure). The fit is excellent, and similar results can be obtained for other solutes, for

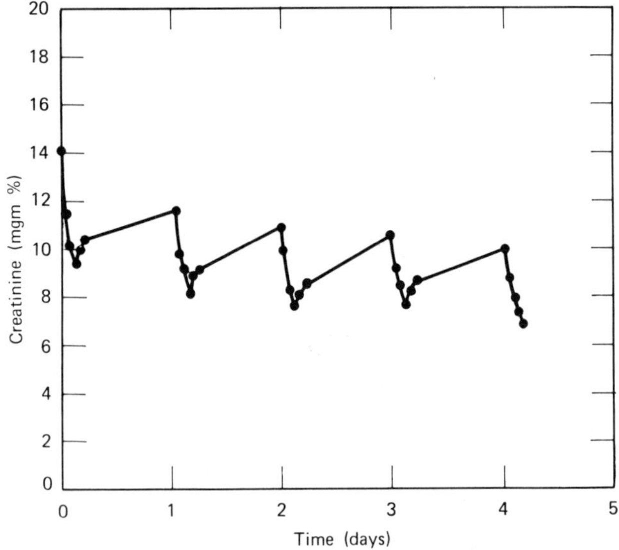


**Figure 6.2.8.** Experimental (dots) and simulated creatinine data (solid line) for patient R. H. At the time of writing it appears that effective creatinine removal is critical to patient well-being. From Bell, Curtiss, and Babb (1965).

* Actually this was done directly from Eqs. 1 and 2 using an analog computer.

example, urea. It can also be seen from the sharp "rebound" in blood solute concentration immediately following dialysis that there is quite an appreciable disequilibrium between body compartments during treatment.

The agreement between simulation and experiment is so good that a word of caution is in order. The model used is an excellent one for predicting the general course of treatment and is thus a great help to the attending physician. However, it is incapable of predicting local solute concentrations with any accuracy and must therefore be used with caution. It is known, for example, that urea leaves the brain relatively slowly during dialysis, and it is widely believed that an appreciable osmotic pressure builds up as a result.

The above approach has since been refined considerably, and a particularly good example is provided by the work of Abbrecht and Prodany (1971). These authors consider the interaction of the patient and the dialyzer in some detail. They also consider the effects of osmotic (diffusional) water transport and of ultrafiltration on the dialyzer. Their analysis provides useful insight into the problem of osmotic imbalance and suggests further investigation of this important and complex problem.

Another defect of this type of model is that the mass-transfer coefficient $K$ is completely empirical. It cannot be predicted *a priori* (although it does appear to correlate well with such accessible parameters as body weight*), and it does not give much insight into the nature of the mass-transfer processes involved. The flow-limited models discussed immediately below offer marked improvement in this respect.

*The Flow-Limited Models of Bischoff and Dedrick.* The extreme rapidity of oxygen transfer in vascular beds, typically completed in less than a second for humans, is in striking contrast with drug elimination, which normally takes place over hours. One would therefore suspect that diffusional limitations are unimportant in determining drug distribution, and this expectation is normally borne out in medical practice. As a result, most studies of drug distribution, or *pharmacokinetics*, are based on the assumption of *flow-limited mass transport*.

A simple representative example, corresponding closely to the Bell model above, is the Bischoff-Dedrick linear two-compartment model, shown schematically in Fig. 6.2.7. Here the body is visualized as two compartments connected by the flowing blood. Each compartment is well mixed, and the blood, circulating at an *effective* volumetric flow rate $Q$, leaves each compartment in equilibrium with it. The solute is assumed to distribute itself linearly between each compartment and the blood. The

* The effect of sudden solute infusion can be treated by a very similar model, as discussed by R. L. Bell, W. T. Gormley, and P. R. Yarnell, *Proc. 7th Annu. Rocky Mt. Bioeng. Symp. and 8th Int. Biomed. Sci. Instrum. Symp.*, Denver, Colo., May 4–6, 1970.

*effective* volume and solute concentrations are $V_1, c_1, V_2, c_2$, respectively.* The system description then takes the form

$$V_1 \frac{dc_1}{dt} = Q(c_2 - c_1) + m_0 g(t) - C_t c_1 \tag{5}$$

$$V_2 \frac{dc_2}{dt} = Q(c_1 - c_2) \tag{6}$$

Here $C_t$ = the total clearance of solute, assumed all from compartment 1.
$m_0$ = total mass of drug, or other solute, absorbed (less than the total dose in case of incomplete absorption).
$g(t)$ = time distribution of drug absorption.

The most commonly assumed forms for $g(t)$ are

1. Pulse input at $t=0$: $g(t)=\delta(t)$.
2. First order absorption:

$$\begin{aligned} g &= 0 \quad (t<0) \\ &= ke^{-k_0 t} \quad (t>0) \end{aligned}$$

or

$$g = ke^{-k_0 t}U(t)$$

where $U(t)$ is the unit step function.

The first-order expression is commonly used as an approximation for the absorption of an oral dose from the gut ("G. I. impulse").

It can be seen that Eqs. 5 and 6 are very similar to Eqs. 1 and 2 used to describe the diffusion-limited model. In fact, except for the physical significance of the terms, these two equations differ only in the form and location of solute input. The evaluation of the parameters in terms of directly measurable quantities, however, proves to be somewhat more straightforward here.

* For simplicity, these have been written in slightly different form from that in Bischoff and Dedrick. The corresponding quantities are as follows:

| This text | Bischoff & Dedrick |
|---|---|
| $c_1$ | $C_1$ |
| $c_2$ | $C_2/K$ |
| $V_1$ | $V_1$ |
| $V_2$ | $KV_2$ |
| $Q$ | $Q_1$ |

For pulse injection at zero time the concentration in compartment 1 is given by

$$c_1 = \frac{m_0}{V_1}\left\{\frac{[r_2+\lambda(1+\alpha)]e^{r_2\tau}-[r_1+\lambda(1+\alpha)]e^{r_1\tau}}{r_2-r_1}\right\} \tag{7}$$

(pulse injection)

where $\tau = kt/(V_1+V_2)$,

$$r_1 = \frac{V_1+V_2}{2V_1}\left\{[1+\lambda(1+\alpha)]+\sqrt{[1+\lambda(1+\alpha)]^2-4\lambda\alpha}\right\}$$

$$r_2 = \frac{V_1+V_2}{2V_1}\left\{[1+\lambda(1+\alpha)]-\sqrt{[1+\lambda(1+\alpha)]^2-4\lambda\alpha}\right\}$$

$$\alpha = V_1/V_2$$

$$\lambda = Q/k$$

Equation 7 and the corresponding result for compartment 2 are shown graphically in Fig. 6.2.9. The limiting curve for $\lambda=\infty$ corresponds to such rapid blood flow that the two compartments are in equilibrium with each other. This "elimination-limiting" behavior, equivalent to a one-compartment model, occurs when the elimination of the drug is slow. Solutions for other input schedules can be obtained from Eq. 7 by superposition.

The relations displayed in Fig. 6.2.9 are useful in scheduling drug dosages, to maintain therapeutically effective levels without exceeding toxic limits. In order to develop such a schedule it is clearly necessary to evaluate the parameters in the basic equations. Satisfactory estimates of effective compartment volumes and blood flow rates can usually be made from readily available physiological parameters, but elimination rates (by both metabolism and excretion) must normally be measured.

The use of physiological, anatomical, and physicochemical data for *a priori* estimation has been discussed* at length by Bischoff and Dedrick and briefly in connection with the two-compartment model in the reference just cited. Closed-form solutions of the type discussed here are generally possible only for drugs distributing linearly between tissues.

* Bischoff, Dedrick, and Zaharko (1970); Dedrick and Bischoff (1968).

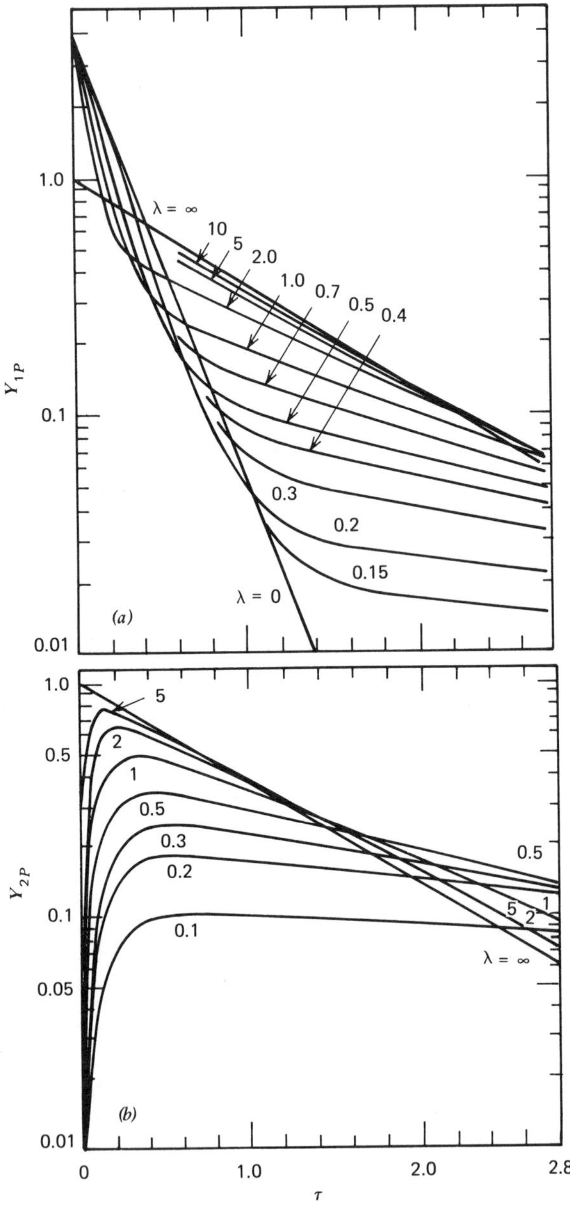


**Figure 6.2.9.** Response to intravenous pulse injection, $\alpha = 0.3$. (*a*) Compartment 1, (*b*) Compartment 2, From Bischoff and Dedrick (1970). Here $Y_{ip} = c_i(V_1 + V_2)/m_0$.

Nonlinear behavior is common, however, particularly as a result of protein binding.†

### EX. 6.2.3. MORE ELABORATE MASS-TRANSFER MODELS

The degree of sophistication needed in mass-transfer modeling depends strongly on the time scale of interest. For times of the order of hours, the two-compartment models discussed above, or even one-compartment approximations, are sufficient to follow the gross distribution of drugs or other solutes. For shorter times, however, it is necessary to use a more detailed picture. Examples of short-time scales include the administration of anesthetics and residence-time distribution studies. One such example has already been given from the work of Nicholes, Warner, and Wood (1964).

A similar one in the chemical engineering literature has been given by Bischoff (1967), who also takes into account tracer dispersion in the major blood vessels. Here individual body regions (see Fig. 6.2.1*b*) are modeled by considering them to consist of discrete subregions or phases. The concentration is considered uniform within a single phase, and each is assumed to be in equilibrium with the effluent blood. Then the mass balance for the system takes the form

$$\frac{d}{dt}\left[c_{iV}\sum_{j=1}^{5}\left(K_j+h_j(c_{iV})\right)V_j\right]=Q_B\left[c_{iV}\left(1+h_B(c_{iV})\right)-c_{iA}\left(1+h_B(c_{iA})\right)\right] \quad (1)$$

where $c_{iA}, c_{iV}$ = concentration of solute $i$ in the arterial and venous blood, respectively; $K_j$ = distribution coefficient for free solute $i$ in subregion $j$, so that free solute concentration in $j$ is $K_j c_{iV}$; $h_j$ = bound solute concentration in subregion $j$; $V_j$ = volume of subregion $j$; $Q_B$ = volumetric blood flow rate. Determination of the parameters appearing in Eq. 1 is discussed by Bischoff and his associates in the previously cited references.

The blood vessels joining the body regions cause both a dispersion of solute and a time delay, and it is suggested by Bischoff that they can be modeled as short chains of mixing tanks, usually two or three to a section.

Discuss the need for models of this type.

† Many drugs and other solutes bind with proteins in reversible complexes. The total concentration in such a case is

$$c_{i,\text{tot}}=c_i[1+h(c_i)]$$

where $h(c_i)$ represents the bound material.

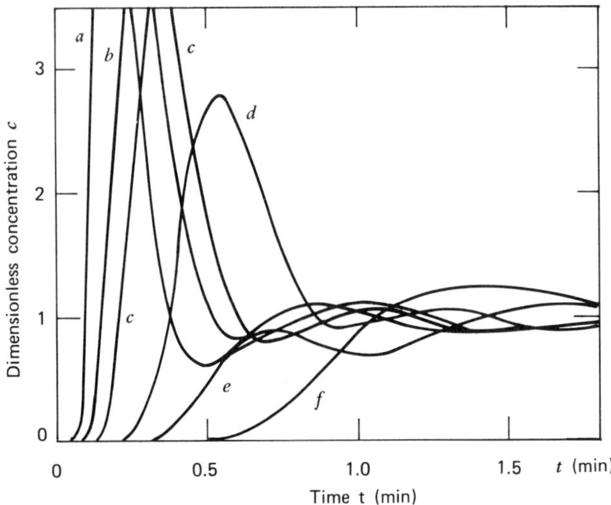


**Figure 6.2.10** Predicted results for 3-sec injection into the right heart of impermeable tracer. (*a*) Pulmonary vein; (*b*) left heart; (*c*) middle aorta; (*d*) femoral artery; (*e*) Middle inferior vena cava; (*f*) lower inferior vena cava. Predictions are based on a compartmentalization such as that pictured in Fig. 6.2.1(*b*) with additional allowance for dispersion in blood vessels. Concentration is normalized so that all organ concentrations approach unity at large time. From Bischoff (1967).

SOLUTION

The properties of this model are shown in Fig. 6.2.10 for 3-sec injection into the right heart of a tracer unable to permeate blood vessel walls. It may be seen here that tracer concentration becomes essentially time independent in all regions shown after a time of the order of a minute.

Transients would persist somewhat longer for a permeable tracer, but the degree of complexity provided by this model is normally not necessary for such relatively slow processes as drug distribution. A particularly simple comparison is available for the distribution of the anticancer agent methotrexate. On the basis of an initial order-of-magnitude based in part on the preliminary experiments represented in Fig. 6.2.11, Bischoff, Dedrick, and Zaharko adopted the moderately simplified pharmacokinetic model shown in Fig. 6.2.12*a*. Here the blood plasma is considered as a single well-mixed compartment, and substantial simplifications are made in the regional models as well. However, clearance of the drug from liver to gut lumen ($k_L$), excretion ($k_F$), and also reabsorption from the gut ($k_0$) had to be considered. The kinetics of these transport processes were approximated by time delays ($D_L$, $D_F$, and $D_L$, respectively). Inspection of

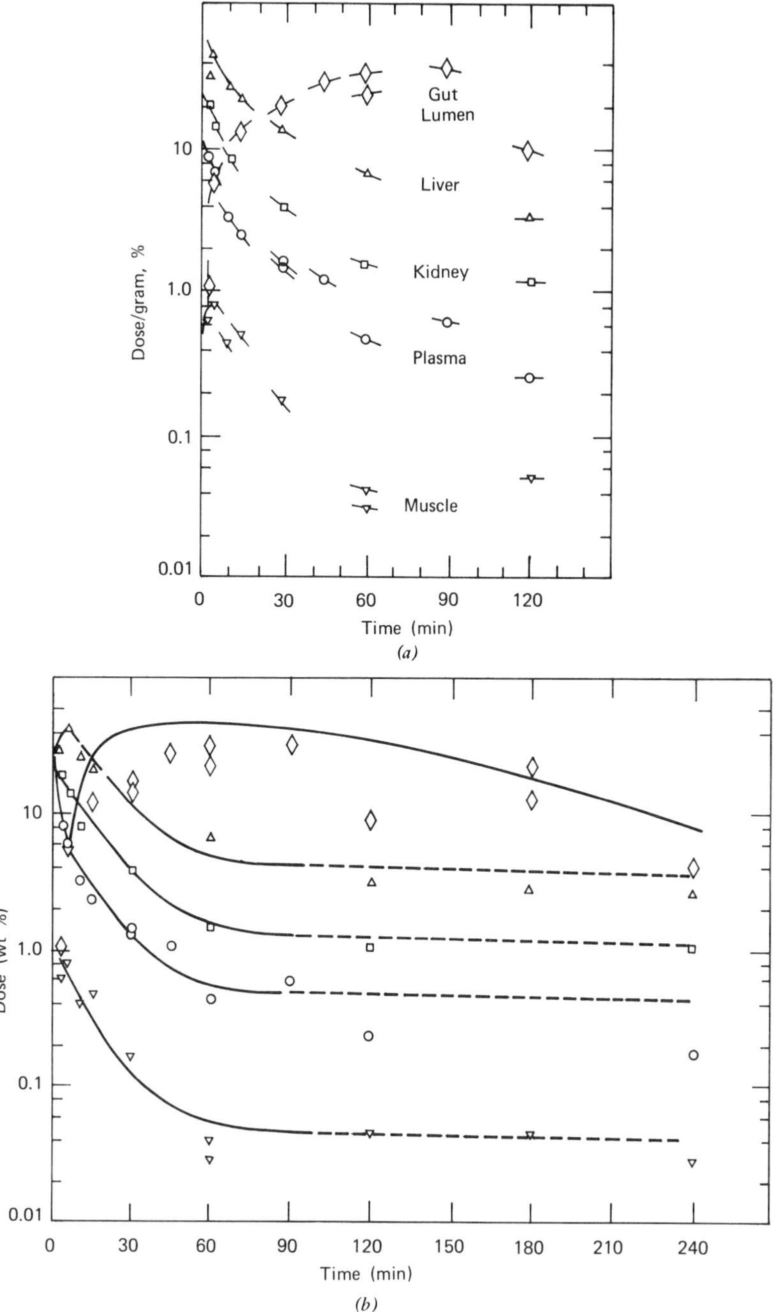


**Figure 6.2.11.** Methotrexate distribution in mice. (*a*) Distribution of MTX in mice following pulse injection in tail vein. (*b*) Comparison between curves predicted by equations and observed data. From Bischoff, Dedrick, and Zaharko (1973).

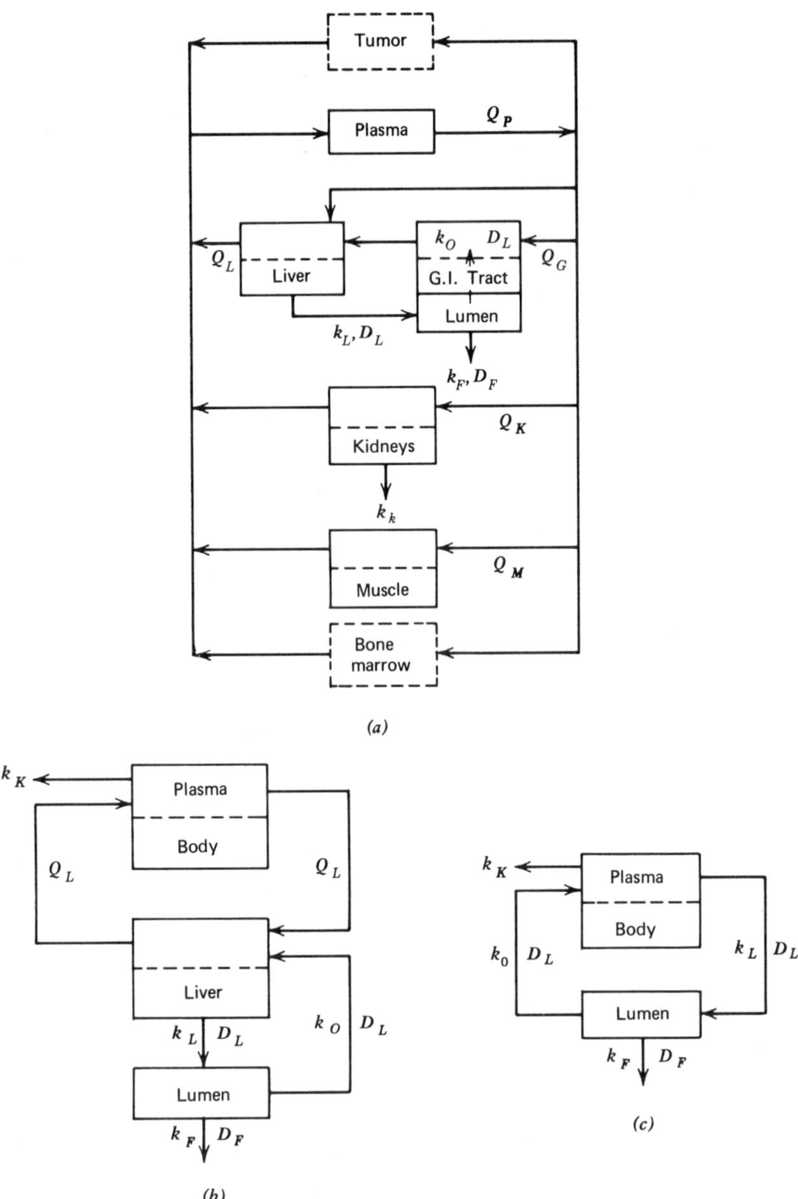


**Figure 6.2.12.** Progressive simplification of pharmacokinetic models for methotrexate (MTX) distribution. From Bischoff, Dedrick, and Zaharko (1970). (*a*) Compartmental model for MTX distribution. The bone marrow and a hypothetical tumor are indicated because these are regions of interest, even though they would normally exert little influence on systemic drug distribution kinetics. (*b*) Two-compartment model plus gut lumen. (*c*) Single-compartment model plus gut lumen.

the resulting mass-balance equations showed, however, that for the behavior observed experimentally for methotrexate (see Fig. 6.2.11*a*), the simple three-compartment model of Fig. 12(*b*) was closely equivalent to the original. Finally it was shown that for times greater than the delay time for liver clearance ($D_L \sim 5$ min), the still simpler model shown in Fig. 6.2.12*c* is sufficient.

Comparison of model predictions with observation is shown in Fig. 6.2.11*b* for a 22-g mouse. It is important to note that no empirical relations parameters are included in the prediction, which is made *a priori* from experimentally measured parameters: organ volumes, equilibrium drug distributions between various tissues, and kinetics of detoxification and elimination.

The results shown in Fig. 6.2.11 are representative of the excellent agreement between prediction and observation obtained for carefully constructed pharmacokinetic models based on the assumption of flow-limited kinetics. This success does, however, depend on careful preliminary investigation of both anatomy and the metabolism of the drug in question. In the case of methotrexate, for example, the simplification in going from Fig. 6.2.12*a* to 6.2.12*b* was made possible by the extensive preliminary investigation, summarized in Fig. 6.2.11*a* and described in more detail in the original paper.

Sometimes, as in setting dosage schedules, lumped-parameter models are sufficient in themselves. In other cases the results of a pharmacokinetic analysis can provide the boundary conditions for more detailed study of a specific region. This is suggested by including compartments for bone marrow and a hypothetical tumor in Fig. 6.2.12*a*, and has since been done by Dedrick, Zaharko, and Lutz (1973).

## 6.3. Distributed-Parameter Models

More elaborate models than the above are needed whenever it is desired to obtain concentration or mass-flux distributions in a specific system. We consider briefly, by way of introduction, three common situations where this type of information is needed:

1. Quantitative description of extracorporeal equipment.
2. Determination of safe operating limits with respect to a toxin or nutrient.
3. Description of complex mass-transfer processes.

These have been chosen here both because of their intrinsic importance and their direct relation to the pharmacokinetic models just discussed.

They also serve to illustrate the need for inventiveness (i.e., a heuristic approach) in modeling. This should already be clear to the perceptive reader from the discussion of Section 6.2; it is touched on more explicitly in the next chapter, which forms the conclusion to this book.

Clearly the macroscopic mass balance, Eq. 6.2.1, contains very little detailed information. A great deal must therefore be supplied by the modeler out of his own judgement and experience. We therefore proceed directly to the examples, where this can be illustrated.

### *EX. 6.3.1. QUANTITATIVE DESCRIPTION OF A BLOOD OXYGENATOR*

For the successful operation of existing mass-transfer devices it is necessary only to relate the performance to such controlling parameters as stream rates and inlet compositions. This information can most easily and reliably be obtained by experiment, perhaps with the aid of dimensional analysis.

If, however, it is hoped to improve performance by some nontrivial modification of configuration or operating procedure, then a deeper understanding is helpful. It is now usually desirable to find out as much as possible about the mechanisms of mass transfer and the flow distribution in the equipment.

The amount of such information which can be obtained, and the most effective means of obtaining it, clearly varies with the type of equipment. However, it is often possible to decompose the overall problem of system description into more manageable components, and the macroscopic mass balance can be very useful for this purpose.

Consider, as a specific example, oxygen transfer in the disc oxygenator shown in Fig. 6.3.1.

SOLUTION

*General discussion.* This device consists essentially of a series of parallel discs mounted on a rotating shaft which is in turn parallel to the axis of the surrounding cylindrical shell. Blood flows in a generally axial direction through the lower half of the shell, countercurrent to an oxygen-rich gas in the upper half. Rotation of the discs brings a continually renewed film of blood, on the disc surface, into contact with the oxygenating gas; oxygen and carbon dioxide are transferred across the blood-gas interface on this film.

In a typical situation the discs turn at 100 rev/min, the blood flow rate is 5 l/min, and the contained blood volume is 2.5 l; the number of discs is

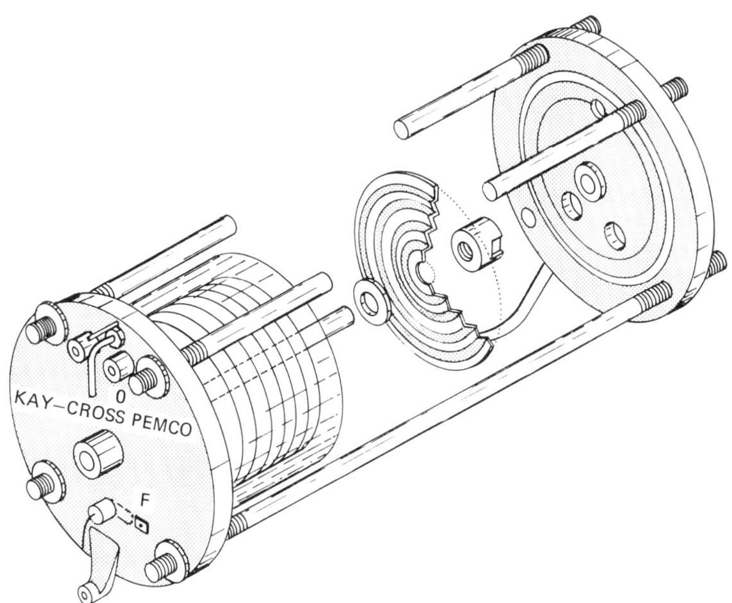


**Figure 6.3.1.** Exploded view of spinning-disc oxygenator.

about 100. We find, from preliminary experiments, that equilibration of blood in the spaces between the discs with that flowing axially outside them is poor, and also that substantial axial dispersion occurs in the outer space.

These conditions and observations suggest modeling the oxygenator in the simplified manner shown in Fig. 6.3.2. Here the system is broken up into units, each representing the blood enclosed between the midplanes of two adjacent discs. These units in turn are subdivided into two blocks, the upper representing the blood between the discs and the lower that between the disc periphery and the cylindrical shell. This must be considered as a first trial model, subject to later modification.

The simplest way to approximate axial dispersion and the mixing resistance between intradisc blood and that flowing outside is via lumped-parameter modeling. We thus define dispersion coefficients $\epsilon_n$ and mass-transfer coefficients $\kappa_n$ (see also Fig. 6.3.2, caption) by

$$\mathcal{W}_n = \left[ Q_B c_n + \epsilon_n \left( \frac{c_n - c_{n+1}}{\delta} \right) \right] (1+m)_{\text{av}} \tag{1}$$

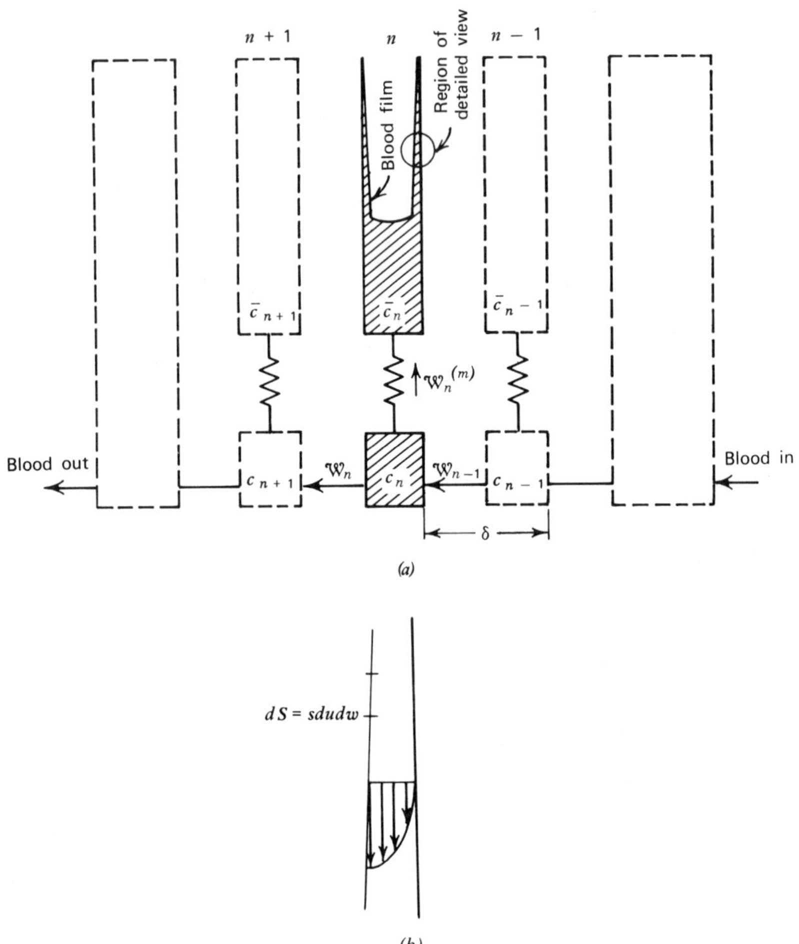


**Figure 6.3.2.** Mass-transfer model of a disc oxygenator. (*a*) Overall view; (*b*) Detailed view. Each unit $n$ represents the blood enclosed between the planes of two adjacent discs. The upper block represents the fluid between the discs, and the lower one that outside the discs (where the axial flow is concentrated). The molar dissolved oxygen concentrations in these regions are denoted by $\bar{c}_n$ and $c_n$, respectively, for unit $n$, and the molar rate at solute leaves unit $n$ is $W_n$. The rate at which solute is transferred to the intradisc region from outside is $W_n^{(m)}$.

and

$$\mathcal{W}_n^{(m)} = \kappa_n(c_n - \bar{c}_n) \tag{2}$$

Here $Q_B$ is the volumetric flow rate of blood, and $(1+m)_{av}$ is the average value of $1+m$ (see Eq. 5.3.4) over the $n$th and $(n+1)$th stages; all other symbols are as defined in Fig. 6.3.2. Equation 1 states that the rate at which the total available oxygen moves from the $n$th to the $(n+1)$th unit depends upon the axial concentration gradient as well as convective transport. Equation 2 states that the rate of solute transfer across the disc periphery is proportional to the difference in solute concentration between the "exterior" and "interior" fluid. The subscript on $\epsilon_n$ and $\kappa_n$ indicates that these quantities may depend upon position in the exchanger. It may be noted that $\epsilon$ and $\kappa$ are measures of flow distribution in the equipment.

Mass transfer across the blood-gas interface in turn may be described in terms of a surface-mean mass-transfer coefficient $(\bar{k}_c)_m$ by

$$\mathcal{W}_n^{(i)} = A_n(\bar{k}_c)_m(\bar{c}_n - \bar{c}_n^*) \tag{3}$$

where $\mathcal{W}_n^{(i)}$ = molar rate of solute transfer across the interface (note that at steady state $\mathcal{W}_n^{(i)} = \mathcal{W}_n^{(m)}$),
$A_n$ = interfacial area in the $n$th unit,
$\bar{c}_n^*$ = molar dissolved oxygen concentration on the blood side of the interface.

Under most conditions of interest $\bar{c}_n$ can be assumed position independent within any stage; this is because the fractional change of $\bar{c}_n$ per stage is small in a 100-stage exchanger. In addition, the total change in $\bar{c}_n$ is small compared to $\bar{c}_n - \bar{c}_n^*$ when (as normally) the oxygen tension in the gas phase is of the order of one atmosphere.

We may next note that the interfacial mass-transfer process meets all the requirements† of the free-surface boundary-layer model introduced in Sect. 5.3. Then from Eq. 5.3.17 we may write

$$(\bar{k}_c)_m \equiv \frac{\overline{\mathrm{Nu}}_m \mathcal{D}_{O_2m}}{D}$$

$$= -\frac{\Pi'(0)}{A_n}\sqrt{\mathcal{D}_{O_2m}}\int_{A_n}\frac{1}{t}\sqrt{\int_{-\infty}^{t}\left(\frac{s'}{s}\right)^2 dt'}\, dS \tag{4}$$

† Note from 5.3.14, with $\mathcal{D}_{O_2m} \sim 1.4\times10^{-5}\mathrm{cm}^2/\mathrm{sec}$ and $t_{exp} \sim 0.3\mathrm{sec}$, that maximum boundary layer thickness for a nonstretching surface is $\delta \sim \sqrt{4\times1.4\times10^{-5}\times0.3}\ cm \doteq 41\ \mu$. This is small relative to any probable film thickness.

Here $t_{loc}$ is the age of any surface element;

$$dS = s\,du\,dw \tag{5}$$

with $s'/s$ equal to the ratio of element surface area at time $t'$ to that at $t=t$; $A_n$ is the total mass-transfer area in the $n$th stage. Evaluation of the integral in Eq. 4 can be difficult, but it is most important to note that it depends only on the velocity profiles within the draining film. The dimensionless concentration gradient $\Pi'(0)$ is obtained from Eq. 5.3.18.

*We then find that the system description requires only a knowledge of flow conditions, entering blood composition, and oxygen tension in the gas.* The velocity profile data need only be sufficient to characterize (1) the dispersion and mixing coefficients ($\epsilon_n$ and $\kappa_n$, respectively) and (2) the surface velocities on the draining blood films. Furthermore, as shown immediately below, considerable useful insight can be gained even without complete flow data.

*Assessing oxygenator potential.* The performance of the oxygenator shown in Fig. 6.3.1 depends in complex ways on both disc geometry and flow path, and its performance in the configuration shown is not entirely satisfactory: it would be highly desirable both to reduce the surface area (hence the blood damage) and the blood volume—if this could be done without decreasing the mass-transfer effectiveness. We seek to determine here if the flow pattern can be improved and whether there is any hope for substantial improvement. Two possibilities are of particular interest:

1. Improve contacting in the present axial flow system by decreasing axial dispersion (lowering $\epsilon$) or improving lateral mixing (raising $\kappa$).
2. Change to cross flow of blood, that is, flow transverse to the shell axis, to improve penetration of blood into the intradisc area.

Before making physical modifications of the exchanger it is clearly desirable to make any possible limiting calculations and to compare the limiting predictions with present performance. This has already been done by Dindorf and Lightfoot (1973); we summarize here their key results.

We begin by assuming rigid rotation of the film surface on the discs, so that there is no surface stretching,* and $s/s_0$ is always unity. Equation 4 then takes the simple form

$$\left(\bar{k}_c\right)_m = -\frac{\Pi'(0)}{A_n}\sqrt{\mathfrak{D}_{O_2m}}\int_{A_n}\frac{dS}{\sqrt{t}} \tag{5}$$

* This assumption has been shown to be highly accurate by Dindorf, Atiemo, and Lightfoot (manuscript in preparation).

where $t$ is the time since formation of area element $dS$. For a rigidly rotating film surface and an oxygenator just half full of blood we may idealize the behavior and write

$$dA_n = \alpha R^2 \, d\theta \tag{6}$$

$$t = \frac{T\theta}{2\pi}$$

where

$$\alpha = \frac{A_n}{\pi R^2} \tag{7}$$

is a correction factor[†] for departures from a flat film surface; $T$ is the period of rotation. Equation 5 may then be directly integrated to give

$$\left(\bar{k}_c\right)_m = -2\Pi'(0)\sqrt{\frac{2\mathfrak{D}_{O_2m}}{T}} \tag{8}$$

The area-mean mass-transfer coefficient $(\bar{k}_c)_m$ on any disc, like $\Pi'(0)$, depends only on local bulk oxygen tensions. Since the dependence of $\Pi'(0)$ on blood and gas oxygen tensions is known (see Dindorf, Lightfoot, and Solen, 1971), one can calculate the oxygenator performance for any given values of $\epsilon_n$ and $\kappa_n$. Of particular interest are the limiting values for

Complete back mixing: $\epsilon_n = \infty$; $\kappa_n = \infty$

Ideal plug flow: $\epsilon_n = 0$; $\kappa_n = \infty$

Note that in both cases good mixing is assumed between the "interior" and "exterior" blood.

For back-mixing conditions the blood composition is uniform at the outlet (arterial) value, and $k_c$ is independent of axial position. Then the amount of oxygen transferred is simply

$$\mathcal{W}_n = \left(\bar{k}_c\right)_m A_{\text{tot}}(c_e^* - c_e) \quad \text{(complete back mixing)} \tag{9}$$

$$= Q_B[c_e(1+m)_e - c_i(1+m)_i] \tag{10}$$

[†] In the device shown in Fig. 6.3.1 the discs are crimped, with each (conic) subsurface at an angle of 60° to the plane of the disc. Thus $\alpha$ is approximately 2 for this system. Note that one must neglect the diameter of the supporting shaft and assume the surface of the blood pool to be flat to obtain Eqs. 6 and 7. The cylindrical coordinates $\theta$ and $r$ are relative to the axis of rotation, and $R$ is the disc radius.

Here the subscripts $e$ and $i$ refer to exit and inlet conditions for the oxygenator. Once again the asterisk indicates the interfacial oxygen concentration; for a completely back-mixed reactor the subscript $e$ on $c^*$ is not really necessary.

For plug flow, bulk oxygen concentrations vary with position—and $k_c$ does also because of its dependence on $m$, the slope of the saturation curve. In principle one must calculate the contribution of each disc and sum over the total number; in practice, however, the change in concentration per disc is small, and one can assume it to vary continuously with axial position. We may then use the conventional description† for a countercurrent mass-transfer apparatus. We may thus write for the mass of oxygen $d\mathcal{W}$ transferred across the mass-transfer area $dA$ in a differential length of the oxygenator

$$d\mathcal{W} = \left(\bar{k}_c\right)_m (c^* - c)\,dA \tag{11}$$

$$= Q_B(1+m)\,dc \tag{12}$$

We may then write

$$A_{\text{tot}} = Q_B \int_{c_i}^{c_e} \frac{\left(\bar{k}_c\right)_m (1+m)\,dc}{(c^* - c)} \quad \text{(Ideal plug flow)} \tag{13}$$

Here $A_{\text{tot}}$ is the total mass-transfer surface, and $m$ is the local slope of the saturation curve (i.e., corresponding to $c$ and $c^*$); $c_i$ and $c_e$ are again the inlet and exit concentrations of dissolved oxygen.

A comparison of Eqs. 10 and 13 with experimental data is shown in Table 6.3.1 for the oxygenator of Fig. 6.3.1. It can be seen here that the actual performance falls considerably short of that predicted for plug flow but not so far from that for back mixing. It should be noted that stretching of the blood-gas interface, as a result of film drainage, should increase the mass-transfer effectiveness above these predictions. There should therefore be quite a potential for improvement in the plug-flow configuration. Cross flow, on the other hand, is not so attractive a possibility.

Studies of flow distribution by Dindorf and Lightfoot showed that there was indeed both back mixing and resistance to lateral blood distribution. This maldistribution becomes much more harmful as the blood volume is decreased. In addition, oxygen-rich blood returning on the disc surface from contact with the gas phase mixes slowly with the rest of the intradisc blood.

† As in *Tr. Ph.*, Ex. 22.5-2.

**Table 6.3.1.** Size Requirements for a Disc Oxygenator [a]

| | | |
|---|---|---|
| Performance specifications: | | |
| $Q_B = 2.6$ l/min | | |
| Disc speed = 120 rev/min | | |
| Oxygen tension = 40 mm Hg (venous) | | |
| = 90 mm Hg (arterial) | | |
| Required length: | | |
| | Observed: | 40.3 cm |
| | Predicted: | 22.5 cm (ideal plug flow) |
| | | 32.4 cm (complete back mixing) |

[a] Representative data of Dindorf and Lightfoot (1973).

These results, along with the effective rigidity of the blood-gas interface, seriously limit the effectiveness of the device. There is therefore very substantial room for improvement. Obtaining these improvements is a major task in equipment design, just starting at the time of writing—and outside the scope of the present text.

### *EX. 6.3.2. RESPIRATORY GAS EXCHANGE IN THE LUNGS*

The mass-transfer system providing exchange of respiratory gases with the environment in intact mammals is both more complex than the exchanger of the last example and better designed for the job. Essentially all of the mass-transfer surface and the bulk of the manifolds are in the lungs themselves, pictured schematically in Figs. 6.3.3 through 6.3.5. Discuss this system from the standpoint of mass-transfer effectiveness.

SOLUTION

Mass transfer in the lungs is complicated by the reciprocating nature of the flow and the small fractional replacement of air per breath. We must therefore consider the problem of *ventilation* in addition to diffusional processes; this corresponds to the determination of the circulation patterns in the previous example.

As indicated in Fig. 6.3.3 (taken from Crandall and Flumerfelt, 1967), much of the gas space in the lung is taken up by ducts with negligible oxygenation capacity: the trachea, bronchi, and bronchioles. Since these are interposed between the mass-transfer surfaces of the alveolar ducts and sacs, they constitute a dead space tending to decrease the effectiveness of ventilation. In a representative adult man under resting conditions the pulmonary gas volume is distributed approximately as follows:

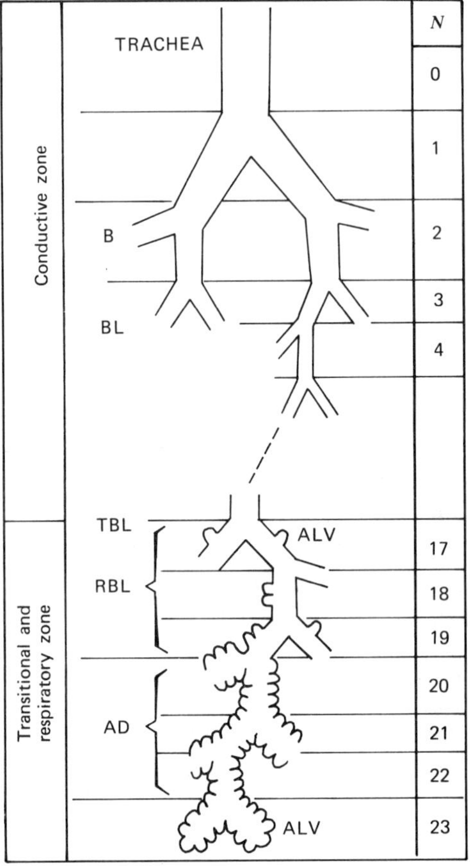


**Figure 6.3.3.** Anatomical segements of the tracheobronchial tree. $N$ = generation number; B = bronchi; BL = bronchioles; TBL = terminal bronchioles; RBL = respiratory bronchioles; AD = alveolar ducts; ALV = alveoli. Redrawn from P. Abbrecht, unpublished notes, U. of Mich., Ann Arbor, Mich. See also Fig. 6.3.5.

| | |
|---|---|
| Dead space: | 150 cm$^3$ |
| Alveolar space: | |
| After exhalation: | 2150 cm$^3$ |
| Tidal volume: | 500 cm$^3$ |
| Average total: | 2400 cm$^3$ |

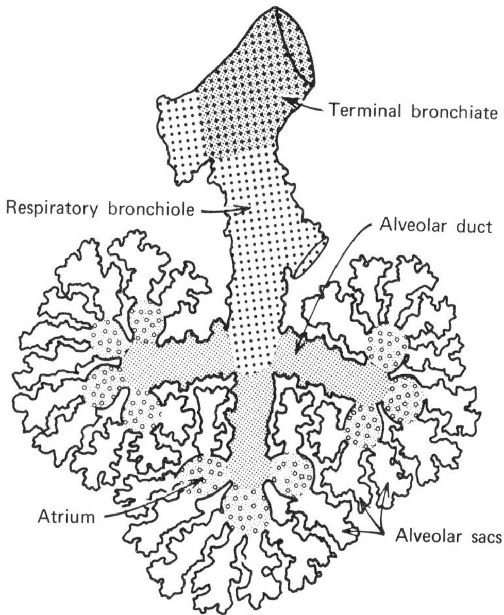


**Figure 6.3.4.** The respiratory lobule. From Miller, (1947).

Of the approximately 500 $cm^3$ inhaled with each breath, only 350 $cm^3$ is fresh air, since the dead space is filled with alveolar air on each exhalation. Thus the fractional replacement of alveolar gas with outside air per breath is only

$$\frac{350}{2400} \doteq 0.15$$

or a little more than one part in seven. This seemingly inefficient ventilation is beneficial in stabilizing respiratory-system control. The body's respiratory control mechanisms can, for example, be upset by hyperventilation. From the standpoint of our present analysis this low rate of exchange means that the average gas concentration in the alveolar space may be considered time independent. Representative gas compositions are:

| | Humidified Atmospheric Air | Alveolar Air | Exhaled Air |
|---|---|---|---|
| $O_2$ | 19.7% | 13.6% | 15.7% |
| $CO_2$ | 0.04% | 5.3% | 3.6% |

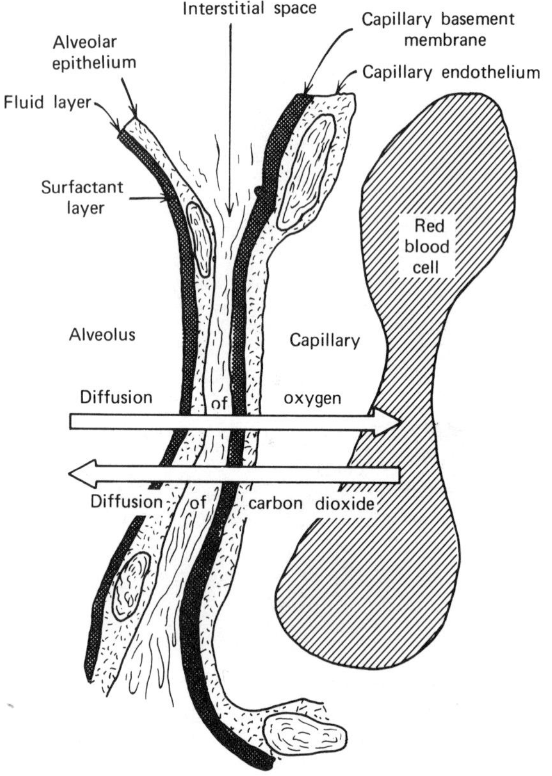


**Figure 6.3.5.** Ultrastructure of the respiratory membrane. From Guyton (1966).

Note that alveolar air is essentially saturated with water vapor. The normal respiration rate is about 12 breaths per minute, so the *alveolar* ventilation rate is about

$$12 \times 350 \text{ cm}^3/\text{min} = 4200 \text{ cm}^3/\text{min}$$

and the fractional rate of replacement of alveolar air is about

$$\left(\frac{350}{2400}\right)(12) = 1.8/\text{min or } 0.03/\text{sec}$$

The pressures required to maintain this air flow in and out are only about $\pm 3$ mm Hg at rest (but up to $\pm 100$ mm Hg during extreme exertion, when both the respiration rate and the tidal volume increase strongly).

The tidal volume is obtained by expansion and compression of the alveoli, shown schematically in Fig. 6.3.4. In addition, it is across the walls of the alveoli that oxygen diffuses to the pulmonary capillaries. There are about 1 million alveolar ducts leading to about 300 million alveolar sacs or alveoli. The alveoli have diameters of about 75 to 300 $\mu$, and the total alveolar surface is about 70 $m^2$—about 40 times the external surface area of the body.

We now turn our attention to diffusion in the pulmonary gas and note that the alveoli are of primary interest here. To obtain a rough approximation to the alveolar response to changes in environment we consider the response of an initially uniform gaseous sphere to a step change in surface oxygen concentration. The Fick number

$$N_{Fi} \equiv \frac{\mathfrak{D}_{O_2m} t}{R^2}$$

is about 0.5 when the approach of the gas to the new equilibrium is 99% complete (see *Tr. Ph.*, Fig. 11.1-3). For

$$\mathfrak{D}_{O_2m} \doteq 0.15\,\text{cm}^2/\text{sec}$$

and

$$R \doteq 0.015 \text{ cm } (150\ \mu)$$

the corresponding time is then

$$t_{0.99} = \frac{(0.5)(2.25 \times 10^{-4})}{(0.15)} \text{ sec}$$

$$= 7.5 \times 10^{-4} \text{ sec}$$

or slightly less than a millisecond. This response is so rapid that one can neglect the gas-phase mass-transfer resistance, just as in the previous example, and consider alveolar composition to be uniform.*

We now turn our attention to blood flow in the lobules, and in particular to the alveolar capillary networks. These capillary networks form a nearly continuous sheet of blood around the alveoli and thus present a highly favorable configuration for gas exchange.† The diffusion path of the exchanging gases is shown in Fig. 6.3.5. Despite its complexity, the region separating alveolar blood and gas is thin—sometimes as little as 0.1 $\mu$, and

* There may, however, be some larger-scale variations of gas concentration, for example, between lobes, due to variations in the gas replacement rate. This aspect of ventilation remains to be thoroughly investigated.

† This blood film is extremely thin. The surface area of the capillary network is that of a 20×30 foot room, and the total blood volume in the capillaries is of the order of 75 $cm^3$. When one thinks of this small amount of blood spread over so large a surface, it is easy to understand why pulmonary gas exchange is so fast (Guyton).

in essentially all areas less than 1 $\mu$. Although extremely permeable to the exchanging gases and such lipid-soluble solutes as alcohol, the respiratory membrane is almost impermeable to sodium and glucose. In general it shows selective permeabilities much like those of the red-cell membrane.

We are now ready to attempt a semiquantitative description of oxygen transfer, and to do this we use many of the approximations of Crandall and Flumerfelt (1967). Specifically, we shall assume that

1. The blood moves in plug flow through a uniform bed of identical parallel channels in contact with alveolar gas of constant composition.
2. The overall mass-transfer resistance can be expressed in terms of a single lumped resistance.

For these assumptions the equations of continuity and motion reduce to

$v = v(t)$

where $v = Q(t)/S$

and $Q$ = volumetric flow rate of blood to the capillaries

$S$ = total capillary cross-sectional area.

The continuity equation for oxygen can be integrated over the capillary cross section to obtain, for any element of moving blood

$$\frac{D}{Dt}\left(c_{O_2} + c_{HbO_2}\right) = \left(\bar{S}K_c\right)\left(c^*_{O_2} - c_{O_2}\right) \tag{1}$$

where $c_{O_2}, c_{HbO_2}$ = bulk oxygen and oxyhemoglobin concentration in the capillary blood,

$c^*_{O_2}$ = the concentration of oxygen in blood equilibrated with the capillary gas,

$\bar{S}$ = specific surface of the capillaries (= $R/2$, where $R$ is the capillary radius),

$K_c$ = overall mass-transfer coefficient.

This equation is to be solved with the assumption of local equilibrium between $O_2$ and hemoglobin and the initial condition

$$\text{At } t = 0, \quad c_{O_2} = c^v_{O_2}$$

where $c^v_{O_2}$ is the oxygen concentration of $O_2$ in the venous blood, and $t$ is time measured from the entry of blood into the alveolar capillaries. We may integrate formally to obtain

$$\left(\bar{S}K_c\right)t = \int_{c^v_{O_2}}^{c_{O_2}(t)} \frac{d\left(c_{O_2} + c_{HbO_2}\right)}{\left(c^*_{O_2} - c_{O_2}\right)} \tag{2}$$

which provides an implicit expression for $c_{O_2}(t)$. This procedure is straightforward and requires no detailed discussion here. The value of $c^*_{O_2}$ may be obtained from the sources listed in the previous chapter. It remains to determine the effect of variable holdup time of the blood, which is in pulsatile flow.

In general one may expect to find a range of residence times for individual volume elements of blood, and it is then necessary to average results for these elements properly. This problem is considered in detail by Crandall and Flumerfelt, but it is not normally of critical importance. This is because the blood residence time in the alveolar capillaries is normally very close to the period of the heart beat. For this situation (which appears to be the only one investigated to date), all blood elements have the same residence time.

The result of integrating Eq. 2 is shown in Fig. 6.3.6*a* for† $(\bar{S}K_c) = 192$ sec$^{-1}$, and the results of a corresponding analysis for $CO_2$ removal are shown in Fig. 6.3.6*b*. It can be seen from these figures, which are widely believed to be reasonable, that healthy human beings have a comfortable "safety factor" under the unstressed conditions shown. Conversely, by the time one feels distress under these resting conditions a very considerable drop in mass-transfer effectiveness has occurred; it is therefore important to anticipate such deterioration, as by a sound preventive-medicine program, rather than to wait for alarming symptoms.

Considerable insight into the effect of progressive deterioration of diffusionally active tissue can be obtained very simply by examination of Fig. 6.3.6. One may see immediately that a given reduction in effective exposure time—resulting, for example, from a reduction in mass transfer area—has a much smaller effect on the exit-gas tension when the exposure time is long. The clinical effect of deterioration thus becomes increasingly serious as the degree of deterioration increases. Thus even a linear drop of mass-transfer area with time would manifest itself as an accelerating rate of change in arterial gas tensions. Such behavior is characteristic of mass-transfer apparatus, and our biological exchangers are no exception.

The foregoing analysis thus illustrates a frequently encountered medical problem as well as providing a semiquantitative description of pulmonary gas exchange. It chief value is in fact to provide insight; considerable further work is indicated to provide a truly reliable quantitative model of the lung.

### *EX. 6.3.3. MASS TRANSFER IN THE MAMMALIAN KIDNEY*

The very success of lumped-parameter models for correlating system performance with operating conditions is their major weakness: one can-

† Compared to the value of 245 sec$^{-1}$ used by Crandall and Flumerfelt.

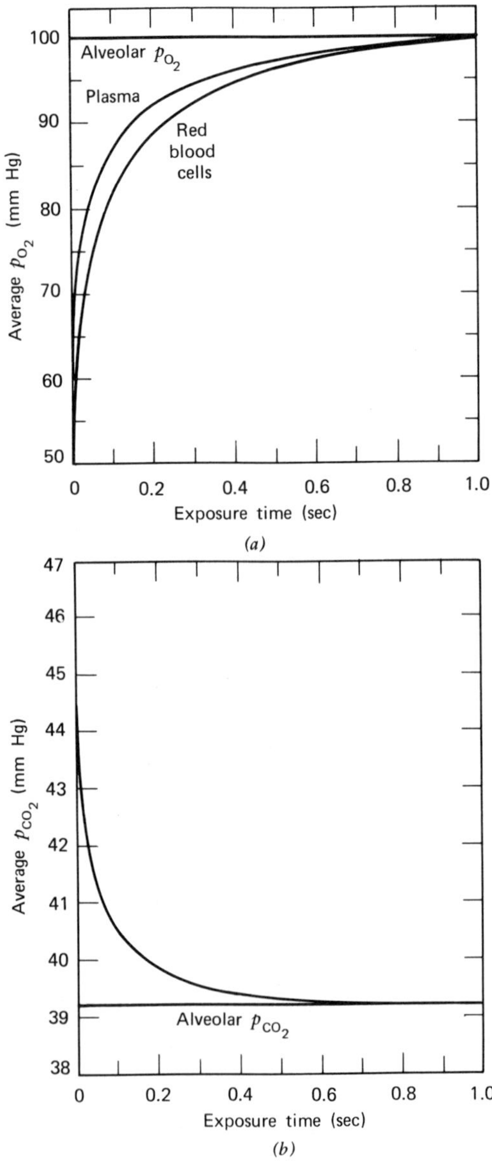


**Figure 6.3.6.** Calculated alveolar gas-exchange effectiveness (adapted from Crandall and Flumerfelt, 1967). Shown here are (*a*) oxygen and (*b*) carbon dioxide tensions at the distal (downstream) end of alveolar networks as a function of exposure time (which is nearly equivalent to effective mass-transfer area).

not easily get much detailed insight into the nature of a complex system from such correlations. This is particularly true for such an organ as the kidney, in which a very wide variety of complex mass transfer processes take place.

Discuss the behavior of the kidney in the light of the entire above text material.

SOLUTION

*General comments.* The kidney performs a number of important functions, but only one of these, the formation of urine, is vital to survival. By means of this process the kidney eliminates soluble end products of body metabolism and controls the concentration of most of the constituents of body fluids. Each kidney is an aggregate of about 1 million nephrons acting in parallel, and for our purposes it is necessary only to consider a single representative nephron.

One of these, a *juxtamedullary nephron*, is shown schematically in Fig. 6.3.7. This nephron differs from the other two types, the outer and the midcortical, primarily in having a longer loop of Henle. The functioning of the nephrons is discussed in detail in standard physiology texts, and we therefore content ourselves with a brief summary.

About one-fifth of the cardiac output, or 1200 ml/min, flows through the kidneys under normal conditions, and of this, about 125 ml/min (or 180 l/day) is ultrafiltered in the glomerulus. The *glomerular filtrate*, which contains all the constituents of plasma smaller than the major plasma proteins,* then passes through the proximal and distal tubules and the collecting duct, from which latter it emerges as urine. During its passage about 99% of the water and most of the desired constituents of the filtrate are reabsorbed, and the undesired solutes are left behind. We now consider the details of this very complex operation briefly.

The glomerular filtration is controlled by the *basement membrane* of the glomerular capillaries, which appears to be a microporous sheet. The endothelial cells on either side are much thicker than the basement membrane, but there appear to be relatively large openings between them. They are the biological counterparts of the mechanical supports required in extracorporeal ultrafiltration. The driving force for this filtration is the difference between the rather high capillary pressure (averaging about 70 mm Hg) and the *colloid osmotic pressure* resulting from the unfilterable proteins (about 32 mm Hg) plus the pressure in Bowman's capsule (about 14 mm). The net average driving force is thus about 24 mm Hg, and the filtration rate is thought to be given by

$$v_w = \kappa[p_c - (p_B + \pi^c)] \tag{1}$$

* And an amount of these proteins as well.

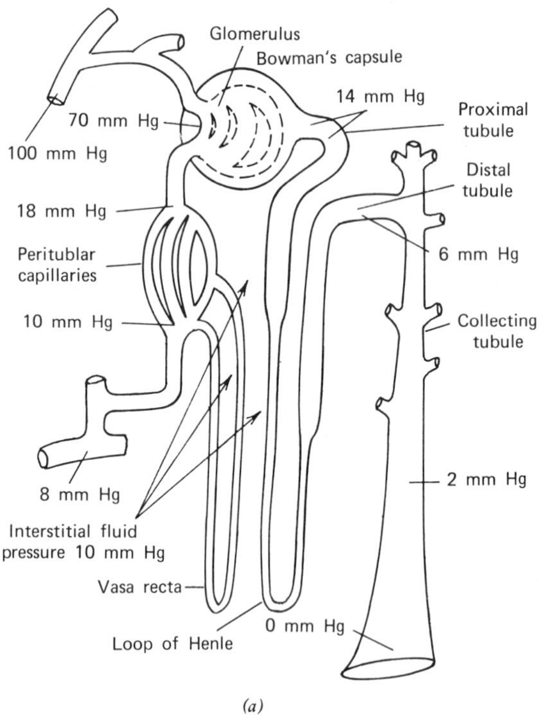


(a)

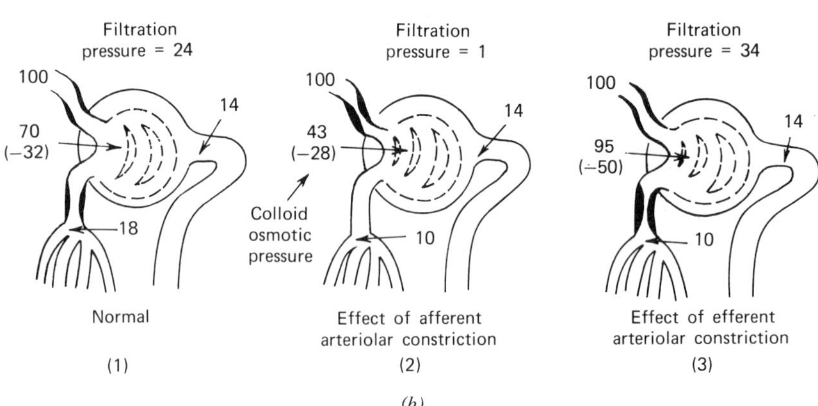


(b)

**Figure 6.3.7.** Salient anatomical and physiological characteristics of a juxtamedullary nephron. (*a*) Pressure at different points in the vessels and tubules of the functional nephron and in the interstitial fluid. Redrawn from Guyton (1966). (*b*) Physiological control of filtration rate: (1) normal pressures at different points in the nephron, and the normal filtration pressure; (2) effect of afferent arteriolar contriction on pressures in the nephron and on filtration pressure; (3) effect of efferent arteriolar constriction on pressures in the nephron and on filtration pressure. From Guyton (1966).

Here $v_w$ is the filtration rate expressed as interfacial water velocity, $p$ is the hydrostatic pressure and $\pi^c$ is the colloid osmotic pressure, $\kappa$ is the membrane hydraulic permeability, and the subscripts $c$ and $B$ refer to the capillary interiors and Bowman's capsule, respectively. The simplicity of Eq. 1 results from the large molecular-weight gap between the serum proteins (molecular weight$>$65,000) and all other solutes of appreciable osmotic activity (the bulk of these are inorganic ions with equivalent weights below $10^2$). The proteins are almost completely excluded, while the small solutes behave essentially like water (see the analysis in Section 3.2). Concentration polarization in the glomerulus and tubules seems not to have been analyzed quantitatively; it is, however, generally thought to be small.

Glomerular filtration is qualitatively similar to that taking place in the arterial ends of all body capillaries, but the total flow is about 7 times that of all other capillary beds of the body combined. The bulk of this fluid is ultimately reabsorbed in the peritubular capillaries, as in the venous ends of other capillary beds. The peritubular capillaries are, however, much more porous than other body capillaries, to accomodate the extremely large flows through their walls. Glomerular filtration rates are regulated quite effectively to respond to body needs, but we do not discuss this subject here.

As the glomerular filtrate passes through the tubules and collecting duct, solutes are selectively reabsorbed or secreted by the tubular epithelium while water moves in response to osmotic gradients. Reabsorption occurs to a much greater extent than secretion, but the latter is especially important in control of $K^+$, $H^+$, and a few other species. Since more than 99% of the water in the glomerular filtrate is normally reabsorbed (see Fig. 6.3.8), unreabsorbed constituents are concentrated more than 100-fold. On the other hand, solutes such as glucose and amino acids are reabsorbed so effectively that their concentration in the urine is very small (See Fig. 6.3.9.)

The tubular absorption mechanisms are the same as in other regions of the body and include both active and passive transport. Active transport of $Na^+$ is illustrated in Fig. 6.3.10. The sodium ions are actively transported across the outer wall of the tubular cell into the peritubular fluid, and the tubular cells are richly supplied with mitochondria to supply the necessary ATP for this energy-consuming process. Transport from the lumen of the tubule into the tubular cells is passive and is aided by a favorable potential difference. Other substances actively transported include glucose, amino acids, and calcium, phosphate, and urate ions. In addition some substances, especially $H^+$ and $K^+$, are secreted actively into the tubule

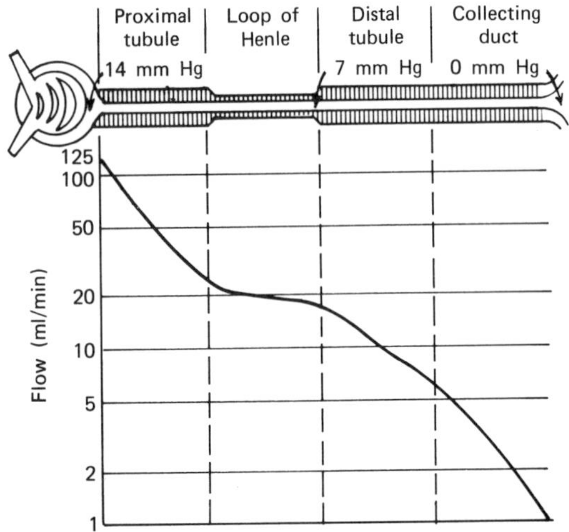


**Figure 6.3.8.** Transport of water at different points in the tubular system, measured by the volume flow of fluid in the tubular system. Note that the flow is plotted on a semilogarithmic scale, illustrating the tremendous decrease of flow with increasing distance from the glomerulus. From Guyton (1966).

lumens. This process is similar to active reabsorption except that it occurs in the opposite direction.

Water is transported passively by osmosis out of the tubules. This in turn concentrates other solutes such as urea, which then also tend to diffuse into the peritubular space. However, the permeability of the tubular membranes is much less for urea than for water, so that much less urea is reabsorbed. The tubular permeability for other solutes, for example creatinine, sucrose, mannitol, and inulin, is effectively zero.

Anions such as $Cl^-$ appear to be transported passively, and $HCO_3^-$ diffuses through the cell membranes as free $CO_2$. The carbon dioxide is formed by the reaction sequence

$$HCO_3^- + H^+ \rightleftarrows H_2CO_3$$

$$H_2CO_3 \rightleftarrows CO_2(aq) + H_2O$$

Therefore the transport of $HCO_3^-$ in as $CO_2$ has the effect of transporting $H^+$ as well. The bicarbonate ion itself does not readily permeate tubule membranes and is therefore not effectively reabsorbed in the absence of $H^+$ (i.e., under alkaline conditions).

Protein reabsorption is a very interesting special case. About 30 g of protein pass into the glomerular filtrate every day, and must be reabsorbed to avoid an excessive metabolic drain on the body. Since protein molecules

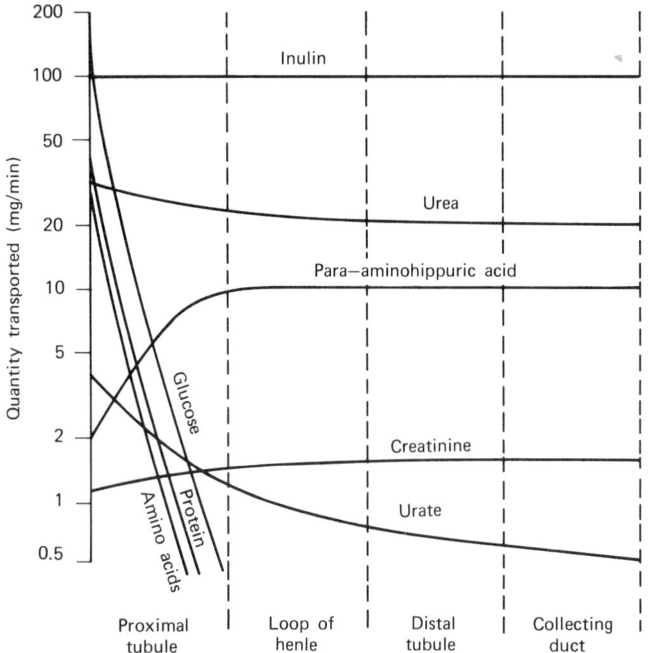


**Figure 6.3.9.** Reabsorption and secretion in the tubular system. Note the reabsorption of the nutritionally important substances in the proximal tubules, and the poor reabsorption of the metabolic end products in all segments of the tubules. Note also the total absence of reabsorption of inulin and the secretion of para-aminohippuric acid into the proximal tubules. From Guyton (1966).

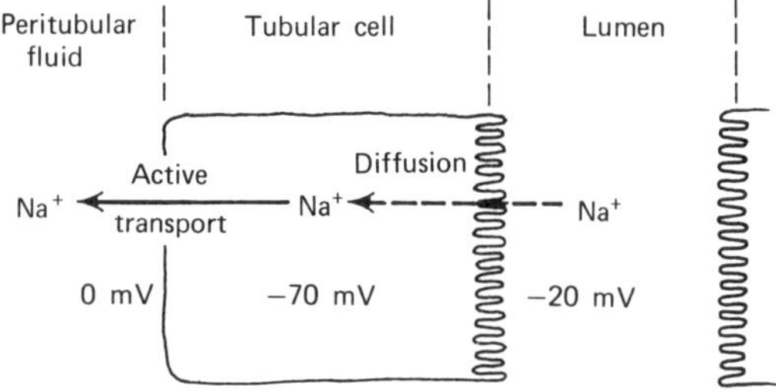


**Figure 6.3.10** Transport of sodium from the tubular lumen to the periturbular fluid. From Guyton (1966).

are too big to diffuse across the tubule membranes they are reabsorbed by *pinocytosis*. That is, they are adsorbed on the *brush border* of the proximal tubule epithelium. This membrane then invaginates to the interior of the cell. Once inside the cell the protein is broken down enzymatically into its constituent amino acids. These in turn are actively transported across the exterior tubule wall to the peritubular fluid and are finally absorbed into the peritubular capillaries.

Figures 6.3.8 and 6.3.9 show that most of the water and desired metabolites of the glomerular filtrate are removed in the proximal tubule. In addition we see that almost no water is removed in the loop of Henle, which has concentration of urine as its primary function.

Fig. 6.3.11 shows* that the reabsorption process for glucose is very effective under normal physiological conditions, but that the transport mechanism can be saturated by an overload. Thus a detectable amount of glucose appears in the urine when the tubular load is about twice normal (threshold), and the maximum rate of reabsorption is of the order of three times the normal load (*transport maximum*). Similar transport maxima exist for many other metabolites, as shown in the accompanying table, but *not* for sodium ions.

It is of interest to note that for any one nephron the glucose loss curve of Fig. 6.3.11 (suitably scaled) would follow the dashed line of the figure. That is, all glucose entering the nephron would be completely recovered up to its transport maximum, and all above this amount would be lost. The presence of a threshold below the transport maximum and the gradual approach to the high-concentration asymptote (or "splay") results from the variability of flow through and reabsorption in the individual nephrons.

*A specific example.* To date no serious attempt seems to have been made to describe kidney function on the basis of the microscopic equations of change. Rather, available descriptions are either essentially stochastic in nature or based on linearized rate expressions. These latter are therefore really extensions of lumped-parameter modeling to local regions. The most elaborate of these, by Jacquez, Carnahan, and Abbrecht (1967), contains a very large number of parameters, and is in principle capable of rather detailed transient descriptions. In practice, however, most of these parameters appear to be inaccessible at present, and we therefore content ourselves with a simpler example. Even it is still speculative.

One of the most interesting properties of the kidney is its ability to

* The *tubular load* is the mass of solute entering the tubules in a unit time. Nephrologists use a highly specialized vocabulary described in detail by Guyton (1966). The most important single term is *plasma clearance*, the flow rate of plasma through the kidney times the fractional rate of removal of the solute in question.

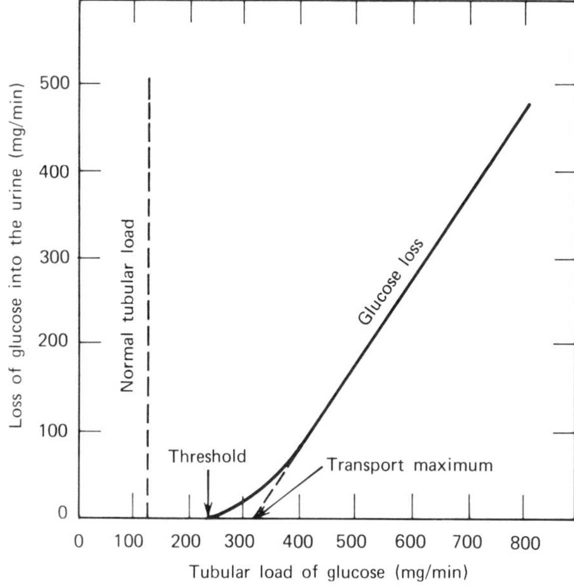


**Figure 6.3.11.** Relationship of tubular load of glucose to loss of glucose into the urine. From Guyton (1966).

produce a highly concentrated (*hyperosmotic*) urine. To do this it carries out an elegant countercurrent mass-transfer operation in the loop of Henle and collecting duct, as described schematically in Fig. 6.3.12. Shown here are the sodium and water transport and the osmolality* of the regions involved in urine concentration. This concentration process occurs primarily in the juxtamedullary nephrons which are about one-fifth of the total number.

The primary concentration process is an active transport of $Na^+$ (and hence NaCl) from the ascending loop of Henle via the intertubular space to the descending loop, which is highly permeable to NaCl. This salt transport produces a local concentration difference of only about 200 milliosmols (about $0.1\,N$), but the countercurrent nature of the flow results in a maximum concentration of about 1200 milliosmols (or 4 times normal)

* The term *osmolality* is widely used in the medical literature and is actually a measure of water activity. Specifically, a one-osmolar solution has the water activity (or partial pressure) of a one-molar solution of an ideal nondissociating solute, or a one-half molar solution of an ideal 1-1 electrolyte. Thus a 300-milliosmolar solution of NaCl chloride corresponds *approximately* to an 0.15 $N$ solution—which is the physiological concentration. This is the approximate equivalent osmotic concentration of the glomerular filtrate, and for our present purposes the solutes in this filtrate may be considered thermodynamically ideal.

**Table 6.3.2.** Transport Maximum of Representative Important Substances Absorbed from the Tubules, or Secreted by Them.[a]

| | | |
|---|---|---|
| Absorption | | |
| Glucose | 320 | mg/min |
| Phosphate | 0.1 | mmole/min |
| Sulfate | 0.06 | mmole/min |
| Amino acids | 1.5 | mmole/min |
| Vitamin C | 1.77 | mg/min |
| Urate | 15 | mg/min |
| Plasma protein | 30 | mg/min |
| Hemoglobin | 1 | mg/min |
| Lactate | 75 | mg/min |
| Acetoacetate | variable | (about 30 mg/min) |
| Secretion | | |
| Creatinine | 16 | mg/min |
| PAH | 80 | mg/min |
| Diodrast | 57 | mg/min (of iodine) |
| Phenol red | 56 | mg/min |

[a] Figures taken from Guyton (1966).

at the "bottom" of the loop (*pelvic tip*). Similar concentrations are produced in the collecting duct and the vasa recta, which are in effective diffusional contact with the peritubular fluid. (It should be noted that only about 5% of the renal blood flow passes through the vasa recta, so that its behavior is of only secondary importance here.)

Since the ascending loop of Henle is only very slightly permeable to water, the active removal of NaCl just described reduces the osmolality to about 100, or one-third of normal at its *distal* (downstream) end. The nephron is now in a position to produce either dilute or concentrated urine.

If a dilute urine is needed (as a result of excessive water intake), the water permeability of the epithelial cells of the distal tubules and collecting ducts is strongly decreased. Then the dilute solution at the distal end of the loop of Henle is passed through these regions unchanged and forms the dilute urine.

If a concentrated urine is needed, on the other hand, the water permeability of the distal tubule and collecting duct is increased. Water then

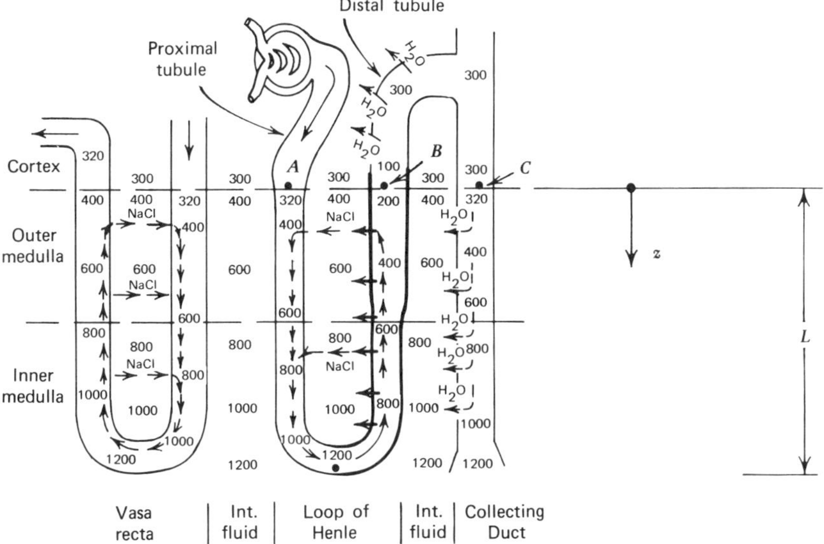


**Figure 6.3.12.** Countercurrent mass transfer and urine production. From Guyton (1966).

diffuses from the hypo-osmotic (dilute) solution, leaving the loop of Henle until it closely approaches the normal physiological level of 300 milliosmols near the end of the distal tubule. (It can be seen from Fig. 6.3.9 that the distal tubule is in intimate diffusional contact with both the efferent arteriole and the proximal tubule, so that a suitable water "sink" does exist as well as a large-mass transfer surface. Fig. 6.3.12 is quite misleading in this respect.) This dewatering process continues in the collecting duct, which is in contact with hyperosmotic (concentrated) solution but is impermeable to NaCl. The success of this process clearly depends for its success upon the remarkably selective permeability of the various epithelial cells. No comparable extracorporeal device has yet been made, or even seriously proposed. The quantitative aspects of urine concentration have been the subject of considerable study but appear to be largely speculative. The model below explains the above-described qualitative behavior, but it is not claimed to be quantitatively reliable.

Consider the urine concentration process as idealized in Fig. 6.3.12. Here 5 $cm^3/min$ of filtrate enters the descending limb of the loop of Henle at a milliosmolarity of 300, which may be assumed to be 0.15 $N$ sodium chloride. Assume that:

1. NaCl is pumped from the ascending limb at a uniform rate of $m_0$ moles per unit time per unit length of limb, so that the total rate of salt

pumped out of this limb is $m_0L$.

2. The ascending limb of the loop of Henle is completely impermeable to water, and the collecting duct is impermeable to salt.

3. The descending limb is sufficiently permeable to salt, and the collecting duct to water, that the fluid in each is in equilibrium with the peritubular fluid at the same "elevation" $z$.

4. In the peritubular fluid, salt and water transport in the $z$ direction are negligible.

5. The distal tubule is permeable only to water, which leaves in sufficient amount that its osmolality $o_S = 300$ at the entrance to the collecting duct.

6. There is no net loss of salt from the system: the amount of salt entering at $A$ is the same as that entering the collecting duct at $c$ and leaving at $E$.

7. *Net* water loss occurs only across the walls (epithelial cells) of the distal tubule.

We now consider the behavior of this model.

The rate at which salt enters the system at $A$ is

$$\dot{m}_s = (5\,\text{ml/min})(0.15\,\text{meq/ml})$$

$$= 0.75\,\text{meq/min} \tag{2}$$

The amount of water which must be removed to produce a 1200-milliosmolar (or 0.6 $N$) urine is

$$Q_{\text{filtrate}} - Q_{\text{urine}} = (5\,\text{ml/min})\left(1 - \frac{300}{1200}\right)$$

$$= 3.75\,\text{ml/min} \tag{3}$$

According to our assumptions all of this water is removed across the wall of the distal tubule.

The volume of solution leaving the distal tubule at $C$ must be the same as that entering our system at $A$, since the salt concentration and flow rate in these two streams are equal. Therefore, at $C$,

$$Q = 5\,\text{ml/min}$$

$$c_s = 0.15\ N$$

The rate of fluid input to the distal tubule at $B$ must be greater than this by the rate of water loss from our system. Therefore at $B$

$$Q = 5 + 3.75\,\text{ml/min} = 8.75\,\text{ml/min} \tag{4}$$

$$c_s = \frac{0.75}{8.75} = 0.0857\ N \tag{5}$$

Since the ascending limb is assumed to be water impermeable, the water flow at $D$ must equal that at $B$. According to assumption 3, however, the salt concentration here must be 1200 milliosmolar. Therefore, at $D$,

$$Q = 8.75\,\text{ml/min}$$

$$\dot{m}_s = (0.6)(8.75) = 5.25\ \text{meq/min}$$

The amount of salt transported out across the wall of the ascending limb is then

$$\Delta\dot{m}_s = 5.27 - 0.75 = 4.50\ \text{meq/min} \tag{6}$$

The intermediate material balances are now complete, and it remains only to determine the concentration profiles.

The salt flow in the ascending limb of Henle is then

$$\dot{m}_s = \left(0.75 + 4.50\frac{z}{L}\right)\ \text{meq/min} \tag{7}$$

according to assumption 1. Since $Q$ in the ascending limb is constant at 8.75 ml/min, the concentration profile in the ascending limb is given by

$$(c_s)_{\text{al}} = \left(\frac{\dot{m}_s}{Q}\right)_{\text{al}} = 0.0857 + 0.514(z/L)\ N \tag{8}$$

The calculated concentration profile in the ascending limb is thus linear.

Because all of the salt pumped out of the ascending limb must be absorbed into the descending limb, it follows that in the descending limb

$$(\dot{m}_s)_{\text{dl}} = (\dot{m}_s)_{\text{al}} = \left(0.75 + 4.50\frac{z}{L}\right)\ \text{meq/min}$$

whereas the salt flow in the collecting duct is constant at

$$(\dot{m}_s)_{\text{cd}} = 0.75\ \text{meq/min}$$

The volumetric flows in the three ducts between $0 < z < L$ are related by

$$\begin{aligned} Q_{\text{dl}} + Q_{\text{cd}} &= Q_{\text{urine}} + Q_{\text{al}} \\ &= 1.25 + 8.75\ \text{ml/min} \\ &= 10\ \text{ml/min} \end{aligned}$$

from a mass balance on water, and

$$\frac{Q_{\text{dl}}}{Q_{\text{cd}}}=\frac{(\dot{m}_s)_{\text{dl}}}{(\dot{m}_s)_{\text{cd}}}$$

$$=\frac{0.75+4.50(z/L)}{0.75}$$

$$=1+6\frac{z}{L}$$

by the assumption that both have the local concentration of the peritubular fluid. It follows from these two equations that

$$Q_{\text{cd}}=\frac{10\ \text{cm}^3/\text{min}}{2+6z/L}$$

and

$$(c_s)_{\text{cd}}=\left(\frac{\dot{m}_s}{Q}\right)_{\text{cd}}$$

$$=(0.075)\left(2+6\frac{z}{L}\right)\ \text{meq/ml}=(c_s)_{\text{dl}}=(c_s)_{\text{ptf}}$$

where the subscript ptf refers to the pertibular fluid. Thus the calculated concentration gradient in these three regions is also constant, but is somewhat less than that in the ascending limb. These predictions agree fairly well with the concentration distribution of the figure.

## 6.4. Macroscopic Balances at the Cellular Level

A wide variety of interesting transient processes takes place at the cellular level in spite of the rapidity of diffusion in systems of such small dimensions. This is possible because of the highly selective permeability of natural membranes and the finite speed of chemical reactions.

The most studied of such processes is almost certainly the transmission of electric impulses along the axons of nerves. This process is, however, much too complex—and too poorly understood from a chemical point of view—for discussion here. Rather the reader is referred to the monograph of Cole (1968), where it is discussed authoritatively and at length.

Another area of great importance is that of periodic chemical reactions, which occur frequently in living systems. A number are understood in detail, and some of the most interesting of these are discussed by Glansdorf and Prigogine (1971). More exciting at the moment, but less well understood, are the cyclic conformational changes accompanying such basic reactions as energy transduction in mitochondria.

We choose here for our single example a simpler, but still speculative, situation which ties in with our previous discussions: the transient response of blood to changes in respiratory gas tension.

### *EX. 6.4.1. OXYGEN UPTAKE BY INTACT RED CELLS*

The transient response of sheep red cells to a sudden increase in oxygen and carbon dioxide tension, shown in Fig. 6.4.1, exhibits a commonly observed "overshoot" phenomenon. That is, the extent of hemoglobin oxygenation* rises rapidly to substantially greater than the equilibrium value and then drops, somewhat more slowly, toward equilibrium. A similar overshoot is thought to occur during oxygenation in the lung.

Suggest a plausible basis for such behavior.

SOLUTION

No definitive explanation is yet available, but the analysis of Sirs (1970), summarized below, appears to be correct in its main features. According to him this behavior results primarily from a combination of the Bohr effect, the localization of all carbonic anhydrase within the red cells, and the impermeability of the red-cell wall to cations. We begin by briefly reviewing these factors.

The Bohr effect, or decrease of hemoglobin oxygen affinity with decrease in pH, is shown qualitatively in Figs. 6.1.7 and 6.1.8; it is also discussed briefly in Section 6.1. This is an important effect physiologically and is described in detail in the *Handbook of Physiology*.

The enzyme carbonic anhydrase (E) which occurs only in the red cell, serves to speed up the hydration of dissolved $CO_2$ to carbonic acid:

$$CO_2 + H_2O \overset{E}{\rightleftharpoons} H_2CO_3 \rightleftharpoons H^+ + HCO_3^- \tag{1}$$

This is the rate-limiting reaction in carbon dioxide transport, and its kinetics depend strongly on the buffering capacity of the ambient solution. Fortunately these kinetics are well understood (and reviewed in some detail by Sirs, 1970). We may note here that for normal physiological conditions the time for 90% completion is of the order of 20 sec in plasma (uncatalyzed) but only a few milliseconds in red cells (catalyzed).

The selective permeability of red-cell membranes for anions has already

* Which may be followed photometrically.

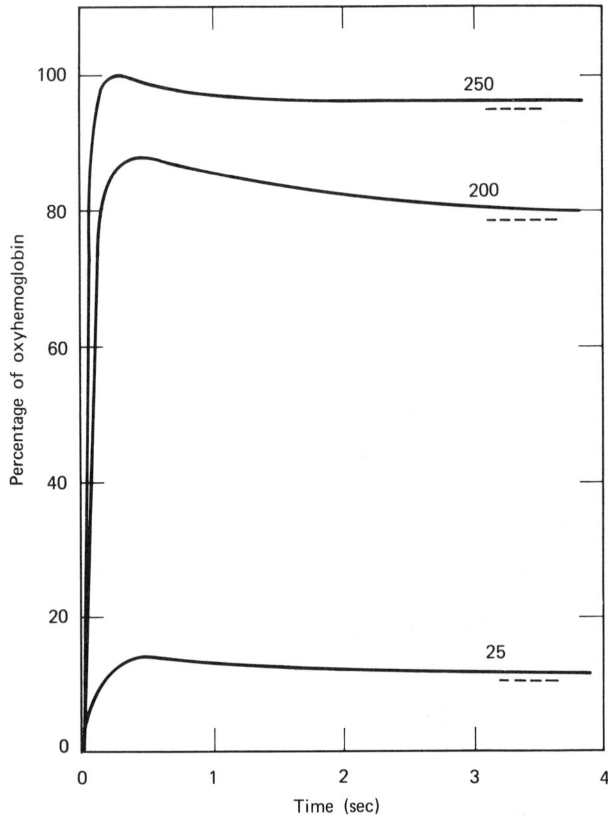


**Figure 6.4.1** The uptake of oxygen by a 1 : 45 oxygen-free suspension of fresh sheep's blood after rapid mixing with an equal volume of oxygenated solution. The three curves were obtained at 18°C from solutions equilibrated with $p_{O_2}=25$, 200, and 250 mm Hg, and with $p_{CO_2}=30$ mm Hg. From Sirs (1970).

been discussed in Section 4.2. It is, however, important to note here that $Cl^-$-$HCO_3^-$ exchange, the Hamburger shift, takes place rapidly—with a half-time on the order of 0.1 sec; $H^+$, on the other hand, permeates this membrane very slowly.

In principle one should also consider the change of hemoglobin pK with degree of oxygenation (the Haldane effect) and the reaction of $CO_2$ with amino groups to form carbamates (see, e.g., Table 2.3.1). These appear, however, to be secondary effects (from our present point of view) in the carbamate case, because these reactions are fast for time scales of seconds. We are therefore now ready to look at the overshoot processes, and we begin with the simpler situation of Fig. 6.4.1.

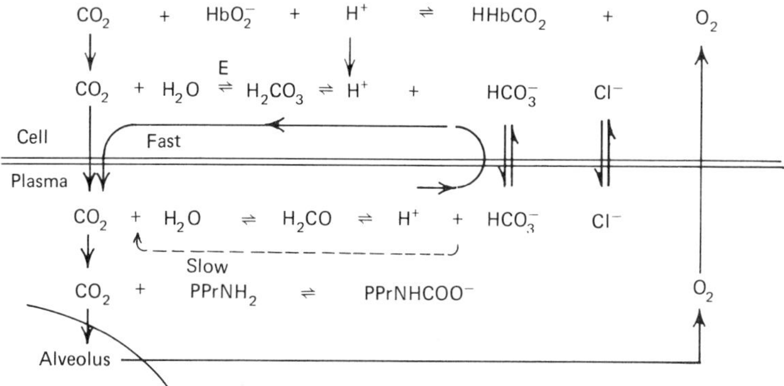


**Figure 6.4.2.** Diagram of the kinetic and diffusional processes accompanying pulmonary gas exchange. From, Sirs (1970).

The red cells in the experiment described here were initially diluted 1:45 with an isotonic solution and then stripped of both $O_2$ and $CO_2$ by $N_2$ flushing. This suspension was then rapidly mixed with a second isotonic solution with a $CO_2$ tension of 30 mm Hg, and $O_2$ tensions of either 25, 200, or 250 mm Hg.

All diffusional processes in this system are most probably fast relative to one second. Therefore, dissolved $CO_2$ equilibrates very rapidly across the membrane, and the intracellular $CO_2$ concentration rises accordingly. In addition, intracellular $CO_2$ hydrates and dissociates effectively instantaneously, and tends to lower the pH. However, since the external $CO_2$ dehydration is slow, partial pressures of $CO_2$ drop substantially below their equilibrium values; meanwhile the external $HCO_3^-$ concentration remains both above the equilibrium value and that in the red cells. This latter effect initiates the Hamburger shift: exchange of intracellular $Cl^-$ for extracellular $HCO_3^-$. This influx of bicarbonate without corresponding hydrogen ions suppresses $H_2CO_3$ ionization, and this limits the formation of $H^+$.

Initially then the intracellular pH surpasses the equilibrium value, and the hemoglobin affinity for oxygen is correspondingly high. Since oxygenation is rapid, a high initial oxyhemoglobin concentration is achieved. At longer times, however, extracellular bicarbonate is converted to $CO_2$, which passes into the cells and rehydrates to carbonic acid. A true transmembrane equilibrium is now established, the intracellular pH drops, and oxygen is partially released from the hemoglobin.

The oxygenation process in the lung is similar except that alveolar gas acts as a $CO_2$ sink. Here $CO_2$ is removed to the gas phase from both cells and plasma, but bicarbonate is consumed appreciably—via $H_2CO_3$ dehydration—only in the cells. The Hamburger shift thus occurs as before, tending to accentuate the increasing intracellular alkalinity. The resulting overshoot in pH again results in an oxygenation overshoot. The chemical and diffusional steps involved, including carbonate formation, are indicated in Fig. 6.4.2. The ultimate drop in intracellular pH accompanying the slower dehydration of extracellular carbonic acid now occurs as the blood moves through the systemic circulation. All of the "excess" oxygen taken up as a result of the overshoot is thus available for body metabolic processes. Sirs suggests that this effect is more pronounced during exercise than at rest and thus doubly beneficial. It remains, however, to determine its quantitative importance.

## BIBLIOGRAPHY

### Books and Reviews

Aris, Rutherford, *Introduction to the Analysis of Chemical Reactors*, Prentice-Hall (1965).

Bassingthwaighte, J. B., "Blood Flow and Diffusion in Mammalian Organs," *Science*, **167**, 1347–1353 (6 March 1970).

Bischoff, K. B., in Hershey (1967).

Cole, Kenneth S., *Membranes, Ions, and Impulses: a Chapter in Classical Biophysics*, Univ. of Calif. Press (1968).

Glansdorf, P., and I. Prigogine, *Thermodynamic Theory of Structure, Stability, and Fluctuations*, Wiley (1971).

Guyton, A. C., *Textbook of Medical Physiology*, Saunders (1966).

Hershey, Daniel, Ed., *Chemical Engineering in Medicine and Biology*, Plenum (1967).

Kleiber, M., *Ann. Rev. Physiol.*, **29**, 1 (1967).

Levenspiel, Octave, and K. B. Bischoff, "Patterns of Flow in Chemical Process Vessels," *Adv. Chem. Eng.*, **4**, 95 (1963).

Lightfoot, E. N., R. L. Safford, and D. R. Stone, "Biological Applications of Mass Transport Phenomena," *CRC Critical Reviews in Bio-engineering.*, **1**, No. 1, 69–138 (1971).

Stahl, W. R., *Physiological Similarity and Modelling. The Application of Dimensional Analysis and Physical Similarity Theory to Mammalian Physiology,* Appleton-Century-Crofts (1970).

### Periodical and Secondary References

Abbrecht, P. H. and N. W. Prodany, "A Model of the Patient-Artificial Kidney System," *I.E.E.E. Trans. Biomed. Eng.* **BME18**, 257 (1971).

Babb, A. L., Lars Grimsrud, and S. B. Layno, "Engineering Aspects of Artificial Kidney Systems," in Hershey (1967).

Bell, R. L., K. Curtiss, and A. L. Babb, "Analog Simulation of Patient-Artificial Kidney Systems," *Trans. Am. Soc. Artif. Intern. Organs*, **11**, 183 (1965).

Bischoff, K. B., and R. L. Dedrick, "Generalized Solution to Linear Two-compartment Open Model for Drug Distribution," *J. Theor. Biol.*, **29**, 63 (1970).

Bischoff, K. B., R. L. Dedrick, and D. S. Zaharko, "Preliminary Model for Methotrexate Pharmaco-kinetics," *J. Pharm. Sci.*, **59**, 149 (1970).

Coulam, C. M., H. R. Warner, H. W. Marshall, and J. B. Bassingthwaighte, "A Steady-State Transfer Function Analysis of the Circulatory System Using Indicator-dilution Techniques," *Comput. Biomed. Res.*, **1**, 124–128 (1967).

Coulam, C. M., H. R. Warner, E. H. Wood, and J. B. Bassingthwaighte, "A Transfer Function Analysis of Coronary and Renal Circulation Calculated from Upstream and Downstream Indicator-dilution Curves," *Circ. Res.*, **29**, 879 (1966).

Crandall, E. D. and R. W. Flumerfelt, in Hershey (1967).

Dedrick, R. L., and K. B. Bischoff, "Pharmaco-kinetics in Applications of the Artificial Kidney," *Chem. Eng. Prog. Symp. Ser.*, **64**, 32 (1968).

Dedrick, R. L., K. B. Bischoff, and D. S. Zaharko, "Interspecies Correlation of Plasma Concentration History of Methotrexate (NSC-740)," *Cancer Ther Rep.*, Part I, **54**, 95 (1970).

Dedrick, R. L. D. S. Zaharko, and R. J. Lutz, "Trransport and Binding of Methotrexate *in Vivo*," *J. Pharm. Sci.* **62**, No. 6, 882–890 (1973).

Dindorf, J. A., E. N. Lightfoot, and K. A. Solen, "Prediction of Blood Oxygenation Rates," Am. Inst. Chem. Engrs., *Chem. Eng. Prog. Symp. Series*, No. 114, 75-87 (1971).

Dindorf, J. A., and E. N. Lightfoot, manuscript in preparation.

Goresky, C. A., W. H. Zeigler, and G. G. Bach, "Capillary Exchange Modelling," *Circ. Res.*, **27**, 739 (1970).

Greenleaf, J. F., T. J. Knopp, C. M. Coulam, and J. B. Bassingthwaighte, *Proc. Ann. Conf. Eng. Med. Biol.*, **10**, 506 (1968).

Hemmingsen, A. M., *Rep. Steno Mem. Hosp. Nord. Insulin Lab.*, **9**, 1 (1960).

Jacquez, J. A., B. Carnahan, and P. Abbrecht, "A Model of the Renal Cortex and Medulla," *Math. Biosci.*, **1**, 227 (1967).

Levitt, D. G., "Theoretical Model of Capillary Exchange Incorporating Interactions between Capillaries," *Am. J. Physiol.*, **220**, 250 (1971).

Miller, W. S., *The Lung*, Chas. C. Thomas (1947).

Nicholes, K. R. K., H. R. Warner, and E. H. Wood, "Study of Dispersion of an Indicator in the Circulation," *Ann. N.Y. Acad. Sci.*, **115**, 721 (1964).

Riggs, A., *J. Gen. Physiol.*, **43**, 737 (1960).

Schmidt-Nielsen, Knud, "Energy Metabolism, Body Size, and Problems of Scaling," *Fed. Proc.*, **29**, No. 4, 1524 (1970).

Sirs, J. A., "The Interaction of Carbon Dioxide with the Rate of Exchange of Oxygen by Red Blood Cells," in *Blood Oxygenation*, Daniel Hershey, Ed., Plenum (1970).

Spalding, D. B., *Chem. Eng. Sci.*, **9**, 74–77 (1958).

Tenney, S. M., and J. E. Remmers, *Nature*, **197**, 54 (1963).

# IV CONCLUSION: THE ROLE OF TRANSPORT PHENOMENA IN ENGINEERING DESIGN

The primary creative activity of engineers is design, and all analytic efforts however important, must play a subordinate role. At the time of writing one of the major challenges facing chemical engineers is the effective incorporation of transport phenomena into a suitable design framework. Available to us for this purpose are both the very extensive literature in transport phenomena, referred to throughout this text, and recent advances in process design (see, e.g., Rudd and Watson, 1968).

The goal of such an effort, which might be called *equipment design*, is to develop specific apparatus and processing procedures. It thus differs from *process design*, which typically deals with the systems aspects of large problems. Equipment design has always been a key activity of chemical engineers, but it has yet to be organized systematically, as we feel it should be. It represents the synthetic, or design, component of unit operations, just as transport phenomena represent the analytic, or descriptive, part.

Equipment design so defined is clearly a key discipline for posing and solving engineering problems in medicine, and it is fitting that we conclude with a brief discussion of it. Since the main emphasis of this text is on transport phenomena, we content ourselves with a single example. Clearly, however, this subject deserves much more extensive treatment elsewhere.

## *EX. 1. BLOOD OXYGENATION*

Temporary replacement of the heart and lungs by a pump-oxygenator system has become increasingly important in surgical procedures and shows promise for the treatment of serious circulatory and respiratory diseases. Furthermore, somewhere in the future lies the possibility of permanently implanted replacements. Discuss the design aspects of blood oxygenation with particular emphasis on the role of transport phenomena.

### SOLUTION

The fundamental nature of this process is indicated schematically in Fig. 1. Here venous blood from the systemic circulation is passed through an oxygenator and pumped back into the arterial end of the system through a heat exchanger. Not shown in this oversimplified diagram are minor (but essential) apparatus such as bubble traps or filters and measurement and control instrumentation. The medical aspects of blood oxygenation are discussed by Galletti (1962) among others, and the engineering aspects have recently been reviewed by Spaeth (1973).

There is clearly no one way to organize research in this area; as in all design problems, one must take a heuristic approach. One very reasonable strategy, inspired by the development of process design, is shown in Table 1.

It is first necessary to define the problem area, as indicated: to decrease both mortality and such sublethal damage as loss of memory, which may occur in present procedures. However, if bypass procedures are to be applied in new areas, for example the treatment of such acute disorders as emphysema attacks or pneumonia, the permissible bypass time must also be increased.

It remains to determine the changes in equipment configuration and processing conditions required to achieve these goals. For such a complex problem as this one it is necessary to begin with a systematic preliminary assessment, involving both a literature survey and preliminary experiments.

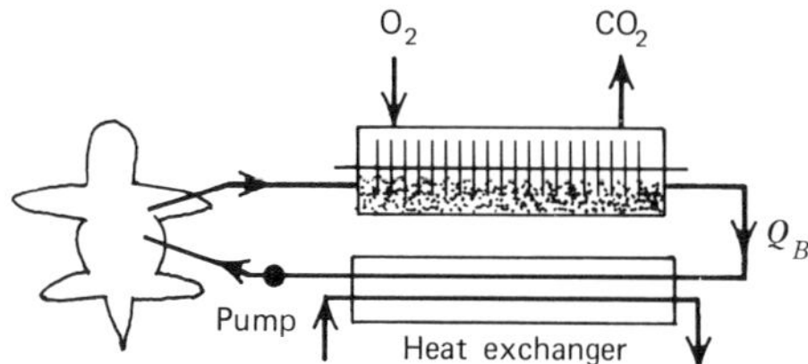


**Figure 1.** Schematic representation of heart-lung bypass procedures.
*Quantitative parameters*; Blood perfusion rate $Q_B$; $P_{O_2}$; $p_{CO_2}$; pH; $T$ (cyclic vs. steady).
*Qualitative parameters*: Surgical procedure; Type of equipment; Anesthetic.

**Table 1.** Organization of Heart-Lung Bypass Research.

| | |
|---|---|
| *Define primitive problem* | |
| Decrease mortality and sublethal damage | |
| Increase permissible bypass time, toward "∞" | |
| *Make a preliminary assessment* | |
| Literature survey ("statistical") | |
| Preliminary experiments | |
| Collective staff judgement | |
| *Identify critical areas* | |
| Physiological: | blood degradation |
| | brain damage |
| | flow maldistribution |
| Technological (device): | insufficient control of parameters |
| | excessive blood volume |
| | flow maldistribution |
| *Select specific research programs* | |
| Flow distribution in "patient" | |
| Blood degradation | |
| Oxygenator analysis and design | |

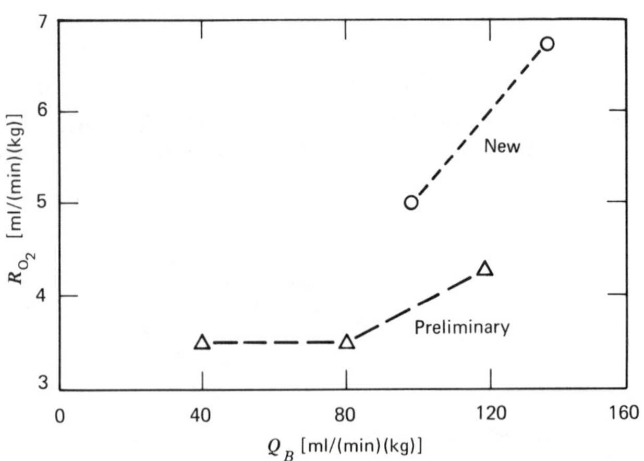


**Figure 2.** Effect of blood perfusion rate on volumetric oxygen consumption.

The medical literature is almost incredibly voluminous by ordinary engineering standards, and it tends to be ambiguous. One must therefore put heavy emphasis on the intangibles of professional judgement and personal contacts. Substantial duplication of effort is inevitable and, given the imperfect reproducibility of medical data, not altogether undesirable. Preliminary experimentation is necessary, both to supplement the literature and to develop experimental techniques, the latter typically being a slow, expensive procedure.

The ordering of experiments is important, and the goals at this preliminary stage are to identify critical areas and to determine the gross effects of major parameters:

1. Quantitative factors such as blood flow rate, arterial oxygen and carbon dioxide tensions, and pH.
2. Qualitative factors such as surgical techniques and the general configuration of the pump and oxygenator.

Both the research capacity and professional contacts of any one laboratory are limited, so specific, relatively narrow, areas must be chosen for detailed investigation, and this choice must be made with care.

Information available at the time of writing indicated that existing heart-lung bypass procedures produced severe blood damage and maldistribution of flow, the latter tending to cause the development of irreversible shock. Blood damage includes formation of small emboli which frequently lodge in the capillaries of the brain and other critical organs. Finally, experiments in the author's laboratory showed that the oxygenator in use there did not permit independent control of major parameters. Partial evidence on these points is given in Figs. 2 and 3.

Figure 2 shows the effect of blood perfusion rate through an oxygenator and test dog on oxygen consumption in the dog and on oxygen tension in the blood returned to him. The increase in oxygen consumption with perfusion rate shown here may result from an increase in the mass of tissue irrigated, since, as indicated in Chapter III, Ex. 2.3.4, aerobic tissue metabolism is zero order with respect to oxygen. This in turn suggests flow maldistribution in the dog, at least for the lower flow rates. Further evidence of maldistribution is obtained from the residence-time distribution of dye injected into the returning blood: mean residence times are typically 5 to 10 sec, whereas under normal conditions they should be on the order of a minute.

It was also found that "arterial" oxygen tensions produced by our oxygenator, which is described in Fig. 6.3.1, drop rapidly with increasing perfusion rate, and that normal physiological tensions cannot be obtained at the highest rates of interest.

Figure 3 shows schematically the effect of bypass time on the physiological condition of test dogs, both for direct return of oxygenated blood to the dog and for prior filtration to remove microemboli. The marked improvement resulting from filtration clearly illustrates the blood damage produced by the oxygenator, here a "direct-contact" device (i.e., one providing for direct contact of blood and oxygenating gas, without an intervening membrane).

Evidence of the type just discussed can now be used to develop a specific research program—large enough, one may hope, to produce significant results without overburdening available personnel. The choice made in the author's laboratory is shown schematically in Table 2. Parallel programs on flow distribution, blood damage, and oxygenator design were felt to be highly desirable, as they tend to reinforce each other quite strongly:

1. Meaningful flow distribution studies can be made only with an adequate oxygenator and on test animals whose condition is not dominated by blood damage.

2. Blood degradation depends strongly, and in a complex way, on both oxygenator configuration and process conditions. It is therefore important

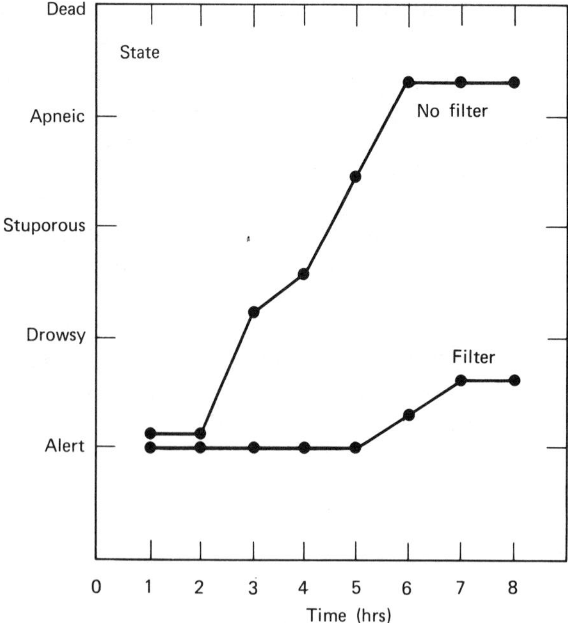


**Figure 3.** The effect of perfusion time on the neurologic state of dogs on heart-lung bypass with a direct-contact oxygenator (from Patterson et al, 1971). The filter removes microemboli formed in the gas-exchange process.

**Table 2.** Schematic flow sheet of bypass research in the author's laboratory in 1972.

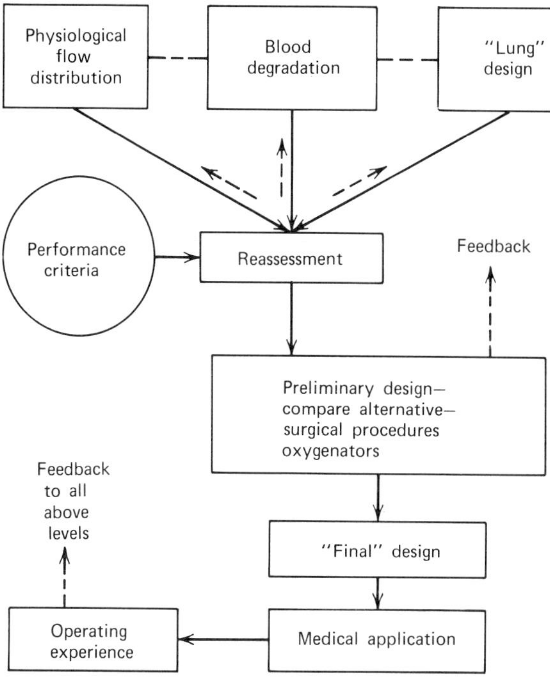

Physiological flow distribution
Blood degradation
"Lung" design
Performance criteria
Reassessment
Feedback
Preliminary design—compare alternative—surgical procedures oxygenators
Feedback to all above levels
"Final" design
Operating experience
Medical application

to carry out degradation studies under conditions of real interest.

3. In oxygenator design, flow-distribution studies provide many of the requirements, whereas blood-damage studies provide some of the most important constraints.

It is clearly not possible for any laboratory to cover all of these areas—or even any one—by itself; these are all very large problems. However, a reasonably broad and active program is required to maintain a realistic research atmosphere.

The remainder of Table 2 shows the normal progression of a (successful) research program through preliminary and final design, and ultimately to practical application. It is important to note the iterative aspects of this process—there is, for example, no true final design except in a dying field—and the need for performance criteria. These latter can be seriously influenced by economic considerations, especially in long-term procedures such as the treatment of end-state kidney disease, and they may even be dominated by them. The most important criteria are medical effectiveness, safety, and reliability. They must clearly be defined primarily by clinicians, and they are not simple to quantify. How much additional blood damage, for example, could one tolerate to reduce by one the required number of units of donor blood?

Transport phenomena enter the design process at many points, but their role is most obvious in the description of oxygenator mass-transfer effectiveness. We therefore concentrate on this aspect and begin by noting that design requirements determine entirely the approach to be taken; these in turn appear at three levels:

1. In the initial stages of the flow distribution studies, where it is necessary to provide adequate gas transfer at all desired flow rates.
2. In the preliminary design stages, where it is necessary to assess a relatively large number of possible configurations before concentrating more deeply on those which appear most promising.
3. In the "final" design where one must make a choice between promising prototypes, which have survived the preliminary comparisons, and size the oxygenators as closely as possible within expectable system requirements.

Clearly the foregoing listing is in the order of increasing required accuracy.

Even in the final design stages, however, highly precise calculations are of relatively little value; there is just too much variability in demand—and there are too many imponderables that can only be investigated experimentally. The judicious use of such approximations as those introduced in Chapter III, Section 5.3 is thus often justified even in the final design.

One must, however, continually review the assumptions on which computational models are based, to be sure they are applicable to any new problem at hand. The assumptions of pseudohomogeneous diffusion and instantaneous equilibration of hemoglobin with oxygen are of particular interest in this respect; any major decrease in boundary-layer thickness below those now normally encountered would probably invalidate both. This would not only greatly complicate design calculations; it would also require extensive new kinetic investigations. The strategy of oxygenator description is thus in itself far from trivial: one should at the same time avoid unnecessary complexities and unjustifiable oversimplifications. Once these factors have been assessed, one has a basis for developing a quantitative description, as was done in Chapter III, Ex. 6.3.1.

A somewhat more subtle problem is to seek qualitative improvements and novel technology without becoming unduly dependent on them. Systematizing design procedures too neatly can easily limit one to pedestrian improvements. One should thus consider such "radical" approaches as oxygenation via hydrogen peroxide along with the optimization of existing configurations. At the same time one must recognize the need for caution in medical procedures and the correspondingly large effort needed for introduction of new concepts.

Transport phenomena, kinetics, and both physical and physiological chemistry play key roles in the evaluation of any such novel procedure. This is true both in preliminary estimates of its potential and in the design of exploratory experiments. Even when *a priori* estimation of system behavior is impossible (and this is normally the case for truly novel situations), much can be done. If we consider, for example, direct hydrogen peroxide infusion to blood, toxicity and bubble nucleation are difficult to predict. It is, however, possible to characterize the mass-transfer behavior with some reliability in properly designed test systems, and to carry out a systematic dimensional analysis.

More generally, a thorough grounding in transport phenomena and experience in its application help in providing "good judgement," the least tangible but most important quality in a good engineer. It is hoped that this monograph has helped the reader to at least some degree in improving his judgement.

## BIBLIOGRAPHY

Galletti, P. M., *Heart-lung Bypass*, Grune and Stratton (1962).

Galletti, P. M., Ed., *Mechanical Devices for Cardio-Pulmonary Assistance*, Karger (1971).

Patterson, R. H., Jack Kessler, and R. M. Bergland, "A Filter to Prevent Cerebral Damage during Experimental Cardiopulmonary Bypass," *Surg., Gynecol., Obstet.*, **132**, 71–74 (1971).

Rudd, D. F., and C. C. Watson, *The Stategy of Process Engineering*, Wiley (1968).

Spaeth, E. E., *CRC Critical Reviews in Bio-engineering* **1**, No. 4, 383–418 (1973).

# APPENDIX SUMMARY OF THE EQUATIONS OF CHANGE

**Table App. 1.** The Equation of Continuity in Several Coordinate Systems.

*Rectangular coordinates* $(x, y, z)$:

$$\frac{\partial \rho}{\partial t} + \frac{\partial}{\partial x}(\rho v_x) + \frac{\partial}{\partial y}(\rho v_y) + \frac{\partial}{\partial z}(\rho v_z) = 0 \tag{A}$$

*Cylindrical coordinates* $(r, \theta, z)$:

$$\frac{\partial \rho}{\partial t} + \frac{1}{r}\frac{\partial}{\partial r}(\rho r v_r) + \frac{1}{r}\frac{\partial}{\partial \theta}(\rho v_\theta) + \frac{\partial}{\partial z}(\rho v_z) = 0 \tag{B}$$

*Spherical coordinates* $(r, \theta, \phi)$:

$$\frac{\partial \rho}{\partial t} + \frac{1}{r^2}\frac{\partial}{\partial r}(\rho r^2 v_r) + \frac{1}{r\sin\theta}\frac{\partial}{\partial \theta}(\rho v_\theta \sin\theta) + \frac{1}{r\sin\theta}\frac{\partial}{\partial \phi}(\rho v_\phi) = 0 \tag{C}$$

**Table App. 2.** The Equation of Motion in Rectangular Coordinates $(x, y, z)$.

In terms of velocity gradients for a Newtonian fluid with constant $\rho$ and $\mu$:

*x component*: $$\rho\left(\frac{\partial v_x}{\partial t} + v_x\frac{\partial v_x}{\partial x} + v_y\frac{\partial v_x}{\partial y} + v_z\frac{\partial v_x}{\partial z}\right)$$

$$= -\frac{\partial p}{\partial x} + \mu\left(\frac{\partial^2 v_x}{\partial x^2} + \frac{\partial^2 v_x}{\partial y^2} + \frac{\partial^2 v_x}{\partial z^2}\right) + \rho g_x \tag{A}$$

*y component*: $$\rho\left(\frac{\partial v_y}{\partial t}+v_x\frac{\partial v_y}{\partial x}+v_y\frac{\partial v_y}{\partial y}+v_z\frac{\partial v_y}{\partial z}\right)$$

$$=-\frac{\partial p}{\partial y}+\mu\left(\frac{\partial^2 v_y}{\partial x^2}+\frac{\partial^2 v_y}{\partial y^2}+\frac{\partial^2 v_y}{\partial z^2}\right)+\rho g_y \quad \text{(B)}$$

*z component*: $$\rho\left(\frac{\partial v_z}{\partial t}+v_x\frac{\partial v_z}{\partial x}+v_y\frac{\partial v_z}{\partial y}+v_z\frac{\partial v_z}{\partial z}\right)$$

$$=-\frac{\partial p}{\partial z}+\mu\left(\frac{\partial^2 v_z}{\partial x^2}+\frac{\partial^2 v_z}{\partial y^2}+\frac{\partial^2 v_z}{\partial z^2}\right)+\rho g_z \quad \text{(C)}$$

---

**Table App. 3.** The Equation of Motion in Cylindrical Coordinates $(r,\theta,z)$.

In terms of velocity gradient for a Newtonian fluid with constant $\rho$ and $\mu$:

*r component*: $$\rho\left(\frac{\partial v_r}{\partial t}+v_r\frac{\partial v_r}{\partial r}+\frac{v_\theta}{r}\frac{\partial v_r}{\partial \theta}-\frac{v_\theta^2}{r}+v_z\frac{\partial v_r}{\partial z}\right)$$

$$=-\frac{\partial p}{\partial r}+\mu\left[\frac{\partial}{\partial r}\left(\frac{1}{r}\frac{\partial}{\partial r}(rv_r)\right)+\frac{1}{r^2}\frac{\partial^2 v_r}{\partial \theta^2}-\frac{2}{r^2}\frac{\partial v_\theta}{\partial \theta}+\frac{\partial^2 v_r}{\partial z^2}\right]+\rho g_r \quad \text{(A)}$$

*θ component*: $$\rho\left(\frac{\partial v_\theta}{\partial t}+v_r\frac{\partial v_\theta}{\partial r}+\frac{v_\theta}{r}\frac{\partial v_\theta}{\partial \theta}+\frac{v_r v_\theta}{r}+v_z\frac{\partial v_\theta}{\partial z}\right)$$

$$=-\frac{1}{r}\frac{\partial p}{\partial \theta}+\mu\left[\frac{\partial}{\partial r}\left(\frac{1}{r}\frac{\partial}{\partial r}(rv_\theta)\right)+\frac{1}{r^2}\frac{\partial^2 v_\theta}{\partial \theta^2}+\frac{2}{r^2}\frac{\partial v_r}{\partial \theta}+\frac{\partial^2 v_\theta}{\partial z^2}\right]+\rho g_\theta \quad \text{(B)}$$

*z component*: $$\rho\left(\frac{\partial v_z}{\partial t}+v_r\frac{\partial v_z}{\partial r}+\frac{v_\theta}{r}\frac{\partial v_z}{\partial \theta}+v_z\frac{\partial v_z}{\partial z}\right)$$

$$=-\frac{\partial p}{\partial z}+\mu\left[\frac{1}{r}\frac{\partial}{\partial r}\left(r\frac{\partial v_z}{\partial r}\right)+\frac{1}{r^2}\frac{\partial^2 v_z}{\partial \theta^2}+\frac{\partial^2 v_z}{\partial z^2}\right]+\rho g_z \quad \text{(C)}$$

---

**Table App. 4.** The Equation of Motion in Spherical Coordinates $(r,\theta,\phi)$.

In terms of velocity gradients for a Newtonian fluid with constant $\rho$ and $\mu$:[a]

*r component*: $$\rho\left(\frac{\partial v_r}{\partial t}+v_r\frac{\partial v_r}{\partial r}+\frac{v_\theta}{r}\frac{\partial v_r}{\partial \theta}+\frac{v_\phi}{r\sin\theta}\frac{\partial v_r}{\partial \phi}-\frac{v_\theta^2+v_\phi^2}{r}\right)$$

$$=-\frac{\partial p}{\partial r}+\mu\left(\nabla^2 v_r-\frac{2}{r^2}v_r-\frac{2}{r^2}\frac{\partial v_\theta}{\partial \theta}-\frac{2}{r^2}v_\theta\cot\theta-\frac{2}{r^2\sin\theta}\frac{\partial v_\phi}{\partial \phi}\right)+\rho g_r \qquad \text{(A)}$$

*θ component*: $$\rho\left(\frac{\partial v_\theta}{\partial t}+v_r\frac{\partial v_\theta}{\partial r}+\frac{v_\theta}{r}\frac{\partial v_\theta}{\partial \theta}+\frac{v_\phi}{r\sin\theta}\frac{\partial v_\theta}{\partial \phi}+\frac{v_r v_\theta}{r}-\frac{v_\phi^2\cot\theta}{r}\right)$$

$$=-\frac{1}{r}\frac{\partial p}{\partial \theta}+\mu\left(\nabla^2 v_\theta+\frac{2}{r^2}\frac{\partial v_r}{\partial \theta}-\frac{v_\theta}{r^2\sin^2\theta}-\frac{2\cos\theta}{r^2\sin^2\theta}\frac{\partial v_\phi}{\partial \phi}\right)+\rho g_\theta \qquad \text{(B)}$$

*φ component*: $$\rho\left(\frac{\partial v_\phi}{\partial t}+v_r\frac{\partial v_\phi}{\partial r}+\frac{v_\theta}{r}\frac{\partial v_\phi}{\partial \theta}+\frac{v_\phi}{r\sin\theta}\frac{\partial v_\phi}{\partial \phi}+\frac{v_\phi v_r}{r}+\frac{v_\theta v_\phi}{r}\cot\theta\right)$$

$$=-\frac{1}{r\sin\theta}\frac{\partial p}{\partial \phi}+\mu\left(\nabla^2 v_\phi-\frac{v_\phi}{r^2\sin^2\theta}+\frac{2}{r^2\sin\theta}\frac{\partial v_r}{\partial \phi}+\frac{2\cos\theta}{r^2\sin^2\theta}\frac{\partial v_\theta}{\partial \phi}\right)+\rho g_\phi \qquad \text{(C)}$$

[a] In these equations

$$\nabla^2=\frac{1}{r^2}\frac{\partial}{\partial r}\left(r^2\frac{\partial}{\partial r}\right)+\frac{1}{r^2\sin\theta}\frac{\partial}{\partial \theta}\left(\sin\theta\frac{\partial}{\partial \theta}\right)+\frac{1}{r^2\sin^2\theta}\left(\frac{\partial^2}{\partial \phi^2}\right)$$

**Table App. 5.** The Equation of Continuity of $A$ for Constant $\rho$ and $\mathfrak{D}_{AB}$

*Rectangular coordinates*:

$$\frac{\partial c_A}{\partial t}+\left(v_x\frac{\partial c_A}{\partial x}+v_y\frac{\partial c_A}{\partial y}+v_z\frac{\partial c_A}{\partial z}\right)=\mathfrak{D}_{AB}\left(\frac{\partial^2 c_A}{\partial x^2}+\frac{\partial^2 c_A}{\partial y^2}+\frac{\partial^2 c_A}{\partial z^2}\right)+R_A \qquad \text{(A)}$$

*Cylindrical coordinates*: (B)

$$\frac{\partial c_A}{\partial t}+\left(v_r\frac{\partial c_A}{\partial r}+v_\theta\frac{1}{r}\frac{\partial c_A}{\partial \theta}+v_z\frac{\partial c_A}{\partial z}\right)=\mathfrak{D}_{AB}\left[\frac{1}{r}\frac{\partial}{\partial r}\left(r\frac{\partial c_A}{\partial r}\right)+\frac{1}{r^2}\frac{\partial^2 c_A}{\partial \theta^2}+\frac{\partial^2 c_A}{\partial z^2}\right]+R_A$$

*Spherical coordinates*:

$$\frac{\partial c_A}{\partial t}+\left(v_r\frac{\partial c_A}{\partial r}+v_\theta\frac{1}{r}\frac{\partial c_A}{\partial \theta}+v_\phi\frac{1}{r\sin\theta}\frac{\partial c_A}{\partial \phi}\right)$$

$$=\mathfrak{D}_{AB}\left[\frac{1}{r^2}\frac{\partial}{\partial r}\left(r^2\frac{\partial c_A}{\partial r}\right)+\frac{1}{r^2\sin\theta}\frac{\partial}{\partial \theta}\left(\sin\theta\frac{\partial c_A}{\partial \theta}\right)+\frac{1}{r^2\sin^2\theta}\frac{\partial^2 c_A}{\partial \phi^2}\right]+R_A \qquad \text{(C)}$$

# NOTATION

Dimensions are given in terms of mass ($M$) or moles (mols), length ($L$), time ($t$), and temperature ($T$). Boldface symbols are vectors or tensors. Symbols that appear infrequently or in one section only are not listed.

$a_i$ = thermodynamic activity of species $i$, dimensionless.
$c$ = total molar concentration, mols$/L^3$.
$c_i$ = molar concentration of species $i$, mols$/L^3$.
$\mathbf{d}_i$ = generalized mass-transfer driving force, $L^{-1}$.
$D$ = characteristic length in dimensional analysis or diameter of sphere or cylinder, $L$.
$\mathfrak{D}_{AB}$ = binary diffusivity for system $A$-$B$, $L^2/t$.
$Đ_{ij}$ = multicomponent diffusivity of the pair $i$-$j$ based on free-energy driving force, $L^2/t$.
$\mathfrak{D}_{im}$ = effective binary diffusivity of $i$ in a multicomponent mixture, $L^2/t$.
$E_v$ = total rate of viscous dissipation of mechanical energy, $ML^2/t^3$.
$e$ = 2.71828....
$\mathbf{e}$ = total energy flux relative to stationary coordinates, $M/t^3$.
$e_v$ = friction loss factor associated with viscous dissipation dimensionless.
$\mathbf{F}$ = force of a fluid on an adjacent solid, $ML/t^2$.
$f$ = friction factor or drag coefficient, dimensionless.
$g$ = gravitational acceleration, $L/t^2$.
$\mathbf{g}_i$ = total body force per unit mass of component $i$, $L/t^2$.
$\mathbf{I}$ = current density, current$/L^2$.

$i = \sqrt{-1}$.

$\mathbf{J}_i$ = molar flux of species $i$ relative to the mass-average velocity, mols$/tL^2$.

$\mathbf{J}_i^\star$ = molar flux of species $i$ relative to the molar average velocity, mols$/tL^2$.

$\mathbf{j}_i$ = mass flux of $i$ relative to mass-average velocity, $M/tL^2$.

$K$ = kinetic energy, $ML^2/t^2$.

$k_x$ = mass-transfer coefficient in a binary system, mols$/tL^2$.

$L$ = length of tube or other characteristic length, $L$.

$M_i$ = molecular weight of species $i$.

$\mathbf{N}_i$ = molar flux with respect to stationary coordinates, mols$/L^2t$.

$\mathbf{n}_i$ = mass flux with respect to stationary coordinates, $M/L^2t$.

$\mathbf{P}$ = momentum, $ML/t$.

$\mathcal{P} = p + \rho g h$ (for constant $p$ and $g$), $M/LT^2$. (More generally $\nabla \mathcal{P} = \nabla p - \rho \mathbf{g}$.)

$p$ = fluid pressure, $M/Lt^2$.

$Q$ = volumetric flow rate, $L^3/t$.

$\mathbf{q}$ = energy flux relative to $\mathbf{v}$, $M/t^3$.

$R$ = gas constant, $ML^2/t^2T$ mols.

$R$ = radius of sphere or cylinder, $L$.

$R_A$ = molar rate of production of species $A$, mols$/tL^3$.

$r$ = radial distance in both cylindrical and spherical coordinates, $L$.

$S$ = cross-sectional area, $L^2$.

$\mathbf{S}$ = vector giving cross-sectional area and its orientation, $L^2$.

$S$ = entropy, $ML^2/t^2T$.

$T$ = *absolute* temperature, $T$.

$t$ = time, $t$.

$U$ = internal energy, $ML^2/t^2$.

$V$ = characteristic speed in dimensional analysis, $L/t$ (or dimensionless velocity).

$V$ = volume, $L^3$.

$\mathbf{v}$ = mass-average velocity, $L/t$.

$\mathbf{v}_i$ = velocity of species $i$, $L/t$.

$\mathbf{v}_\infty$ = approach velocity, $L/t$.

$\mathbf{v}^\star$ = molar-average velocity, $L/t$.

$W$ = rate of doing work on surroundings, $ML^2/t^3$.

$x$ = rectangular coordinate, $L$.

$x_i$ = mole fraction of species $i$, dimensionless.

$y$ = rectangular coordinate, $L$.

$y_i$ = mole fraction of species $i$, dimensionless.

$z$ = rectangular coordinate, $L$.

$\Gamma(x)$ = the gamma function of $x$.

$\delta$ = film thickness, penetration thickness, or boundary-layer thickness, $L$, occasionally dimensionless.
$\zeta$ = dimensionless position variable, variously defined.
$\eta$ = dimensionless position variable, variously defined.
$\theta$ = angle in cylindrical or spherical coordinates (radians).
$\mu$ = viscosity, $M/Lt$.
$\mu_i$ = chemical potential (e.g., partial molal free energy) of species $i$, $L^2/t^2$.
$\nu = \mu/\rho$ = kinematic viscosity, $L^2/t$.
$\xi$ = dimensionless position variable, variously defined.
$\Pi$ = dimensionless profiles.
$\pi$ = 3.14159... ; osmotic pressure $M/t^2L$.
$\boldsymbol{\pi}$ = pressure tensor, $M/t^2L$.
$\rho$ = fluid density, $M/L^3$.
$\rho_i$ = mass concentration of species $i$, $M/L^3$.
$\boldsymbol{\tau}$ = shear stress tensor, $M/t^2L$.
$\tau_0$ = magnitude of shear stress at fluid-solid interface, $M/t^2L$.
$\Phi$ = potential energy, $ML^2/t^2$.
$\Phi$ = dimensionless electrostatic potential.
$\boldsymbol{\phi} = \rho\mathbf{v}\mathbf{v} + \boldsymbol{\pi}$ = total momentum flux relative to stationary coordinates, $M/t^2L$.
$\phi$ = angle in spherical coordinates, radians.
$\phi$ = electrostatic potential, volts.
$\phi_i$ = volume fraction of species $i$ in a mixture.
$\psi$ = stream function; dimensions depend on coordinate system.
$\Omega$ = angular velocity, radians$/t$.
$\omega_i$ = mass fraction of $i$, dimensionless.

DIACRITICAL MARKS

$\tilde{}$, per mole.
$\hat{}$, per unit mass.
$\bar{}$, partial molal.
$\bar{}$, time-smoothed.

BRACKETS

$\langle a \rangle$, average value of $a$ over a flow cross section.
$[=]$, has the dimensions of.

SUPERSCRIPTS

$^0$, value at 1 atm or other standard state.
$'$, deviation from time-smoothed value.

SUBSCRIPTS

$A, B$, species in binary systems.
$b$, bulk or "cup mixing" value for enclosed stream.
$i, j, k, m, n$, species in multicomponent systems.
loc, local transfer coefficient.
$m$, mean transfer coefficient for a submerged object.
tot, total quantity in a macroscopic system.
0, quantity evaluated at a surface.
1,2, quantity evaluated at cross sections "1" and "2".

COMMONLY USED DIMENSIONLESS GROUPS

Nu = Nusselt number for heat transfer.
$\mathrm{Nu}_{AB}$ = Nusselt number for mass transfer.
Pr = Prandtl number.
Re = Reynolds number.
Sc = Schmidt number.

MATHEMATICAL OPERATIONS

$\frac{D}{Dt}$ = substantial derivative, $= \frac{\partial}{\partial t} + \mathbf{v} \cdot \boldsymbol{\nabla}$.
$\operatorname{erf} x = \frac{2}{\sqrt{\pi}} \int_0^x e^{-t^2}\, dt$ = the error function of $x$.
$\ln x$ = the logarithm of $x$ to the base $e$.
$\log_{10} x$ = the logarithm of $x$ to the base 10.
$\Gamma(x) = \int_0^\infty t^{x-1} e^{-t}\, dt$ = the (complete) gamma function.
$\nabla$ = the "del" or "nabla" operator.

# PROBLEMS

## PROBLEMS* FOR CHAPTER I

### I.$A_D$. *ENGINEERS AND EVOLUTION*

Compare critically engineering solutions to several characteristic momentum and mass-transfer problems with their counterparts in biological systems. (See e.g., Bugliarello, 1966.)

### I.$B_D$. *FOUNDATIONS OF BIOMEDICAL ENGINEERING*

Look up biographies of some early contributors to applied biology, for example those cited in the text, and comment critically on their contributions. How were they affected by the technological limitations of their time?

* It is suggested that the D-class problems of these chapters be used as topics for short but carefully researched and prepared term papers. The approach intended for text problems is indicated by the subscript following the problem number: D refers to generally non-mathematical discussion; 1 to numerical problems requiring no more than substitution into available formulas; 2 to those requiring only elementary analysis; 3 to those requiring mature judgement, and in some cases, sources outside the text; 4 to more challenging problems, particularly those requiring mathematics not generally mastered by undergraduates.

## I.$C_D$. *THE BASIS OF INVENTION*

Search for the origins of an important medical device (e.g., thermometers, stethoscopes, proctoscopes) or procedure (e.g., electrocardiography). Seek, in particular, information on motivation, the qualifications of those responsible for development, and the role of engineering science. Also do not forget to explain the technical foundation on which the development rests.

## I.$D_D$. *THE IMPACT OF "ENGINEERING SCIENCE"*

(a) Look for evidence that basic developments in technology motivated medical oriented engineers and scientists to produce medically useful developments.

(b) Suggest possible medical applications for facts and concepts you have acquired during your education. (Do not be too critical here—look for possibilities.)

## I.$E_D$. *WHAT DID THEY REALLY DO?*

The tendency in textbooks is to put important developments into well-organized and easily understood form. Leaf through the text for results of interest to you and trace them back to the original work. Comment on your experience.

## I.$F_D$. *THE BODY AS A UNIT*

Look up alternative references to overall characteristics of the body and compare your results with the text. Compare the variability of various measures—for example, body mass, energy conversion, fluid composition, and the make-up of blood.

## I.$G_D$. *BIOENERGETICS*

Expand Table 2.1.2 by more detailed analysis of body energy metabolism. Discuss in particular the efficiency of the phosphate-linked energy conversion (see, e.g., Lehninger, 1971).

## I.$H_D$. *THE CELL AS A CHEMICAL PLANT*

Compare individual cells and industrial chemical processing units in terms of complexity of function, the effects of scale, and the degree of organization. (It will take a lot of thought and background reading to do this one right.)

### I.I$_D$. *CARDIOVASCULAR MODELING*

(a) Compare the "pump characteristics" shown in Fig. 2.2.5 with those you would expect for a positive displacement pump in a system of rigid pipes.

(b) Compare the hydrodynamic behavior shown for the major arteries in Fig. 2.2.3 with that of the smaller vessels.

(c) Show that the *Windkessel* analogs of Fig. 2.2.58 are valid. Can you think up any others?

(d) Describe the behavior of the system shown in Fig. 2.2.6 in terms of its use as a flow model.

### I.J$_3$. *CHARACTERISTICS OF THE WINDKESSEL MODEL*

The outflow from the heart can be written in complex form as a Fourier series:

$$Q = \sum_{n=0}^{\infty} \tilde{Q}_n e^{in\omega t} \tag{1}$$

where $\tilde{Q}_n = \tilde{Q}_n(r) + iQ_n(i)$ is a complex constant and $\omega$ is the frequency. The individual Fourier elements $Q_n$ can be separately calculated for the *Windkessel* model (Eq. 2.2.2), since this relation is linear.

(a) Calculate the pressure-flow relation for the first harmonic ($n = 1$).

(b) Discuss the effect of arterial elasticity. In particular, how does it moderate arterial pulses (given that cardiac output is independent of aortic pressure). How does it affect the cyclic and steady flow components to the microcirculation?

(c) What is the relation for $Q_\omega = Q_0 \cos \omega t$?

(d) What is the outflow rate to the microcirculation for (c)?

PARTIAL ANSWER

(a) $p_\omega = [R/(1 + iN)]Q_\omega$; $N = R\omega/k$
(c) $p_\omega = [RQ_0/(1 + N^2)](\cos \omega t + N \sin \omega t)$
(d) $Q_{\text{out}} = [Q_{\text{in}}/(1 + N^2)](\cos \omega t + N \sin \omega t)$

## PROBLEMS FOR CHAPTER II

### II.1.A$_D$. *HYDRODYNAMIC EFFECTS OF SERUM PROTEINS*

(a) Compare the effects of shear rate, temperature, and protein composition on the apparent viscosity of blood and plasma.

(b) A large number of diseases are accompanied by elevated globulin concentrations. What effect would you expect this to have on flow in the microcirculation? (You will have to speculate a bit here.)

### II.1.$B_D$. *VISCOMETRIC MEASUREMENTS*

(a) Look up the Cokelet review and any pertinent references to the problems of interpreting viscometric measurements. Summarize your findings.

(b) Is the yield stress of blood a meaningful concept in the sense that it represents a realizable low-shear-rate limit? Discuss in the light of Fig. 1.1.7.

### II.1.$C_D$. *CASSON VERSUS NEWTON*

It is stated in the text that although blood follows the Casson equation reasonably well at low shear rates, it should be considered as Newtonian at high shear. Is this a necessary distinction in view of the fact that the Casson equation becomes identical with Newton's law of viscosity in the high-shear limit? Consider Fig. 1.1.9 in preparing your answer, and justify your conclusions.

### II.1.$D_D$. *CONTROVERSIAL RED-CELL RHEOLOGY*

Rheological measurements on red cells are necessarily ambiguous, and the results depend—often quite strongly—on measurement procedures. Discuss this problem after reading the pertinent literature. Include the general references in the Bibliography as well as those dealing specifically with red-cell membranes.

### II.1.$E_2$. *RABINOWITSCH AND CASSON*

Does a fluid whose rheological behavior is described by the Casson relation (Eq. 1.1.8) meet the requirements of Eq. 1.1.6? You may wish to use the integrated Casson equation, as described in Problem II.3-G.

### II.1.$F_2$. *THE RABINOWITSCH EQUATION*

(a) Show that an approach similar to the Rabinowitsch development can be used to investigate flow between two *close-fitting* coaxial cylinders.

(b) How would you expect entrance effects to affect a Rabinowitsch plot?

(c) Can the Rabinowitsch approach be used for any other geometries? Which? Why?

## II.1.$G_2$. *THE FÅHREUS-LINDQVIST EFFECT*

(a) How would the high-shear anomaly shown in Fig. 1.1.5 look in Fig. 1.1.7? (Only a qualitative answer is wanted here.)

(b) How would you expect the effect shown in Fig. 1.1.8 to affect the behavior of a capillary viscometer in which the bulk of the upstream reservoir is emptied during viscosity determinations? Be as quantitative as you can.

## II.1.$H_2$. *HYALURONATE RHEOLOGY*

(a) Show that curves $E$ and $F$ of Fig. 1.2.1 are qualitatively consistent with $G''(\omega)$ in Fig. 1.2.2. Equation 1.2.6 may prove useful to you.

(b) How would you expect the apparent viscosity of synovial fluid in your knee joints to vary with the speed of walking? How about energy dissipation?

(c) Look up some of the listed references on synovial rheology and discuss (1) the relation between rheological properties and effectiveness of lubrication; (2) the importance of lubrication theory in medicine.

## II.1.$I_2$. *RED-CELL RHEOLOGY À LA RAND*

(a) Develop limiting relations between pressure differences required to rupture a red cell as a function of rupture time for: (1) long times (greater than a minute) and (2) short times (less than a second).

(b) Describe the differences in red-cell behavior under these two sets of conditions.

(c) Calculate some actual pressures for selected times.

(d) Recalculate them in terms of (tensile) membrane stresses for a membrane thickness of 80 Å.

## II.1.$J_2$. *THE MACRORHEOLOGY OF SKELETAL STRUCTURES*

Carl Hirsch and Lars Sonnerup [*J. Biomech.*, **1**, 13–18 (1968)] suggest that the stress-strain behavior of bone can be described by the simple visco-elastic model

(a) Show that stress-strain behavior is described by

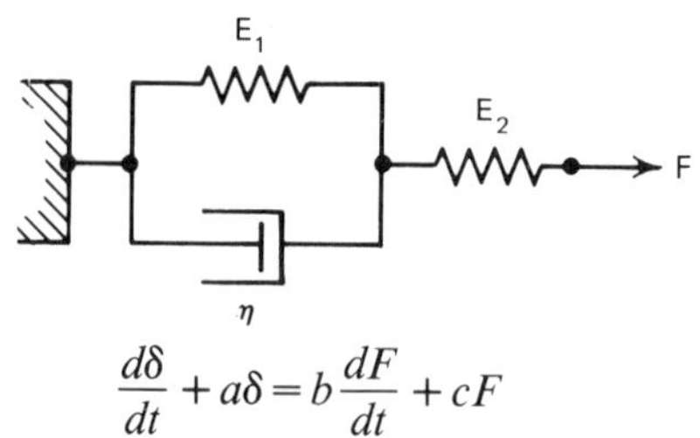


$$\frac{d\delta}{dt} + a\delta = b\frac{dF}{dt} + cF$$

and identify the coefficients a, b and c with the reheological parameters $E_1$, $E_2$, and $\eta$.

(b) Describe the response of this model to the three common experimental tests: (1) The sudden imposition of a constant force F, (2) The sudden impositon of a constant strain $\delta$, and (3) The imposition of a cyclic strain $\delta = \delta_0 \cos \omega t$ where $\delta_0$ and $\omega$ are constants. See also A. Viidik, *J. Biomech.*, **1**, 3–11, (1968).

## II.2.A$_1$. *THEORETICAL SEDIMENTATION RATES*

(a) Estimate the sedimentation rate of an isolated red cell, assuming it to be an oblate spheriod with major and minor axes of 8.5 and 2.4 $\mu$ (see Fig. 1.1.1) filled with a solution of 35% hemoglobin by weight. For the purposes of this problem all proteins may be considered to have a partial specific volume of 0.75 cm$^3$/g.

(b) Try to find characteristic sedimentation rates for human blood in various conditions of health. Compare these with your above calculation, and discuss.

## II.2.B$_1$. *STOKES'S LAW IN THE LUNG*

The mean residence time of air in the adult human lung is about $\frac{1}{2}$ min (see Chapter III, Section 6). Calculate the distance a suspended aerosol particle of density 3 g cm$^{-3}$ will fall relative to air in that time, using a kinematic viscosity of 0.17 cm$^2$/sec. Assume the particles to be spheres in the diameter range of 50 $\mu$ to zero. These calculations are of very considerable practical importance. The larger particles fall rapidly on to the mucous-lined walls of the larger ducts, and are pumped up out of the lungs by ciliary actions. The very small particles, on the other hand, do not depart appreciably from the gas streamlines (except for Brownian motion) and are not deposited at all; neither these nor the large ones are dangerous. However, particles of intermediiate size, on the order of 2 $\mu$, fall fast enough to be deposited in the alveoli and the smaller ducts, which to not have cilia; these remain in the lung and produce permanent impairment. There is a large literature in this important area.

## II.2.$C_1$. *EFFECTIVE VISCOSITY OF SUSPENSIONS*

(a) Compare Eqs. 8, 9, and 10 of Ex. 2.3.1 with the data of Fig. 1.1.3 for rigid spheres.

(b) Why should rigid discs and sickled cells give higher apparent viscosities than rigid spheres? Why should emulsions give less?

(c) How meaningful is the concept of effective viscosity?

## II.2.$D_1$. *BROWNIAN MOTION AND HYDRODYNAMIC SHEAR*

The relative effects of Brownian motion and hydrodynamic shear can be expressed in terms of the dimensionless ratio

$$N = \frac{t_B}{T}$$

where $T$ is the rotational period of a hydrodynamic particle in a shear field, given by Eq. 2.3.10. The Brownian time constant $t_B$ is a measure of the time required for Brownian motion to randomize the orientation of particles; it is described more completely in Problem II.2.N. Hydrodynamic shear has a predominant effect on particle behavior when $N \gg 1$; this is, for example, the region of shear degradation and non-Newtonian viscosity. It has a negligible effect when $N \ll 1$.

Typical values of $t_B$ are given in Table II.2.$D_1$.1. (See next page.)

The quantity $R_\omega$ is the ratio of torque required to rotate the particle divided by the resulting angular velocity. Estimate $t_B$ for:

(a) Red cells, approximated both as discs of radius 4 $\mu$ and spheres of radius of 2.5 $\mu$.

(b) Serum albumin approximated as a prolate ellipsoid with $a = 75$ Å and $b = 25$ Å.

(c) Fibrinogen as a cylinder with $a = 250$ Å and $b = 10$ Å. (See also Problem II.2.N). Discuss the relative importance of hydrodynamic and Brownian forces.

APPROXIMATE ANSWERS

$\sim 10^2$, $10^{-6}$, $10^{-4}$ sec, respectively.

## II.2.$E_2$. *RED-CELL SETTLING*

(a) Calculate the steady sedimentation rate of a red cell, and compare with that of a sphere of equal volume. Assume the red cell to be an oblate ellipsoid of revolution with an axis ratio of 4. Assume that the red-cell interior is a 35% hemoglobin solution and that all serum proteins have a partial specific volume of 0.75 $cm^3/g$.

(b) Calculate the effect of aggregation on sedimentation rates assuming spherical aggregates of (1) 50% by volume of occluded plasma, and (2) no occluded plasma.

(c) Compare the sedimentation rates of rouleaux and spherical aggregates of the same volume. Assume for these purposes that rouleaux act like prolate elliposiids with an axis ratio equal the actual length-to-diameter ratio. Discuss this assumption briefly.

## II.2.$F_2$. *FLOW ENTRAINED BY A MOVING PARTICLE: SIGNIFICANCE OF THE STREAM FUNCTION*

It is common practice, for example in discussing so-called zeta potentials, to talk about liquid moving with a suspended particle.

(a) Use Eq. 2.2.4 along with Eq. 1 of Ex. 2.1.1 to show how the amount of "entrained" fluid varies with the distance from the particle surface for a sphere. For simplicity make your calculation for $\theta=\pi/2$. Define entrainment as the volumetric rate of fluid flow across the plane $\theta=\pi/2$ resulting from motion of the sphere relative to the fluid.

(b) Discuss the significance of your result.

(c) What does Eq. 2.2.7 say about the corresponding result for nonspherical particles. (See also Eqs. 2.2.9 and 2.2.10).

**Table II.2.$D_1$.1.** Typical Values of the Brownian Time Constant $t_B$[a]

| Shape | $R_\omega=(\kappa T)t_B$ | Comment |
|---|---|---|
| Sphere | $8\pi R^3\mu$ | Radius $R$ |
| Disc | $\frac{32}{3}R^3\mu$ | Radius $R$, rotation about diameter *or* axis of symmetry. |
| Prolate ellipsoid | $\dfrac{\frac{8}{3}\pi\mu a^3}{\ln(2a/b)-0.5}$ | Approximate, $a\gg b$, where $a$ and $b$ are half axes; rotation about transverse axis |
| Cylinder | $\dfrac{\frac{8}{3}\pi\mu a^3}{\ln(2a/b)-0.8}$ | Approximate, $a\gg b$, where $a=$ half length and $b=$ radius; rotation about transverse axis. |

[a] See Happel and Brenner (1965) and Burgers (1938) for a more complete listing and discussion. Note that $\kappa=$ Boltzmann's constant $=1.380\times10^{-16}$ dyn cm/(molecule)(°K).

ANSWER

(b) Since the entrainment so defined is not finite, the concept is at best ambiguous.

### II.2.$G_2$. *TRANSPORT PROPERTIES OF PROTEIN SOLUTIONS*

Use the results of Ex. 2.2.1 and Section 2.3 to develop an expression for the Schmidt numbers, $\text{Sc} = \mu/\rho\mathfrak{D}_{AB}$, of suspensions of ellipsoids of revolution. Show how Sc changes with the molecular shape of a prolate globular protein.

### II.2.$H_{2,D}$. *EFFECTIVE VISCOSITY OF CONCENTRATED SUSPENSIONS*

(a) Compare Eqs. 8, 9, and 10 of Ex. 2.3.1 with the data of Fig. 1.1.3, for rigid spheres and discs.

(b) Why should the curve for sickled cells lie above the curve for rigid spheres?

(c) Why should that for emulsions lie below (at least for high volume fractions)?

(d) Why should normal red cells give the lowest viscosities of all? What does this have to do with the shape of the relaxed red cell?

### II.2.$I_2$. *PARTICLE SHAPE FROM VISCOSITY MEASUREMENTS*

(a) Show how the length of rouleaux could be estimated from viscosity measurements in dilute suspensions subjected to a uniform shear rate. How sensitive is this technique over the range of $L:D$ between 1:4 (oblate ellipsoids containing one red cell) and 10:1 (prolate ellipsoids roughly comparable to rouleaux of forty red cells). What are the major sources of ambiguity in this technique? [You will need to look up one of the basic references, for example, Tanford (1961), to answer this question precisely.]

(b) It is often said that the decrease in the effective viscosity of blood with increasing shear is due to rouleaux breakup. The high-shear asymptotic viscosity corresponds in this argument to isolated red cells. How sound do you consider this theory?

### II.2.$J_2$. *BROWNIAN "TORQUES"*

In Ex. 2.2.2 Brownian forces tending to randomize the orientation of the dumbbell were neglected. These are most simply expressed as the ratio $R_\omega = \mathcal{T}/\omega$, where $\mathcal{T}$ is the torque required to spin the dumbbell at an angular velocity $\omega$. See Table II. 2.D. Redo Problem II.2.D for fibrinogen considering it to be a dumbbell with $2l = 475$ Å and $2R = 65$ Å.

PARTIAL ANSWER

$$R_\omega = 12\pi\mu R l^2$$

## II.2.K$_3$. *DUMBBELLS AND RED-CELL ADHESION*

A rouleaux of red cells closely resembles a row of spheres, which in turn is a simple generalization of the dumbbell of Fig. 2.2.2. It has been shown by Burgers (1938) that to a reasonable first approximation the drag forces between individual spheres are unaffected by sphere-sphere interactions. Generalize Eq. 8 of this example for such a chain of spheres. Show how one can calculate the adhesive forces holding rouleaux together by relating rouleaux length to the shear field in which they find themselves.

PARTIAL ANSWER

$$F_{\max} = 3\pi R l \dot{\gamma}\left(1 + \frac{l}{D}\right) \doteq \tfrac{3}{2}\pi\mu l^2 \dot{\gamma}$$

## II.2.L$_3$. *SUSPENSION RHEOLOGY SIMPLIFIED*

Calculate the instantaneous effective viscosity of a suspension of Kuhn dumbbells at a concentration of $n$ dumbbells per unit volume *all aligned at the orientation of maximum force* corresponding to Eq. 7, Ex. 2.2.1. Define

$$(-\boldsymbol{\tau} : \nabla \mathbf{v}) = \mu_{eff}\Phi_v^0$$

(see, e.g., *Tr. Ph.*, Table 3.4-8, p. 91), where $\Phi_v^0$ is the dissipation function in the absence of dumbbells. Assume $(-\boldsymbol{\tau} : \nabla\mathbf{v}) = \mu\Phi_v^0 + 2n\mathbf{F}\cdot(\mathbf{v}_f - \mathbf{v}_s)$, where $\mathbf{F}$ and $(\mathbf{v}_s - \mathbf{v}_f)$ are given in the text. (Why is this reasonable?) Note that $\mathbf{F} = \boldsymbol{\delta}_r F_r$ so that $\mathbf{F}\cdot(\mathbf{v}_f - \mathbf{v}_s) = F_r(v_f)_r$. (Why?) Would the instantaneous $\mu_{\text{eff}}$ be greater or less for other orientations?

## II.2.M$_3$. *LOW-REYNOLDS-NUMBER PROPULSION*

Consider Ex. 2.5.1 and remember that for any point on the tail $(y_p, z_p)$,

$$y = b \sin kz$$

$$v_y = \frac{dy_p}{dt} = -Q\sin\theta$$

$$v_z = \frac{dz_p}{dt} = -Q\cos\theta$$

in the $z$-$y$ coordinate system. Assume $bk \ll 1$, and note that

$$\theta = \tan^{-1}\frac{dy_p}{dz_p}$$

$$\frac{dy_p}{dz_p} = bk\cos kz$$

(a) Show that

$$\frac{dz_p}{dt} = -Q[1 - \tfrac{1}{2}b^2k^2\cos^2 kz + O(b^4k^4)]$$

by expanding both $\tan^{-1}u$ and $\cos u$ in a truncated series.

(b) Show* that the time $T$ required for a point to move one cycle ($0 < kz < 2\pi$) is

$$T = 2\pi\frac{1 + \tfrac{1}{2}b^2k^2}{kQ} \qquad (\text{ to order } b^2k^2)$$

(Note that $z$ is decreasing.)

(c) The distance moved by this same point relative to the head, $\Delta x(T)$, is zero. Then from the definition $kz = k(x - ct)$ it follows that

$$T = \frac{2\pi}{kc}$$

Show from these two relations for $T$ that

$$Q = c(1 + \tfrac{1}{2}b^2k^2)$$

(d) From intermediate results in part (b) you should (if neither you nor I made a mistake) be able to show that

$$kx_p = b^2k^2\sin kz\cos kz$$

From this and the expression

$$ky_p = bk\sin kz$$

* Use the expansion

$$[1 - b^2k^2\cos kz]^{-1} = 1 + b^2k^2\cos kz \text{ to order } b^2k^2$$

it is possible to relate $x_p$ and $y_p$, the horizontal and vertical positions in coordinates fixed to the organism head. Make such a plot for $bk=0.458$.

### II.2.N$_4$. *BROWNIAN MOTION AND ROTATIONAL DIFFUSION*

Any particle suspended in a fluid is subject to Brownian forces which tend to change its orientation, (i.e., the angles $\theta$ and $\phi$ in Fig. 2.3.2) as well as the position of its center. In both cases the effect is to make all possible orientations and positions equally likely, and the rotational diffusion can be described by a two-dimentional analog of Fick's law:

$$\frac{\partial P}{\partial t}=\left(\frac{\mathfrak{D}^{(\omega)}}{R^2}\right)\frac{1}{\sin\theta}\left\{\frac{\partial}{\partial\theta}\sin\theta\frac{\partial P}{\partial\theta}+\frac{\partial}{\partial\phi}\frac{1}{\sin\theta}\frac{\partial P}{\partial\phi}\right\}$$

where $P$ is the probability of any orientation $(\theta,\phi)$, and $(\mathfrak{D}^{(\omega)}/R^2)$ is the angular equivalent to the binary diffusivity $\mathfrak{D}_{AB}$. See Table II.2.D, p. 434.

Describe $P(\theta,\phi)$ as a function of time for an axisymmetric particle in a stagnant fluid initially at $\theta=0$. Use a normalized definition so that $P$ approaches unity for all orientations at lung times. Note that $\partial P/\partial\phi$ is always zero for this situation.

(Note that the two ends of the particle are treated as distinct.)

### II.3.A$_1$. *NATURE OF THE CASSON EQUATION*

(a) Calculate $\tau_y$ and $s^2$ for the type-O banked blood referred to in the text for $H=0.45$ (i.e., in volume fraction rather than volume percent) and a fibrinogen concentration of 0.3g/100 cm$^3$. The viscosity of the suspending plasma is 1.3 cp at the temperature of interest [1cp$=10^{-2}$g/(cm)(sec)], and the exponent $K$ may be taken as unity.

(b) Calculate the velocity profile for this blood in a tube with an inside diameter of 1.0 cm at a wall shear stress of 15 dyn/cm$^2$. (This is roughly representative of the steady flow component in major arteries. See Problem II.3.G for the nature of the velocity profile.)

(c) Compare the above velocity profile with that for Newtonian flow at the same maximum velocity. What does this tell you about the Newtonian approximation?

### II.3.B$_1$. *ARTERIAL RESPONSE TIMES*

Check the entries in Table 3.4.1 against solutions for unsteady tube flow, which are available in such standard references as R. B. Bird et al. (1971), *Tr. Ph.*, or H. Schlichting, *Boundary Layer Theory*, 4th ed., McGraw-Hill (1960).

## II.3.$C_{1,D}$. *ENTRANCE EFFECTS IN SLIT FLOW*

Use the result of Ex. 3.4.1 and the data of Problem II.3.H, to determine the importance of entrance effects in the dialyzer discussed in the example.

(a) Calculate the entrance length.

(b) How important would you expect hydrodynamic entrance effects to be for a dialyzer about 1m long?

ANSWER

(a) $Lo/B \sim \frac{3}{4}$. (The calculation technique is not very accurate in this Reynolds-number range; why is this not important?)

## II.3.$D_1$. *WAVE CHARACTERISTICS*

(a) Calculate the length, phase velocity, and damping characteristics of a sinusoidal disturbance for

$$\omega = \frac{2\pi}{T} = 2.5\,\text{sec}^{-1}$$

$$R = 1.25 \text{ cm}$$

$$\nu = 0.04\text{cm}^2/\text{sec}$$

$$h = 0.2 \text{ cm}$$

$$E = 6.0 \times 10^6 \text{ dyn/cm}^2$$

$$\sigma = \tfrac{1}{2}$$

$$\rho_{\text{wall}} \approx \rho_{\text{blood}} \approx 1\text{g/cm}^3$$

These conditions are characteristic of the fundamental frequency in the aorta. (See footnote in connection with Eq. 3.5.37.)

(b) State qualitatively how the higher harmonics will compare with the above calculations.

## II.3.$E_1$. Wave Propagation

Consider a pressure wave of the form

$$\frac{P}{P_0} = e^{-\zeta}\,\text{Re}\left\{ e^{i(\tau - \zeta)} \right\}$$

where $\zeta = z/R$; $t = \omega t$; $e^{i\chi} = \cos\chi + i\sin\chi$

(a) Sketch the shape of this wave at $t=0$ and state the direction in which it is traveling.

(b) Calculate:

(1) the change in amplitude over a length equal to one tube diameter $(=2R)$; (2) the celerity, or wave speed; (3) the wavelength.

(c) Compare the above wave with the fundamental pressure wave of the aorta for damping characteristics and wavelength.

### II.3.F$_1$. *HARMONIC ANALYSIS*

Carry out the calculations required in Ex. 3.5.4, and check your results against Fig. 3.5.6.

### II.3.G$_2$. *VELOCITY PROFILES ACCORDING TO THE CASSON EQUATION*

(a) Develop an expression for the velocity profile for a Casson fluid in steady tube flow by noting that

$$\tau_{rz} = \tau_w(r/R) \qquad \text{(neglecting end effects)} \tag{1}$$

$$-s^2\frac{dv_z}{dr} = \tau_{rz} - 2\sqrt{\tau_{rz}\tau_y} + \tau_y \tag{2}$$

(b) Why is the second of the above expressions necessary?

ANSWER

$$v_z = \frac{R\tau_w}{s^2}\left\{\frac{1}{2}\left[1-\left(\frac{r}{R}\right)^2\right] - \frac{4}{3}\sqrt{\frac{\tau_y}{\tau_w}}\left[1-\left(\frac{r}{R}\right)^{3/2}\right] + \left(\frac{\tau_y}{\tau_w}\right)\left(1-\frac{r}{R}\right)\right\}$$

for $\dfrac{\tau_y}{\tau_w} \leqslant \dfrac{r}{R}$

$$v_z = \left(\frac{R\tau_w}{s^2}\right)\left\{\frac{1}{2}\left[1-\left(\frac{\tau_y}{\tau_w}\right)^2\right] - \frac{4}{3}\sqrt{\frac{\tau_y}{\tau_w}}\left[1-\left(\frac{\tau_y}{\tau_w}\right)^{3/2}\right] + \frac{\tau_y}{\tau_w}\left[1-\frac{\tau_y}{\tau_w}\right]\right\}$$

for $\dfrac{\tau_y}{\tau_w} \geqslant \dfrac{r}{R}$

## II.3.H$_2$. *WALL EFFECTS IN A DIALYZER CHANNEL*

(a) In a typical dialyzer of the type shown in Fig. 3.2.1*b*, the total volumetric flow rate is about 350 cm$^3$/min, $B \doteq 0.1$ mm, and $W \doteq 10$ cm. Note from Section 1 that the kinematic viscosity of blood is about 0.04 cm$^2$/sec.

Use the velocity profile of Problem II.3.R to determine the importance of wall effects (i.e., departures from the slit flow result near $x = \pm w$). Sketch this profile. What is your conclusion?

(b) Show that the velocity profile you began with does meet the pertinent differential equation and boundary conditions.

## II.3.I$_2$. *EFFECT OF DISTORTION ON FLOW DISTRIBUTION*

Consider the situation shown at the right in Fig. 3.2.1*b*, where the blood channel has been distorted by unbalanced pressures.

(a) Develop an expression for the velocity distribution and volumetric flow per unit width if the half channel thickness becomes

$$\frac{B}{B_0} = 1 + \alpha\left(\frac{x}{W}\right)^2$$

where $B_0$ is local half thickness, and $\alpha$ is another constant. Use the Hele-Shaw approximation.

(b) What do the results look like for $\alpha$ equal to unity?

## II.3.J$_2$. *STEADY FLOW THROUGH AN ELASTIC TUBE*

Develop an expression for the shape of an elastic tube subjected to internal pressures $p_0$ and $p_L$ at its upstream and downstream ends, respectively, and the resultant steady flow rate through the tube. (See the figure, and assume that the tube radius is proportional to pressure.)

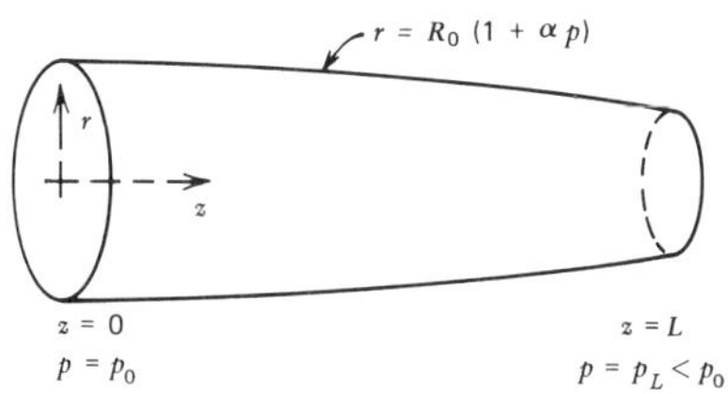


Steady flow through an elastic tube.

II.3.K$_2$. *ULTRAFILTRATION IN SLIT FLOW*

Parallel Ex. 3.3.1 for a linear variation of seepage rate with position. Show that this is a degenerate case of the example.

II.3.L$_2$. *INVISCID TUBE FLOW*

(a) Develop an expression for the velocity profile in a long rigid tube of length $L$ filled with an incompressible inviscid fluid for the following boundary conditions:

$$\text{At } z=L, \quad p=0 \qquad (\text{all time})$$

$$\text{At } z=0, \quad p=0 \qquad (t<0)$$

$$(1)\ \text{At } z=0, p=p_0 \cos \omega t\, t>0$$

and

$$(2)\ \text{At } z=0, \quad p=p_0 \sin \omega t>0$$

(b) Discuss the importance of transients and the behavior for arbitrary time dependence of the terminal pressures.

II.3.M$_2$. *PULSATILE FLOW IN A STRAIGHT DUCT*

Modify Ex. 3.5.2 by using the following alternative "downstream" boundary condition:

$$\text{At} \quad z=L, \quad \langle v_z \rangle = 0$$

PARTIAL SOLUTION:

$$\tilde{P}=P_0\left(\cosh \sqrt{iCZ}\, \zeta - \sinh \sqrt{iCZ}\, \zeta \tanh \sqrt{iCZ}\, \lambda\right)$$

II.3.N$_2$. *BURSTING A BLOOD VESSEL*

Consider an elastic tube of relaxed radius $R_0$ and volume $V$ extended to radius $R_0+\Delta R$ by an internal gauge pressure $p$. Assume Hooke's law,

$$\text{stress} = E \cdot (\text{fractional elongation})$$

to hold, *and V to be constant.* Calculate $\Delta R$ as a function of $p$, and discuss the need for collagen in blood vessels. What for example would be the maximum pressure that a tube with the dimensions of the aorta in Fig. 3.1.2 could resist if it exhibited an effective Young's modulus of (a) $6 \times 10^6$ dyn/cm$^2$ (typical of elastin; see Table 3.1.2) or (b) $10^8$ dyn/cm$^2$ (typical of collagen)? (Note: the results of this highly simplified analysis should not be relied on quantitatively.)

## II.3.O$_3$. *ENTRANCE LENGTHS IN TUBE FLOW*

Extend Ex. 3.4.1 to the analogous case of tube flow and discuss your result in the light of the discussion in Section 3.4. If you have time, look up the paper of Lee and Fung (1970) for a more detailed comparison.

## II.3.P$_3$. *SEPARATION OF VARIABLES IN OSCILLATORY RIGID-DUCT FLOW*

(a) Parallel the development given in Eqs. 3.5.1 through 3.5.15 for a Casson fluid.

(b) To what extent can this development be generalized with respect to duct geometry and fluid rheological behavior?

## II.3.Q$_3$. *FLOW OF A SPHERE IN A CLOSE-FITTING TUBE: LUBRICATION THEORY*

A sphere of radius $R$ is flowing with a Newtonian liquid in a tube of radius $(R+\delta)$ where $\delta \ll R$.

(a) Show to a first approximation that in the region of appreciable viscous stresses,

$$\Delta = \delta + \frac{Z^2}{2R} \tag{1}$$

where $\Delta$ is the distance measured perpendicularly from the tube wall to the sphere surface, and $Z$ is the distance measured axially from the sphere center.

(b) Show that if the sphere is moving axially at a velocity $V$, the corresponding fluid velocity profile is

$$\frac{v_z}{V} = \frac{\eta}{\Delta} - \frac{\Delta^2(\partial \mathcal{P}/\partial z)}{2\mu V}(\eta - \eta^2) \tag{2}$$

where $\eta = y/\delta$, and $y$ is distance from the tube wall.

(c) *Show how one can* calculate the pressure drop across the sphere and the flow of liquid relative to the sphere by using (1) a force balance on the sphere surface, neglecting buoyancy effects and acceleration of the sphere, (2) the conservation of mass between any two planes of constant $z$, and (3) Eqs. (1) and (2) for $-\infty < z < \infty$.

(d) Sketch velocity profiles for this system and comment qualitatively on the pressure drop resulting from presence of the sphere. How would the deformability of the red cell affect its behavior in a similar situation?

*Note*: For a more thorough discussion of this problem see R. M. Skalak, "Mechanics of the Micro-circulation," in *Foundations of Biomechanics,* Y. C. Fung, Max Anliker, and N. Perrone, Eds., Prentice-Hall (1972).

## II.3.R$_4$. *FLOW IN A RECTANGULAR CHANNEL*

(a) Parallel the general approach of Section 3.2 to describe the velocity profile in the idealized (rectangular) dialyzer blood channel of Fig. 3.2. 1(*b*). (Note: since $W \gg B$, there is no need for a symmetrical relation. Rather it is simplest to write

$$\frac{v_z}{-(1/\mu)d\mathcal{P}/dz} = \tfrac{1}{2}(B^2 - y^2) + f(x,y)$$

where $\nabla^2 f = 0$ and $f$ is zero at $y^2 = B^2$.

(b) Discuss the problem of determining the velocity profile in the dialysate channels, which to a first approximation have an equilateral triangular cross section bounded by the surfaces $y = H$, and $y = \pm\sqrt{3}\,x$. The velocity profile

$$v_z = \left(\frac{\mathcal{P}_0 - \mathcal{P}_2}{4\mu LH}\right)(y - H)(3x^2 - y^2)$$

is surprisingly simple. An extensive discussion of duct flows is given by R. Berker, *Handbuch der Physik*, Vol. III/2, Springer, Berlin (1963), pp 67-77.

ANSWER

$$\text{(a)}\quad f = -2B^2\left(\frac{2}{\pi}\right)^3 \sum_{n=0}^{\infty} \frac{(-1)^n}{(2n+1)^3} \frac{\cosh[(2n+1)\pi z/2B]}{\cosh[(2n+1)\pi W/2B]} \cos\left[(2n+1)\frac{\pi y}{2B}\right]$$

## II.4.$A_1$. *KINETIC AND HYDRODYNAMIC ENERGY*

(a) Use Fig. 2.2.2(*b*) of Chapter I to compare the *orders of magnitude* of the kinetic energy flux $\frac{1}{2}\langle v_i^3\rangle$ with the corresponding hydrodynamic work term $p_i\langle v_i\rangle/\rho$ for the major arteries under resting conditions. Assume flat velocity profiles and neglect the differences in phase between pressure and velocity, but discuss the probable effects of these assumptions briefly.

(b) How would you expect this comparison to look at twice normal cardiac output? (This is quite an open-ended question and will take considerable thought.)

PARTIAL ANSWER

The ratio of kinetic to hydrodynamic energy varies from the order of $10^{-1}$ at the ascending aorta to $10^{-2}$ at the saphenous artery.

## II.4.$B_2$. *HARMONIC ASPECTS OF VISCOUS DISSIPATION*

A typical small artery has a diameter of 1 mm and a time-average blood velocity $\langle\bar{v}\rangle$ of 10 cm/sec.

(a) Calculate the pressure gradient required to produce this steady-flow component and the resulting rate of viscous dissipation per unit length, averaged over one heartbeat.

(b) Calculate the corresponding dissipation for the first three harmonics if

$$\frac{\langle v\rangle}{\langle\bar{v}\rangle}=\sum_{n=0}^{\infty} a_n\cos n\omega t+b_n\sin n\omega t$$

with $a_0=1$, $a_2=6$, $b_1=-0.3$, $a_2=-5$, $b_2=9$, $a_3=-1.5$, and $b_3=-3$. (This corresponds roughly to the conditions of Fig. 3.5.6.)

## II.4.$C_2$. *PRESSURE RISE IN AN IDEALIZED ANEURYSM*

Compute the pressure rise *as a function of time* for an enlargement of the type considered in Ex. 4.2.2. Assume that the upstream diameter is 2.5 cm and that the downstream one is much larger; use the simplified flow pulse

$$Q=(5\ \text{l/min})\left(1+\sin\frac{2\pi t}{\text{sec}}\right)$$

and consider the oscillatory flow to be flat. How are your predictions changed by a doubling of cardiac output?

### II.4.$D_2$. *PRESSURE LOSS AT A STENOSIS*

A stenosis, or reduction in cross section, corresponds to an orifice in piping systems and results in a loss of velocity head. As a first approximation, the contraction to the minimum cross-sectional area can be considered frictionless, and the subsequent expansion can be assumed to give *no* pressure recovery.

Parallel Ex. 4.2.2 to show how one can describe the behavior of a stenosis in a major artery.

### II.4.$E_2$. *THE IMPULSIVE NATURE OF THE HEART*

Use Fig. 2.5.6 of Chapter I to estimate the *order of magnitude* of the reaction force required to pump blood from the heart at the peak of systolic injection. Assume the aorta to be 2.5 cm in internal diameter, and the velocity profile to be flat.

ANSWER

On the order of $10^5$ dyn.

### II.4.$F_2$. *ENERGY DISSIPATION AND THE WINDKESSEL MODEL*

The *Windkessel* model (Eq. 2.2.2 of Chapter I) provides a good overall description of the pressure-flow relations in the arterial system. Show how it can be used to calculate viscous energy dissipation as follows.

(a) Calculate the rate of energy dissipation for:

$$Q_i = (5\ \mathrm{l/min})(1 + \sin T)$$

$$R = \frac{5\ \mathrm{l/min}}{100\ \mathrm{mm\ Hg}}$$

where $T = 2\pi/(1\,\mathrm{sec})$ and $t$ is the time. (See Problem I.J for the needed pressure-flow relation.)

Begin your calculations for $k=0$ and repeat for a value of $k$ which halves the peak oscillatory pressure.

(b) Why does arterial elasticity affect viscous dissipation even though all calculated dissipation takes place in the resistive part of the circuit (primarily the microcirculation)?

### II.4.G$_2$. *VISCOUS DISSIPATION IN A RIGID DUCT*

(a) Calculate the viscous dissipation rates per unit length of duct for representative entries in Table 3.1.1 using the rigid-tube theory (e.g., Eq. 3.5.42), the average velocities $\langle \bar{v} \rangle$ (given as blood velocity in the table), and the simplified velocity-time relation

$$\langle v \rangle = \langle \bar{v} \rangle \left( 1 + \sin \frac{2\pi t}{1\,\text{sec}} \right)$$

(b) Use the results of (a) and the remaining data in the table to estimate the distribution of viscous dissipation in the circulatory system. How do your calculations compare with the remarks made in the text?

## PROBLEMS FOR CHAPTER III

### III.1.A$_1$ *DIFFUSION POTENTIALS*

Consider the cell of Fig. 1.3.2 with $M^+$ equal to $Na^+$, $c_1=0.1\ N$ and $c_2=1\ N$.

(a) Calculate the potential across the diffusion path using the diffusivities of Problem III.1-C and assuming constant activity coefficients for the ions.

(b) Calculate the voltage measured by the potentiometer, and compare with the result of (a). Are the potentials such as (a) "real"?

### III.1.B$_1$. *STEADY ULTRACENTRIFUGATION*

A dilute aqueous solution containing both sucrose (molecular weight 342.3) and a typical globular protein (molecular weight 69,000) is placed in a centrifuge tube with an effective length of 10 cm and with the free liquid

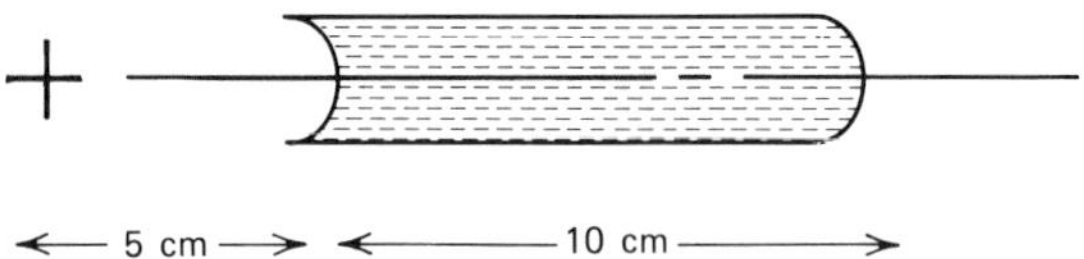


surface 5 cm from the axis of rotation (see figure). For dilute solutions at 298°K E. A. Moelwyn-Hughes [*Physical Chemistry*, 2nd ed., Macmillan (1961)] gives the following data:

$$\tilde{V}_{H_2O} = 29.8 \times 10^{-24}\,\text{ml/molecule}$$

$$\tilde{V}_{\text{sucrose}} = 350.0 \times 10^{-24}\,\text{ml/molecule}$$

$$\tilde{N} = 6.024 \times 10^{23}\,\text{molecules/mole}$$

$$C_{H_2O} \approx 55.5\,\text{moles/l}$$

It is harder to measure the partial molal or specific volumes of proteins because it is hard to get an accurate dry weight. However, nearly all soluble proteins have partial specific volumes very near

$$\frac{\tilde{V}}{M} = 0.75 \pm 0.01\ \text{ml/g}$$

[E. J. Cohn and J. T. Edsall, *Proteins, Amino Acids, and Peptides*, Reinhold (1943).]

Calculate the concentration profiles for these two solutes as a function of the angular velocity of the centrifuge, $\Omega$. Comment on the relative separability of these two solutes.

### III.1.C$_1$. *IONIC DIFFUSIVITIES FROM LIMITING CONDUCTANCE*

The limiting equivalent conductances at zero concentration, $\lambda_{i0}$, at 25°C for the following ions are [see Robinson and Stokes (1965) Table 6.1]: $Na^+$, 50.10 cm$^2/\Omega$; $K^+$, 73.5 cm$^2/\Omega$; $Cl^-$, 76.35 cm$^2/\Omega$. Estimate the corresponding ionic diffusivities from the formula

$$Ð_{iW} = \frac{RT}{F^2}\frac{\Lambda_{i0}}{\nu_i}$$

(see Ex. 1.3.1). Note that 1 joule = 1 volt coulomb = 1 volt ampere second.

ANSWERS:

Approximately, to the nearest percent, in units of $10^{-5}$ cm$^2$/sec,

$$\text{Na}^+,\ 1.33,\ K^+;\ 1.95,\ \text{Cl}^-;\ 2.00$$

III.1.D$_1$. *SALT DIFFUSIVITIES FROM LIMITING IONIC CONDUCTANCES*

Calculate the diffusivity $\mathfrak{D}_{SW}$ of sodium chloride in water at 25°C from Eq. 8 of Ex. 1.3.1 and the following data: ionic diffusivities from Problem III.1 – C and the expression

$$\log_{10}\gamma_s = -\frac{0.5115\sqrt{I}}{1+1.316\sqrt{I}} + 0.055\ I$$

where $I$ is the ionic strength, here equal to molar salt concentration in gram-moles per liter. Compare with the following experimental values, taken from Robinson and Stokes (1965). How important is the thermodynamic correction in this system?

| $c$ (mole/1) | 0 | 0.001 | 0.01 | 0.1 | 1.0 |
|---|---|---|---|---|---|
| $\mathfrak{D}$sw(exptl) | 1.610 | 1.585 | 1.545 | 1.483 | 1.484 |

III.1.E$_1$. *OSMOTIC PRESSURES*

(a) The ion-exchange membrane of Ex. 1.5.3 contains about 1 mole of sulfonic acid to 13 moles water, and is thus about 4.3 molal. Estimate membrane-phase osmotic pressure assuming $(p_{W1}/p_{W2}) \sim 0.75$, which is typical for corresponding sulfonic-acid solutions.

(b) The total osmotic pressure of blood plasma is normally taken to be about 7.9 atm, the osmotic pressure of the isotonic saline solution (0.154$N$ NaCl, 0.9 g/100 ml). Estimate the term

$$e^{\pi V_W/RT}$$

for plasma.

ANSWERS

(a) 400 atm.
(b) 1.0058.

## III.1.F$_1$. *MEMBRANE EQUILIBRIA*

Estimate the thermodynamic equilibrium constant $K_\gamma$ for the membrane of Ex. 1.5.3, assuming an osmotic pressure of 400 atm and a solute partial molal volume of 90 $cm^3$/g-mole.

ANSWER

$K_\gamma \sim 0.15.$

## III.1.G$_2$. *EFFECTIVENESS OF A SALT BRIDGE*

Microelectrodes of the type discussed in Ex. 2.2.2 are normally filled with concentrated KCl to minimize the diffusion potentials (see Exs. 1.3.1 and 1.3.2): Since the diffusivities of $K^+$ and $Cl^-$ are very nearly equal, KCl diffusion causes only very small potentials; by using concentrated KCl in the bridge, the diffusion potentials of other salts present are believed to be reduced. Test this belief for a *continuous-mixture junction* [see Ex. 2.4.1 and Section 2.4(c)] between pure 0.15 *N* NaCl and pure 3.0 *N* KCl. Neglect all ion-ion diffusional interactions and variations in diffusivities and activity coefficients. Use diffusivities as calculated in Problem III.1.C, and proceed as follows:

(a) Show that the potential gradient

$$\nabla\phi = \frac{(RT/\mathfrak{F})[(\mathcal{D}_{KW}-\mathcal{D}_{CW})\nabla x_K + (\mathcal{D}_{NW}-\mathcal{D}_{CW})\nabla x_N]}{x_K + x_N + x_C}$$

for arbitrary geometry. Here K, N, C, and W refer to potassium, sodium, chloride, and water, respectively.

(b) Show that for the system of interest the potential drop between the electrode body and external solution is

$$\Delta\phi = \frac{RT}{\mathfrak{F}}\left[\frac{3(\mathcal{D}_{KW}-\mathcal{D}_{CW}) - 0.15(\mathcal{D}_{NW}-\mathcal{D}_{CW})}{3-0.15}\right]\cdot \ln\frac{3}{0.15}$$

Calculate this potential.

(c) Consider the case of arbitrary terminal concentrations but $\mathcal{D}_{KW} = D_{CW}$, and discuss the effectiveness of KCl in reducing junction potentials.

## III.1.H$_2$. *THERMODYNAMIC IDENTITIES*

(a) Show that Eq. 1.2.2 follows directly from the first postulate and the basic relation

$$T\,dS = dU - p\,dV - \sum_{i=1}^{n} \frac{\mu_i}{M_i}\,dm_i$$

where $m_i$ is mass of species $i$. To obtain the desired relation start by considering a unit mass of fluid following the mass-average velocity.

(b) Obtain Eq. 1.2.3 by putting

$$\rho \frac{D\hat{U}}{Dt} = (\nabla \cdot \mathbf{q}) - p(\nabla \cdot \mathbf{v}) - (\tau : \nabla \mathbf{v}) + \sum_{i=1}^{n} (\mathbf{j}_i \cdot \mathbf{g}_i)$$

$$\rho \frac{D\omega_i}{Dt} = -(\nabla \cdot \mathbf{j}_i) + r_i$$

into Eq. 1.2.2.

Now take advantage of the identity

$$\left(\nabla \cdot \frac{\mathbf{q}}{T}\right) = \frac{1}{T}(\nabla \cdot \mathbf{q}) - \frac{1}{T^2}(\mathbf{q} \cdot \nabla T)$$

and similar expressions for $\mathbf{j}_i$ and $\mu_i$ to obtain Eq. 1.2.3.

(c) Show in detail how Eq. 1.2.11 is obtained.

### III.1.I.$_2$ *DIFFUSION VELOCITIES*

Two large electrode compartments of the type shown in Fig. 1.3.2 are joined by a short cylindrical tube representing the diffusion path.

(a) Develop an expression for the mass- and molar-average velocities, $\mathbf{v}$ and $\mathbf{v}^*$, respectively, as a function of position. Assume total molar concentration $c$, salt diffusivity $D_{SW}$, and terminal concentrations to be constant;† assume also that the water velocity $\mathbf{v}_W$ is zero.

(b) Calculate the maximum value of $\mathbf{v}$ and $\mathbf{v}^*$ for a tube 1 mm long if the terminal solutions are 0.2 *N* sodium chloride at 25°C. Under these conditions $D_{SW} = 1.48 \times 10^{-5}\,\text{cm}^2/\text{sec}$.

(c) Estimate the effect of water movement on these maximum velocities by assuming the volume-average velocity to be zero. For present purposes the partial molal volumes of sodium chloride and water may be considered constant at $18\,\text{cm}^3/\text{g-mole}$, for both.‡

(d) Discuss the problem of determining molar flux ratios, assuming that hydrostatic equilibrium is maintained.

† The effect of transients is discussed in *Tr. Ph.*, Prob. 19.L.

‡ See G. Kortum and J. O'M. Bockris, *Textbook of Electrochemistry*, Elsevier (1951), p. 60. Partial molal volumes are not significantly concentration independent in this system.

III.1.J$_2$. *FICK'S LAW IN BINARY SYSTEMS*

Show that Eq. 1 of Ex. 1.2.1 also leads to

$$\mathbf{j}_A = \rho_A(\mathbf{v}_A - \mathbf{v}) = -\rho \mathfrak{D}_{AB} \nabla \omega_A \tag{1}$$

where $\mathbf{v} = \omega_A \mathbf{v}_A + \omega_B \mathbf{v}_B$ is the mass-average velocity, and to

$$c_A(\mathbf{v}_A - \mathbf{v}^\circ) = -\mathfrak{D}_{AB} \nabla c_A \tag{2}$$

where

$$\mathbf{v}^\circ = \left(c_A \bar{V}_A\right)\mathbf{v}_A + \left(c_A \bar{V}_B\right)\mathbf{v}_B$$

This latter expression is, however, limited to processes taking place at constant temperature and pressure.

III.1.K$_2$. *COMPARISON OF BODY FORCES*

The component body forces on membrane matrices are given by Eqs. 1.4.3 and 1.4.8. Calculate the electrostatic potential gradients in volts per centimeter and pressure gradients in dynes per cubic centimeter required for electrical and "support" forces to equal the gravitational force $g = 980\,\text{cm/sec}^2$. Assume the density $\rho_m$ of the membrane matrix to be $1\,\text{g/cm}^3$, and the equivalent weight $|M_i/\nu_i|$ to be 200 g/eq. Discuss the significance of your answer.

ANSWER

$$2.05 \times 10^{-7}\,\text{V/cm};\ 9.7 \times 10^{-3}\,\text{atm/cm}.$$

III.1.L$_2$. *PRESSURE DIFFUSION*

(a) Calculate the equilibrium difference in atmospheric oxygen mole fraction at Denver, Colorado (elevation 5000 ft) and Madison Wisconsin (elevation 1000 ft).

(b) In actuality this difference is not observed. Why?

(c) Discuss the feasibility of producing oxygen-enriched air at a bedside from compressed air available at 80 psi gauge without the use of additional energy.

III.1.M$_2$. *TRANSIENT ULTRACENTRIFUGATION*

Extend Problem III.1.B to the early stages of ultracentrifugation, during which relative motion of protein, sucrose, and water occurs.

(a) Develop expressions for protein and sugar velocities at times so short that appreciable concentration gradients have not yet developed.

(b) Contrast the types of information which can be obtained from transient and steady experiments. What simplifications would you expect to be useful in these experiments?

## III.1.$N_2$. *SALT BRIDGES*

Show how the presence of a salt $M^+X^-$ for which $Ð_{MW} \doteq Ð_{XW} \equiv Ð_{SW}$, such as $K^+Cl^-$, can be used to suppress junction potentials. Consider the salt bridge to represent the diffusion path in Fig. 1.3.2 and that concentration changes take place primarily in the *salt bridge*;* assume that the concentration of $M^+X^-$ is very high here, compared to the salt $M^+Y^-$ in the two compartments joined by the bridge. Use the Nernst-Planck approximations and neglect convection.

ANSWER

$$\nabla\phi = \frac{RT}{\mathfrak{F}}\left(1 - \frac{Ð_{YW}}{Ð_{SW}}\right)\left[\frac{\nabla x_Y}{x_M + x_X + x_Y(Ð_{YW}/Ð_{SW})}\right]$$

## III.1.$O_{D,2}$. *IONIC INTERACTIONS*

The equivalent conductivity $\Lambda$ of an electrolyte is equal to the sum of the equivalent conductivities $\lambda_i$ of the constituent ions—for example,

$$\Lambda = \lambda_+ + \lambda_-$$

for a simple salt; these in turn are equal to

$$\lambda_i = \frac{\kappa_i}{c}$$

where $\kappa_i$ is the specific conductance. The equivalent conductance of NaCl at 25°C, shown below, is seen to be much more concentration dependent than the salt diffusivity.

(a) Discuss qualitatively.

(b) Estimate the ionic interaction term $D_{+-}$, assuming the ion-water diffusivity and activity coefficients to be constant.

* It is common practice to use saturated KCl immobilized by agar in these bridges.

| $c$ (mole/l) | 0 | 0.001 | 0.01 | 0.1 | 1.0 |
|---|---|---|---|---|---|
| $\Lambda$ [cm$^2$/($\Omega$)(eq)] | 126.45 | 123.74 | 118.53 | 106.74 | 85.76 |

### III.1.P$_2$. *MEMBRANE BODY FORCES*

What is the equivalent to Eq. 1.4.9 for the membrane matrix $m$? Is this equation needed? Useful?

### III. 1.Q$_2$. *MEMBRANE STRESSES*

It may be seen that the colloid osmotic pressure, that is, that resulting from the presence of hemoglobin, is about an atmosphere in red cells. If it were unopposed, what would be the resulting membrane stress? What effect would you expect such a stress to have on the human red-cell membrane (refer back to Fig. 1.3.3)?

### III.1.R$_2$. *EFFECTS OF THERMODYNAMIC ACTIVITY ON DONNAN EXCLUSION*

The equilibrium constant of Ex. 1.5.3 can be decomposed into "osmotic" and "chemical" components $K_\Pi$ and $K_\gamma$ defined by

$$K_s = K_\Pi K_\gamma \tag{1}$$

with

$$K_\Pi = e^{\Pi \bar{V}_s/RT} \tag{2}$$

where $\Pi$ is the osmotic pressure. Estimate $K_\gamma$ as follows:

(a) Estimate the osmotic pressure $\Pi$ from Eq. 3a of Ex. 1.5.1 by assuming that the high interior ionic strength reduces the water vapor pressure by 25% but that the partial molal volume of water is essentially equal to its molal volume of 18 cm$^3$.

(b) Evaluate $K_\Pi$ using this osmotic pressure and a solute partial molal volume of 90 cm$^3$/g-mole. Note that for 25°C one may use $RT \doteq 2.45 \times 10^4$ cm$^3$ atm/g-mole.

ANSWER

$K_\gamma \doteq 0.146.$

III.1.S$_3$. *A SIMPLE APPROACH TO THE KEDEM-KATCHALSKY EQUATIONS*

Consider a pseudosteady transfer of one solute and water across a membrane. Show that the rate of entropy increase accompanying such transport is given by

$$T\frac{d}{dt}\frac{dS}{dA}=N_{\mathrm{W}}\left[RT\Delta\ln(a_{\mathrm{W}})_{T,p}+\bar{V}_{\mathrm{W}}\Delta p\right]$$

$$+N_{\mathrm{S}}\left[RT\Delta\ln(a_{\mathrm{S}})_{T,p}+\bar{V}_{\mathrm{S}}\Delta p\right] \tag{1}$$

where $(d/dt)dS/dA$ is the rate of entropy increase resulting from fluxes across a differential area element $dA$. To write this equation we must assume the partial molal volumes to be constant. Note also that the activities and pressures are those in the external solutions.

Further limit consideration to thermodynamically ideal binary external solutions and show that for this special case,

$$T\frac{d}{dt}\frac{dS}{dA}=\Delta pJ_v+RT\Delta c_{\mathrm{S}}J_D \tag{2}$$

To complete the development assume the second and third postulates of irreversible thermodynamics to apply under these conditions. Show that one thus obtains Eqs. 1.6.7 and 1.6.8, with $L_{PD}=L_{DP}$.

III.1.T$_3$. *FORMAL DESCRIPTION OF A SALT BRIDGE*

Show that for Eq. 2.4.21 one may write for monovalent ions

$$\alpha_{\mathrm{N}}\nu_{\mathrm{N}}=(r_{\mathrm{MM}}r_{\mathrm{XX}}-r_{\mathrm{MX}}^2)+(r_{\mathrm{MM}}r_{\mathrm{NX}}-r_{\mathrm{MN}}r_{\mathrm{XM}})-(r_{\mathrm{XX}}r_{\mathrm{MN}}-r_{\mathrm{MX}}r_{\mathrm{NX}})$$

and

$$\alpha_{\mathrm{X}}\nu_{\mathrm{X}}=(r_{\mathrm{NN}}r_{\mathrm{MM}}-r_{\mathrm{MN}}^2)+(r_{\mathrm{MM}}r_{\mathrm{NX}}-r_{\mathrm{NM}}r_{\mathrm{MX}})+(r_{\mathrm{NN}}r_{\mathrm{MX}}-r_{\mathrm{MN}}r_{\mathrm{NX}})$$

where

$$r_{ij}=\frac{1}{Ð_{ij}}$$

Calculate the remaining $\alpha_i \nu_i$ and show that in the absence of ion-ion interaction,

$$\frac{\alpha_M \nu_M}{A} = \frac{x_M Ð_{WM}}{x_M Ð_{WM} + x_N Ð_{WN} + x_X Ð_{WX}}$$

What are the other corresponding terms?

## III.1.U$_4$. *A MICROSCOPIC APPROACH TO THE KEDEM-KATCHALSKY RELATIONS*

Complete the development sketched out in Section 3.3, beginning with Eqs. 3.3.6 and 3.3.7, to obtain Eqs. 1.6.3 and 1.6.4. Discuss in your own words the approximations involved to obtain the reciprocal relations $L_{PD} \doteq L_{DP}$ for these latter equations.

## III.2.A$_D$. *MAGNITUDES OF DIFFUSION COEFFICIENTS*

(a) Look up the diffusivities of common small electrolytes and comment both on the range and concentration dependence. Report representative values.

(b) Compare mutual and tracer diffusivities for any systems you can find data on—for example in Robinson and Stokes (1965).

(c) Look up and report diffusivities of representative proteins, available in any standard work on proteins. If you have time, report on the use of viscosity and diffusivity for determining protein structure.

## III.2.B$_1$. *DIFFUSION OF RED CELLS*

Estimate the diffusivity of red blood cells, as diffusing hydrodynamic particles, assuming:

(a) They are spheres containing the actual red-cell volume, which can be calculated by assuming them to be oblate ellipsoids with axes of 8 and 2.4 $\mu$.

(b) They are oblate ellipsoids, and the effective drag force is the average of that for motion along the $x$,$y$, and $z$ axes of a rectangular coordinate system aligned with the cell axis of rotation.

## III.2.C$_2$. *ALVEOLAR DIFFUSION*

(a) Determine the binary pair diffusivities, under alveolar conditions, of $O_2$, $N_2$, and $CO_2$, for example from *Tr. Ph.*, Chapter 10. Note that these are essentially insensitive to mole fraction.

(b) Develop an expression for the *effective binary diffusivity* $\mathfrak{D}_{O_2m}$ for oxygen using the formula

$$\mathfrak{D}_{O_2m}^{-1} = \frac{\sum_{j=1}^{n} (x_j\mathbf{N}_i - x_i\mathbf{N}_j)/\mathfrak{D}_{ij}}{\mathbf{N}_i - x_i \sum_{j=1}^{n} N_j}$$

Show that this is consistent with Eq. 2.1.2 and assume the fluxes are always in the same ratio as at the alveolar wall.* Note that $\mathfrak{D}_{O_2m}$ is sensitive to mole fraction.

(c) Calculate $\mathfrak{D}_{O_2O_2}$ from Eq. 1.2.23. Note both sign and composition dependence.

PARTIAL ANSWER

(A) $\mathfrak{D}_{O_2N_2} = 0.217\ \text{cm}^2/\text{sec}$, $\mathfrak{D}_{O_2CO_2} = 0.162\ \text{cm}^2/\text{sec}$, $\mathfrak{D}_{N_2CO_2} = 0.163\ \text{cm}^2/\text{sec}$.

## III.2.D$_2$. *OXYGEN TRANSPORT TO A RED CELL*

Develop an expression for the rate of oxygen uptake by a red cell initially at 40 mm Hg tension placed suddenly in quiescent surroundings at a tension of 100 mm Hg. Assume the red cell to be a sphere of 3-$\mu$ radius, and for convenience use the linear approximation to the equilibrium curve. Neglect diffusional resistance within the cell.

Base your calculations on the results of Ex. 2.2.2 and:

(a) Assume $t \gg 1$ and calculate the time to approach within a factor $1/e$ of saturation.

(b) Calculate $t$ for this situation, and discuss the utility of the pseudo-steady approximation of part (a). Also compare with the ellipsoidal approximation of Problem III.2.E.

## III.2.E$_2$. *EQUILIBRATION RATES FOR A RELAXED RED CELL*

Estimate the response of an isolated red cell in an unstirred solution by using (1) the pseudo-steady expressions developed in Problem III.2.K for the diffusion rate and (2) the linear approximation to the hemoglobin

* Note that the normal respiratory quotient of 0.8 yields $N_{CO_2} = -0.8N_{O_2}$. Why not worry about water vapor? Should you?

dissociation curve.* For this approximation,

$$\pi = e^{-\bar{t}}$$

where

$$\pi = (c_{O_2} - c_\infty)(c_O - c_\infty)$$

$$\bar{t} = \frac{t}{t_e}$$

Here $c_O$ and $c_\infty$ are the initial and final oxygen concentrations in the red cell, and $t_e$ is a time constant for the process. Calculate $t_e$ for normal red cells with initial and final oxygen tensions of 95 and 40 mm Hg. Use as geometric models:

(a) An oblate ellipsoid with half axes of 1 and 4 $\mu$. Note that the minor and major axes are $a \sinh \eta_0$ and $a \cosh \eta_0$ in the nomenclature of Problem III.2.K (which see).

(b) A flat disc 2 $\mu$ thick and 8 $\mu$ in diameter [neglecting disc thickness in calculating *diffusional flux*; note that this is a degenerate case of (a)].

ANSWER

(a) ~0.09 sec.
(b) ~0.16 sec.

## III.2.F$_2$. *DIFFUSION OF OXYGEN IN A SPHERED RED CELL*

Repeat Problem III.2.D for a sphered red cell 8 $\mu$ in diameter.

APPROXIMATE ANSWERS

(a) $10^{-1}$ sec.
(b) $\frac{1}{4}$ sec.

## III.2.G$_2$. *EFFECTIVENESS OF A SALT BRIDGE (SIMPLIFIED)*

Microelectrodes of the type discussed in Ex. 2.2.2 are normally filled with concentrated KCl to minimize the diffusion potentials (see Exs. 1.3.1 and 1.3.2): Since the diffusivities of $K^+$ and $Cl^-$ are very nearly equal, KCl diffusion causes only very small potentials; by using concentrated KCl in the bridge, the diffusion potentials of other salts present are believed to be reduced. Test this belief for a *continuous-mixture junction* [see Ex. 2.4.1 and

* The effect of diffusional transients is duscussed in Problem III. 2.J; the use of the linear approximation is only a minor convenience and can easily be relaxed.

Section 2.4(c)] between pure NaCl and pure KCl. *Neglect all ion-ion diffusional interactions and variations in diffusivities and activity coefficients*, and proceed as follows:

(a) Show from the results of Section 2.4 that the potential gradient

$$\nabla\phi = \frac{(RT/\mathfrak{F})[(Đ_{KW} - Đ_{CW})\nabla x_K + (Đ_{NW} - Đ_{CW})\nabla x_N]}{(x_K Đ_{KW} + x_N Đ_{NW} + x_C Đ_{CW})}$$

for arbitrary geometry. Here K, N C, and W refer to potassium, sodium, chloride, and water, respectively. To do this, use Eq. 2.4.19 with the help of Eq. 2.4.5.

(b) Show that for the system of interest the potential drop between the two solutions on either side of the diffusion path can be written as follows, if at point 1′ the solution is pure KCl at concentration $x_{K0}$, while at 2′ it is pure NaCl at concentration $x_{N0}$:

$$-(\phi_2' - \phi_1') = \frac{RT}{\mathfrak{F}} \frac{x_{K0}(Đ_{KW} - Đ_{CW}) + x_{N0}(Đ_{NW} - Đ_{XW})}{x_{K0}(Đ_{KW} + Đ_{CW}) + x_{N0}(Đ_{NW} + Đ_{CW})} \ln \frac{x_{N0}(Đ_{NW} + Đ_{XW})}{x_{K0}(Đ_{KW} + Đ_{XW})}$$

(c) Calculate this junction potential $\phi_2' - \phi_1'$ for $x_{N0}$ and $x_{K0}$ corresponding to 0.15 *N* NaCl and 4.0 *N* KCl, respectively, if the ion-water diffusivities are as given in Problem III.1.C. Assume, *for purposes of this problem,* that mole fractions are proportional to normality over this range.

(d) Calculate to the same order of accuracy the potential $\phi_2 - \phi_1$ measured by reversible chloride electrodes.

ANSWER

(c) $-1.73$ mV.
(d) $-45$ mV.

## III.2.H.$_2$. *ZERO-ORDER CHEMICAL REACTION IN SPHERES*

Parallel the development of Ex. 2.3.4 for spheres, *but* limit yourself to steady operation unless you have time for the full development. Obtain the spherical analog to Eq. 12. Note that for steady operation it is simpler to work directly with $c_i$:

$$\mathfrak{D}_{im} \frac{1}{r^2} \frac{\partial}{\partial r} r^2 \frac{\partial c_i}{\partial r} = -R_i$$

## III.2.I$_3$. *DIFFUSION AND THE VOLUMETRIC FLUX*

Justify Eq. 5 of Ex. 2.1.1 as follows, starting with the more familiar expression

$$\mathbf{J}_{\mathrm{P}}^{\bigstar} = -c\mathfrak{D}_{\mathrm{PW}}\nabla x_{\mathrm{P}} \tag{1}$$

(a) Show that

$$\mathbf{J}_{\mathrm{P}}^{\bigstar} = c x_{\mathrm{P}} x_{\mathrm{W}}(\mathbf{v}_{\mathrm{P}} - \mathbf{v}_{\mathrm{W}}) \tag{2}$$

and

$$\mathbf{J}_{\mathrm{P}}{}^{0} = c x_{\mathrm{P}} c_{\mathrm{W}} \bar{V}_{\mathrm{W}}(\mathbf{v}_{\mathrm{P}} - \mathbf{v}) \tag{3}$$

(b) Show that, for constant temperature and pressure, one may write

$$\nabla c_{\mathrm{P}} = c^2 \bar{V}_{\mathrm{W}} \nabla x_{\mathrm{P}} \tag{4}$$

This equation essentially completes the development. To obtain it note that

$$d\tilde{V} = d\frac{1}{c} = \sum_{i=1}^{n} \bar{V}_i dx_i + \frac{\partial \tilde{V}}{\partial T} dT + \frac{\partial \tilde{V}_{\mathrm{P}}}{\partial p} dp$$

It follows that for our system

$$(\nabla c)_{T,p} = -c^2\left(\bar{V}_{\mathrm{P}} - \bar{V}_{\mathrm{W}}\right)\nabla x_{\mathrm{P}}$$

This with the relation $\nabla c_{\mathrm{P}} = \mathrm{x}_{\mathrm{P}} \nabla c + c \nabla x_{\mathrm{P}}$ gives Eq. 4.

## III.2.J$_3$. *DIFFUSION OF OXYGEN IN A RED CELL*

A red cell initially in a normal arterial state is suddenly placed in effective diffusional contact with an oxygen-free fluid, so that the red-cell surface is maintained at zero oxygen tension. (For a discussion of external diffusional resistance see Problem III.2.E.) Approximate the red cell as a slab 2 $\mu$ in thickness and estimate the time required for it to equilibrate with its new environment by extending Ex. 2.3.2 to the new boundary conditions:

(a) Calculate the time for complete deoxygenation by use of the step-function approximation.

(b) Calculate the time for dissolved oxygen concentration to reach 10% of its original value at the midplane of the red cell using the linear approximation. (Suggestion: work by analogy with the nonreactive problem: see, e.g., *Tr. Ph.,* Fig. 11.1-3.

Note that the two calculations are no longer comparable, because the terminal conditions differ.

ANSWER

(b) About 0.05 sec.

### III.2.K$_3$. *PSEUDOSTEADY DIFFUSION FROM AN OBLATE ELLIPSOID*

An oblate ellipsoid of revolution with minor and major axes $a \sinh\eta_0$, and $a\cosh\eta_0$, respectively, is placed in a unbounded quiescent medium which is solute free

(a) Calculate the steady-state rate of diffusion of solute to the fluid if the ellipsoid surface is maintained at unit dimensionless concentration $\phi$. See the figure for an explanation of the coordinate system, and note that for $\phi$ independent of $\theta$ and $\psi$,

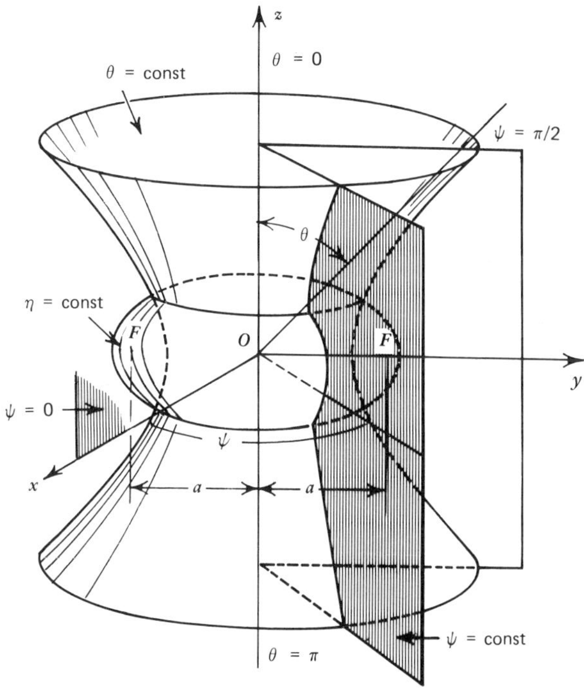


Oblate spheroidal coordinates $(\eta,\theta,\psi)$. Coordinate surfaces are oblate spheroids ($\eta=$const), hyperboloids of revolution ($\theta=$const), and half planes ($\psi=$const).

$$\nabla^2\phi = \frac{d^2\phi}{d\eta^2} + \tanh\eta\frac{d\phi}{d\eta} = 0$$

$$\text{grad}\,\phi = \delta_\eta \frac{1}{a\sqrt{\cosh^2\eta - \sin^2\theta}}\frac{d\phi}{d\eta}$$

The metric coefficients are defined by

$$dl^2 = g_{\eta\eta}(d\eta)^2 + g_{\theta\theta}(d\theta)^2 + g_{\psi\psi}(d\psi)^2$$

with

$$g_{\eta\eta} = g_{\theta\theta} = a^2(\cosh^2\eta - \sin^2\theta)$$

$$g_{\psi\psi} = a^2\cosh^2\eta\sin^2\eta$$

Here $dl$ is the distance between points whose coordinates differ by $(d\eta, d\theta, d\psi)$.

(b) Calculate the volume of the elliposoid.

ANSWER

(a) Total rate of diffusion of solute $i$,

$$M_i = \mathfrak{D}_{im}\Delta c_i\left\{8a\left[1 - \frac{2}{\pi}\tan^{-1}(\sinh\eta_0)\right]\right\}$$

where $\mathfrak{D}_{im}$ and $\Delta c_i$ are the effective binary diffusivity and the concentration driving force.

(b) Volume $= \frac{4}{3}\pi a^3\sinh\eta_0\cosh^2\eta_0$.

## III.2.$L_4$. *TERNARY DIFFUSION IN A MICROELECTRODE*

Generalize Eq. 6 of Ex. 2.2.2 to the interdiffusion of two salts, starting with the matrix formulation of Ex. 2.4.1. Write your answer explicitly by the use of Sylvester's theorem.

## III.3.$A_1$. *POTABLE WATER FROM THE SEA*

Typical seawater, containing 3.45% by weight of dissolved salts, has a vapor pressure 1.84% below that of pure water. Estimate the minimum osmotic pressure to produce pure liquid water from this solution at 25°C (where $cRT = 1354$ atm).

ANSWER

$\pi_{\min} \doteq 25$ atm.

### III.3.$B_1$. *OSMOTIC CONTRIBUTION TO THE DISTRIBUTION COEFFICIENTS*

The total osmotic pressure of blood plasma is normally taken to be about 7.0 atm. Calculate the magnitude of

$$e^{\pi \bar{V}_W/RT}$$

where $\pi$ is the osmotic pressure. Assume $\bar{V}_W$ to be 18 ml/g-mole, the value for pure water. How much does this osmotic pressure affect the interphase distribution coefficients? To what degree does it affect hemodialysis?

PARTIAL ANSWER

About 0.6%.

### III.3.$C_2$. *MEMBRANE SELECTIVITY*

(a) A membrane impermeable to protein but permeable to inorganic ions separates two solutions, initially of the following compositions:

| | Left | Right |
|---|---|---|
| Sodium | 0.15 *N* | 0.06 *N* |
| Chloride | 0.15 *N* | 0 |
| Protein (anion) | 0 | 0.06 *N* |

What is the equilibrium composition in the right compartment, approached at large times, if (1) the left compartment is large enough that its composition remains essentially constant, and (2) the activity coefficients of all species are unity. Would you expect an osmotic pressure to develop in this system?

(b) Discuss the significance of "membrane potentials", $V_2 - V_1$, as measured in the apparatus shown in Fig. 1.3.2.

### III.3.$D_2$. *THE RELATIVE TRANSFER RATES OF SOLUTE AND WATER AT ZERO PRESSURE DROP*

During hemodialysis water is removed from the patient's blood by ultrafiltration: by maintaining the blood at a higher hydrodynamic pressure than the isoosmotic dialyzate solution on the other side of the membrane. It has occasionally been suggested that this pressure difference can be reduced, or even entirely eliminated, by making the dialyzate hyperosmotic, as by adding glucose to an otherwise isoosmotic dialyzate.

Calculate the reduction in pressure drop permitted as a function of dialyzate glucose concentration if the water flux is to be maintained at that for an isoosmotic dialyzate solution at a pressure difference of 200 mm Hg. Neglect the glucose concentration in the blood. Use the conditions of Ex. 3.2.1, and assume

$$cRT = 1360\,\text{atm}$$

$$\gamma_{\text{glucose}} = \text{const}$$

for the purposes of this problem. (Note that the boundary conditions differ from those for ultra-filtration. As a matter of practical interest, water fluxes are of the order of 2 $\text{ml/hrm}^2\text{mm}$ Hg for "typical" membranes.)

Do you consider the suggestion to have any merit?

### III.3.E$_2$. *MEMBRANE POTENTIALS*

Extend the discussion of Ex. 1.3.2 to a diffusion path consisting of a highly charged cation-exchanging membrane. Use the Nernst-Planck approximation, neglect convection, and assume no current is flowing. Show that in the limit of an ideally selective membrane (i.e., one completely excluding mobile anions) that

$$\phi_2 - \phi_1 = \frac{RT}{\mathfrak{F}} \ln\left(\frac{a_{M1}}{a_{M2}}\right)$$

What is the voltage read by the potentiometer?

### III.3.F$_3$. *UNIMPORTANCE OF TRANSIENTS IN MEMBRANE TRANSPORT*

Consider a membrane of area $A$ and thickness $\delta$ in contact with two well-stirred solutions, each of volume $AL$. A solute of interest, $i$, distributes itself linearly into the membrane, so that at equilibrium its concentration within the membrane phase is $K_i$ times that in the external solution.

(a) Show by reference to the literature, for example *Tr. Ph.* (Chapter 11) or—better—H. S. Carslaw and J. C. Jaeger, *Conduction of Heat in Solids*, 2nd ed., Oxford Univ. Press (1959), that the response of the membrane to an abrupt change in interfacial compositions will be largely completed in a characteristic response time

$$t_m = \delta^2 \mathfrak{D}_{im}$$

where $\mathfrak{D}_{im}$ is the effective diffusivity of $i$ in the membrane.

(b) Show that the response time of the *system to a change of concentration in one solution is*

$$t_s = \frac{\delta L}{\mathfrak{D}_{im} K_i}$$

if $t_m \ll t_s$, that is, if

$$\delta \ll \frac{L}{K_i}$$

It is this last inequality which justifies the neglect of membrane-phase transients in many applications.

## III.4.$A_1$. *SHORT-ANSWER QUESTIONS*

(a) When is diffusion too fast, and what is done about this problem?

(b) In ultrafiltration of a multicomponent solution across a semipermeable membrane, the water flux can be expressed by

$$N_W = -\left(\Delta p - \sum_{\substack{i=1 \\ \neq W}}^{n} \sigma_i \pi_i\right)$$

What is the significance of the $\pi_i$? what values can be taken by the $\sigma_i$ in the absence of diffusional interactions? What are some reasonable values of $\sigma_i$ for important metabolites in the glomerulus of the kidney?

(c) In general there is a measurable electrostatic potential difference between the interiors of red blood cells and the surrounding plasma. What do you think might be responsible for this? In general what is the origin of diffusion potentials?

(d) Why do red cells sometimes burst if their metabolic activity ceases?

## III.4.$B_1$. *OSMOTIC PRESSURE ACROSS SEMIPERMEABLE MEMBRANES*

When kidneys are stored between "harvest" from a donor and transplant into a recipient they must be perfused with a solution containing both oxygen and such nutrients as sugar. If an *isotonic* saline solution (i.e., one having the same equilibrium partial pressure of water as the kidney tissue) is used for this purpose, the kidney swells to an unacceptable degree and becomes rigid; if isotonic blood plasma is used, no swelling occurs. Why?

Consider both electrolyte and water transport in your answer.

## III.4.$C_2$. *INTERPRETATION OF MEMBRANE PERMEABILITIES*

Since cell membranes are very thin, measured diffusional permeabilities $P_{ma}$ must always be defined in terms of the bulk compositions of the solutions separated by the membrane.

(a) Show that the true permeability of the membrane

$$P_m = \left( \frac{1}{P_{ma}} - \frac{1}{k_{c_i}} - \frac{1}{k_{c_0}} \right)^{-1}$$

where $k_{c_i}$ and $k_{c_0}$ are the boundary-layer permeabilities defined by

$$N_s = k_{c_i} \Delta c_i = k_{c_0} \Delta c_0$$

Here $\Delta c_i$ and $\Delta c_0$ are the solute concentration differences across the "inside" and "outside" boundary layers.

(b) Correct the apparent permeability shown in Table 4.2.1 for the red cell to water using $\mathfrak{D}_{Wm} = 10^{-5} \text{cm}^2/\text{sec}$ and $D = 6\ \mu$ if the Nusselt number in the external phase has a value of 2.

ANSWER

$64 \times 10^{-4}$cm/sec. Allowance for the inside diffusional resistance is more difficult. It appears likely, however, that it will account for much of the remaining difference between observed diffusional and hydraulic permeabilities.

## III.4.$D_2$. *ANOMALOUS SOLVENT DRAG*

Urea added to the outside of toad skin in Ringer's solution increases the permeability of the skin to urea and sucrose (as well as many other solutes) and most particularly creates a net inward flux of sucrose: where there is no sucrose concentration difference across the skin, the sucrose influx can be 3 to 4 times the efflux. Stender, Kristensen, and Skadhauge (*J. Mem. Biol.*, **11**, 377–398 (1973)) suggest the following explanation (see figure). Transport is via pores (intra-cellular spaces) closed at the outside end by a tight junction (TJ), freely permeable to solutes but not water, and lined by a membrane (SPM) permeable to water and inorganic salts but not organic solutes. Urea diffusing down the pore pulls water from the surrounding porous tissue by osmosis; the result is an increasing convection of water

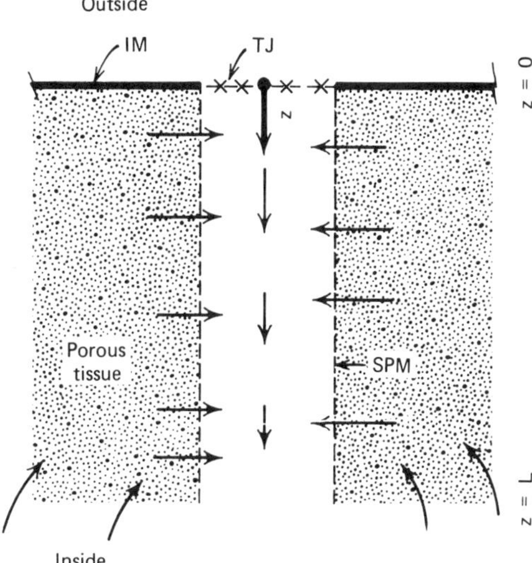


Anomalous-solvent-drag model of toad skin. IM, totally impervious membrane. TJ, tight junction: permeable to solutes as urea, but not water. SPM, semipermeable membrane permitting only water transport. Arrows indicate movement of water.

down the tube. This convection aids the unidirectional flux of any solute in the positive $z$ direction and retards the reverse flux.

(a) Show that for any solute,

$$N_i = \mathfrak{D}_{im}\frac{dc_i}{dz} + c_i v_{\mathrm{W}}$$

while for water,

$$\frac{dv_{\mathrm{W}}}{dz} = Kc_u$$

where $u$ refers to the driving solute urea.

(b) Show that for urea and the forward flux of any driven solute,[†]

$$\frac{c_i}{c_{i0}} = e^{v_i(h)}\left(1 - N_i^* \int_0^h e^{-U_i} de\right)$$

where $c_{i0} = c_i(z=0)$, $h = z/L$, $N_i^* = N_i L/\mathfrak{D}_{im} c_{i0}$, and

[†] As in Ex. 3.4.1, the "downstream" concentration for each unidirectional flux is considered to be zero.

$$U_i = \int_0^h \frac{v_W L}{\mathfrak{D}_{im}} dh$$

What is the corresponding result for the reverse flux of driven solute?

(c) Show that the ratio of the unidirectional fluxes is

$$\frac{N_{iF}}{N_{iR}} = -\frac{c_{iF}}{c_{iR}} \exp\left(\int_0^1 \frac{v_W L}{\mathfrak{D}_{im}} dh\right)$$

where $F$ and $R$ refer to forward and reverse fluxes.

(d) Discuss the system behavior in the light of Problem III.4.G.

### III.4.E$_{2,3}$. *MULTICOMPONENT CARRIER TRANSPORT*

Extend Ex. 2.3.3 to show that the flux of a given solute, $S_1$, in the $z$ direction can be speeded by addition of a competing solute (i.e., one complexing with the same carrier) $S_2$ at $z=\delta$.

(a) For simplicity you may assume if you like that $S_1$ and $S_2$ have identical thermodynamic and diffusional properties and that $S_1(\delta)=S_2(0)=0$.

(b) Work out the more general case of $n$ competing solutes with arbitrary boundary concentrations.

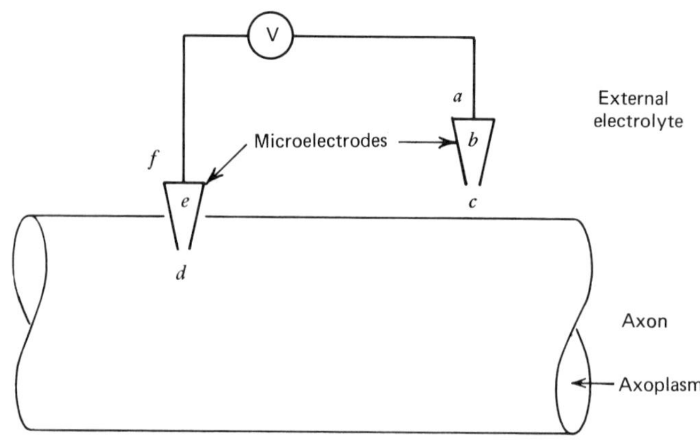


Membrane potential measurements in a squid nerve.

III.4.F$_2$. *INTERPRETATION OF MEMBRANE POTENTIALS IN COMPLEX SYSTEMS*

A typical situation for which membrance potential measurements are made is shown in the figure. The composition of the axoplasm is given in the table; a typical external electrolyte might be 0.5 $N$ in NaCl and 0.01 $N$ in KCl. Such a complex mixture defies exact analysis, but a reasonable approximation can be obtained by using the Nernst-Planck approximations, neglecting convection and activity coefficients, and assuming similar concentration profiles for the ions (as suggested by Ex. 2.4.1). Show that for this situation,

$$\phi_c - \phi_d = \frac{RT}{\mathfrak{F}}\left(\frac{u_c - u_d}{v_c - v_d}\right)\ln\left(\frac{v_d}{v_c}\right)$$

This result is known as the Henderson relation. Here

$$u = \sum_{i=1}^{n} \nu_i x_i \mathcal{D}_{i\mathrm{W}} \text{ and } v = \sum_{i=1}^{n} \nu_i^2 \mathcal{D}_{i\mathrm{W}}.$$

Typical Axoplasm Composition

| Ion | Conc. (meq/l) | | Ion | Conc. (meq/l) | |
|---|---|---|---|---|---|
| $Cl^-$ | 140 ± 20 | 509 ± 20 | $K^+$ | 344 ± 20 | 520 ± 20 |
| $HPO_4^{2-} + H_2PO_4^-$ | 24 ± 4 | | $Na^+$ | 65 ± 20 | |
| Aspartate | 65 ± 3 | | $Ca^{++}$ | 7 ± 5 | |
| Glutamate | 10 ± 3 | | $Mg^{++}$ | 20 ± 10 | |
| Fumarate, Succinate | 15 ± 5 | | Organic bases | 84 ± 20 | |
| Sulfonate | 35 ± 10 | | | | |
| Isethionate | 220 ± 20 | | | | |

III.4.G$_3$. *VELOCITY PROFILES IN ANOMALOUS SOLVENT DRAG*

Urea and water transport for the conditions of Problem III.4.D can be described by

$$\frac{d\pi}{d\zeta} - u\pi = -N^* \qquad \frac{d\phi}{d\zeta} = \beta\pi$$

$$\pi(0) = 1 \qquad \pi(1) = 0 \qquad \phi(0) = 0 \tag{1,2}$$

where $\pi$, $N^*$, $\phi$, $\zeta$, and $\beta$ are the suitably defined dimensionless urea concentration, urea flux, water velocity, distance, and permeability.

(a) Justify these expressions.

(b) Develop successive low-$\beta$ approximations for $\pi$ and $\phi$ as functions of $e$ by Picard's method: (i) Assume $u=0$ in Eq. 1 and calculate a first approximation $\pi_1$ for $\pi$. (ii) Use this in Eq. 2 to determine $\phi_1$. (iii) Put $\phi_1$ in Eq. 1 to obtain $\pi_2$, and $\pi_2$ in Eq. 2 to obtain $\phi_2$. One can proceed to higher approximations, but calculations by Stender, Kristensen, and Skadhauge suggest that $\phi_1$ is sufficient.

(c) For high $\beta$ the above procedure is awkward. Show that here expressions of the form

$$\phi=\beta\left(\zeta-\frac{1}{r}\zeta^u\right) \qquad 1<u<2$$

can be used to obtain reasonable upper and lower bounds to $\phi$ for any $\beta$. Discuss the limiting behavior for high and low values of $\beta$.

PARTIAL ANSWER

$$\phi_1=\beta(\zeta-\tfrac{1}{2}\zeta^2)$$

## III.5.A$_1$. *SHORT-ANSWER QUESTIONS*

(a) If oxygen has to diffuse across stagnant air and blood films of the same thickness in series, which will provide the greater mass-transfer resistance?

(b) In the above situation, how does the mass-transfer resistance of the blood film vary with dissolved oxygen concentration and film thickness?

(c) How do you suppose oxygen and glucose are getting into your brain cells across the surrounding cell membranes?

(d) How does the rate of oxygen consumption in these cells depend on dissolved oxygen concentration?

## III.5.B$_1$. *CONVECTIVE MASS TRANSFER*

(a) Why is convective mass transfer necessary, and how is it accomplished?

(b) Compare the structures and functions of the major arteries, microcirculation, and large veins.

## III.5.$C_1$. *DIRECT-CONTACT OXYGENATION*

Show how the relative mass-transfer rates for direct-contact oxygenation can be determined for

(a) Change of venous and gas partial pressures of oxygen without change in geometry or flow conditions.
(b) Change of geometry or flow conditions without change in these partial pressures.

Assume in (b) that the mass-transfer coefficients are known in the absence of reaction.

## III.5.$D_2$. *CONTROLLING MASS-TRANSFER RESISTANCE IN DIRECT-CONTACT OXYGENATORS*

Check the assumption made in the text that blood-phase mass-transfer resistance "controls" during oxygenation of blood—that is, that mass-transfer resistance in the gas can be ignored. To do this, note that the penetration theory gives the *smallest possible* mass-transfer coefficient for the gas phase: any convective mixing in the boundary layer can only increase the effectiveness of transfer. It will thus suffice to compare resistances for this special case.

(a) Begin by showing that

$$\frac{\mathrm{Nu}_G}{\mathrm{Nu}_L}=\sqrt{\frac{\mathfrak{D}_{O_2L}}{\mathfrak{D}_{O_2G}}}\,\frac{\Pi'(0)_G}{\Pi'(0)_L} \tag{1}$$

with $\Pi'(0)_G=\sqrt{4/\pi}$ .
(b) Evaluate $\Pi'(0)_L$ for $p_0=700$ mm Hg and $p_0=30$ from Eq. 5.3.20, and check your result against Fig. 5.3.4 to see if it is reasonable.
(c) Calculate the ratio of drops in oxygen concentration across the two boundary layers from the relations

$$N_{O_2}(y=0)=k_c\Delta c_{O_2}; \qquad \mathrm{Nu}=\frac{k_cD}{\mathfrak{D}_{O_2m}}$$

and the continuity of oxygen flux across the interface. Obtain the blood-phase diffusivity from Table 5.3.1 and assume that $\mathfrak{D}_{O_2G}=0.2\ \mathrm{cm^2/sec}$.
(d) Calculate the corresponding ratio of drops in oxygen tension, which may be considered as a realistic measure of the relative mass-transfer resistances. (See Table 2.3.1 for the Henry's-law constant of oxygen in blood.)

PARTIAL ANSWER

Only about $\frac{1}{2}\%$ of the drop in oxygen tension takes place in the gas phase, even under the calculated conditions. The gas-phase resistance is thus truly negligible—provided the *ventilation* is adequate.

### III.5.E$_2$. *CONVECTIVE MASS TRANSFER WITH CONSTANT WALL FLUX*

Obtain Eq. 5.4.14a as follows. Begin with

$$v_z(1+m)\frac{\partial c_{O_2}}{\partial z}=\frac{1}{r}\frac{\partial}{\partial r}\mathfrak{D}_{O_2m}r\frac{\partial c_{O_2}}{\partial r}$$

and put $c_{O_2}=f(r)+\alpha z$. Proceed as in *Tr. Ph.*, Section 9.8, and note that the oxygen concentration profile is symmetric about the $z$ axis. Assume $m$ to be a constant during your development, but discuss this assumption critically.

### III.5.F$_2$. *BLOOD OXYGENATION*

A proposed blood oxygenator consists of a falling blood film exposed to an oxygen-rich gas as shown in the figure. The surface at $y=0$ is kept at "arterial" oxygen partial pressure $p_0$, whereas for $y \gg 0$, and at all $y$ for $z=0$, the "venous" partial pressure $p_\infty$ is maintained.

(a) Determine the functional dependence of the oxygen absorption rate on $W$, $L$, and $v_{max}$ (assuming $v_z \approx v_{max}$ throughout the concentration boundary layer).

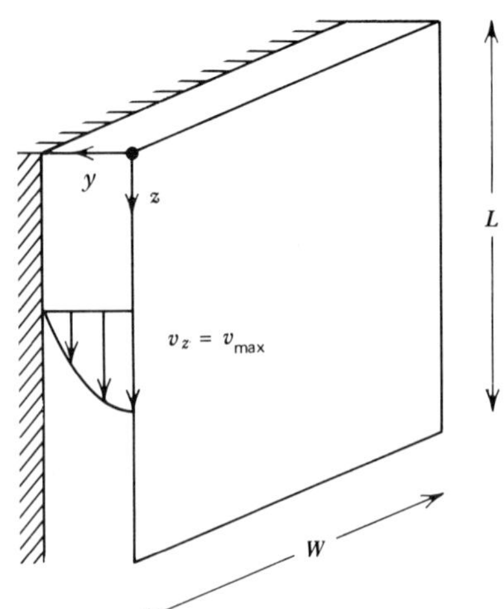

(b) Obtain an estimate of the coefficient appearing in your expression using an effective diffusivity of $0.8 \times 10^{-5}$ $\text{cm}^2/\text{sec}$ for oxygen in blood, and typical heart-lung bypass conditions:

$$p_0 = 700\,\text{mm Hg}$$

$$p_\infty = 40\,\text{mm Hg}$$

III.5.G$_2$. *ESTIMATION OF MASS-TRANSFER COEFFICIENTS*

Develop an expression for local Nusselt numbers in terms of Re and Sc for two-dimensional stagnation flow of a high-Schmidt-number fluid past an impermeable wall. This type of flow is characterized by

$$\tau_0 = \tau_{yx}\big|_{y=0} = 1.2326x\sqrt{k^3\mu\rho}$$

where outside the momentum-transfer boundary layer $V_x = kx$ and $k$ is a constant. (Use the distance $x$ from the stagnation locus as your reference length.)

III.5.H$_2$. *AXIAL DIFFUSION IN CAPILLARY TRANSPORT*

(a) Show that the Péclet number

$$\frac{\mathfrak{D}_{\mathrm{O}_2 m}}{L(1+m_{av})\langle v\rangle} \sim \frac{\text{axial capillary diffusion of oxygen}}{\text{convective oxygen transport}} \tag{1}$$

for the conditions of Section 5.4. To obtain this result one must make the approximation

$$\frac{\partial^2 p_{\mathrm{O}_2}}{\partial z^2} = \frac{[p_{\mathrm{O}_2}(0) - p_{\mathrm{O}_2}(L)]|_0}{L^2} \tag{2}$$

How reasonable do you consider Eq. 1? Should it overestimate or underestimate the diffusion flux?

(b) Calculate this ratio for $m = 40$ and the conditions in the text.

ANSWER

(b) $\text{Pé} \sim 10^{-3}$.

III.5.I$_2$. *OXYGEN TRANSFER ACROSS A BLOOD FILM IN COUETTE FLOW*

Gas of two different oxygen partial pressures is separated by a thin cylindrical film of blood in shear flow, as indicated in the diagram.

(a) Show that the continuity equations can be reduced to

$$(1+m)\left(\frac{\partial c_{O_2}}{\partial t}+v_\theta\frac{1}{r}\frac{\partial c_{O_2}}{\partial\theta}\right)=\mathfrak{D}_{O_2m}\frac{1}{r}\frac{\partial}{\partial r}r\frac{\partial c_{O_2}}{\partial r}$$

List the assumptions you must make to get this result and discuss them briefly.

(b) Make what further assumptions you consider reasonable and develop an expression for the rate of oxygen transfer at steady state. Discuss the validity of your result.

(c) How would you estimate the importance of transients?

III.5.J$_3$. *AN ALTERNATIVE EXPRESSION FOR MASS-TRANSFER COEFFICIENTS*

For any flow which is steady in a given coordinate system, Eq. 5.3.17 may be written more conveniently as

$$k_c=\frac{\pi'(0)}{c\sqrt{\mathfrak{D}_{O_2m}}}\frac{h_z v_{x0}}{\left[\int_0^x h_x h_z^2 v_{x0}dx\right]^{1/2}} \tag{i}$$

where $\pi'(0)$ is the same as in Eq. 5.3.17 and $c$ is the total molar concentra-

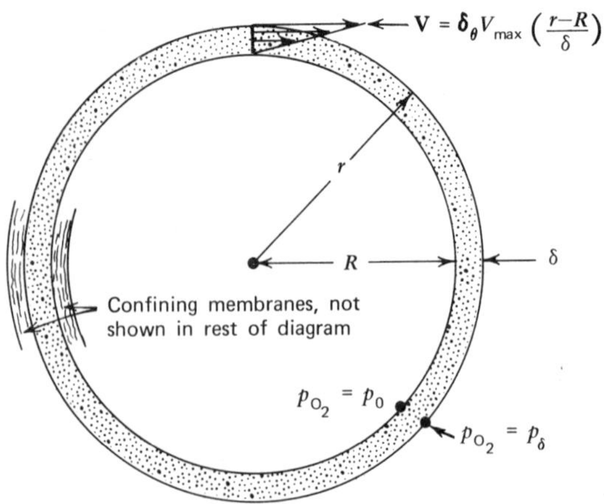

tion of blood (about 55 g-mol/l). Here $x$ and $z$ are orthogonal coordinates on the blood surface parallel and perpendicular, respectively, to the surface velocity $v_{x0}$ ($v_s$ of Eq. 3, Ex. 5.3.1). The scale factors

$$h_x = \frac{\partial \mathbf{r}}{\partial x}, \qquad h_z = \frac{\partial \mathbf{r}}{\partial z}$$

are the distances in the $x$ or $z$ directions corresponding to a unit change in $x$ or $z$, respectively (see also Stewart, 1963 or Lightfoot, 1969).

Repeat Ex. 5.3.1 using Eq. i.

SUGGESTION

Define $x$ as the subtended distance along the cone axis and $z$ as the angle about the axis. Then

$$h_x = (\sin\alpha)^{-1}$$

and

$$h_z = x\tan\alpha$$

if $z$ is in radians.

### III.5.K$_3$. *THE BOUNDARY-LAYER TRANSFORMATION IN THREE DIMENTIONS*

Extend Eq. 5.2.20 to three-dimensional systems by paralleling the development of Stewart (1963). For this situation the governing equations take the form

$$\text{Continuity:} \quad \frac{\partial}{\partial x} h_z v_x + h_x h_z \frac{\partial v_y}{\partial y} = 0$$

$$\text{Diffusion:} \quad \frac{v_x}{h_x}\frac{\partial x}{\partial x} + v_y \frac{\partial x}{\partial y} = \frac{\partial}{\partial y} D_{im} \frac{\partial x}{\partial y}$$

$$\text{Motion:} \quad \frac{\partial v_x}{\partial y} = \frac{\mu_\infty}{\mu}\gamma_\infty(x,z)$$

Here $h_x x$ is the distance measured along the body in the direction of immediately adjacent streamlines (the $x$ direction), $y$ is measured into the fluid perpendicular to the interface, and $h_z z$ is measured in the surface perpendicular to $x$ (i.e., in the $z$ direction).

ANSWER

Eq. 5.2.20 is still valid, but with

$$\delta = \frac{\left(9D_{\text{PW}\infty}\int_0^x h_x h_z \sqrt{h_z \gamma_\infty}\, dx\right)^{1/3}}{\sqrt{h_z \gamma_\infty}}$$

## III.5.L$_3$. *THE EFFECT OF A FIRST-ORDER REACTION ON CONCENTRATION PROFILES*

Convective mass transfer with first-order reaction can be described by

$$\frac{\partial c_A}{\partial t} + (\mathbf{v} \cdot \nabla c_A) = (\nabla \cdot \mathfrak{D}_{AM} \nabla c_A) + k_1 c_A \tag{1}$$

to the general level of approximation used in this chapter. This relatively simple expression is useful for a surprising range of situations, for example the dispersion of actively multiplying (log phase) bacteria in a complex flow system such as the mammalian circulation. Here $\mathfrak{D}_{AM}$ is an effective dispersion coefficient, and both it and $k_1$ will normally be functions of position.

Consider the case of homogeneous boundary conditions

$$\mathbf{n} \cdot (\nabla c_A + R c_A)|_s = 0 \tag{2}$$

at all bounding surfaces, which is a generalization of the expression

$$[\mathbf{n} \cdot (\mathbf{v} c_A - \mathfrak{D}_{AM} \nabla c_A)]|_s = 0 \tag{3}$$

valid for a system with boundaries impermeable to species $A$. Here $\mathbf{n}$ is a unit normal to the bounding surface $s$. Show that for this situation,

$$c_A = u e^{k_1 t} \tag{4}$$

where $u$ is the solution to Eqs. (1) and (2) for $k_1 = 0$. Note that neither $\mathfrak{D}_{AM}$ nor $k_1$ may be time varying if Eq. (4) is to be valid. Similar but less simple generalizations may be made for other boundary conditions by use of the principle of superposition. For an introduction to other aspects of first-order reaction see Lightfoot (1969).

## III.5.M$_3$. *ELECTROPHORETIC MIGRATION OF PROTEIN*

A thin layer of a protein solution is introduced into an electrolyte stream passing into a rectangular channel, as shown in the figure. Concentration levels are such that the protein has little effect on the conductivity, viscosity, or density of the electrolyte in which it is suspended. The protein is convected in the $x$ direction and simultaneously migrates vertically in the electric field provided by the electrodes indicated in the figure. The

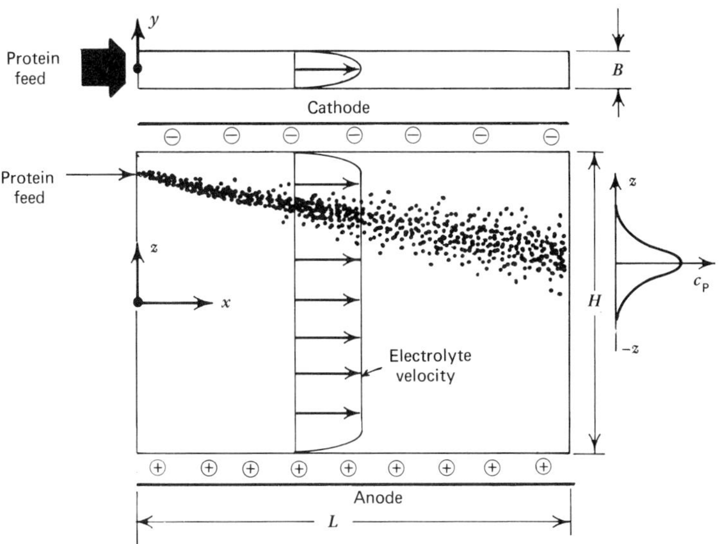


Electrophoretic migration of protein.

initially sharp protein band also spreads by concentration diffusion. To obtain a first approximation to protein distribution it may be assumed that: (1) $B \ll H, L$ so that protein diffusion in the $y$ direction essentially eliminates concentration gradients in this direction; (2) the electric field is essentially uniform, so that

$$\nabla \phi = \boldsymbol{\delta}_z E_0$$

where $E_0$ is a constant; end effects in the flow field at $z = \pm H$ are negligible. In addition it may be noted that protein diffusion is effectively binary, that is, dominated by protein-water interactions, and that the solution is primarily water on a molar basis.

(a) Show that

$$x_P(\mathbf{v}_W - \mathbf{v}_P) = \mathcal{D}_{PW}\left[\frac{\partial \ln a_P}{\partial \ln x_P}\nabla x_P + x_P \nu_P \boldsymbol{\delta}_z E_0\right]$$

with

$$\mathbf{v}_W = \boldsymbol{\delta}_x \langle v \rangle,$$

where $\langle v \rangle$ is the flow-average velocity in the cell.

(b) Next note that in the absence of diffusion,

$$v_P = \delta_x \langle v \rangle - \delta_z \mu_P E_0$$

where

$$\mu_P = \mathcal{D}_{PW} \nu_P = \text{protein mobility}$$

Define rotated coordinates $(u, w)$ by

$$\delta_u = \delta_x \cos\theta - \delta_z \sin\theta$$

$$\delta_W = \delta_z \cos\theta + \delta_x \sin\theta$$

where $\tan\theta = \mu_P E_0 / \langle v \rangle$. Show that in the absence of diffusion protein moves in the $u$-direction at a speed $v^0 = \sqrt{\langle v \rangle^2 + (\mu_P E_0)^2}$ and that diffusion is primarily in the $w$ direction if the convective transport is reasonably rapid.

(c) Put the above results into the continuity equation for protein to obtain

$$v_u \frac{\partial c_P}{\partial u} = \mathfrak{D}_{PW} \frac{\partial^2 c_P}{\partial u^2}$$

where $\mathfrak{D}_{PW} = \mathcal{D}_{PW}(\partial \ln a_P / \partial \ln x_P)$. Show that for negligible wall effects this equation yields

$$c_P = (c_0 \delta) \sqrt{\frac{v^0}{4u \mathfrak{D}_{PW}}} \exp\left( \frac{-w^2}{4u \mathfrak{D}_{PW} / v^0} \right)$$

where $B c_0 \delta \langle v \rangle$ is the rate of protein input to the cell. (Note that the distribution of protein in the $w$ direction is thus described by the normal error curve with a standard deviation of $\sigma = \sqrt{2u \mathfrak{D}_{PW} / v^0}$.)

## III.6.A$_1$. *STOCHASTIC DESCRIPTION OF A RECIRCULATING SYSTEM*

Shown in the table in digitized form is the response of a Kay-Cross blood oxygenator to pulse injection of "cardiac green" dye. The signal is complicated by recirculation of the dye via the dog being supported by the oxygenator. The basic data are:

$$m_0 = 2.2 \text{ mg of dye}$$

$$c_{\text{exit}} = 0.054 \text{ mg/l per division of ordinate}$$

Calculate the following:

| Time, sec: | 0 | 10 | 14 | 18 | 22 | 26 | 30 | 34 |
|---|---|---|---|---|---|---|---|---|
| Ordinate, No. of divisions: | 0 | 0 | 0.2 | 0.5 | 1.0 | 1.8 | 2.8 | 3.8 |
| $t$ | 38 | 42 | 46 | 50 | 54 | 58 | 62 | 66 |
| Ordinate | 5.2 | 6.8 | 9.2 | 10.0 | 11.8 | 12.4 | 13.5 | 14.0 |
| $t$ | 70 | 74 | 78 | 82 | 86 | 90 | 94 | 98 |
| Ordinate | 14.5 | 14.5 | 14.2 | 14.0 | 13.6 | 13.0 | 12.5 | 11.7 |
| $t$ | 102 | 106 | 110 | 114 | 118 | 122 | 126 | 130 |
| Ordinate | 11.2 | 10.7 | 9.8 | 9.0 | 8.3 | 8.0 | 8.0 | 8.0 |

(a) The exit concentration corrected for circulation by extrapolation of $\ln c_{\text{exit}}$ vs. $t$.

(b) The volumetric flow rate $Q$ through the oxygenator (compare with actual rate of 1360 $\text{cm}^3/\text{min}$).

(c) The oxygenator volume $V$ (which is believed to be 2100 $\text{cm}^3$) and mean residence time $\bar{t}$.

(d) The second and third moments of $c_{\text{exit}}$ with respect to $(t-\bar{t})$. How reliable do you consider these?

### III.6.B$_1$. *PHARMACOKINETICS*

Describe quantitatively the distribution of methotrexate MTX in a laboratory mouse using the flow models in Fig. 6.2.12 and the parameters in the table (both from Bischoff, Dedrick, and Zaharko, 1970).

(a)(i) Convince yourself that Fig. 6.2.12*b* is described by*

$$V_T \frac{dc_P}{dt} = Q_L\left(\frac{c_L}{R_L} - c_P\right) - K_K c_P + Mg(t) \tag{1}$$

$$V_L \frac{dc_L}{dt} = Q_L\left(c_P - \frac{c_L}{R_L}\right) - K_L \frac{c_L}{R_L} U(t-D_L) + K_0 U(t-D_L) \tag{2}$$

* No written explanation is required, but this would be a good test question.

Pharmacokinetic Parameters for MTX Distribution in a 22-g Mouse

---

$V_i$ = Volumes (ml):
- $V_P = 1.$ (plasma)
- $V_G = 1.5$ (gut wall and contents)
- $V_K = 0.34$ (kidney)
- $V_M = 10.$ (muscle)
- $V_H = 1.3$ (liver)

$R_i$ = equilibrium (tissue concentration)/(plasma concentration)
- $R_G = 0.15$
- $R_K = 3.$
- $R_M = 0.15$
- $R_L = 10$
- $R_B = 300$ (bile)

$k_i$ = clearances (ml/min)
- $k_K = 0.26$
- $k_L = 1.$

$k_0 = 0.39 \times 10^{-6}$ g MTX/min for conditions of this problem.
$k_F = 7.5 \times 10^{-3}\,\text{min}^{-1}$ = reciprocal of fecal residence time.

$Q_L = 1.1$ ml/min = hepatic (liver) plasma flow rate.

$D_i$ = delay times (min)—highly approximate
- $D_L = 5$
- $D_F = 180$

The unit step function is defined by

$$U(t) = 0 \text{ for } t < t_0$$

$$= 1 \text{ for } t > t_0$$

---

$$\frac{dM_{\text{GL}}}{dt} = -K_0 U(t - D_L) - K_F M_{\text{GL}} U(t - D_F) + K_L \frac{c_L}{R_L} U(t - D_L) \quad (3)$$

Note that clearances are calculated on the basis of plasma concentrations. In the above equations:

$$M_{\text{GL}} = \text{total mass of MTX in gut lumen}$$

$$V_T = V_P + R_M V_M + R_K V_K + R_G V_G$$

$$c_P = \text{MTX concentration in plasma}$$

Other symbols are defined in the table or in the text.

(ii) Put numbers into the above equations *for the special case* of the instantaneous injection of $66 \times 10^{-6}$ g of MTX into the blood at $t=0$. (This is equivalent to eliminating $Mg(t)$ from Eq. 1 and using it as an initial condition. At $t=0$, $c_P=(66\times 10^{-6}/V_T)$, g/ml $c_L=M_{GL}=0$).

(b) Note that for times less than $D_L=5$ min, Eqs. 1 and 2 can be solved independently of Eq. 3. Obtain a formal solution for $c_P$, for this short-time situation, by eliminating $c_L$ between these two equations and integrating.*

(c) For times greater than 5 min. it is reasonable to assume that $c_P=c_L/R$ so that Eqs. 1 and 2 can be combined. Calculate the *fraction* of the original dose in each body compartment (*P, G, K, M, L*) as a function of time for these longer times. What is the significance of the calculated long-time limiting behavior? Do you believe it? To what extent?

## III.6.C$_2$. *DISCUSSION OF THE HUMAN KIDNEY*

(a) Sketch a juxtamedullary nephron and label its major regions.

(b) Discuss briefly the process of glomerular filtration and the factors determining the filtration rate. What is meant by colloid osmotic pressure?

(c) Compare the behavior of the kidney with respect to glucose, water, and urea.

## III.6.D$_2$. *A SINGLE-POOL MODEL FOR DRUG ELIMINATION*

Consider the elimination of a water-soluble drug removed from the body entirely by kidney clearance. Assume the drug to be ingested fairly rapidly at time zero to a "standard American male" as described by Tables 2.1.1 and 2.1.4 of Chapter I.

(a) Estimate the fraction remaining in the body as a function of time if (1) the patient is consuming 500 cm$^3$/min of oxygen; (2) the blood hematocrit is raised from 43 to 70 during glomerular filtration; (3) the drug in question is neither fat soluble nor protein bound, moves with the water during filtration, and is not reabsorbed appreciably; and (4) the time scale for elimination is small compared to that for drug equilibration in the body.

* Note that these equations can be put in the form

$$\frac{dc_1}{dt}=a_{11}c_1+a_{12}c_2 \qquad a_{11}\frac{dc_1}{dt}+a_{12}\frac{dc_2}{dt}=\frac{d^2c_1}{dt^2} \tag{4}$$

$$\frac{dc_2}{dt}=a_{21}c_1+a_{22}c_2 \qquad a_{11}\frac{dc_1}{dt}+a_{12}(a_{21}c_1+a_{22}c_2)=\frac{d^2c_1}{dt^2} \tag{5}$$

The remaining $c_L$ can be eliminated by use of Eq. 1. As an alternative one can use matrix notation, as in the Bell model in the text.

(b) Compare your predictions with the MTX elimination curve, Fig. 6.1.10.

(c) Develop an expression for the concentration history if the drug binds to body proteins according to the overall approximation

$$\bar{c}/\bar{c}_0 = \frac{c/c_0}{1+\alpha c/c_0}$$

Here $\bar{c}$ and $c$ are the concentrations of the protein-bound and unbound drug, respectively; $\bar{c}_0$, $c_0$, and $\alpha$ are constants characteristic of the system. The *free* protein concentration after injection and equilibration with body tissues may be taken as $c_i$.

(d) In part (a) is it reasonable to assume instant equilibration of the drug between body compartments? Discuss briefly but carefully.

ANSWER

(a) The drug is removed exponentially with a time constant of about 80 min. This is much faster than MTX removal except perhaps at very short times.

III.6.E$_2$. *MEAN RESIDENCE TIME IN SIMPLE SYSTEMS*

It has been found that dispersion of a dye tracer in the major blood vessels is similar to that produced by a chain of ideal mixing tanks. Describe the response of such a system of $n$ tanks in series, each of volume $V$, to a pulse input at time zero of a mass $M$ of inert tracer to the first tank. Assume the volumetric flow $Q$ of solvent to be steady, and proceed as follows:

(a) Show that the effluent concentration from the first tank is $(M/V)e^{-\tau}$ where $\tau = tQ/V$.

(b) Show that the concentration in the $n$th tank is

$$C_n = e^{-\tau}\int_0^{\tau} C_{n-1}(\lambda)e^{+\lambda}d\lambda$$

Use this result and that of part (a) to obtain an explicit expression for $C_n$.

(c) Finally, show that Eq. 3 of Ex. 6.2.1 holds for this system. (The generality of Eq. 3 is examined in Problem III.6.H.)

PARTIAL SOLUTION

$$C_n = \frac{\tau^{n-1}e^{-\tau}}{(n-1)!}$$

## III.6.$F_2$. *SEPARATION OF PROTEINS BY ELECTROPHORESIS*

Use the results of Problem 5.M to estimate the separability of two proteins of differing mobilities $\mu_1$ and $\mu_2$, but essentially the same diffusivity, in the apparatus described there. Assume that $|\theta_1-\theta_2|\ll\theta_1$ for present purposes. It is suggested that you proceed as follows:

(a) Show that separation will be about 95% complete when the difference in migration distances is about 4 times the standard deviation. Thus

$$w_1-w_2=E_0(\mu_1-\mu_2)t=(4\sqrt{2})\sqrt{\frac{u\mathfrak{D}_{\mathrm{PW}}}{v^0}} \tag{1}$$

For small migration angles $\theta$ this degree of separation at $x=L$ is then described by

$$t\doteq\frac{L}{\langle v\rangle},\quad v^0\doteq\langle v\rangle,\quad u\doteq L$$

or

$$\left|E_0(\mu_1-\mu_2)\sqrt{\frac{L}{\mathfrak{D}_{\mathrm{PW}}\langle v\rangle}}\right|=4\sqrt{2}=\sqrt{2}n \tag{2}$$

where $n$ is the number of standard deviations between peaks.

(b) Rewrite Eq. 2 in terms of the volumetric flow rate of electrolyte,

$$Q=BH\langle v\rangle$$

to show that the degree of separation depends only on the cell volume and not on the individual cell dimensions (always provided the geometry-dependent assumptions of Problem III.5.M are satisfied).

(c) Show that the temperature rise for a given degree of separation depends only on the chemical nature of the proteins and the electrolyte.*

(d) Calculate the adiabatic temperature rise for 95% separation of serum albumin and $\alpha$-globulin if:

$$\mu_{\mathrm{alb}}=8.15\times10^{-5}\ (\mathrm{cm/sec})/(\mathrm{Vcm})$$

$$\mu_{\mathrm{glob}}=6.39\times10^{-5}$$

* This result was first obtained explicitly by J. St. L. Philpot, *Trans. Faraday Soc.*, **36**, 38–46 (1940); it is implicit in the paper of Arne Tiselius, *ibid*, **33**, 524 (1937).

$$\mathfrak{D}_{PW} = 6.1 \times 10^{-7}\,\text{cm}^2/\text{sec}$$

$$K = 0.0129\ \text{mho/cm}$$

where $K$ is the specific conductivity of the electrolyte, here about that of 0.1 $N$ KCl solution at 25°C. Note that the local rate of dissipation of electrical energy is $E_0^2 K$.

PARTIAL ANSWER

(c) For $2m$ standard deviations between the positions of the peak concentrations of the proteins,

$$\Delta T = \frac{8m^2 K \mathfrak{D}_{PW}}{\rho c_P (\mu_1 - \mu_2)^2}$$

For $m = 2$ and the above system, $\Delta T \sim 195°C$. Note that the separation temperature could be reduced by using a more dilute carrier solution.

### III.6.G$_3$. *MASS TRANSFER DURING FILM WITHDRAWAL*

Consider the vertical withdrawal of a flat sheet at velocity $v_0$ from a quiescent liquid, as in the figure. The surface velocity of the liquid is $v_s$, and the distance along the surface in the direction of motion is $x$. Assume for this problem that the surface is nearly flat, so that $x$ is essentially the vertical distance from the liquid surface.

(a) Show from Problem III.5.J (or Eq. 5.3.17) that the local mass-transfer coefficient if the boundary-layer approximations of Section 5.3 are valid is given by

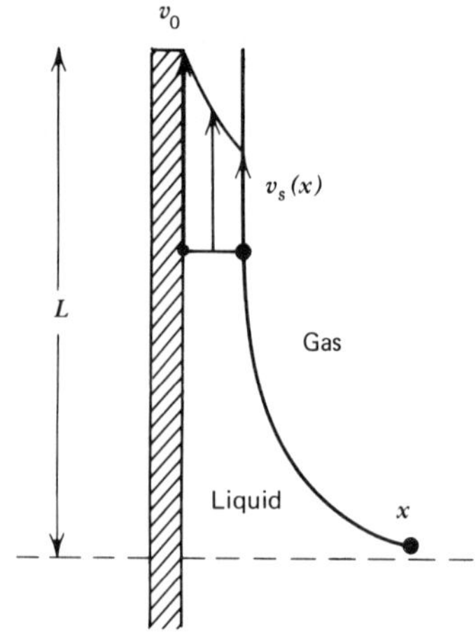

$$k_{c,\mathrm{loc}}=\sqrt{\frac{\mathfrak{D}_{AB}}{\pi c^2}}\;\frac{v_s}{\sqrt{\int_0^x v_s\,dx}}$$

(b) Show that the ratio of mass transferred in the actual case to that if $v_s=v_0$ is

$$\frac{k_{c,m}(v_s)}{k_{c,m}(v_0)}=\sqrt{\int_0^1 (v_s/v_0)\,dx/L}$$

(c) Discuss the effectiveness of disc oxygenators in the light of this example.

III.6.H$_4$. *MEASUREMENT OF MEAN RESIDENCE TIME**

To determine the generality of Eq. 3, Ex. 6.2.1, consider steady flow through a container with impermeable walls and with a single inlet and a single outlet. A nonreactive dye is injected into the feed stream in such a way that no dye reaches the inlet before zero time. Assume that the dye concentration is uniform over the inlet and outlet cross sections and that diffusional transport through these cross sections is negligible relative to convection. These conditions are closely approximated, for example, in tracer flow through the lung.

(a) Begin with the diffusion equation in form

$$\frac{\partial c_i}{\partial t}=(\nabla\cdot(\mathfrak{D}_{im}\nabla c_i-c_i\mathbf{v})) \tag{1}$$

and integrate with respect to time from zero to infinity to obtain

$$Q=(\nabla\cdot(\mathfrak{D}_{im}\nabla C-C\mathbf{v})) \tag{2}$$

with

$$C=\int_0^\infty c_i\,dt \tag{3}$$

Note that Eq. 2 has the form of the steady-state diffusion equation and that for the inlet and outlet

* This problem is based on the original development of D. B. Spalding, *Chem. Eng. Sci.* **9**, 74–77 (1958).

$$C = \frac{M_0}{Q} \tag{4}$$

where $M_0$ is the mass of tracer injected, and Q is the volumetric flow rate through the system. Show that $C$ must have this value at all points in the system.

(b) Now show from the macroscopic mass balance (Eq. 6.2.1) that

$$V\frac{\partial \bar{c}_i}{\partial t} = Q[(C_d)_i - (C_d)_0] \tag{5}$$

where $V$ is the system volume and $c.\mathrm{ea92}_i$ is the volume-average concentration of contained solute at any time. Multiply Eq. 5 by $t\,dt$ and integrate from zero to infinity. Integrate the left side by parts to obtain

$$\frac{VM_0}{Q} = Q\int_0^\infty t[(c_d)_0 - (c_d)_i]\,dt \tag{6}$$

which is the desired relation. (For all physically interesting systems, $tc_d$ will approach zero at large $t$, as required by this development.)

This result can also be applied to turbulent systems by writing the diffusion equation in time-averaged form and replacing $\mathfrak{D}_{im}$ by $\boldsymbol{\varepsilon} = \mathfrak{D}_{im}\,\boldsymbol{\delta} + \mathfrak{D}_{im}^{(t)}$. (See *Tr. Ph.,* Chapter 20. Note that in generalizing that discussion to three-dimensional systems, $\mathfrak{D}_{im}^{(t)}$ becomes a second-order tensor; it is, however, time independent for our conditions.)

# INDEX